Methods in Enzymology

Volume 323
ENERGETICS OF BIOLOGICAL MACROMOLECULES
Part C

METHODS IN ENZYMOLOGY

EDITORS-IN-CHIEF

John N. Abelson   Melvin I. Simon

DIVISION OF BIOLOGY
CALIFORNIA INSTITUTE OF TECHNOLOGY
PASADENA, CALIFORNIA

FOUNDING EDITORS

Sidney P. Colowick and Nathan O. Kaplan

Methods in Enzymology

Volume 323

Energetics of Biological Macromolecules

Part C

EDITED BY

Michael L. Johnson

CENTER FOR BIOMATHEMATICAL TECHNOLOGY
UNIVERSITY OF VIRGINIA HEALTH SYSTEM
CHARLOTTESVILLE, VIRGINIA

Gary K. Ackers

DEPARTMENT OF BIOCHEMISTRY AND MOLECULAR BIOPHYSICS
WASHINGTON UNIVERSITY SCHOOL OF MEDICINE
ST. LOUIS, MISSOURI

ACADEMIC PRESS

San Diego   London   Boston   New York   Sydney   Tokyo   Toronto

Academic Press
A Harcourt Science and Technology Company
525 B Street, Suite 1900, San Diego, California 92101-4495, USA

http://www.academicpress.com

Academic Press Limited
32 Jamestown Road, London NN1 7BY

International Standard Book Number: 0-12-182224-9

PRINTED IN THE UNITED STATES OF AMERICA
00 01 02 03 04 05 06 MM 9 8 7 6 5 4 3 2 1

100 2000513

This volume honors Serge N. Timasheff for his fundamental and far-reaching contributions to the understanding of biological macromolecules. His pioneering research has elucidated the self-assembly of brain microtubules and its regulation through the binding of small ligands to the protein subunits.[1-3] His research on the thermodynamics of protein–solvent thermodynamics[4-5] has provided an important conceptual basis for understanding physiological adaptation by organisms in extreme environments. As an editor and university professor he has contributed importantly to the dissemination of new discoveries and methodologies among several generations of scientists.

JAMES C. LEE
MICHAEL L. JOHNSON
GARY K. ACKERS

[1] J. C. Lee and S. N. Timasheff, *Biochemistry* **14,** 5183–5187 (1975).
[2] K. E. Shearwin, B. Perez-Ramirez, and S. N. Timasheff, *Biochemistry* **33,** 885–893 (1994).
[3] G. C. Na and S. N. Timasheff, *Biochemistry* **25,** 6214–6222 (1986).
[4] S. N. Timasheff, *Biochemistry* **31,** 9857–9864 (1992).
[5] S. N. Timasheff, *Adv. Prot. Chem.* **51,** 356–432 (1998).

Table of Contents

Contributors to Volume 323

Article numbers are in parentheses following the names of contributors.
Affiliations listed are current.

BERNARD Y. AMEGADZIE (9), *Eli Lilly and Company, Indianapolis, Indiana 46225*

L. MARIO AMZEL (8), *Department of Biophysics and Biophysical Chemistry, The Johns Hopkins University School of Medicine, Baltimore, Maryland 21205-2185*

CAMMON B. ARRINGTON (5), *Department of Biochemistry, University of Iowa, Iowa City, Iowa 52242*

RODNEY L. BILTONEN (15), *Departments of Biochemistry and Molecular Genetics, and Pharmacology, University of Virginia Health Sciences Center, Charlottesville, Virginia 22908*

MICHAEL N. BLACKBURN (10), *Department of Structural Biology, SmithKline Beecham Pharmaceuticals, King of Prussia, Pennsylvania 19406-0939*

MICHAEL R. BRIGHAM-BURKE (9, 10), *Department of Structural Biology, SmithKline Beecham Pharmaceuticals, King of Prussia, Pennsylvania 19406-0939*

IAN S. BROOKS (10), *Wolfram Research Inc., Champaign, Illinois 61820-7237*

JONATHAN B. CHAIRES (16), *Department of Biochemistry, University of Mississippi Medical Center, Jackson, Mississippi 39216-4505*

BABUR Z. CHOWDHRY (16), *School of Chemical and Life Sciences, The University of Greenwich, Woolwich, London SE18 6PF, United Kingdom*

JOHN J. CORREIA (4), *Department of Biochemistry, University of Mississippi Medical Center, Jackson, Mississippi 39216*

BARRIE E. DAVIDSON (11), *Department of Biochemistry and Molecular Biology, University of Melbourne, Parkville 3052, Victoria, Australia*

MICHAEL L. DOYLE (9, 10), *Department of Structural Biology, SmithKline Beecham Pharmaceuticals, King of Prussia, Pennsylvania 19406-0939*

MAURICE R. EFTINK (20), *Department of Chemistry, University of Mississippi, University, Mississippi 38677*

CONNIE L. ERICKSON-MILLER (9), *Department of MVHD, SmithKline Beecham Pharmaceuticals, Collegeville, Pennsylvania 19426-0989*

KAREN G. FLEMING (3), *Department of Molecular Biophysics and Biochemistry, Yale University, New Haven, Connecticut 06520-8114*

CHARLES A. GRIFFIN (9), *Department of Downstream Process Development, SmithKline Beecham Pharmaceuticals, King of Prussia, Pennsylvania 19406-0939*

IGNACY GRYCZYNSKI (21), *Department of Biochemistry and Molecular Biology, Center for Fluorescence Spectroscopy, University of Maryland School of Medicine, Baltimore, Maryland 21201*

IHTSHAMUL HAQ (16), *Krebs Institute for Biomolecular Sciences, Department of Chemistry, University of Sheffield, Sheffield S3 7HF, United Kingdom*

PRESTON HENSLEY (9, 10), *Pfizer Inc., Central Research Division, Groton, Connecticut 06340-9979*

GEOFFREY J. HOWLETT (11), *Department of Biochemistry and Molecular Biology, University of Melbourne, Parkville 3052, Victoria, Australia*

TERENCE C. JENKINS (16), *Yorkshire Cancer Research Laboratory of Drug Design, Cancer Research Group, University of Bradford, Bradford, West Yorkshire BD7 1DP, United Kingdom*

MICHAEL L. JOHNSON (6, 7, 21), *Departments of Pharmacology and Internal Medicine, Center for Biomathematical Technology, General Clinical Research Center, NSF Center for Biological Timing, University of Virginia Health System, Charlottesville, Virginia 22908-0735*

CHRISTOPHER S. JONES (9), *MedImmune Inc., Process Biochemistry, Gaithersburg, Maryland 20878*

DONALD W. KUPKE (18), *Department of Chemistry, University of Virginia, Charlottesville, Virginia 22908*

JOSEPH R. LAKOWICZ (21), *Department of Biochemistry and Molecular Biology, Center for Fluorescence Spectroscopy, University of Maryland School of Medicine, Baltimore, Maryland 21201*

JAMES C. LEE (v), *Department of Human Biological Chemistry and Genetics, University of Texas Medical Branch, Galveston, Texas 77555-1055*

SHARON LOBERT (4), *School of Nursing & Department of Biochemistry, University of Mississippi Medical Center, Jackson, Mississippi 39216*

LAURIE MACKENZIE (9), *Department of Screening Sciences, SmithKline Beecham Pharmaceuticals, King of Prussia, Pennsylvania 19406-0939*

LUIS A. MARKY (18), *Departments of Pharmaceutical Sciences, Biochemistry and Molecular Biology, University of Nebraska Medical Center, Omaha, Nebraska 68198-6025*

DEAN E. MCNULTY (9), *Department of AND, SmithKline Beecham Pharmaceuticals, King of Prussia, Pennsylvania 19406-0939*

DAVID P. MILLAR (19), *Department of Molecular Biology, The Scripps Research Institute, La Jolla, California 92037*

DAVID G. MYSZKA (9, 14), *Protein Interaction Facility, Department of Oncological Sciences, Huntsman Cancer Institute, University of Utah, Salt Lake City, Utah 84132*

RAJESH NAIR (21), *Department of Biochemistry and Molecular Biology, Center for Fluorescence Spectroscopy, University of Maryland School of Medicine, Baltimore, Maryland 21201*

ROLAND NEWMAN (10), *IDEC Pharmaceuticals Corporation, San Diego, California 92121-1104*

KAZIMIERZ NOWACZYK (21), *Department of Biochemistry and Molecular Biology, Center for Fluorescence Spectroscopy, University of Maryland School of Medicine, Baltimore, Maryland 21201*

SHAWN P. O'BRIEN (9), *Department of Molecular Biology, SmithKline Beecham Pharmaceuticals, King of Prussia, Pennsylvania 19406-0939*

DANIEL J. O'SHANNESSY (10), *Alchemia Pty. Limited, Indooroopilly 4068, Queensland, Australia*

SUSAN PEDIGO (12), *Department of Biochemistry, University of Iowa, College of Medicine, Iowa City, Iowa 52242-1109*

GRZEGORZ PISZCZEK (21), *Department of Biochemistry and Molecular Biology, Center for Fluorescence Spectroscopy, University of Maryland School of Medicine, Baltimore, Maryland 21201*

GEORGE P. PRIVALOV (2), *Applied Thermodynamics, Hunt Valley, Maryland 21031-0157*

PETER L. PRIVALOV (2), *Department of Biology, Biocalorimetry Center, Johns Hopkins University, Baltimore, Maryland 21218-2685*

MITCHELL REFF (10), *IDEC Pharmaceuticals Corporation, San Diego, California 92121-1104*

JINSONG REN (16), *Department of Biochemistry, University of Mississippi Medical Center, Jackson, Mississippi 39216-4505*

ANDREW D. ROBERTSON (5), *Department of Biochemistry, University of Iowa, Iowa City, Iowa 52242*

M. DOMINIC RYAN (9), *Department of Physical and Structural Chemistry, SmithKline Beecham Pharmaceuticals, King of Prussia, Pennsylvania 19406-0939*

MADELINE A. SHEA (12), *Department of Biochemistry, University of Iowa, College of Medicine, Iowa City, Iowa 52242-1109*

THOMAS M. SMITH (10), *Department of Downstream Process Development, SmithKline Beecham Pharmaceuticals, King of Prussia, Pennsylvania 19406-0939*

BRENDA R. SORENSEN (12), *Department of Biochemistry, University of Iowa, College of Medicine, Iowa City, Iowa 52242-1109*

CHARLES H. SPINK (17), *Chemistry Department, State University of New York-Cortland, Cortland, New York 13045*

WALTER F. STAFFORD III (10, 13), *Analytical Ultracentrifugation Research Laboratory, Boston Biomedical Research Institute, Boston, Massachusetts 02114*

MARTIN STRAUME (7), *NSF Center for Biological Timing, Center for Biomathematical Technology, Department of Internal Medicine, Division of Endocrinology and Metabolism, University of Virginia, Charlottesville, Virginia 22903*

ISTVÁN P. SUGÁR (15), *Departments of Biomathematical Sciences and Physiology/Biophysics, Mount Sinai School of Medicine, New York, New York, 10029*

RAYMOND W. SWEET (10), *Department of Immunology, SmithKline Beecham Pharmaceuticals, King of Prussia, Pennsylvania 19406*

LEAH TOLOSA (21), *Department of Biochemistry and Molecular Biology, Center for Fluorescence Spectroscopy, University of Maryland School of Medicine, Baltimore, Maryland 21201*

ALEMSEGED TRUNEH (10), *Department of Immunology, SmithKline Beecham Pharmaceuticals, King of Prussia, Pennsylvania 19406*

MARC R. VAN GILST (1), *Department of Cellular and Molecular Pharmacology, University of California, San Francisco, California 94143-0450*

AMY S. VERHOEVEN (12), *Department of Biochemistry, University of Iowa, College of Medicine, Iowa City, Iowa 52242-1109*

PETER H. VON HIPPEL (1), *Institute of Molecular Biology and Department of Chemistry, University of Oregon, Eugene, Oregon 97403*

THOMAS P. WHITE (17), *Chemistry Department, State University of New York-Buffalo, Buffalo, New York 14260*

ROBERT W. WOODY (9), *Department of Biochemistry and Molecular Biology, Colorado State University, Fort Collins, Colorado 80523*

PETER R. YOUNG (9), *Department of Cardiovascular Disease Research, DuPont Pharmaceuticals, Wilmington, Delaware 19880-0400*

Preface

Understanding the molecular mechanisms underlying a biological process requires detailed knowledge of the structural relationships within the system and an equally detailed understanding of the energetic driving (i.e., thermodynamic) forces that control the structural interactions. Understanding the energetic driving forces is concomitant with understanding (1) the source of the driving energies, (2) the distribution of the energies among the individual steps of the structural changes, and (3) which of the structural pathways are of biological significance.

Unfortunately, the most common examples used in the teaching of thermodynamics are exercises in heat engines. Living systems are best understood as "chemical potential engines" where the functional driving forces are generated by changes in chemical potentials. This and our previous *Methods in Enzymology* volumes (259 and 295) are devoted to describing some modern thermodynamic techniques currently being utilized to study these energetic driving forces in biological systems. This volume will be useful as a reference source and as a textbook for scientists and students whose goal is to understand the energetic (i.e., thermodynamic) relationships between macromolecular structures and biological functions.

MICHAEL L. JOHNSON
GARY K. ACKERS

METHODS IN ENZYMOLOGY

xvii

VOLUME XVII. Metabolism of Amino Acids and Amines (Parts A and B)
Edited by HERBERT TABOR AND CELIA WHITE TABOR

VOLUME XVIII. Vitamins and Coenzymes (Parts A, B, and C)
Edited by DONALD B. MCCORMICK AND LEMUEL D. WRIGHT

VOLUME XIX. Proteolytic Enzymes
Edited by GERTRUDE E. PERLMANN AND LASZLO LORAND

VOLUME XX. Nucleic Acids and Protein Synthesis (Part C)
Edited by KIVIE MOLDAVE AND LAWRENCE GROSSMAN

VOLUME XXI. Nucleic Acids (Part D)
Edited by LAWRENCE GROSSMAN AND KIVIE MOLDAVE

VOLUME XXII. Enzyme Purification and Related Techniques
Edited by WILLIAM B. JAKOBY

VOLUME XXIII. Photosynthesis (Part A)
Edited by ANTHONY SAN PIETRO

VOLUME XXIV. Photosynthesis and Nitrogen Fixation (Part B)
Edited by ANTHONY SAN PIETRO

VOLUME XXV. Enzyme Structure (Part B)
Edited by C. H. W. HIRS AND SERGE N. TIMASHEFF

VOLUME XXVI. Enzyme Structure (Part C)
Edited by C. H. W. HIRS AND SERGE N. TIMASHEFF

VOLUME XXVII. Enzyme Structure (Part D)
Edited by C. H. W. HIRS AND SERGE N. TIMASHEFF

VOLUME XXVIII. Complex Carbohydrates (Part B)
Edited by VICTOR GINSBURG

VOLUME XXIX. Nucleic Acids and Protein Synthesis (Part E)
Edited by LAWRENCE GROSSMAN AND KIVIE MOLDAVE

VOLUME XXX. Nucleic Acids and Protein Synthesis (Part F)
Edited by KIVIE MOLDAVE AND LAWRENCE GROSSMAN

VOLUME XXXI. Biomembranes (Part A)
Edited by SIDNEY FLEISCHER AND LESTER PACKER

VOLUME XXXII. Biomembranes (Part B)
Edited by SIDNEY FLEISCHER AND LESTER PACKER

VOLUME XXXIII. Cumulative Subject Index Volumes I–XXX
Edited by MARTHA G. DENNIS AND EDWARD A. DENNIS

VOLUME XXXIV. Affinity Techniques (Enzyme Purification: Part B)
Edited by WILLIAM B. JAKOBY AND MEIR WILCHEK

VOLUME XXXV. Lipids (Part B)
Edited by JOHN M. LOWENSTEIN

VOLUME LV. Biomembranes (Part F: Bioenergetics)
Edited by SIDNEY FLEISCHER AND LESTER PACKER

VOLUME LVI. Biomembranes (Part G: Bioenergetics)
Edited by SIDNEY FLEISCHER AND LESTER PACKER

VOLUME LVII. Bioluminescence and Chemiluminescence
Edited by MARLENE A. DELUCA

VOLUME LVIII. Cell Culture
Edited by WILLIAM B. JAKOBY AND IRA PASTAN

VOLUME LIX. Nucleic Acids and Protein Synthesis (Part G)
Edited by KIVIE MOLDAVE AND LAWRENCE GROSSMAN

VOLUME LX. Nucleic Acids and Protein Synthesis (Part H)
Edited by KIVIE MOLDAVE AND LAWRENCE GROSSMAN

VOLUME 61. Enzyme Structure (Part H)
Edited by C. H. W. HIRS AND SERGE N. TIMASHEFF

VOLUME 62. Vitamins and Coenzymes (Part D)
Edited by DONALD B. MCCORMICK AND LEMUEL D. WRIGHT

VOLUME 63. Enzyme Kinetics and Mechanism (Part A: Initial Rate and Inhibitor Methods)
Edited by DANIEL L. PURICH

VOLUME 64. Enzyme Kinetics and Mechanism (Part B: Isotopic Probes and Complex Enzyme Systems)
Edited by DANIEL L. PURICH

VOLUME 65. Nucleic Acids (Part I)
Edited by LAWRENCE GROSSMAN AND KIVIE MOLDAVE

VOLUME 66. Vitamins and Coenzymes (Part E)
Edited by DONALD B. MCCORMICK AND LEMUEL D. WRIGHT

VOLUME 67. Vitamins and Coenzymes (Part F)
Edited by DONALD B. MCCORMICK AND LEMUEL D. WRIGHT

VOLUME 68. Recombinant DNA
Edited by RAY WU

VOLUME 69. Photosynthesis and Nitrogen Fixation (Part C)
Edited by ANTHONY SAN PIETRO

VOLUME 70. Immunochemical Techniques (Part A)
Edited by HELEN VAN VUNAKIS AND JOHN J. LANGONE

VOLUME 71. Lipids (Part C)
Edited by JOHN M. LOWENSTEIN

VOLUME 72. Lipids (Part D)
Edited by JOHN M. LOWENSTEIN

VOLUME 320. Cumulative Subject Index Volumes 290–319 (in preparation)

VOLUME 321. Numerical Computer Methods (Part C)
Edited by MICHAEL L. JOHNSON AND LUDWIG BRAND

VOLUME 322. Apoptosis
Edited by JOHN C. REED

VOLUME 323. Energetics of Biological Macromolecules (Part C)
Edited by MICHAEL L. JOHNSON AND GARY K. ACKERS

VOLUME 324. Branched-Chain Amino Acids (Part B) (in preparation)
Edited by ROBERT A. HARRIS AND JOHN R. SOKATCH

VOLUME 325. Regulators and Effectors of Small GTPases (Part D: Rho Family) (in preparation)
Edited by W. E. BALCH, CHANNING J. DER, AND ALAN HALL

VOLUME 326. Applications of Chimeric Genes and Hybrid Proteins (Part A: Gene Expression and Protein Purification) (in preparation)
Edited by JEREMY THORNER, SCOTT D. EMR, AND JOHN N. ABELSON

VOLUME 327. Applications of Chimeric Genes and Hybrid Proteins (Part B: Cell Biology and Physiology) (in preparation)
Edited by JEREMY THORNER, SCOTT D. EMR, AND JOHN N. ABELSON

VOLUME 328. Applications of Chimeric Genes and Hybrid Proteins (Part C: Protein–Protein Interactions and Genomics) (in preparation)
Edited by JEREMY THORNER, SCOTT D. EMR, AND JOHN N. ABELSON

VOLUME 329. Regulators and Effectors of Small GTPases (Part E: GTPases Involved in Vesicular Traffic) (in preparation)
Edited by W. E. BALCH, CHANNING J. DER, AND ALAN HALL

VOLUME 330. Hyperthermophilic Enzymes (Part A) (in preparation)
Edited by MICHAEL W. W. ADAMS AND ROBERT M. KELLY

VOLUME 331. Hyperthermophilic Enzymes (Part B) (in preparation)
Edited by MICHAEL W. W. ADAMS AND ROBERT M. KELLY

VOLUME 332. Regulators and Effectors of Small GTPases (Part F: Ras Family I) (in preparation)
Edited by W. E. BALCH, CHANNING J. DER, AND ALAN HALL

VOLUME 333. Regulators and Effectors of Small GTPases (Part G: Ras Family II) (in preparation)
Edited by W. E. BALCH, CHANNING J. DER, AND ALAN HALL

VOLUME 334. Hyperthermophilic Enzymes (Part C) (in preparation)
Edited by MICHAEL W. W. ADAMS AND ROBERT M. KELLY

VOLUME 335. Flavonoids and Other Polyphenols (in preparation)
Edited by LESTER PACKER

[1] Quantitative Dissection of Transcriptional Control System: N-Dependent Antitermination Complex of Phage λ as Regulatory Paradigm

By MARC R. VAN GILST and PETER H. VON HIPPEL

Introduction

All living organisms depend on precise regulation of complex patterns of gene transcription. Remarkably, the same basic transcriptional machinery is responsible for producing the immense diversity of RNA transcripts at a variety of defined levels. Therefore, in order to satisfy the requirement for specific and timely gene expression, it is necessary for this machinery to be sensitive to an abundance of regulatory mechanisms. To cope with this challenge, transcription complexes often assemble to the point where the decision "to transcribe, or not to transcribe" rests on the sharp edge of an "energetic fulcrum"; consequently, small modifications can push the complex down one pathway or another. This balanced state presents an excellent opportunity for regulation by "transcription factors," protein or nucleic acid components that are responsible for the specific activation or repression of transcription.

As a means of directing spatial and temporal specificity, transcription factors frequently require the assistance of additional protein or nucleic acid components (accessory factors), all of which assemble along with the transcription factor into a transcriptional regulatory complex. The dependence of a transcription factor on *cis*-acting nucleic acid components directs the sequence specificity of this regulation, and dependence on *trans*-acting protein components can control the timing and intensity of a regulatory response.

In this chapter, we use basic kinetic and thermodynamic principles to illustrate how a transcription complex can achieve a state that is sensitive to regulation. Next, we present a working model for the assembly of a specific regulatory complex. Our method involves addressing quantitatively the individual contributions of accessory factors and defining them as "layers of specificity." We then describe how these layers of specificity can control the level of function of a transcription factor at distinct steps of the overall regulatory pathway. Finally, to provide a set of concrete illustrations of these principles, we describe physicochemical and molecular biological studies of the mechanisms whereby the N protein-dependent antitermination system activates the delayed early genes of bacteriophage λ.

METHODS IN ENZYMOLOGY, VOL. 323   0076-6879/00 $30.00

We stress that although our focus here is primarily on this single prokaryotic regulatory system, the principles embedded in this approach are applicable to regulatory processes in general. In the last section of this chapter we discuss the extension of these ideas to other systems that employ multilayered mechanisms to activate specific responses to cellular signals.

Transcriptional Antitermination in Phage λ

Biological Significance of λ-Mediated Antitermination

Bacteriophage λ proceeds from the lysogenic to the lytic cycle through a well-timed pattern of gene expression. Genes located downstream of the early operons of phage λ are responsible for switching the λ system into the lytic mode and synthesizing many copies of the phage genome. After sufficient copies of the genome have been made, the late genes are activated and the proteins responsible for packaging the λ genome into phage heads are produced. This delay between early and late gene expression is essential because it allows many copies of the phage genome to be made prior to packaging it into phage particles.

The N and Q proteins are responsible for regulation of the delayed-early and late genes, respectively. Both of these proteins activate transcription of their target genes by modifying the transcription complex to allow it to read through terminators that would otherwise prevent synthesis of these genes. These "antitermination" mechanisms are crucial to the timely expression of the lytic phase of λ development. Strikingly, the N protein is able to function only on terminators located upstream of the early genes, while the Q protein functions only on terminators located upstream of the late genes. This terminator specificity is not due to the composition of the terminators themselves, but to regulatory elements located between the promoter and the terminator site of action. The λ phage-encoded N protein works with a DNA element, called *nut,* and with a number of *Escherichia coli* host factors, to enhance the transcription of the early genes by modifying RNA polymerase of the *E. coli* transcription complex ($RNAP_{elong}$[1]) into a termination-resistant form.[2-5] The Q protein works with the *qut*

---

[1] In this chapter we are concerned primarily with interactions between transcription factors and the elongation complex, as opposed to interactions between N and RNA polymerase free in solution. Therefore we designate the elongating RNA polymerase within the transcription complex as $RNAP_{elong}$.

[2] A. Das, *J. Bacteriol.* **174,** 6711 (1992).

[3] A. Das, *Annu. Rev. Biochem.* **62,** 893 (1993).

[4] J. Greenblatt, J. Nodwell, and S. Mason, *Nature (London)* **364,** 401 (1993).

[5] J. W. Roberts, *Cell* **72,** 653 (1993)

DNA element to modify the transcription complex by a somewhat different mechanism.[4,5] Because the N-dependent antitermination mechanism utilizes both *trans* and *cis* "accessory factors" to form a specific regulatory complex, this system provides an excellent model for quantitative study of the assembly of a biological regulatory complex.

Assembly of N-Dependent Antitermination Complex

The assembly of a stable antitermination complex involves several components. As shown in Fig. 1, the Nus proteins of *E. coli* (NusA, B, E, and G), and the *nut* DNA element in the early operons of phage λ, all participate in the formation of the N-dependent antitermination system.[6] Transcription through the *nut* sequence produces two RNA elements, boxA and boxB, that function as specific protein-binding sites on the nascent RNA transcript.[7-11] The N protein binds to an RNA stem–loop structure (hairpin) formed by the boxB element of the nascent transcript, and the NusB/E heterodimer interacts with the boxA element. The NusA component is not a nucleic acid-binding protein, but rather serves as an RNAP-associated factor that also binds to the N protein.[12-14] Little is known about how NusG works to assist N function, but it appears to stabilize the N–RNAP$_{elong}$ interaction within the context of a full antitermination complex.[7,15,16] While the impairment (by mutation) of any one of these factors may severely compromise N function *in vivo, in vitro* experiments designed to bypass the requirement for one or more of these factors have provided information about how each contributes to the construction of a terminator specific and properly regulated antitermination mechanism.

In Vitro Antitermination

Figure 1 shows three sets of N-dependent antitermination complexes characterized *in vitro*. Complex 1 contains all of the known components

[6] S. W. Mason and J. Greenblatt, *Genes Dev.* **5,** 1504 (1991).

[7] J. Mogridge, T. F. Mah, and J. Greenblatt, *J. Biol. Chem.* **273,** 4143 (1998).

[8] J. R. Nodwell and J. Greenblatt, *Cell* **72,** 261 (1993).

[9] S. Chattopadhyay, J. Garcia-Mena, J. DeVito, K. Wolska, and A. Das, *Proc. Natl. Acad. Sci. U.S.A.* **92,** 4061 (1995).

[10] J. Mogridge, T. F. Mah, and J. Greenblatt, *Genes Dev.* **9,** 2831 (1991).

[11] M. R. Van Gilst, W. A. Rees, A. Das, and P. H. von Hippel, *Biochemistry* **36,** 1514 (1997).

[12] S. C. Gill, S. E. Weitzel, and P. H. von Hippel, *J. Mol. Biol.* **220,** 307 (1991).

[13] J. Greenblatt and J. Li, *J. Mol. Biol.* **147,** 11 (1981).

[14] M. R. Van Gilst and P. H. von Hippel, *J. Mol. Biol.* **274,** 160 (1997).

[15] S. W. Mason, J. Li, and J. Greenblatt, *J. Biol. Chem.* **5,** 1504 (1992).

[16] J. DeVito and A. Das, *Proc. Natl. Acad. Sci. U.S.A.* **91,** 8660 (1994).

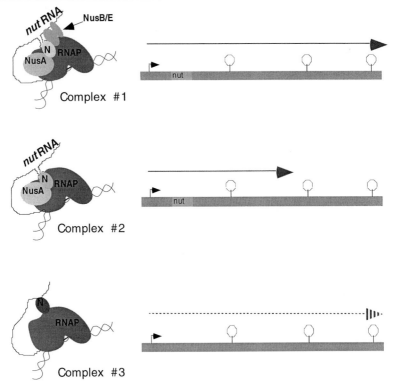

Fig. 1. Three classes of antitermination complexes and the template distances over which they function. The solid arrows represent the distances (from the *nut* site to the terminator) over which the antitermination complex can induce readthrough. Complex 1 contains all the factors genetically identified as being necessary for N-mediated antitermination to support λ growth. This complex can also function *in vitro* on termination signals located as far as 7000 nucleotides downstream of the *nut* site.[15] Complex 2 can function *in vivo* and *in vitro* at terminators located relatively near the *nut* site.[16,18] N alone can induce antitermination in the absence of all the accessory factors (complex 3) at elevated concentrations of N and somewhat reduced salt concentrations.[16,19] The dashed arrow here indicates that the terminator sequence specificity of N function (which depends on the *nut* site) is lost under these conditions, since antitermination can be induced in this system by the binding of N protein to any nonspecific RNA-binding site on the nascent transcript that is located within effective "RNA looping" distance (see below) of a transcription complex at an intrinsic terminator.

of the *in vivo* antitermination complex described above, self-assembles *in vitro* at physiologically relevant concentrations of proteins and salts, and forms an antitermination system that achieves efficient readthrough of terminators located up to 7000 bp downstream of the *nut* site.[15,16] This complex is specific, in that it is *nut* RNA dependent and, as a consequence, operates only on terminators that are located downstream of the *nut* site on the same RNA transcript. This complex also has a long "range" of action,[17] in that the N-dependent antitermination function is maintained through multiple terminators within a polycistronic operon while the $RNAP_{elong}$ complex transcribes across several thousand base pairs of template.[15,16]

Complex 2 of Fig. 1 corresponds to a reduced N-dependent antitermination complex that can operate *in vitro* at physiological salt and protein concentrations. This reduced complex contains only N protein, $RNAP_{elong}$, NusA, and a transcript that contains boxB, yet still regulates transcription in a template sequence-specific manner, in that it requires the *nut* DNA site for synthesis of the required boxB element. However, this complex has significantly less range than complex 1, in that the template distance over which complex 2 exerts effective N-dependent antitermination extends only to terminators located less than a few hundred base pairs downstream of the *nut* site.[15,16,18,19]

By lowering the salt concentration, or by elevating the concentration of N protein, antitermination can be achieved in the absence of all the antitermination accessory factors.[19] This minimal complex, represented by complex 3 in Fig. 1, is not terminator specific (i.e., it has no range of action), and depends neither on *cis*-acting upstream regulatory signals nor on *trans*-acting protein components.

"Layers of Specificity" Concept

The ability of N to bring about antitermination without the participation of other factors is consistent with the hypothesis that N is solely responsible for the modification of the elongation complex into a termination-resistant state. Thus, in this model, the other factors merely serve to facilitate a productive $N–RNAP_{elong}$ interaction.[16,19,20] Alternatively, the other antitermination components could contribute directly to the N-dependent modification of $RNAP_{elong}$ via parallel pathways. However, since these factors do not induce antitermination of intrinsic terminators in the absence of N (in

---

[17] The term "range" is introduced to represent the template distance (in base pairs from the specific *nut* site to the terminator) over which N-dependent antitermination can occur. This term replaces the less appropriate term "processivity," which has commonly been used to describe the template distance over which N acts.

[18] W. Whalen, B. Ghosh, and A. Das, *Proc. Natl. Acad. Sci. U.S.A.* **85**, 2494 (1988).

[19] W. A. Rees, S. E. Weitzel, A. Das, and P. H. von Hippel, *Proc. Natl. Acad. Sci. U.S.A.* **93**, 342 (1996).

[20] W. Whalen and A. Das, *New Biol.* **2**, 975 (1990).

fact, they play different roles in the host[12,21]), we postulate that the N protein alone accounts for the fundamental modification of $RNAP_{elong}$, and that the manifestation of the antitermination phenotype is the result of $N\text{-}RNAP_{elong}$ association. Furthermore, we suggest that N protein alone cannot bind effectively to $RNAP_{elong}$ under physiological salt and protein concentrations and that the overall role of the *nut* site and the Nus factors is to increase the occupancy of $RNAP_{elong}$ by N.

In terms of the usage introduced in this chapter, we will define the contribution of each component to the net stability of the fundamental $N\text{-}RNAP_{elong}$ interaction as representing a "layer of specificity." We use this term because the dependence of N activity on these stabilizing layers controls the specific timing, the template location, and the efficacy of the N-dependent antitermination response. The fundamental $N\text{-}RNAP_{elong}$ interaction is defined as the first layer of specificity. This layer involves the *specific modification* of $RNAP_{elong}$ into a termination-resistant form, and we postulate that the binding of N to $RNAP_{elong}$ is uniquely responsible for triggering this modification. The interaction of N with the boxB RNA hairpin represents the second layer of specificity. This layer introduces *sequence specificity* into N-dependent antitermination, because the requirement for boxB on the nascent RNA transcript ensures that N can function only at terminators located at limited distances downstream of the transcribed *nut* site. The N–NusA interaction adds a third layer of specificity. This layer can provide *host–factor specificity,* as antitermination can occur only when there is sufficient concentration of NusA in the host to enhance the binding of N to $RNAP_{elong}$. The association of the NusB/E heterodimer with the boxA element of the *nut* RNA contributes a fourth layer of specificity. This layer contributes both *sequence specificity* and *host–factor specificity,* because it depends not only on the concentration of host proteins but also, because these proteins must bind to an upstream RNA element, their contribution to antitermination is exclusively targeted to terminators located downstream of the *nut* site. Finally, we point out that NusG also appears to participate in the antitermination mechanism.[7,15,16] However, because its mechanism of action at the molecular level is still not understood, we will not consider this factor in our present analysis. When further information about the role of NusG is available, this approach can be extended to accommodate this factor.

---

[21] In the absence of N, the Nus factors, together with a boxA element similar to the *nut* boxA, can facilitate "readthrough" of many of the rho-dependent terminators that punctuate ribosomal operons. However, these components do not function on intrinsic terminators, and it appears that an unknown factor, which may be "N-like," is required for this activity [see C. L. Squires, J. Greenblatt, J. Li, and C. Condon, *Proc. Natl. Acad. Sci. U.S.A.* **90,** 970 (1993)].

Thermodynamics and Kinetics of Antitermination

In this section we consider the function and assembly of the antitermination complex in kinetic and thermodynamic terms. To this end we propose, as supported both theoretically and experimentally, that the elongation-termination decision is controlled by a simple kinetic competition between the overall elongation and termination reaction processes (Fig. 2A).[22–25] In addition, we postulate that the formation of the antitermination complex is fast compared with either of these overall reactions, and thus the assembly and regulation of the antitermination complex at any particular template position constitute an equilibrium-controlled process within the overall kinetic competition.[26] Further detailed justification for this equilibrium assumption, which is central to our thermodynamic treatment of N-dependent antitermination complex assembly, is presented elsewhere.[27]

The use of these postulates, in combination with the "layering" concept described above and existing ideas,[20,28] provides a working thermodynamic model that we can use to investigate transcriptional control by N-dependent antitermination within intrinsic terminators.[29] More complicated versions of this simple model may need to be formulated in the future, but for now this framework appears sufficient to describe the mechanisms that accessory factors use to control the function of the N protein. In addition, we point out that this model can serve as a guide for the design and interpretation of quantitative investigations into N-mediated antitermination.

Perturbation of Kinetic Competition by Antitermination. As indicated above, the observed termination efficiency (TE) at a given template position is dependent on a kinetic competition between elongation (defined as the overall process of adding the next template-required nucleotide residue to the nascent transcript) and termination (defined as the controlled disruption of the transcription complex, leading to the release of the nascent RNA and polymerase components into solution).[22–25] Figure 2B presents simplified Eyring diagrams that describe the kinetic competition between these two processes at nonterminator and terminator positions along the template. At nonterminator positions, the activation barrier for disruption of the

---

[22] P. H. von Hippel and T. D. Yager, *Science* **255,** 809 (1992).

[23] P. H. von Hippel and T. D. Yager, *Proc. Natl. Acad. Sci. U.S.A.* **88,** 2307 (1991).

[24] K. S. Wilson and P. H. von Hippel, *J. Mol. Biol.* **244,** 36 (1994).

[25] W. A. Rees, S. E. Weitzel, A. Das, and P. H. von Hippel, *J. Mol. Biol.* **273,** 797 (1997).

[26] T. D. Yager and P. H. von Hippel, *Biochemistry* **30,** 1097 (1991).

[27] M. R. Van Gilst, C. A. Conant, W. A. Rees, S. E. Weitzel, and P. H. von Hippel, in preparation (2000).

[28] J. R. Nodwell and J. Greenblatt, *Genes Dev.* **5,** 2141 (1991).

[29] N also inhibits termination at rho-dependent terminators (see Ref. 21). However, most mechanistic investigations to date (and thus also those summarized in this chapter) have focused on regulation at intrinsic terminators.

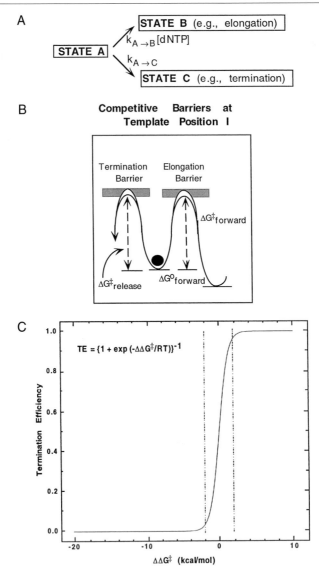

Fig. 2. A kinetic competition determines termination efficiency. (A) The two competing rates and their regulation. (B) Simplified free energy of activation diagram, showing the relative magnitudes of the rate-limiting steps for the overall elongation and termination processes at terminator and nonterminator positions along the template.[23] (C) Plot of the observed termination efficiency as a function of $\Delta\Delta G^{\ddagger}$ at a given template position.[23] This plot shows that small changes in $\Delta\Delta G^{\ddagger}$ can have significant effects on termination efficiency only when the heights of the competing free energy of activation barriers are comparable.

transcription complex is high ($\Delta G_{term}^{\ddagger} \approx +30$ kcal/mol), and elongation ($\Delta G_{forward}^{\ddagger} \approx +16$ kcal/mol) is favored. On the other hand, at sites within the zone of opportunity for intrinsic termination, structural changes in the elongation complex that either destabilize the ground state (ΔG_f^0) or stabilize the termination transition state ($\Delta G_{term}^{\ddagger}$) can result in a decrease in the magnitude of $\Delta G_{term}^{\ddagger}$ to a point at which the rates of the elongation and termination processes (and thus the heights of the free energy of activation barriers to termination and elongation—see Fig. 2B) are similar.[24,26] Therefore, small modifications to the heights of the free energy of activation barriers (2–3 kcal/mol) can result in dramatic differences in the observed efficiency of termination [see Eq. (1) and Fig. 2C].

If we apply this model to N function, N may operate to bring about antitermination in two distinct ways: (1) N could decrease the rate of disruption of the transcription complex ($\Delta\Delta G_{term}^{\ddagger}$), either thermodynamically by stabilizing the ground state (ΔG_f^0), or kinetically by destabilizing the transition state leading to termination ($\Delta\Delta G_{term}^{\ddagger}$) (Fig. 2B); (2) alternatively, N could increase the rate of elongation ($\Delta\Delta G_{forward}^{\ddagger}$) by lowering the activation barrier for the addition of the next nucleotide residue. N could also function by a combination of both mechanisms. For this analysis, we are primarily concerned with the total free energy of activation contribution of N to antitermination ($\Delta\Delta G_{at}^{\ddagger}$), and we will define this parameter [Eq. (1)] as the algebraic sum of the effect of N on each of these competing pathways:

$$\Delta\Delta G_{at}^{\ddagger} = \left| \Delta\Delta G_{forward}^{\ddagger} - \Delta\Delta G_{term}^{\ddagger} \right|$$
$$= \left| RT \ln[(1 - TE_0)/TE_0] - RT \ln[(1 - TE_N)/TE_N] \right| \qquad (1)$$

where TE_0 is equal to the termination efficiency in the absence of N, TE_N is equal to the termination efficiency in the presence of N, and TE itself is defined as

$$TE = [RNA_{term}/(RNA_{term} + RNA_{runoff})] \qquad (2)$$

where RNA_{term} and RNA_{runoff} correspond, for a given experiment, to the amounts of RNA released at the terminator and at the runoff position of the template, respectively. Note that the change in TE (ΔTE_{obs}) that results from N function depends both on the value of TE for a given terminator in the absence of N (TE_0) and on the additional free energy of activation ($\Delta\Delta G_{at}^{\ddagger}$) contributed to the antitermination process at that terminator by N.

Viewed in this way, the binding of N to a particular transcription complex does not ensure that it will avoid termination; it merely increases the probability that it will read through. Thus, the value of $\Delta\Delta G_{at}^{\ddagger}$ determines the change in the fraction of a population of complexes that read through a terminator when 100% of the transcription complexes are bound by N.

We will now define the parameter $\Delta\Delta G_{at}^{\ddagger}$ as determining the "efficacy of antitermination," and the maximum value of ΔTE that applies at a given terminator when all the functional $RNAP_{elong}$ is complexed with N will be ΔTE_{max}. We note also, because ΔTE is not linearly proportional to $\Delta\Delta G_{at}^{\ddagger}$, that ΔTE_{max} is terminator specific and must be calculated from $\Delta\Delta G_{at}^{\ddagger}$ and ΔTE_0 at different terminators according to Eqs. (1) and (2).

Because many of the antitermination accessory proteins may also bind to $RNAP_{elong}$, it is important to define two types of binding interactions. First, there are interactions that contribute free energy to the antitermination mechanism ($\Delta\Delta G_{at}^{\ddagger}$). We will consider interactions of this type to represent fundamental modifications of $RNAP_{elong}$ that result in an increase in the efficacy of the antitermination modification (ΔTE_{max}) at a given terminator.

Importantly, in our model, N is the sole antitermination protein that can modify $RNAP_{elong}$ into a termination-resistant form. Therefore, if only a fraction of transcription complexes are bound by N, then the observed antitermination response (ΔTE_{obs}) will be equal to ΔTE_{max} multiplied by the fraction of complexes bound by N. As a consequence, accessory factors may employ a second type of binding interaction that serves to increase the fraction of transcription complexes bound by N by somehow increasing the affinity of N for the transcription complex. This type of $RNAP_{elong}$ binding interaction will increase the observed antitermination response without affecting the efficacy of the antitermination modification (i.e., without moving ΔTE_{obs} closer to ΔTE_{max}). For accessory factors of this type, their free energy of binding to the complex is not directly coupled to the antitermination modification, and thus they have no effect on the efficacy of antitermination (ΔTE_{max}). A notable conclusion from this model is that these accessory factors will have no effect on termination efficiency if N is not present and thus there is no factor with which to modify $RNAP_{elong}$ ($\Delta TE_{max} = 0$), or under conditions where 100% of the transcription complexes are already bound by N (and thus $\Delta TE_{max} = \Delta T_{obs}$).

Layer 1, Part A: Interaction of N with $RNAP_{elong}$

If the direct interaction between N and $RNAP_{elong}$ ultimately suffices to induce antitermination, then under conditions where N is functioning without the Nus factors and the *nut* site, and at high enough concentrations of N to ensure full occupancy of the elongating RNAP, the free energy provided by this fundamental N–$RNAP_{elong}$ binding interaction must be equal to or greater than the free energy required for antitermination. Therefore, we can write

$$\Delta\Delta G_{at}^{\ddagger} + \Delta G_{other} = \Delta G_{N-P} \tag{3}$$

TABLE I

Changes in Termination Efficiency at Intrinsic $t_{R'}$ Terminator for Nonspecific and Specific Antitermination in Presence and Absence of NusA[a]

Antitermination components	TE_N (%)	TE_O (%)	ΔTE (%)	$\Delta\Delta G_{at}^{\ddagger}$ (kcal/mol)
N–boxB	15	85	70	−2.1
N–nonspecific RNA	15	85	70	−2.1
N–boxB–NusA	10	90	80	−2.6
N–nonspecific RNA–NusA	10	90	80	−2.6

[a] TE data were obtained from Ref. 29a, and $\Delta\Delta G_{at}^{\ddagger}$ was calculated according to Eq. (1).

where ΔG_{other} is equal to the binding free energy component of the N–RNAP$_{elong}$ interaction that is not directly coupled to the antitermination modification induced by N (see description of the kinds of interactions of regulatory factors with RNAP$_{elong}$ above). Because the dissociation constant for the N–RNAP$_{elong}$ interaction ($K_{d(N-P)}$) is related to ΔG_{N-P} by Eq. (4):

$$K_{d(N-P)} = e^{-\Delta G_{N-P}/RT} \qquad (4)$$

we can calculate an upper limit for $K_{d(N-P)}$ by assuming that all of the N–RNAP$_{elong}$ binding free energy is coupled to antitermination, meaning that $\Delta G_{other} = 0$ and therefore that $\Delta\Delta G_{at}^{\ddagger} = \Delta G_{N-P}$ [see Eq. (3)].

Previous results characterizing the effect of N on termination at the intrinsic $t_{R'}$ terminator are summarized in Table I.[29a] We note that, in the absence of all other components of the N-dependent antitermination complex (i.e., experiments with complex 3 of Fig. 1), the observed TE for this terminator changes from $TE_O = 0.85$ to $TE_N = 0.15$ on saturation of the transcription complex with N protein.[19,27] This shift in ΔTE_{obs} (which is close to ΔTE_{max} since RNAP$_{elong}$ should be fully complexed at saturating N concentrations[27]) corresponds to a value of $\Delta\Delta G_{at}^{\ddagger}$ of $\sim$−2.1 kcal/mol. Therefore, according to Eq. (4) the value of $K_{d(N-P)}$ must be at least $10^{-2}\ M$.

The observation that $K_{d(N-P)}$ could be as high as $10^{-2}\ M$ in order to provide enough free energy for the antitermination mechanism has important biological implications. First, since the direct binding affinity of N for RNAP$_{elong}$ could be weak, effective binding to RNAP$_{elong}$ cannot occur without stabilizing contributions from additional (accessory) factors. Consequently, these accessory factors can serve, as we will show, to introduce terminator specificity into the N-dependent antitermination reaction. In contrast, if larger changes in the binding free energy of N to RNAP$_{elong}$ were required to permit terminator readthrough, then the binding of N

[29a] W. A. Rees, Ph.D. thesis. University of Oregon, Eugene, Oregon, 1996.

alone would have to be much tighter, the accessory factors would not be needed for binding, and the opportunity for introducing terminator specificity through these other regulatory factors would be lost.

Antitermination Support by Other Layers of Specificity. Our model postulates that the fraction of ΔTE_{max} achieved is equal to the fraction of elongating RNA polymerases bound by the N protein ($f_{P,B}$). For a fixed concentration of elongation complexes, the level of antitermination observed is then dependent on the concentration of the N protein and the K_d of the N–RNAP$_{elong}$ association. This dependence is represented by Eqs. (5A) and (5B):

$$f_{P,B} = \frac{(K_d + [N_0] + [RNAP_0]) \pm \sqrt{(K_d - [N_0] - [RNAP_0])^2 - 4[N_0][RNAP_0]}}{2[RNAP_0]}$$

$$\tag{5A}$$

$$\Delta TE_{obs} = f_{P,B} \cdot \Delta TE_{max} \tag{5B}$$

where $f_{P,B}$ is equal to the fraction of RNAP$_{elong}$ that is bound by N, $[N_0]$ is equal to the total N concentration, K_d is the dissociation constant for the binding of N to RNAP$_{elong}$ ($K_{d(N-P)}$), and $[RNAP_0]$ is equal to the total concentration of RNAP$_{elong}$.

Figure 3 displays the fraction of RNA polymerase bound by N as a function of N concentration for different values of $K_{d(N-P)}$. From this plot it is evident that $K_{d(N-P)}$ must be larger (weaker binding) than 10^{-6} M if N is not to interact directly with RNAP$_{elong}$ under physiological conditions.[27] Physical measurements and functional assays both show that the interaction of N with RNAP$_{elong}$ is weaker than this in the absence of accessory factors[19] (M. R. Van Gilst and P. H. von Hippel, unpublished results, 1999). Combining this lower estimate for $K_{d(N-P)}$ with the upper estimate of 10^{-2} M rationalized above yields a range of $K_{d(N-P)}$ values that can both supply the amount of binding free energy required for antitermination and result in a net N–RNAP$_{elong}$ binding affinity that is weak enough to require assistance from other factors under physiological salt and protein concentrations.[30]

Figure 3 and Eq. (5) show that antitermination accessory factors can work to promote the association of N with RNAP$_{elong}$ by two possible mechanisms. (i) These factors could operate by increasing the affinity of N for RNAP$_{elong}$, resulting in a shift of the $K_{d(N-P)}$ curves of Figure 3 to the right. (ii) Alternatively, these factors could operate by increasing the concentration of the N protein at RNAP$_{elong}$ (i.e., at the transcription com-

---

[30] For purposes of these calculations we assume that the free concentration of N protein in the *E. coli* cell under lytic phage λ growth conditions is ~0.1 μM.

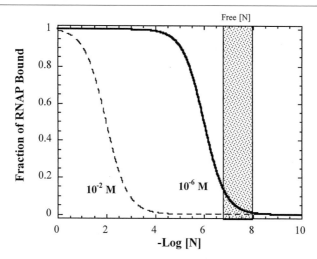

FIG. 3. Fraction of RNAP bound as a function of the free N concentration at different values of K_d for the N–RNAP interaction. The fraction of RNAP that would be bound by N as a function of N concentration for dissociation constants $[K_{d(N-P)}]$ of $10^{-2}\,M$ (dashed line) and $10^{-6}\,M$ (solid line). The stippled region represents the range of N concentrations generally used for antitermination experiments *in vitro* (10–200 nM).[15,16,19]

plex), resulting in a shift in the stippled rectangle (representing the free N concentration) to the left. To fulfill the requirements of mechanism (ii), accessory factors could work to increase the concentration of N directly by stimulating the production of N, or indirectly by recruiting N to the vicinity of the elongation complex. Measurement of the binding constant of N for $RNAP_{elong}$ will lead to refined estimates of how much free energy accessory factors must provide to promote the effective association of N with $RNAP_{elong}$ under physiological conditions.

Layer 2, Part A: Binding of N to boxB Hairpin and RNA Looping

Occupancy of $RNAP_{elong}$ by N Increased by RNA Looping. The interaction of N protein with the boxB RNA hairpin is responsible for the specific localization of N-mediated antitermination.[9,31] As outlined above, the accessory factor boxB can contribute to N function by strengthening the fundamental N–$RNAP_{elong}$ interaction, by increasing the concentration of N relative to the elongating RNA polymerase, or by a combination of both effects. However, boxB must be present on the nascent RNA transcript (rather than added in *trans*) in order to facilitate antitermination, suggesting

---

[31] M. Rosenberg, S. Weissman, and B. DeCrombrugghe, *J. Biol. Chem.* **250,** 4755 (1975).

that boxB may operate to increase the local (effective) concentration of N by tethering it in the vicinity of the transcription complex.[11,19,20,28]

In the preceding section we suggested that N cannot associate with $RNAP_{elong}$ in the absence of other antitermination factors at physiological concentrations of salt and N protein. However, N will bind to the boxB RNA hairpin under these conditions since the dissociation constant of N for boxB $[K_{d(N-boxB)}]$ is $\sim 10^{-8}\ M$.[11,27] Therefore, in the presence of physiological concentrations of N, nearly all elongation complexes that have transcribed through the *nut* site will carry an N protein bound to boxB on the nascent transcript. This tight binding of N to boxB will tether N to the transcription complex, reducing the volume through which this fraction of the N protein is free to diffuse to a sphere centered on the elongation complex. The volume of this sphere depends on the contour length and flexibility of the RNA between boxB and the elongating RNAP. Therefore tethering N to the transcription complex allows the RNA-bound N to function at $RNAP_{elong}$ by "*cis*-RNA looping." The consequent reduction in the volume through which N is free to diffuse results in an increase in the local concentration of N at a transcription complex localized at the terminator.[11,20,28]

If we consider the N–boxB complex and $RNAP_{elong}$ to represent two ends of a freely jointed (random coil) RNA chain, we can calculate the mean end-to-end distance of the chain and thus determine $[j_N]$, which we define as the local concentration of tethered N at the transcription complex that results from RNA looping. This parameter depends on the properties of the connecting RNA chain as described by Rippe *et al.*,[32] who have summarized the relevant equations needed to calculate the local concentrations $[j]$ of transcriptional activator proteins that result from *cis* DNA looping. We have used versions of these equations that have been modified for the parameters of single-stranded RNA to calculate $[j_N]$, the local concentration of N at $RNAP_{elong}$, as a function of the contour length (in nucleotides) and flexibility of the RNA transcript between the N–boxB complex and $RNAP_{elong}$.

The results of these calculations, using 6.5 Å as the average internucleotide distance for single-stranded RNA and 40 Å for the persistence length of single-stranded RNA, are presented in Fig. 4. This value of the persistence length parameter was measured experimentally for poly(rU) in solution.[33] The calculations used to determine $[j_N]$ in Fig. 4 for short ($< \sim 30$ nucleotides) lengths of RNA chain are not straightforward, since the relevant equations must be solved numerically for these RNA lengths using a Porod–Kratky wormlike chain model.[32] Equation (6A) describes the

[32] K. Rippe, P. H. von Hippel, and J. Langowski, *Trends Biochem. Scis.* **20**, 500 (1995).
[33] L. D. Inners and G. Felsenfeld, *J. Mol. Biol.* **50**, 373 (1970).

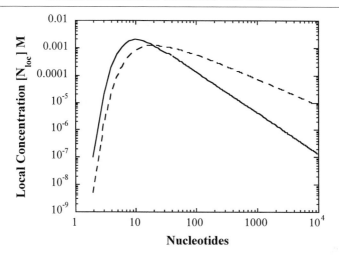

Fɪɢ. 4. The local concentration of N protein at RNAP for boxB-bound N and nonspecifically bound N. The local concentration of N at $RNAP_{elong}$ as a function of the distance between boxB on the transcript and $RNAP_{elong}$ located at the terminator, calculated by a freely jointed chain model.[11,32] The solid line represents the local concentration of N at the terminator that would result from binding to a specific site on the nascent transcript [Eqs. (6A) and (6B)]. The dashed line (see Ref. 11) represents the local concentration that would result from an average of one N protein bound nonspecifically and randomly to the nascent transcript. The distance (in nucleotides) shown on the x axis for the nonspecific antitermination plot corresponds to the distance along the template from the promoter to $RNAP_{elong}$ at the terminator, since N can bind at random to the nascent transcript in this mode and thus the N-to-$RNAP_{elong}$ distance is variable at any given template position.

probability density, $W(r)$, that the two ends of a random coil are separated by an average end-to-end distance, r:

$$W(r) = \left(\frac{3}{2\pi n l}\right) e^{-3r^2/2\pi l^2} \qquad (6A)$$

Equation (6A) is derived from the classic freely jointed chain model of polymer chain statistics. Here n is equal to the number of chain segments, and l is equal to the statistical segment length. For r values at contour lengths greater than 20 nm ($n > \sim30$ nucleotides), the exponential term of this equation goes to one; therefore, by converting $W(r)$ to units of concentration (moles per liter), a simpler expression can be derived to calculate the local concentration of N ($[N_{loc}]$) at a terminator where N is bound to the nascent transcript at a position located more than 30 nucleotides from $RNAP_{elong}$. Using the poly(rU) parameter for l (40 Å), and taking n as the contour length (in nucleotides per 6.5 Å) for the nascent

RNA chain between the boxB hairpin and the 3′ terminus of the transcript, we may write (see Ref. 32):

$$[j_N] = [N_{loc}] = 0.131(nts)^{-3/2} \qquad (6B)$$

for such chains, where (nts) represents the contour length of the intervening RNA transcript in nucleotides. This simple relationship provides an estimate of the binding density of N at $RNAP_{elong}$ for most of the physiologically relevant portions of the curves in Fig. 4.

Figure 4 shows that the local concentration of N induced by binding to the boxB RNA hairpin is 1000-fold higher than the estimated cellular concentration of N at an RNA contour length (from boxB to $RNAP_{elong}$) of 100 nucleotides, and that this local concentration is still 10-fold higher at an RNA contour length of 5000 nucleotides. Therefore, if N is bound tightly to the RNA transcript, the local (i.e., tethered) concentration of N overwhelms the solution concentration of N, and the binding equilibrium of N to $RNAP_{elong}$ is effectively determined by this *cis*-RNA looping component of the overall N concentration.[30] The effective concentration of N present, $[N_{eff}]$, is therefore

$$[N_{eff}] = [N_{loc}] + [N_{free}] \qquad (7)$$

where $[N_{free}]$ is the free solution concentration of N and $[N_{loc}]$ is $[j_N]$. If $[N_{loc}] \gg [N_{free}]$, then $[N_{eff}] = [N_{loc}]$, and $[N_{loc}]$ can be substituted into Eq. (5A) for both $[N_0]$ and $[RNAP_0]$ (since N and $RNAP_{elong}$ are tethered to each other the local concentration applies to both binding partners) to determine the fraction of $RNAP_{elong}$ that is complexed with a tethered N protein. We note that by calculating the effective concentration of N at $RNAP_{elong}$ in this way we are essentially considering the $N–RNAP_{elong}$ binding interaction as a bimolecular reaction, even though N is tethered to $RNAP_{elong}$ by a flexible (random chain) linker.

Figure 5 shows that RNA looping can provide the concentrations of N that are needed to bind $RNAP_{elong}$ efficiently at physiologically relevant separations between the *nut* site and intrinsic terminators. *In vitro* studies have shown that effective N-dependent antitermination can occur, in the absence of other Nus factors, at intrinsic terminators located 100 to 200 bp downstream of the *nut* site.[16,19] The local concentration of N bound to a boxB hairpin at these distances from $RNAP_{elong}$ is calculated to be $\sim 10^{-4} M$ (indicated by the vertical dashed line in Fig. 5). N cannot induce antitermination at terminators located thousands of base pairs (or even 600 bp[16]) downstream of *nut* in the absence of the Nus factors.[15] We estimate that these distances correspond to local concentrations of tethered N of $\sim 10^{-7} M$ (indicated by the solid vertical line Fig. 5). A value of $K_{d(N–P)}$ [or more specifically, since N is now bound to the nascent transcript, a value of

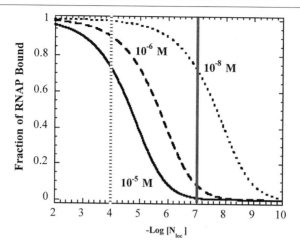

FIG. 5. The local concentration effect increases the fraction of RNAP$_{elong}$ that is complexed with N. The fraction of RNAP$_{elong}$ complexed with N is plotted as a function of N concentration. The local concentrations of N that result from binding to a boxB site on the nascent transcript, either ~7000 bp (shaded vertical line) or ~100 bp (dashed vertical line) upstream of the terminator along the template, are shown. The fractions of RNAP$_{elong}$ that would be complexed by N for a $K_{d(NR-P)}$ value of 10^{-5} M (solid curve), 10^{-6} M (dashed curve), and 10^{-8} M (dotted curve) are indicated. To account for the fact that N can function without Nus factors at a *nut* site to terminator separation of ~100 bp, but not at a separation of several thousand base pairs, the K_d for the N–RNAP$_{elong}$ interaction must be ~10^{-5} to 10^{-6} M. However, for the Nus factors to facilitate antitermination at a *nut* site to terminator separation of ~7000 bp, $K_{d(NR-P)}$ must be ~10^{-8} M.

K_d for the interaction of the N–boxB complex with RNAP$_{elong}$ of the transcription complex, $K_{d(NB-P)}$; see below] of ~10^{-5} M, can account for these *in vitro* observations. An accurate experimental measurement of $K_{d(NB-P)}$ under the conditions used, and more systematic analysis of the RNA length dependence of the local concentration of N for real transcripts, are needed to define the actual values of the local concentration parameter ($[j_N]$) obtained as a consequence of looping of real mRNAs. Such measurements are currently in progress in our laboratory.

[N$_{loc}$] can also be calculated as described above for N-dependent nonspecific antitermination; i.e., for antitermination that occurs in the absence of a boxB RNA hairpin on the transcript. Antitermination then depends on the nonspecific binding of N to the nascent transcript.[11,19] A calculation also presented in Fig. 4 represents the estimated local concentration of N as a consequence of RNA looping for N bound to the nascent transcript, nonspecifically and randomly, at an average binding density of one N per transcript. Since the binding of N can, in principle, occur anywhere on the nascent transcript, $[j_N]$ (or [N$_{loc}$]) is plotted (dashed line in Fig. 4) as a function of the length (in nucleotides) of the nascent RNA transcript from the transcription start site to the transcription complex.[11]

It is important to emphasize that these calculations have been performed using poly(rU) as a model for a nascent RNA chain, which means that the RNA transcript has been assumed to be both unstacked and free of elements of local secondary structure (i.e., stem–loop hairpins). Real transcripts will not only carry significant amounts of secondary structure, but will also show base–base (especially purine–purine) stacking. Both effects will shorten the effective contour length of the overall nascent transcript, and also decrease the average (per unit length) flexibility of the transcript relative to the poly(rU) model. A rigorous implementation of such looping calculations for specific transcripts will require better estimates or measurements of the RNA persistence lengths and flexibilities of such transcripts.

N-Dependent Antitermination Dependent on Two Binding Equilibria

We now suggest that two distinct equilibria control the interaction of N with the elongating RNA polymerase. These equilibria are shown schematically in Fig. 6A and B. The first equilibrium describes the binding of N to (specific or nonspecific) sites on the nascent transcript, which is a prerequisite for the induction of the local concentration effect by *cis*-RNA looping. N cannot associate with $RNAP_{elong}$ under physiological conditions if it does not bind to the nascent transcript.[19] As a consequence, antitermination is strictly dependent on the presence of a high-affinity RNA-binding site for N on the transcript. Since a tight-binding boxB site is present only if the wild-type *nut* sequence has been transcribed, this equilibrium is the controlling feature that establishes the terminator specificity of the N-dependent antitermination mechanism.

Successful binding of N to boxB sets up the second required binding equilibrium; this one between the RNA-tethered N protein and $RNAP_{elong}$. Once N is bound to boxB, the fraction of transcription complexes that are modified by N depends on the affinity of the N–boxB complex for RNAP and on the contour length and flexibility of the nascent RNA chain between the boxB hairpin and the transcription complex at a terminator. The important conclusion from this model is that this second binding equilibrium determines the distance downstream from boxB that N can function (i.e., the range of the antitermination effect) and thus the terminator specificity of N-dependent antitermination.

Regulation of Three Parameters of N-Dependent Antitermination by RNA Looping. The RNA looping effect has significant implications for how the remaining antitermination (accessory) factors affect N function. To address the potential mechanisms by which the various "layers of speci-

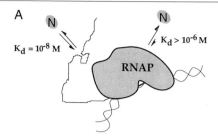

$$[N_{tot}] = 10^{-7}\ M$$

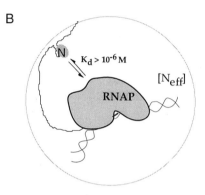

FIG. 6. Two equilibria control N-dependent antitermination. (A) The concentration of the N protein and the affinity of N for a site on the nascent RNA determine whether antitermination can occur, since N cannot interact with $RNAP_{elong}$ directly unless N is present at much higher than physiological concentrations. Thus binding of N to the high-affinity (boxB) site on the nascent RNA determines the specificity of antitermination. (B) Once N is bound to the nascent RNA transcript, the affinity of N for $RNAP_{elong}$ and the local concentration that results from RNA looping determine whether N can bind and modify $RNAP_{elong}$. Since the local concentration of N in this system is dependent on the length and flexibility of the nascent RNA, changes in the affinity of N for $RNAP_{elong}$ affect the range (template distance from the *nut* site to the terminator) over which N can function.

ficity" involved in N function might assist the N-dependent antitermination reaction, it is important to define clearly the parameters that these layers can affect. Using the models presented in this chapter, we define three different parameters of N-mediated antitermination.

"EFFICACY" OF ANTITERMINATION MODIFICATION BY N. As described previously, the amount of interaction free energy that N can contribute to the modification of the elongation complex ($\Delta\Delta G_{at}^{\ddagger}$) determines the maximum change in TE (ΔTE_{max}) that can occur if all the elongation complexes are fully bound by N. This parameter, which we call the "efficacy of antitermination,'' reflects the amount of binding free energy of the N–RNAP$_{\text{elong}}$ interaction that is coupled to the antitermination modification. An accessory factor could, in principle, alter the efficacy of antitermination either by directly contributing to the value of $\Delta\Delta G_{\text{at}}^{\ddagger}$ itself or by altering the amount of free energy that N binding couples to the antitermination modification. However, since our model assumes that N alone is responsible for the antitermination modification, we assume that N-dependent antitermination accessory factors will alter the efficacy of N function by the latter mechanism.

FRACTION OF TRANSCRIPTION COMPLEXES THAT ARE SUSCEPTIBLE TO N-DEPENDENT ANTITERMINATION. Since we have shown that the affinity of N for RNAP$_{\text{elong}}$ is too low to permit direct binding under physiological conditions, only elongation complexes that have an N protein bound to the nascent RNA transcript are susceptible to N modification via RNA looping. Therefore the fraction of complexes that can receive the fundamental antitermination modification is strictly dependent on the fraction that contains an N protein bound to the RNA transcript ($f_{\text{RNA,B}}$). This parameter is defined by Eq. (8):

$$f_{\text{RNA,B}} = \frac{(K_d + [\text{N}_0] + [\text{RNA}_0]) \pm \sqrt{(K_d - [\text{N}_0] - [\text{RNA}_0])^2 - 4[\text{N}_0][\text{RNA}_0]}}{2[\text{RNA}_0]}$$

(8)

where $[\text{RNA}_0]$ is the total concentration of boxB RNA-binding sites, K_d is the affinity of N for the boxB RNA site, and $[\text{N}_0]$ is the total N concentration.

Equation (8) can also be used to calculate the potential for antitermination through nonspecific binding to the nascent transcript[11,19,27] if $[\text{RNA}_0]$ corresponds to the number of potential nonspecific RNA binding sites on the nascent transcript from which N can function through looping.[34]

Accessory factors can assist this step of antitermination by improving the binding of N to the nascent transcript (the boxB hairpin itself is an example of such an accessory factor), thereby lowering the total concentration of N protein that is required to manifest N-dependent antitermination activity. Importantly, the presence of high-affinity N-binding sites in the nascent RNA that raise the value of the f_{RNA} can result in the specific localization of N function to terminators located downstream of these sites.

---

[34] The number of nonspecific RNA-binding sites for N along the transcript that appears in Eq. (8) represents sites that can effectively place N into the proper interaction with RNAP$_{\text{elong}}$ by RNA looping. If N cannot function from certain nonspecific binding sites on the transcript (e.g., because these sites lie too close to the transcription complex for effective delivery of N to RNAP$_{\text{elong}}$), then those sites will serve merely to compete with functional nonspecific transcript sites for N binding, rather than as facilitators of N-dependent antitermination.

"RANGE" OF N-DEPENDENT ANTITERMINATION. Once N is bound to the boxB RNA hairpin on the nascent transcript, the fraction of elongation complexes that subsequently do receive the antitermination modification is determined by the net affinity of the N–boxB complex for $RNAP_{elong}$ [$K_d(_{NB–P})$] and by the contour length and flexibility of the RNA between boxB and the elongating RNAP. Therefore modifying the affinity of the N–boxB subassembly for the transcription complex will change the template (*nut* site to terminator) distance (range) over which N can act. Antitermination factors that increase the affinity of the N–boxB subassembly for $RNAP_{elong}$ would allow modification of transcription complexes at terminators located further downstream of the *nut* site. Thus the $K_d(_{N–B})$ parameter will control the "range" of the antitermination modification along the template.

Interactions between Three Antitermination Parameters. The preceding arguments lead to the conclusion that the actual value of ΔTE_{obs} that is realized under a given set of experimental conditions is determined by the product of the three parameters defined above. Thus we can write

$$\Delta TE_{obs} = (f_{RNA,B})(f_{P,B})(\Delta TE_{max}) \qquad (9)$$

where ΔTE_{max} is the efficacy of the antitermination mechanism calculated from Eqs. (1) and (2), $f_{RNA,B}$ is calculated by Eq. (8) using [N_{tot}], and $f_{P,B}$ is calculated according to Eq. (5A), using [N_{loc}] and [$RNAP_{loc}$]. It is apparent from the form of Eq. (9) that simply measuring the change in the observed TE does not tell us which of these mechanisms a particular antitermination factor may be affecting. However, the simple approach outlined here allows us to separate the various contributions to the N-dependent antitermination phenotype and to test them by working under conditions where one or more of the parameters is held at a fixed value (e.g., saturating N concentrations would lead to $f_{RNA,B} = 1$).

Finally, it is important to note that the preceding three parameters of the N-dependent antitermination process are not necessarily independent of one another, in that a given antitermination accessory factor may influence more than one of them.

Layer 1, Part B: Other Functional Aspects of N–RNAP$_{elong}$ Interaction

As shown in Fig. 5, we estimate that, in the absence of any Nus factors, the dissociation constant of N for $RNAP_{elong}$ [$K_{d(N–P)}$] must be greater than $\sim 10^{-5}$ M; i.e., the (favorable) binding free energy of the fundamental N–P interaction ($\Delta G_{N–P}$) must be ~ -6.9 kcal/mol or less. Since the "efficacy" of non-factor-dependent antitermination, as defined by $\Delta\Delta G^{\ddagger}_{at}$, is ~ -2.1 kcal/mol, ΔG_{other} could be as high as ~ -4.8 kcal/mol [Eq. (3)]. Although the ΔG_{other} term [Eq. (3)] does not provide a direct contribution to the fundamental N-dependent modification of $RNAP_{elong}$, it can affect the range of antitermination (as defined above) since it increases the binding affinity of N for $RNAP_{elong}$. This suggests that the interactions of the N protein with $RNAP_{elong}$ that are utilized to produce the antitermination modification may be separable from the interactions that merely affect the range of antitermination. Therefore, it may be possible to isolate mutations in N that affect the magnitude and range parameters of antitermination differently, thus helping to define domains of the N protein that are responsible for these activities.

Layer 2, Part B: Other Functional Interactions of boxB RNA Hairpin

If we set aside the role of the boxB hairpin in controlling the local concentration of N ($[N_{loc}]$) at the transcription complex, we can now ask whether boxB plays an additional role in N function that results in an effect on one of the other parameters of antitermination. To begin this dissection, we will consider N-dependent antitermination in the absence of any of the Nus factors of *E. coli.*

boxB and Fundamental Modification of $RNAP_{elong}$. N can function in the absence of all the other antitermination factors, suggesting that N alone can ultimately modify the elongation complex if conditions are made favorable for the N–$RNAP_{elong}$ association.[16,19] However, the boxB RNA hairpin could, in principle, still participate in the control of the efficacy of the N-dependent antitermination modification (i.e., the ΔTE_{max} parameter) by enhancing $\Delta \Delta G_{at}^{\ddagger}$.

boxB-independent antitermination can occur at moderate salt concentrations ($\sim$50 mM KCl) when one N protein is bound nonspecifically to the nascent RNA transcript.[19,27] Under these conditions, nonspecific antitermination differs from boxB-dependent antitermination only in the presence or absence of a boxB RNA hairpin bound to the amino-terminal domain of N (Fig. 7). Experimental measurements have shown that under conditions where transcription complexes are saturated with N, the efficacy of termination (ΔTE_{max}) is identical with boxB-containing and ΔboxB templates (Table I).[19,27] These results strongly argue that the participation of the boxB RNA hairpin in the N–boxB–$RNAP_{elong}$ complex does not improve the

ARM		NusA		RNAP

N   22   34   51   70   C

FIG. 7. The interaction domains of N protein. The positions along the N polypeptide chain of the proposed RNA, NusA, and $RNAP_{elong}$ interaction domains.[35]

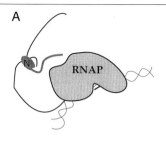

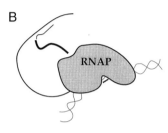

FIG. 8. Models for specific and nonspecific N-dependent antitermination. (A) boxB provides a high-affinity binding site for N. (B) An N protein bound nonspecifically to the nascent transcript presents the antitermination domain of N protein to $RNAP_{elong}$ in the same conformation as does N bound specifically to the boxB site on the nascent transcript.

efficacy of the fundamental $N–RNAP_{elong}$ modification. These functional considerations are consistent with biophysical findings that suggest that N has (at least) two structurally independent domains, one of which (the amino-terminal domain) is responsible for RNA binding, while the other (the carboxy-terminal domain) participates in the modification of $RNAP_{elong}$.[11,14,20,27,35] Therefore we postulate that the domain of the N protein that is responsible for the antitermination modification is presented to $RNAP_{elong}$ in the same conformation, regardless of whether N is bound specifically or nonspecifically to the nascent RNA transcript (Fig. 8).[27]

boxB and Range of N-Dependent Antitermination. Several mutations in the boxB RNA hairpin significantly compromise antitermination *in vivo*.[9,10] Some of these mutations abolish specific N binding, abrogating the capacity of boxB to render complexes susceptible to antitermination through the local concentration effect (i.e., $f_{RNA,B} = 0$). However, there also exist antitermination-deficient mutations in boxB that do not significantly affect N binding.[9–11,27] Whether or not these small binding deficiencies are large

[35] J. Mogridge, P. Legault, J. Li, M. D. Van Oene, L. E. Kay, and J. Greenblatt, *Mol. Cell* **1**, 265 (1998).

enough to abrogate N–RNA binding *in vivo* is still not clear; however, for the sake of the present discussion we will assume that the weakened binding affinities of these hairpins are still sufficient to support N–boxB association *in vivo* (therefore $f_{RNA,B}$ is ≈ 1). Because we have demonstrated that the involvement of boxB does not alter the fundamental N-dependent modification of $RNAP_{elong}$ (ΔTE_{max}), it is most likely that these mutations operate by perturbing the role of boxB in determining the range of the antitermination modification (i.e., $f_{P,B}$), since λ growth depends on readthrough of terminators located several thousand bases downstream of the *nut* site.

boxB could increase the range of N function directly by enhancing the ability of the N–boxB complex to bind to $RNAP_{elong}$, or indirectly by assisting other antitermination factors in increasing the affinity of the N–boxB complex for $RNAP_{elong}$. In the direct mechanism, boxB could carry out its effect by inducing an allosteric change in N that increases the affinity of N for $RNAP_{elong}$, or simply by interacting with the elongating transcription complex itself to enhance the binding affinity of the N–RNA complex for $RNAP_{elong}$ (we recall that this latter binding does not alter ΔTE_{max}). As a consequence, the binding free energy of the N–boxB complex for $RNAP_{elong}$ can be written

$$\Delta G_{(NB-P)} = \Delta G_{N-P} + \Delta G_{B-P} + \Delta G_{coop} \qquad (10)$$

where ΔG_{N-P} is the free energy of the N–$RNAP_{elong}$ binding interaction, ΔG_{B-P} is the binding free energy for the boxB–$RNAP_{elong}$ interaction, and ΔG_{coop} represents any cooperative interactions between boxB and N that may affect the overall affinity of the N–boxB complex for $RNAP_{elong}$.

A Nus-independent contribution of boxB to the antitermination range parameter could be experimentally demonstrated either by a direct measurement of the dissociation constant of N for $RNAP_{elong}$ in the presence and absence of the boxB RNA hairpin, or by measuring the ability of N to induce readthrough of terminators located at various distances downstream of wild-type and mutant boxB sites. If the boxB mutants that do not affect N binding decrease the range of antitermination, then they should exhibit both an increased $K_d(_{NB-P})$ and function less efficiently than wild-type boxB hairpins at terminators located greater distances downstream of boxB RNA.

Layer 3: Contribution of NusA to Antitermination

N-mediated antitermination in phage λ is also dependent on the presence of the NusA protein. In fact, N, together with the boxB RNA hairpin and NusA, suffices to modify $RNAP_{elong}$ into a metastable termination-resistant form that can read through terminators located relatively short distances downstream of the *nut* site *in vitro* (i.e., at template lengths of ~200–300 bp from boxB to the terminator).[15,16]

NusA as Elongation Factor That Binds to N Protein. NusA binds to RNAP$_{elong}$ with a $K_{d(NusA-RNAP_{elong})}$ of ~10^{-7} M.[12] In the absence of N, NusA operates to increase the termination efficiency at intrinsic terminators, possibly due to a decrease in the rate of transcript elongation.[24,36] NusA also binds directly to the N protein with a $K_{d(NusA-N)}$ of ~$10^{-7}M$, suggesting that NusA might help to increase the overall affinity of N for the RNAP$_{elong-NusA}$ complex.[14]

NusA Enhancement of Efficacy of N-Dependent Modification of RNAP$_{elong}$. Under *in vitro* conditions in the presence of NusA, saturating concentrations of N protein change the observed termination efficiency at the $t_{R'}$ terminator from ~0.90 to ~0.10 ($\Delta TE_{max} = 0.80$) (Table I).[19,27] This corresponds to a $\Delta\Delta G_{at}^{\ddagger}$ value of ~-2.6 kcal/mol [see Eqs. (1) and (2)], which differs from the value of $\Delta\Delta G_{at}^{\ddagger}$ of ~2.1 kcal/mol that applies in the absence of NusA (Table I). Therefore, NusA appears to increase the amount of N–RNAP$_{elong}$ binding free energy that is coupled to the N-dependent antitermination modification by ~-0.5 kcal/mol. An alternative explanation could state that NusA directly contributes its own NusA–RNAP$_{elong}$ binding free energy to the antitermination modification; however, this explanation is less likely because it would necessitate a complete switch of the effect of NusA on the transcription complex on the addition of N. Thus it is much simpler to suggest that NusA merely enhances the ability of N to carry out its antitermination modification.

The ability of NusA to enhance the efficacy of the N-dependent antitermination modification does not require boxB. *In vitro* experiments performed with NusA under saturating N concentrations reveal that ΔTE_{max} (and thus $\Delta\Delta G_{at}^{\ddagger}$), is the same whether or not a boxB hairpin is present on the nascent RNA transcript (Table I).[27]

NusA Inhibition of Nonspecific Interactions of N with Nascent RNA Transcript. Higher levels of N protein are required to induce antitermination (at terminators located 100–200 bp downstream of the transcription start site) in the absence of boxB when NusA is present. Since we have established that the total concentration of N determines whether or not N will bind to the RNA transcript, this result suggests that NusA somehow interferes with the nonspecific binding of N to the nascent transcript. Biophysical experiments have demonstrated that NusA does not affect the nonspecific binding of N to RNA in the absence of RNAP$_{elong}$. However, it is known that NusA is in close contact with the nascent RNA in the transcription complex. Therefore, it is possible that NusA could reduce the number of nonspecific sites along the RNA transcript from which N can

[36] T. D. Yager and P. H. von Hippel, in *"E. coli* and *S. typhimurium:* Cellular and Molecular Biology" (F. Neidhardt, ed.), Vol. 1, p. 1241. Am. Soc. Microbiol. Washington, D.C., 1987.

effectively induce antitermination, either by occluding potential binding sites or by restricting the number of sites that can bring N into the proper position for interaction with $RNAP_{elong}$ by looping.

Contribution of NusA to Range of Antitermination. If we consider the NusA protein to be tightly bound to the transcription complex, then the binding of the N-boxB complex to NusA could contribute to $\Delta G_{(NB-P)}$ as follows:

$$\Delta G_{(NB-PA)} = \Delta G_{NB-NusA} + \Delta G_{NB-P} + \Delta G_{coop} \qquad (11)$$

where $\Delta G_{(NB-PA)}$ represents the binding free energy of the association of the N–boxB complex with $RNAP_{elong}$–NusA, $\Delta G_{NB-NusA}$ is the binding free energy of the N–boxB complex for NusA, ΔG_{NB-P} corresponds to the affinity of N–boxB for $RNAP_{elong}$, and ΔG_{coop} is a general term that includes any positive or negative cooperativity between components that may result from geometric constraints or conformational changes.

If we insert the value of $\Delta G_{NB-NusA}$ measured for the N–BoxB–NusA binding interaction (~ -9.7 kcal/mol)[14] into Eq. (11), and estimate ΔG_{NB-P} to be ~ -6.9 kcal/mol [for an estimated $K_{d(NB-P)}$ of $\sim 10^{-5}$ M; see Fig. 5]; then, if the cooperativity term of Eq. (11) is equal to zero, the value of $\Delta G_{(NB-PA)}$ is ~ 16.6 kcal/mol. This free energy of interaction translates to a $K_d(NB-PA)$ of $\sim 10^{-12}$ M. Therefore, the majority of the N–NusA binding free energy is not utilized to enhance the N–boxB–$RNAP_{elong}$ binding interaction. In fact, Fig. 5 shows that since the presence of NusA does not result in readthrough of terminators located 7000 bp downstream of the *nut* site, the contribution of NusA to the antitermination modification of the transcription complex must be much less than -4.2 kcal/mol. These results suggest that either the binding of N to NusA is severely compromised in the context of the transcription complex, or that the NusA-binding site is not positioned both to permit N to bind efficiently to NusA and to effectively induce N-dependent antitermination.

Cooperative Interactions between boxB and NusA. In the absence of the transcription complex, the affinity of N for the boxB RNA hairpin is the same, whether or not NusA is also bound to N.[14] However, functional experiments reveal that NusA prevents boxB RNA added in *trans* from competing for N binding with the boxB hairpin on the transcription complex.[19,27] This same stabilization is observed at higher salt concentrations. Therefore, NusA must somehow improve the interaction of N with the boxB RNA hairpin within the context of the transcription complex. NusA also appears to stabilize nonspecific N–RNA binding, although to a lesser degree than the specific N–boxB interaction. Thus, NusA must stabilize the N–RNA interaction within the context of the elongation complex.

We conclude that NusA serves both to enhance the efficacy of the

N-dependent modification of $RNAP_{elong}$ and to stabilize the N–boxB–$RNAP_{elong}$ interaction. However, the complete establishment of the importance of N–NusA binding for N function requires more detailed study, including an exploration of the effect of this factor on the range of the N-dependent antitermination process.[27]

Layer 4: Role of NusB/E Heterodimer and BoxA

The NusB and NusE proteins form a heterodimer that interacts with the boxA element of the *nut* site, enabling antitermination to occur at terminators located thousands of base pairs downstream of *nut*.[7,8,10,15,16] However, there are numerous interactions that must be accounted for before we can estimate the effects of the NusB/E heterodimer on N-dependent antitermination. The NusB/E heterodimer binds efficiently to the consensus boxA sequence of ribosomal operons in the absence of the transcription complex. However, this binding interaction is weaker in the N-antitermination system than in the ribosomal system and may require the assistance of other antitermination components for stabilization.[7,8] BoxA binding does appear to be important in N-dependent antitermination, since using a template containing the consensus boxA element can suppress the N-mediated antitermination-deficient *nusA1* mutation. These results suggest the existence of a complex network of cooperative interactions between the *nut* site, the Nus proteins, N, and $RNAP_{elong}$ in the formation of the full antitermination complex.[7,8,10] Quantitative data on antitermination effects of NusB/E and boxA RNA are sparse, and the effect of these factors on the efficacy of antitermination has not been measured. Nonetheless, the addition of boxA and NusB/E does result in an overall increase in the stability of the N–RNA–$RNAP_{elong}$ complex and thus on the range of the N-dependent antitermination modification, and allows N to function several thousand base pairs downstream of the *nut* site.[15,16] It remains to be shown whether the formation of this larger N-dependent antitermination complex can still be treated in a totally equilibrium context, or whether some kinetic considerations will need to be introduced.[8,10,27]

Effect of NusB/E–boxA on Range of Antitermination. A reduced antitermination complex consisting of N, NusA, boxB, and $RNAP_{elong}$ can function at terminators located 200–300 bp downstream of the *nut* site.[15,16] The NusB/E heterodimer and the boxA RNA element can increase this range up to 7000 bp.[16] By examining Fig. 5 it is apparent that antitermination at these distances requires that the affinity of the N–boxB RNA complex for $RNAP_{elong}$ correspond to a value of $K_d \sim 10^{-8}\ M$. Therefore, if the addition of NusB/E and boxA RNA to the system drives the required value of the overall K_d from $\sim 10^{-5}$ to $\sim 10^{-8}\ M$, then, using Eq. (8), we can calculate that these factors stabilize the interaction of N with $RNAP_{elong}$ by ~ -4.2 kcal/mol.

If we assume that the NusB/E heterodimer forms a tight complex with the N–boxB/boxA RNA, then the affinity of this complex for $RNAP_{elong}$ is determined by

$$\Delta G_{(NBBEA-PA)} = \Delta G_{NB-PA} + \Delta G_{BEA-PA} + \Delta G_{coop} \qquad (12)$$

where $\Delta G_{(NBBEA-PA)}$ corresponds to the binding free energy of the N–boxB–NusB/E–boxA complex for RNAP–NusA, ΔG_{NB-PA} is equal to the binding free energy of the N–boxB complex for $RNAP_{elong}$–NusA [Eq. (11)], ΔG_{BEA-PA} is equal to the binding free energy of the NusB/E–boxA complex for $RNAP_{elong}$–NusA, and ΔG_{coop} is a general term that includes the binding free energies of all the various cooperative interactions (positive and negative) that may occur within this complex.

Although we can estimate a value of $\Delta G_{(NBBEA-PA)}$, just how NusB/E–boxA does contribute to the range of antitermination is not currently clear. The most important question is whether much of this contribution depends on direct interaction of the NusB/E–boxA complex with $RNAP_{elong}$ (ΔG_{BEA-PA}), or whether this effect is mediated predominantly through cooperative interactions within the overall complex (ΔG_{coop}). Biophysical experiments to probe these interactions have not been done, and although some functional experiments that systematically "drop out" the various antitermination components at this level have been performed,[15,16] such experiments cannot be used to distinguish between the preceding mechanistic possibilities unless the changes in the range over which N-dependent antitermination is effective in the presence and absence of each of these factors is determined.

Summary and Perspectives

In this chapter, we have presented a detailed approach to the assembly of the N-dependent antitermination complex of phage λ, and have attempted to exploit this methodology to estimate the contributions of the individual components to the various parameters of the antitermination mechanism. Where possible, we have also incorporated and interpreted existing results within the context of this approach. Our models reveal that "layers of specificity" may control three different characteristics of N-mediated antitermination. These are (1) the efficacy of the N-dependent antitermination modification, which reflects the actual mechanism of N, (2) the fraction of complexes susceptible to antitermination, which depends on the total concentration of N and the affinity of N for a binding site on the nascent transcript to provide higher local concentrations of N to RNAP$_{elong}$ via RNA looping, and (3) the range of the antitermination modification, which reflects the functional stability of the N–RNAP$_{elong}$ interaction and its relation to the length of the RNA transcript.

We have also provided a quantitative description of how accessory factors may affect each of these characteristics of the antitermination mechanism. Although some quantitative data have been obtained that can be used to implement and test this model, it is also apparent that several of the ideas presented are based largely on theoretical calculations or assumptions that have not been fully proven. Additional biophysical measurements and functional experiments are needed to further define the contributions of antitermination factors to the mechanisms and stability of N-mediated antitermination.

Future experiments clearly need to focus on careful measurement of the effects of the antitermination factors on the range of N-dependent antitermination, perhaps by means of a systematic study of the capacity of reduced antitermination complexes to operate as a function of template distance between the *nut* site and the terminators at which N acts, and a correlation of such effects with studies of the contour lengths and flexibilities of real RNA transcripts. In addition, parallel biophysical experiments are needed to measure the N–RNAP$_{elong}$ binding interaction in the presence and absence of antitermination components, and in the absence of local concentration effects due to specific and nonspecific RNA looping.

Finally, the approach that we have taken here focuses on well-characterized *in vitro* antitermination complexes. *In vivo,* other cellular processes are likely to be involved in the regulation of antitermination, including the degradation of N by the *lon* protease, the autoregulation of N translation, inhibitors of boxA binding, etc. (e.g., see Ref. 37). We assert, however, that after the basic assembly process of the N-dependent antitermination complex is well understood, this type of model should be extendable to incorporate other types of regulation as well.

Multilayered Regulatory Processes

As we have seen in our consideration of the N-dependent antitermination complex of λ, the assembly of such a complex often represents what an engineer might describe as a "sloppy" process. Thus, although the individual interactions that contribute in the antitermination complex might suggest an amazingly stable outcome, this is not observed. Rather, much

---

[37] S. Gottesman, M. Gottesman, J. E. Shaw, and M. L. Pearson, *Cell* **24,** 225 (1981).

of the potential free energy available for assembling the complex is not utilized, suggesting that the complex is makeshift and not optimized for stability. However, the resulting complex is optimal for specific regulation of the activation of the delayed early genes of λ. This follows because the balance of all the contributions that make up an antitermination complex has evolved so that phage growth is dependent on several weak interactions between factors that, collectively, control the crucial timing and location of an antitermination event. We have shown here, quantitatively, that because the N-dependent antitermination complex exists in such a balanced state, subtle changes in the interactions that form this complex can have large effects on the ability of N to activate expression of the delayed early genes in phage λ.

The approach that we have presented here can be applied to other types of "multilayered" regulatory processes, with the general method of analysis involving several steps: (1) The action of the regulatory factor under study is first described theoretically to determine how much "modification free energy" must be contributed to the overall complex to make the regulatory function "perform" within the biological system, as well as to determine how this contribution can be recognized and dissected out from other interactions; (2) the function under study is then examined to describe how it depends on other (accessory) factors, which we have denoted as the "specific layers" of interaction that control specificity of function. Specificity factors may involve RNA, DNA, or even protein looping, or other types of localization, such as sliding, tracking, bending, and recruitment; and (3) the "durability" (or range) of the function must be analyzed, in order to determine how regulatory factors work together to assure that the system under study holds together in the presence of adequate concentrations of the other regulatory factors involved in the system. In principle, the system should be "open-ended," so that additional regulatory factors and interactions can be introduced to permit adaptation to complexities or intrusions that may develop as a consequence of evolution or the interaction of the organism with the environment.

We have used a completely thermodynamic approach in considering the design of the N-dependent antitermination system of phage λ, since this regulatory subsystem seems to operate totally "within" the overall kinetic competition between elongation and termination. However, it is clear that in N-dependent antitermination processes, as well as in other systems, kinetic steps may be "woven in" to the story as well, and analysis of new mechanisms should always focus initially on the dissection of the system into its thermodynamic and kinetic parts, with due attention being paid to a possible shift in the rate-limiting step of the overall process as a consequence of regulatory changes. In principle, this approach should be

useful for designing and interpreting experiments to discover the mechanisms by which any regulatory complex assembles and operates.

Acknowledgments

This work was submitted to the Graduate School of the University of Oregon (by M.R.V.G.)
in partial fulfillment of the requirements for the Ph.D. in Chemistry. The studies described
were supported in part by NIH Research Grants GM-15792 and GM-29158 (to P.H.v.H.).
M.R.V.G. was a predoctoral trainee on USPHS Institutional Training Grant GM-07759.
P.H.v.H. is an American Cancer Society Research Professor of Chemistry. We thank Karsten
Rippe for help with the RNA looping calculations and for the preparation of Fig. 4, and Feng
Dong and Mark Young for assistance with some of our theoretical approaches. Finally, we
thank Asis Das (University of Connecticut), Jonathon Weisman (UCSF), William Rees, and
other colleagues in our laboratory at Oregon and elsewhere for many helpful and stimulating discussions.

[2] Problems and Prospects in Microcalorimetry of Biological Macromolecules

By GEORGE P. PRIVALOV and PETER L. PRIVALOV

Introduction

Interest in calorimetric studies of biological macromolecules, in particular proteins, nucleic acids, and their complexes, is rapidly growing with the
realization of the importance of information on the energetics of formation
of macromolecular structures and their interactions with partner molecules.
Correspondingly, the demands placed on calorimetric instruments and experiments are increasing, as well as on the methods of analyzing calorimetric
data. In the past calorimeters were used only for measuring the heat effects
of sharp temperature-induced transitions; now the main interest is concentrated on studying the heat effects of complicated changes in macromolecules and their complexes under varying conditions, on measuring the
absolute values of thermodynamic parameters specifying the states of these
molecules, and on correlating this thermodynamic information with structural information. This has led to the development of more sensitive calorimetric instruments, so-called nanocalorimeters, and special programs for
analysis of calorimetric and structural information. The question now is
how to use these instruments and programs effectively and how to utilize
these two pools of information to gain a deeper understanding of biological
macromolecules.

METHODS IN ENZYMOLOGY, VOL. 323

Calorimetry versus Equilibrium Analysis in Studying
Intramacromolecular Processes

A study of the thermodynamics of the unfolding/refolding of macromolecules under varying external conditions by analyzing the equilibrium, as observed by some "sensitive to the state" parameter (e.g., optical), is much easier and less expensive than by calorimetry, which requires rather expensive instrumentation. From an equilibrium analysis one can obtain the equilibrium constant as a function of some variable parameter and this can then be converted into a Gibbs energy specifying the studied process under standard conditions. This, however, is possible only if (1) the observed process is reversible, (2) it involves only two states, and (3) we know how to convert the equilibrium constant determined under the real conditions into a Gibbs energy under standard conditions. The unfolding/refolding of proteins induced by denaturants (urea or guanidinium chloride) is a reversible process in many cases, but it is unclear how closely this process is approximated by a two-state transition. Even more so, it is unclear how to extrapolate the equilibrium constant determined at relatively high concentrations of denaturant to zero concentration, so as to obtain the Gibbs energy specifying the stability of the protein in the absence of denaturant. Usually this is done by assuming that the logarithm of the equilibrium constant is a linear function of denaturant concentration,[1] but this is only an assumption that does not have any theoretical justification. The temperature-induced unfolding/refolding of proteins is reversible under some solvent conditions preventing aggregation, but again it is not always clear how close it is to a two-state transition. The usual test of the two-state transition, i.e., full cooperativity of unfolding, is the synchronous change of all parameters sensitive to the state of the protein on variation of the external conditions. However, if the protein consists of several domains, which have similar stabilities, these domains can unfold independently over the same range of external conditions, mimicking a two-state transition for the whole molecule.

There is only one way to test how closely the process is approximated by a two-state transition: this is to compare the effective thermodynamic parameter of the process derived by equilibrium analysis of the changes observed under varying conditions (assuming that it is a two-state process) with the same thermodynamic parameters directly measured. However, if we are studying protein unfolding/refolding as a function of ligand concen-

---

[1] C. N. Pace, *Trends Biochem. Sci.* **8**, 93 (1990).

tration, the parameter that we can determine by equilibrium analysis is the number of ligands, $\Delta \nu^{\text{eff}}$, bound by protein on unfolding:

$$\Delta \nu^{\text{eff}} = \frac{\partial \ln K^{\text{eff}}}{\partial \ln a_i}$$

where K^{eff} is the effective equilibrium constant and a_i is the activity of the ligand (cosolute). If the ligand is a proton, we can measure potentiometrically the number of protons bound by the protein on acid-induced denaturation and check the assumption that it is a two-state process. If the ligand is urea or guanidinium hydrochloride, we do not have an experimental method to measure the number of denaturant molecules bound on unfolding. If we are studying a temperature-induced process, the equilibrium analysis can give us the effective, i.e., the so-called van't Hoff, enthalpy of this process:

$$\Delta H^{\text{eff}} = RT^2 \frac{\partial \ln K^{\text{eff}}}{\partial \ln T}$$

This effective enthalpy we can compare with the calorimetrically measured enthalpy and decide whether the observed process is a two-state transition.[2] Furthermore, we can determine the effective, van't Hoff enthalpy from the sharpness of the heat absorption peak, i.e., its relative height:

$$\Delta H^{\text{eff}} = 4RT^2 \frac{\Delta C_p(T_t)}{Q}$$

where $\Delta C_p(T_t)$ is the heat capacity peak height at the transition temperature T_t and Q is the total heat absorbed in the peak.[2] A much more efficient method of analysis is, however, to compare the experimental heat capacity curve with the computer-simulated heat capacity function.[2-4]

$$\text{Cp}(T) = \text{Cp}^0 + \frac{\partial}{\partial T}\left(\sum_{i=1}^{N} P_i \Delta H_i\right)$$

where Cp^0 is the partial molar heat capacity of the reference state. The second term accounts for any temperature-induced transitions and the possibility that the transition involves an arbitrary number of states, N. In this equation P_i represents the population of molecules in the ith state and ΔH_i its relative enthalpy. Expanding the above equation and bearing in

[2] P. L. Privalov and S. A. Potekhin, *Methods Enzymol.* **131**, 4 (1986).
[3] E. Freire, *Methods Mol. Biol.* **40**, 191 (1995).
[4] E. Freire, *Methods Enzymol.* **259**, 144 (1995).

mind that the population of each state is a function of the Gibbs free energy of the state, ΔG_i, we have

$$\mathrm{Cp}(T) = \mathrm{Cp}^0 + \left\{ \left[\sum_{i=1}^{N} \Delta H_i^2 \exp(-\Delta Gi/RT)/Q \right] \right.$$
$$\left. - \left[\sum_{i=1}^{N} \Delta H_i \exp(-\Delta Gi/RT)/Q \right]^2 \right\} \bigg/ RT^2$$

where Q is the partition function and $\Delta G_i(T) = \Delta H_i(T) - T\Delta S_i(T)$,

$$\Delta H_i(T) = \Delta H_i(T_t) + \int_{T_t^i}^{T} \Delta C_p^i(\mathrm{T}) \, dT$$

and $\Delta S_i(T)$ for a monomolecular transition is

$$\Delta S_i(T) = \frac{\Delta H_i(T_t^i)}{T_t^i} + \int_{T_t^i}^{T} \Delta C_p^i(T) \, d\ln T$$

Here $\Delta H_i(T_t^i)/T_t^i = \Delta S(T_t)$ is the entropy change at the transition point T_t, since for a monomeric reaction $\Delta G(T_t) = \Delta H(T_t) - T_t\Delta S(T_t) = 0$.
For the multimeric reaction, $X \rightleftharpoons m_1 + m_2 + \cdots + m_k$, the entropy

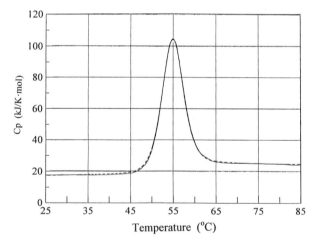

FIG. 1. The partial molar heat capacity function of barnase in 20 mM phosphate buffer, pH 5.5, measured by Nano-DSC (Calorimetry Science Corporation, SCS). The solid curve is the calorimetrically determined function; the dashed curve is computer simulated, assuming that protein unfolding represents a two-state transition with the parameters: $T_t = 54.8°$, $\Delta H_t = 541$ kJ mol^{-1}, $\Delta C_p = 6.43$ kJ K^{-1} mol^{-1}.

change at the transition point, T_t, with half of the molecules unfolded, is determined as

$$\Delta S_i(T_t) = \frac{\Delta H_i(T_i) + RT \ln K}{T_t}$$

where $K = (N/2)^{n-1} \prod_{i=1}^{k} (m_i)^{(m_i)}$, $N = N^0/N^{st}$ is the dimensionless initial concentration of the complex relative to the standard concentration and $n = \sum_{i=1}^{k} m_i$ is the order of reaction.[2,5]

Figure 1 shows the results of a calorimetric study of barnase. The experimental heat capacity function is almost perfectly simulated by the function calculated by assuming that the temperature-induced process represents a single two-state transition. It appears that unfolding of barnase is an extremely cooperative process. This means that the population of all intermediate states on temperature induced unfolding of barnase is small.

Figure 2 shows the results of a study of hen egg white lysozyme. In this case the heat absorption peak corresponding to its temperature-induced unfolding is not well simulated by a single two-state transition, but is perfectly simulated by three-state transitions.[6] The unfolding of this molecule therefore proceeds through noticeably populated intermediate states. This is not surprising because in contrast to barnase, lysozyme is a two-domain protein and the fact that its unfolding is not a perfect two-state transition might mean that its domains unfold in a not fully cooperative fashion. Indeed, if we take the closely related protein, equine lysozyme, which does not differ much from hen egg white lysozyme, we will see that the two domains in this protein unfold quite independently and that the stability of one of them depends on the concentration of calcium in solution,[7] demonstrating that this domain specifically binds calcium ions (Fig. 3). It is notable that human lysozyme unfolds cooperatively, like hen egg white lysozyme, but after replacement of only two amino acid residues the domains in the mutant protein lose cooperativity and unfold at different temperatures and concentrations of urea.[8] The smallest changes in structure can thus result in significant changes in the thermodynamic properties of proteins, particularly as regards the cooperativity of unfolding.

[5] A. Tamura and P. L. Privalov, *J. Mol. Biol.* **273**, 1048 (1997).

[6] G. P. Privalov, V. Kavina, E. Freire, and P. L. Privalov, *Anal. Biochem.* **232**, 79 (1995).

[7] Y. V. Griko, E. Freire, G. Privalov, H. van Dael, and P. L. Privalov, *J. Mol. Biol.* **252**, 447 (1995).

[8] D. R. Booth, M. Sunde, V. Bellotti, C. V. Robinson, W. L. Hutchinson, P. E. Fraser, P. N. Hawkins, C. M. Dobson, S. E. Radford, C. C. F. Blake, and M. B. Pepys, *Nature (London)* **385**, 787 (1997).

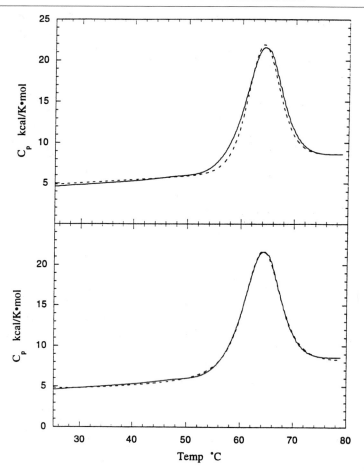

FIG. 2. The calorimetrically determined partial molar heat capacity of hen egg white lysozyme in 20 mM glycine buffer, pH 2.5 (the solid curve and the computer-simulated curve assume that the transition is two-state (dashed line, *top*) and three-state (dashed line, *bottom*).[6]

A similar situation was observed with T$_4$ lysozyme.[9] This protein also has two domains and in the wild-type protein these domains strongly cooperate. However, the cooperation of these domains can be decreased dramatically by a single point mutation: Fig. 4 shows the result of replacing serine by lysine at position 44 in T$_4$ lysozyme. This not only destabilizes the protein and decreases the temperature range of its unfolding, but makes this unfolding much broader and less cooperative. Thermodynamic analysis of the

[9] J. H. Carra, E. C. Murphy, and P. L. Privalov, *Biophys. J.* **71**, 1994 (1996).

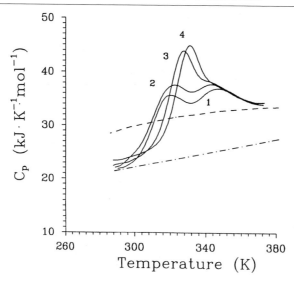

Fɪɢ. 3. Temperature dependence of the partial heat capacity of equine lysozyme in 10 mM sodium acetate (pH 4.5) in the presence of various concentrations of $CaCl_2$: curve 1, 0.00 mM; curve 2, 0.10 mM; curve 3, 0.75 mM; and curve 4, 1.50 mM $CaCl_2$. The dashed line represents the heat capacity function calculated for the completely unfolded polypeptide chain of equine lysozyme. The dash-dot line represents the heat capacity function of the native state of equine lysozyme obtained by linear extrapolation with the slope specific for the native state of globular proteins.[7]

heat capacity function shows that unfolding of this mutant protein proceeds through a highly populated intermediate state in which one of the domains is unfolded but the other is not, i.e., the domains of the mutant protein unfold almost independently.

Another example of this type is staphylococcal nuclease, Anfinsen's classic object for studying protein folding. This protein also has two domains, but unfolding of the wild-type protein is close to a two-state transition, which means that the domains cooperate almost perfectly. The unfolding of many mutants of this protein, however, is not a two-state transition (Fig. 5).[10,11] It is thus clear that substitution of some amino acid residues can significantly affect the interaction between the domains, resulting in loss of their cooperativity.

These examples show how dangerous it is to judge the effects of a mutation on the thermodynamics of protein folding if the protein is studied by equilibrium analysis of indirect parameters sensitive to the state of the

[10] J. H. Carra and P. L. Privalov, *Biochemistry* **34**, 2034 (1995).
[11] J. H. Carra and P. L. Privalov, *FASEB J.* **10,** 67 (1996).

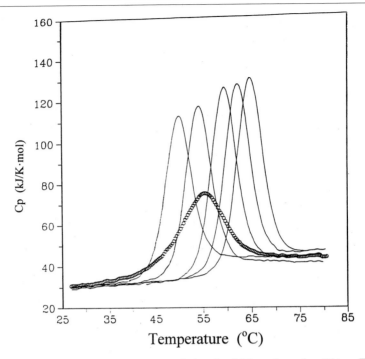

FIG. 4. Calorimetric study of temperature-induced unfolding of pseudo-wild-type T_4 lyso-zyme (solid lines) at pH values of 2.8, 3.0, 3.5, and 3.7 and of the S44A mutant protein at pH 4.0.[9]

protein. A change in folding/unfolding curves caused by replacements of amino acid residues, which is usually assigned to the enthalpic or entropic contributions of the replaced amino acid to the stabilization of protein structure, can simply be just the result of changes in the cooperativity of the unfolding process. Even a small change in the cooperativity can appear as a significant change in the van't Hoff enthalpy of temperature-induced unfolding, a parameter that is derived on the basis of the assumption that the cooperativity of unfolding of the mutant protein is the same as that of the wild-type protein and that the unfolding of both is close to a two-state transition. In the case of denaturant-induced unfolding, the change in cooperativity will result in a change in the dependency of the apparent Gibbs energy on the concentration of denaturants (i.e., in the change of parameter m) and, correspondingly, in a significant change in the Gibbs energy value at zero concentration of denaturant, which is used to specify the protein stability.[1]

Such small changes in the cooperativity of unfolding can be detected

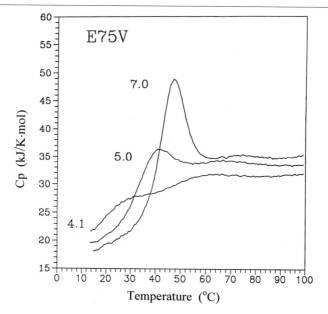

Fig. 5. Calorimetric study of temperature-induced unfolding of the E75V mutant of staphylococcal nuclease at pH 7.0, 5.0, and 4.1, showing that domains in this mutant unfold over separate temperature ranges and in different pH-dependent ways. [Reproduced from J. H. Carra and P. L. Privalov, *Biochemistry* **34,** 2034 (1995).]

only by scanning calorimetry, in which the stability of the protein is studied by temperature variation. If the stability of the protein is studied by variation of denaturant concentration (urea or guanidinium chloride) there is no experimental test to detect small changes in the cooperativity of protein unfolding. This is one of the reasons why numerous publications on the evaluation of the contributions of various amino acids to the stabilization of protein structures, obtained by equilibrium studies of unfolding of mutants, have not been much help in understanding the mechanisms of protein structure stabilization.

The Need to Measure Absolute Partial Heat Capacity of Macromolecules

The absolute value of the heat capacity of proteins and nucleic acids provides essential information about the state of the molecule. The heat capacity of a native compact protein is significantly lower than that of the unfolded protein[12]; the heat capacity of complexes is significantly lower

[12] P. L. Privalov, *Adv. Protein Chem.* **33,** 167 (1979).

than the sum of the heat capacities of the compact components.[13] This is because the absolute heat capacity depends both on the degrees of freedom of the macromolecule and on the extent of hydration of its groups. The degrees of freedom do not contribute much to the heat capacity of proteins,[14] but the heat capacity effect of hydration is significant (see below). The heat capacity change accompanying a protein structural transformation therefore provides information about the hydration changes occurring in this process, i.e., about the extent of exposed surfaces.

On the other hand, the heat capacity change in any process determines the temperature dependence of the enthalpy and entropy of this process, since

$$\left(\frac{\partial \Delta H}{\partial T}\right)_p = \Delta C_p(T) \qquad \left(\frac{\partial \Delta S}{\partial T}\right)_p = \frac{\partial C_p(T)}{T}$$

Therefore, if we know the heat capacity difference we can determine the enthalpy and entropy of the process at any other temperature, and if we know the enthalpy and entropy as a function of temperature we can determine the Gibbs energy difference between the considered states at any temperature:

$$\Delta G(T) = \Delta H(T) - T\Delta S(T) = \frac{T_t - T}{T_t} \Delta H(T_t)$$
$$+ \int_{T_t}^{T} \Delta C_p(T)\, dT - T \int_{T_t}^{T} \Delta C_p(T)\, d \ln T$$

The most important point here is that the Gibbs energy value is derived with experimentally determined parameters, i.e., it is not based on some assumed extrapolation procedure.

If the unfolding of a protein takes place over a short temperature range, one can determine the heat capacity difference by extrapolating the heat capacity of the native and unfolded states to the transition zone. This does not require knowledge of the absolute values of the heat capacities of either state. However, to extrapolate the heat capacity of the native and unfolded states into the transition zone we need to know the temperature dependencies of these heat capacities outside this zone, which are not necessarily the same.[15] We must therefore determine the partial heat capacities of these states over the pre- and posttransition temperature ranges. However, extrapolation of the apparent heat capacities of the pre- and postdenatur-

[13] R. S. Spolar and M. T. Reckord, *Science* **263**, 777 (1994).
[14] J. M. Sturtevant, *Proc. Natl. Acad. Sci. U.S.A.* **74**, 2440 (1977).
[15] P. L. Privalov, E. I. Tiktopulo, S. Y. Venyaminov, Y. V. Griko, G. M. Makhatadze, and N. N. Khechinashvili, *J. Mol. Biol.* **205**, 737 (1989).

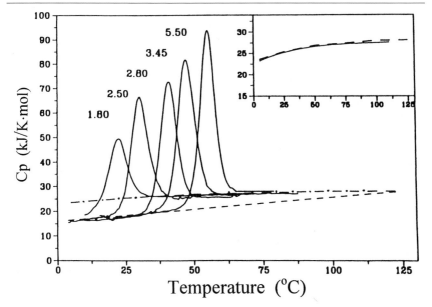

FIG. 6. Temperature dependence of the partial molar heat capacity of barnase in solutions with different pH values, indicated above the curves.[19] The dashed line represents the partial molar heat capacity of native barnase. The dash-dot line represents the partial molar heat capacity of unfolded barnase. *Inset:* Comparison of the temperature dependencies of the heat capacity of barnase obtained for irreversibly denatured protein by heating to 100° (solid line), with the heat capacity of the unfolded polypeptide chain (dashed line), calculated by summing the heat capacities of individual amino acid residues.[20]

ation ranges can result in significant error because we actually do not know where the transition starts and where it ends. Furthermore, the unfolding sometimes results in aggregation, which usually proceeds with a negative heat effect. If, however, the aggregation is not extensive, we cannot recognize its presence but it will nevertheless significantly change the slope of the heat capacity function of the unfolded protein. The best way to check the correctness of the heat capacity function of an unfolded protein is to compare it with that expected for ideally unfolded polypeptide chains, a function that can be calculated from the known partial heat capacities of amino acid residues.[16–18] This is illustrated in Fig. 6 for barnase.[19]

[16] G. I. Makhatadze and P. L. Privalov, *J. Mol. Biol.* **213,** 375 (1990).

[17] P. L. Privalov and G. I. Makhatadze, *J. Mol. Biol.* **213,** 385 (1990).

[18] P. L. Privalov and G. I. Makhatadze, "Physical Properties of Polymers Handbook" (E. Mark. ed.), p. 91. AIP Press, Woodbury, New York, 1996.

[19] Y. V. Griko, G. I. Makhatadze, P. L. Privalov, and R. W. Hartley, *Protein Sci.* **3,** 669 (1994).

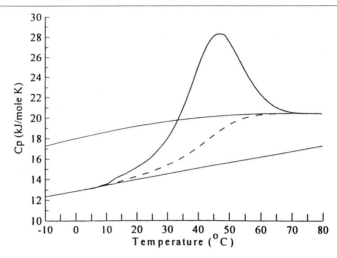

Fig. 7. The partial molar heat capacity of the HMG box from mouse Sox-5 protein.[22] The thin lines indicate the heat capacity functions expected for the protein with stable, compact native structure (lower line) and for the completely unfolded polypeptide chain (upper line).

The situation with the native state is more complicated because the heat capacity of this state depends not simply on the content of amino acid residues, but on the three-dimensional structure of the protein, i.e., on the water-exposed groups, and also on the fluctuations of the native structure, which can accumulate significant thermal energy with temperature rise.[20,21] However, the specific partial heat capacities (i.e., the heat capacity calculated per gram) of compact native globular proteins appeared to be similar.[21] They are on the order of (1.33 ± 0.13) J K^{-1} g^{-1} at 25°, and their temperature dependencies are between 5×10^{-3} and 7×10^{-3} J K^{-2} g^{-1}. If the heat capacity of a protein at 25° is significantly larger than 1.33 J K^{-1} g^{-1}, this is a clear indication that the protein is not fully folded. If the initial heat capacity slope of the protein is larger than 7×10^{-3} J K^{-2} g^{-1}, this is an indication that there is some temperature-induced process that proceeds with diffuse heat absorption, i.e., a process that has either a low enthalpy or a low cooperativity.

Let us consider as an example a small protein domain, the HMG box from mouse Sox-5 protein.[22] The molecular mass of this domain is about 10 kDa. On heating it shows significant heat absorption that is associated with its unfolding (Fig. 7). How can we determine the enthalpy of this

[20] P. L. Privalov and G. I. Makhatadze, *J. Mol. Biol.* **224,** 715 (1992).

[21] G. I. Makhatadze and P. L. Privalov, *Adv. Protein Chem.* **47,** 307 (1995).

[22] C. Crane-Robinson, C. M. Read, P. D. C. Cary, P. C. Driscoll, A. I. Dragan, and P. L. Privalov, *J. Mol. Biol.* **281,** 705 (1998).

process? If we interpolate the initial and final slopes of the heat capacity function to the transition zone we conclude that unfolding of this protein results in a heat capacity increment on the order of 1 kJ K^{-1} mol^{-1} and the enthalpy (the area of the heat absorption peak above the extrapolated lines) is about 150 kJ mol^{-1}. However, we can see that the heat capacity of this protein at 25° is about 16 kJ K^{-1} mol^{-1}, i.e., the specific heat capacity is 1.72 J K^{-1} g^{-1}, which is significantly larger than that expected for this size of compact globular protein. Also, the slope of the heat capacity curve is about 15 × 10^{-3} J K^{-2} g^{-1}, which is almost twice that expected for stable compact globular proteins. On the other hand, at 5° and lower temperatures, the partial heat capacity of this protein is close to the value expected of a compact globular protein at that temperature. It appears thus that the temperature-induced unfolding of this protein starts from low temperatures. The enthalpy of unfolding should then be calculated from the area above the reference slope specific for compact globular proteins of rigid structure. For the HMG box the total enthalpy is then calculated to be 232 kJ K^{-1} mol^{-1}, i.e., much larger than obtained by straightforward extrapolation of the initial slope. However, statistical–thermodynamic analysis of the excess heat effect shows that the heat absorption profile does not correspond to a two-state transition. In fact, this small protein domain unfolds in two stages, each of which is associated with a heat capacity increment (Fig. 8).[22]

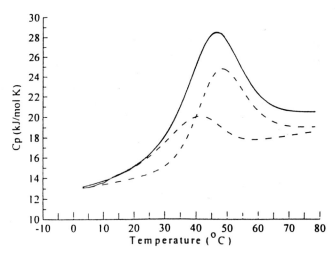

FIG. 8. Deconvolution of the partial specific heat capacity of the HMG box of mouse Sox-5 protein (solid line) into two components, each corresponding to a simple two-state transition (dashed lines).[22] The experimental functions and the function synthesized with the two-component model are almost indistinguishable on this scale. The thermodynamic parameters found for the component transitions are as follows: (1) T_t = 33.6°, $\Delta_t H$ = 95 kJ mol^{-1}, $\Delta_t C_p$ = 1.6 kJ K^{-1} mol^{-1}; (2) T_t = 44.7°, $\Delta_t H$ = 142 kJ mol^{-1}, $\Delta_t C_p$ = 3.0 kJ K^{-1} mol^{-1}.

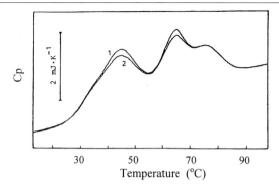

Fig. 9. Original calorimetric recordings of the apparent excess heat capacity of plasminogen on two consecutive heatings of the same solution (30 mM glycine, pH 3.4).[23] Reproducibility of the heat effect shows that the baseline is stable and that the process of unfolding is to a large extent reversible. The complex profile of the curves makes it clear that unfolding of plasminogen proceeds in several stages, each shifted in temperature.

With an increase in the number of domains in a protein or in the number of subunits in a complex, the situation with regard to the heat capacity becomes more complicated. In such cases the temperature-induced unfolding proceeds in several overlapping and rather diffuse stages (because of their small individual transition enthalpies), each with a characteristic $T_{t,i}$. The overall heat absorption curve appears complex, and it is far from clear at what temperature unfolding starts and at what temperature it ends (Fig. 9). Determination of the excess heat effect under such circumstances requires knowledge of the absolute heat capacity function in the temperature range from at least 0 to 100°. To determine it, an instrument with an extremely stable baseline over the whole of this temperature range is needed. By analyzing this absolute heat capacity function we can deconvolute it on the basis of several simple, two-state components of the observed complex process of heat absorption. In the case of plasminogen, analysis of the excess heat capacity curve showed that this protein unfolds in seven stages. Calorimetric study of the fragments of plasminogen showed that each stage corresponds to unfolding of individual domains, found to be seven in this molecule.[23]

The same problem is faced in studying α-helical peptides. The α helix is one of the basic structural elements in the proteins. Understanding the thermodynamics of its formation is thus of prime importance. It is assumed that the α helix is largely stabilized by hydrogen bonding but the enthalpy of hydrogen bonds is still a matter of debate. There have been many

[23] V. V. Novokhatny, S. A. Kudinov, and P. I. Privalov, *J. Mol. Biol.* **179**, 215 (1984).

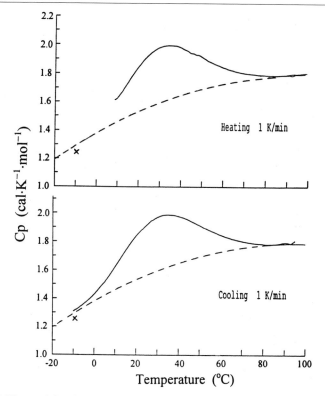

Fɪɢ. 10. The partial molar heat capacity of the α helix forming peptide, measured in the heating and cooling experiments.[24] The dashed line is the calculated heat capacity of the unfolded peptide. This 29-amino acid residue peptide with covalently closed terminal loops is 100% helical at $-10°$ and unfolds completely on heating to 100°.

attempts to measure the enthalpy of unfolding of the α helix calorimetrically and correspondingly evaluate the energetics of hydrogen bonding. These attempts, however, were not successful because an individual α helix unfolds over too extended a temperature range: at 20° it is partly unfolded, whereas at 100° it is still not completely unfolded. Measuring the enthalpy of its unfolding therefore requires an instrument with an extremely stable baseline and that can measure the absolute heat capacity below 20° and above 100°. The heat effect of unfolding an α helix with covalently closed terminal loops studied by differential scanning calorimetry (Nano-DSC; Calorimetry Science Corporation, SCS) is presented in Fig. 10.[24] This instrument can measure the heat capacity of aqueous solutions in a supercooled state down

[24] J. W. Taylor, N. J. Greenfield, B. Wu, and P. L. Privalov, *J. Mol. Biol.* **291,** 965 (1999).

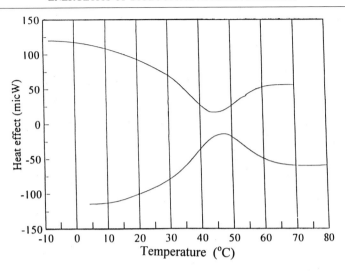

FIG. 11. Fragment of a calorimetric recording of the heat effects on scanning up and down of the HMG box from Sox-5 protein in a pH 5.5 solution.[22] The heat effects associated with unfolding/refolding of the protein appear as mirror images, somewhat shifted in temperature because of the different kinetics of unfolding and refolding. Heating and cooling rates: 1 degree (K) per minute.

to −10°, and up to 125° under excess pressure. Also, by closing covalently terminal loops we increased considerably the cooperativity of α-helix unfolding. Calorimetric experiments showed that the unfolding of such a helix consisting of 29 amino acid residues represents a two-state transition that proceeds with a significant heat effect and results in a heat capacity increase (Fig. 10). This means that unfolding of an α helix proceeds with exposure of nonpolar groups. It may therefore be concluded that folding of the helix is associated with the tight packing of the nonpolar groups of the backbone, i.e., van der Waals interactions play a significant role in its stabilization.[24]

The Need to Measure Heat Capacity of Proteins Not Only on Heating but Also on Cooling and Below 0°

By scanning up and down in temperature we can check the reversibility of the temperature-induced process (Fig. 11; see also Fig. 10). This is important for the thermodynamic analysis of the considered process. If the heat effect is not reproducible on cooling this means that the process is irreversible and should be analyzed with caution. This does not always mean that one cannot analyze it thermodynamically. The irreversibility might be due to some secondary effects (e.g., aggregation, chemical degradation of groups at high temperatures), processes slower than protein unfolding. In that case the calorimetrically recorded melting profile is not significantly affected if the scanning rate is fast enough and small deformations of the melting profile due to irreversibility can easily be corrected.[25]

The other reason for recording the refolding profile is to measure the heat capacity below 0° and thereby expand the operational temperature range. A dust-free aqueous solution can readily be supercooled below 0° without freezing[26] and although the aqueous solution is then in a metastable state, this does not change its thermodynamic properties. One should, however, bear in mind that freezing of a supercooled solution can start at any moment and that this can damage the instrument. Calorimetric studies of aqueous solutions below 0° can therefore be done only with an instrument equipped with an automatic system for the immediate switch-on of fast heating on the initiation of freezing. The formation of ice in supercooled solutions usually starts on dust particles that serve as nucleation centers. The simplest way to remove dust from the solution is to filter it in microcentrifuge tubes by centrifugation.

Expansion of the operational range below 0° is especially important for studying cold denaturation phenomena, which usually take place at low temperatures (Fig. 12; for review, see Ref. 27). It is also important for studying processes that end at low temperatures. An example of this is the folding of an α helix, which was considered above (Fig. 10): only below 0° is it completely folded. We can therefore measure the total enthalpy of its formation only in a supercooling experiment. Another example is the temperature-induced melting of residual structure in single-strand oligonucleotides. The oligonucleotides do not have residual structure at temperatures above 80°. This residual structure appears as the temperature decreases. The enthalpy of its formation is not large and it is a rather diffuse process that ends below 0° (Fig. 13). Consequently, measurements of the residual structure enthalpy in the oligonucleotides are not easy, but without such measurements one cannot determine the energetics of double helix formation.[30]

[25] B. I. Kurganov, A. E. Lyubarev, J. M. Sanchez-Ruiz, and V. L. Shnyrov, *Biophys. Chem.* **69,** 125 (1997).
[26] P. L. Privalov, Y. V. Griko, S. Y. Venyaminov, and V. P. Kutyshenko, *J. Mol. Biol.* **190,** 487 (1986).
[27] P. L. Privalov, *CRC Crit. Rev. Biochem. Mol. Biol.* **25,** 281 (1990).
[28] Y. V. Griko, P. L. Privalov, S. Y. Venyaminov, and V. P. Kutyshenko, *J. Mol. Biol.* **202,** 127 (1988).
[29] Y. V. Griko, P. L. Privalov, J. M. Sturtevant, and S. Y. Venyaminov, *Proc. Natl. Acad. Sci. U.S.A.* **85,** 3343 (1988).
[30] I. Jelesarov, C. Crane-Robinson, and P. L. Privalov, *J. Mol. Biol.* **294,** 981 (1999).

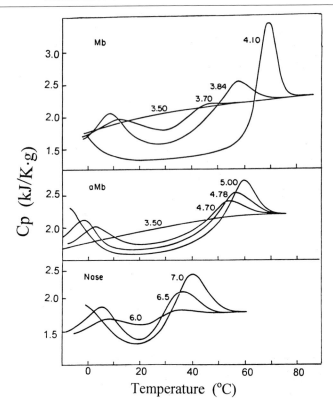

Fig. 12. The temperature dependence of the partial specific heat capacity of metmyoglobin (Mb),[26] apomyoglobin (aMb),[28] and staphylococcal nuclease (Nase)[29] in solutions with different pH values. The excess heat absorption at low temperatures results from refolding of the heat-denatured proteins; the heat absorption at higher temperatures is associated with their heat denaturation.[27]

The Need to Measure Heat Capacity of Proteins Above 100°

Under an excess pressure of 3 atm one can heat aqueous solutions to about 130° without boiling and without the appearance of the microscopic gas bubbles that would be damaging for the calorimetric experiment. Calorimetric measurements at such high temperatures are required because there are many extremely stable macromolecular structures that melt above 100°. These include more than proteins from thermophiles—even the unfolding of an α helix ends above 100° (Fig. 10). The main problem in such studies is that these temperatures induce irreversible degradation of the chemical

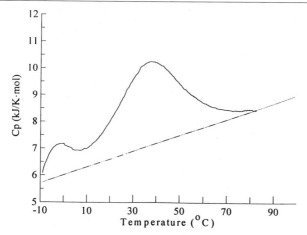

Fig. 13. The temperature dependence of the partial heat capacity of the 12-base single-strand deoxyoligonucleotide 5'-GCGAACAATCGG-3'. Concentration of oligonucleotide, 1.88 mg/ml; volume of the cell, 0.328 ml. It is remarkable that unfolding of the oligonucleotide proceeds with a heat capacity increment on the order of 1 kJ K^{-1} mol^{-1}. The enthalpy of total unfolding, which starts at $-10°$ and ends at $80°$, amounts to 77 kJ mol^{-1}. [Reproduced from I. Jelesarov, C. Crane-Robinson, and P. L. Privalov, *J. Mol. Biol.,* in press (1999).]

structure of proteins and nucleic acids.[31] Incubation of proteins at temperatures above $80°$ results in a decrease in their ability to fold back into the native conformation; the higher the temperature and the longer the time of incubation, the lower is the probability of refolding. This damaging action of high temperatures can be reduced by increasing the heating rate to shorten the exposure to high temperatures and by eliminating oxygen from the solution. The content of oxygen can be reduced by placing the protein solution prior to the experiment under vacuum and also by flushing nitrogen through the solution. This, however, must be done with caution so as not to change the concentration of cosolutes (salts) in the protein solution from that of the buffer solution, which is used as a reference in the calorimetric experiment.

Measuring Heat Effects of Association at Various Fixed Temperatures

By measuring the heat effect of protein–ligand, protein–protein, or protein–DNA/RNA association by isothermal titration microcalorimetry

---

[31] A. M. Klibanov, *Adv. Appl. Microbiol.* **29,** 1 (1983).

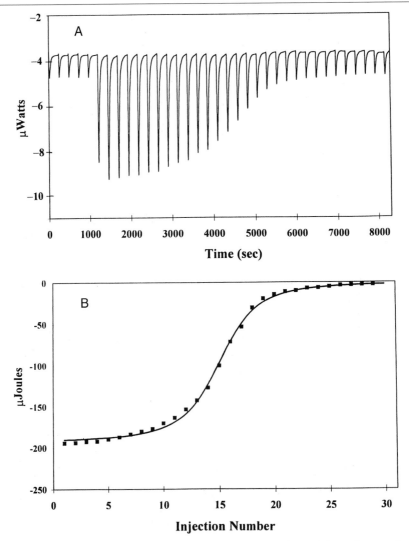

FIG. 14. (A) Isothermal titration of pancreatic ribonuclease A (RNase) by cyclic monophosphate (2'-CMP) at 25°, using a Jet-Titration calorimeter (Calorimetry Science Corporation, SCS). Concentrations: RNase, 68.8 μM; 2'-CMP, 0.5 mM. The volume of the calorimetric cell is 0.6 ml, and the volume of each injection is 5 μl. (B) The integral heat effect of RNase association with 2'-CMP as a function of injection number. Analysis of this function shows that the enthalpy of binding is -72.0 kJ mol^{-1}, the association constant is 1.10×10^{-6}, and the number of binding sites is 0.95.

at a standard temperature (e.g., 25°) one can determine the enthalpy of formation of these complexes (Fig. 14). If the heat effect of association is large and the instrument is sensitive enough, one can also determine the binding constant in the isothermal calorimetric titration experiment if that is not too high.[32-34] The equilibrium constant gives us the Gibbs energy of complex formation $\Delta G = -RT \ln K$, and since the Gibbs energy and enthalpy are then known, we can also determine the entropy of complex formation as $\Delta S = (\Delta H - \Delta G)/T$.

If the binding constant is high, in order to determine it by a titration experiment, one must use low concentrations of reagents and the heat effect of their mixing becomes too low for accurate determination. In that case it is necessary to use some other method, e.g., the fluorescence titration or the gel-shift assay.

The enthalpy and entropy, as well as the Gibbs free energy, are important parameters specifying the energetics of complex formation. However, to understand the nature of this process and the forces responsible for such energetics, we need to know these parameters not only at a standard temperature but also at other temperatures, i.e., we need to know their dependence on temperature. In the case of macromolecules in aqueous solution, their partial heat capacity is largely determined by hydration effects, i.e., the surfaces of polar and nonpolar groups exposed to water (see below). If we know the heat capacity change on association of the components into the complex, we can judge how large the surfaces are that were dehydrated on complex formation, i.e., the significance of the hydrophobic and polar interactions involved in this process. The temperature dependence of the enthalpy of complex formation can give us this information, but only if temperature change does not induce alterations in the states of the components and the complex. If temperature induces gradual changes in any of the components, the heat effect that results will be reflected in the apparent heat capacity changes with temperature. Therefore, without studying the heat capacities of the components and the complex in the temperature range over which we measured the heat effect of complex formation, we cannot draw any conclusions regarding the heat capacity effect and the enthalpy of this process.

Consider as an example the association of a DNA-binding protein domain with its specific DNA recognition site.[35] Figure 15 shows the en-

[32] T. Wiseman, S. Wiliston, J. Brandts, and L. N. Lin, *Anal. Biochem.* **179,** 131 (1989).

[33] E. Freire, O. L. Mayorga, and M. Straume, *Anal. Chem.* **62,** 950A (1990).

[34] K. Breslauer, E. Freire, and M. Straume, *Methods Enzymol.* **211,** 533 (1992).

[35] P. L. Privalov, I. Jelesarov, C. M. Read, A. Dragan, and C. Crane-Robinson, *J. Mol. Biol.* **294,** 997 (1999).

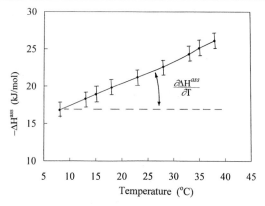

FIG. 15. Plot of the enthalpies of specific binding of the HMG box of Sox-5 protein with a 12-base pair DNA duplex containing the recognition sequence. The initial slope of this dependence equals -6 kJ K^{-1} mol^{-1} and increases with temperature rise.

thalpy of association of the HMG box with the DNA duplex measured at different temperatures by isothermal titration calorimetry (ITC). From the slope of this function it may be concluded, as is usually done in the literature, that the heat capacity change on DNA association with protein is about -6.0 kJ K^{-1} mol^{-1}. However, analysis of the surface areas that are screened from water on complex formation shows that the heat capacity change on association should be at least two times smaller (for the analysis, see below). The reason for this discrepancy is that both components, the HMG box and the DNA duplex, are changing their state in the temperature range over which the titration experiment has been carried out. As discussed above, the HMG box starts to unfold on heating from 5° (see Fig. 7). Therefore, in the titration temperatures the HMG box is partly unfolded, but on association with DNA it refolds, releasing heat. An increase in temperature also results in a significant increase in thermal fluctuations in the DNA duplex, which manifests itself as a heat capacity rise prior to unfolding. A temperature rise also results in an increase in thermal fluctuations in the complex, and this is also seen as a heat capacity increase before the extensive heat absorption due to its cooperative dissociation/unfolding (Fig. 16). All these effects contribute to the ITC-measured enthalpy of complex formation and they all need to be taken into account if we want to estimate the net heat capacity effect and the net enthalpy of fully folded protein and DNA association into fully folded complex.[35]

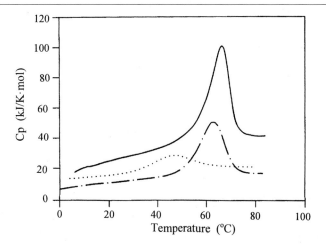

Fig. 16. The molar partial heat capacities of the 12-bp DNA duplex (dash-dot line), Sox-5 HMG box (dotted line) and their complex (solid line).[35]

Analysis of Complexes

The fact that the thermal properties of protein complexes with other proteins or nucleic acids change gradually on heating and that this occurs in the temperature range over which the formation of the complexes is usually studied by ITC considerably complicates the analysis of the energetics of formation of these complexes. The situation becomes especially complicated if the binding is not strong and the heating results in partial dissociation of the complex, as shown schematically for the protein–DNA complex in Fig. 17. The equilibrium in that case is shown in the following set of equations:

$$[PD] \xrightarrow{K_A} [P] + [D]$$

$$[P] \xrightarrow{K_P} [P']$$

$$[D] \xrightarrow{K_D} [D'] + [D'']$$

where [PD] is the concentration of the protein–DNA complex, [P] is the concentration of the free folded protein, and [D] is the concentration of the free DNA duplex. The concentration of free unfolded protein is [P'] and the concentration of free DNA strands is [D'] and [D''], which are

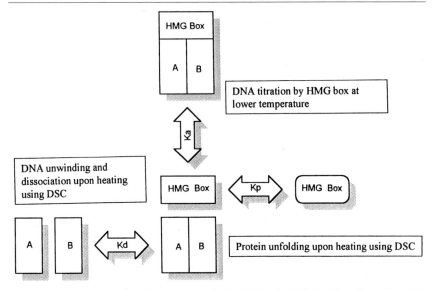

Fig. 17. Thermodynamic pathways of heat-induced dissociation/unfolding of protein–DNA complex. A, Protein–HMG box; B, strands of DNA duplex.

usually equal. The equilibrium constants can be expressed by the following set of equations:

$$K_A = \frac{[PD]}{[P][D]}$$

$$K_P = \frac{[P']}{[P]}$$

$$K_D = \frac{[D'][D'']}{[D]}$$

$$[P] + [P'] + [PD] = P_{\text{total}}$$
$$[D] + [D'] + [PD] = P_{\text{total}}$$

By solving this set of equations, the free two-stranded DNA concentration can be determined from

$$[D] + ([D]K_D)^{1/2} + K_A[D]\frac{P_{\text{total}}}{1 + K_P + K_A[D]} - P_{\text{total}} = 0$$

while the concentrations of the rest of the components are

$$[P] = \frac{P_{\text{total}}}{1 + K_P + K_A[D]}$$

$$[PD] = K_A[P][D]$$

Binding parameters $K_a(T)$ and $\Delta H_a(T)$ can be evaluated over a wide temperature range using values determined at a given temperature T_0[36]:

$$K_A(T) = K_A(T_0) \cdot \exp\left\{-\frac{\Delta H_A}{R}\left(\frac{1}{T} - \frac{1}{T_0}\right) + \frac{\Delta Cp_A}{R}\left[\ln\left(\frac{T}{T_0}\right) + \frac{T_0}{T} - 1\right]\right\}$$

For K_P and K_D we have

$$K = \exp\left[-\frac{\Delta H(T) - T \cdot \Delta S(T)}{RT}\right]$$

The apparent heat capacity is determined by the concentration of intermediates:

$$Cp(T) = \Delta H_a \frac{\partial F_C(T)}{\partial T} + \Delta H_U \frac{\partial[1 - F_P(T)]}{\partial T}$$

$$+ \Delta H_{\text{DNA}} \frac{\partial[1 - F_D(T)]}{\partial T} + \Delta C_{\text{pb}}(T)$$

where ΔH_a is the enthalpy of the DNA–protein association, ΔH_U is the enthalpy of protein unfolding, ΔH_{DNA} is the enthalpy of DNA unwinding, and $\Delta C_{\text{pb}}(T)$ is the baseline heat capacity:

$$\Delta Cp(T) = Cp_P^N(T) \cdot F_P^N(T) + Cp_D^N \cdot F_D^N(T)$$
$$+ Cp_P^U(T) \cdot F_P^U(T) + Cp_D^U(T) \cdot F_D^U(T)$$
$$+ Cp_C^F(T) \cdot F_C(T)$$

where $Cp_P^N(T)$ is the heat capacity of protein in the native state, $F_P^N(T)$ is the concentration of free protein in the native state, $F_c(T)$ is the concentration of the protein–DNA complex, $C_D^N(T)$ is the heat capacity of double-stranded DNA, $F_D^N(T)$ is the concentration of free double-stranded DNA, $Cp_P^U(T)$ and $F_P^U(T)$ are the heat capacity and concentration of unfolded protein, $Cp_D^U(T)$ and $F_D^U(T)$ are the heat capacity and concentration of single-stranded DNA, and $Cp_C^F(T)$ is the heat capacity of the complex.

By modeling the observed heat capacity functions of the complex and its free components, all the thermodynamic characteristics of these systems can be evaluated. Particularly in the case of Sox-5 HMG box association with the cognate DNA, this analysis revealed that dissociation of protein

[36] J. F. Brandts and L.-N. Lin, *Biochemistry* **29,** 6927 (1990).

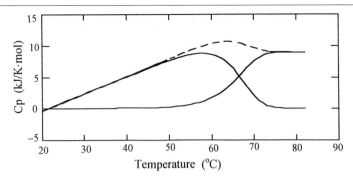

FIG. 18. The apparent intrinsic heat capacity of the DNA and HMG box complex consists of two overlapping function: (1) the heat capacity of the complex, which increases gradually on heating because of increasing fluctuations of its structure (dashed line), and (2) the heat capacity effect of the cooperative dissociation of the complex and unfolding of its components, the protein and DNA (solid lines).[35]

starts before cooperative disruption of the complex and that there are two stages in the temperature-induced changes of the complex (Fig. 18). The first stage is characterized by a gradual increase in heat capacity, i.e., gradual accumulation of the thermal energy by the fluctuating structure of the complex. The second stage represents the highly cooperative dissociation of the complex with unfolding of its components, the protein and the DNA duplex. This second stage proceeds with a significant increase in the heat capacity, which then drops due to the sharp decrease of the native complex population in the temperature zone of the cooperative transition.

Utilizing Calorimetric Information

Accurate calorimetric data are essential for the quantitative description of the energetic bases of molecular structures and their transformations in various processes. In other words, the main goal of calorimetric studies is to find connections between thermodynamic and structural parameters specifying proteins, nucleic acids, and their complexes.

As discussed above, by knowing the sequence of a polypeptide chain, or even just the total amino acid content, it is possible to calculate the partial heat capacity of the unfolded polypeptide chain in aqueous solution. This is done simply by summing the partial heat capacities of amino acid residues, the values of which are known over the temperature range from 5 to 125°.[16–18] The simple summation of the heat capacities of the amino acid residues in fact means that in the unfolded polypeptide all residues are fully exposed to water. It might be expected that there would be some screening effects between neighboring side chains. However, the heat capacities of polypeptide chains calculated by assuming that the contributions of amino acid residues are additive were found to be in good correspondence with the calorimetrically measured values. So if there are screening effects, they must be smaller than the error in the calorimetric determination of the partial heat capacity, which is on the order of 10%.

In the case of native proteins, we cannot calculate their absolute partial heat capacity from the sequence because it depends on the water-exposed groups and the rigidity of the protein structure, the fluctuations of which can contribute significantly to the heat capacity.[37] What is usually calculated therefore is the heat capacity difference between the folded and unfolded protein states caused by changes in hydration of groups as a result of unfolding.

There have been many attempts to evaluate hydration effects on protein unfolding, based on studying the transfer into water of various low molecular weight compounds from the liquid state, solid state, or gaseous state (for a review see Ref. 21). The solid state may be considered the most appropriate reference state for the densely packed interior of the native protein in studying the transfer of its internal groups into water on unfolding. However, if we take this state, or the liquid state, as a reference in considering transfer of protein internal groups into water, then we cannot determine the energies of interactions between these groups in the native protein. We also cannot determine the configurational entropy of polypeptide chain unfolding using transfer from the liquid state because in the native protein the groups are fixed, in contrast to the liquid state. To evaluate the internal interactions in proteins and the configurational entropy of the polypeptide chain we need to know the net hydration effects, not including the interactions between the groups considered and the effects associated with thermal motion of these groups. These net hydration effects of protein groups can be determined only by considering transfer of model compounds from a fixed position in the gaseous phase to a fixed position in water.[38]

It was found that in the first approximation the hydration effects of various groups are additive, i.e., proportional to the accessible surface area, $\Delta ASA_{k,j}$. Therefore, using the specific, surface-normalized hydration effects of the given type k of the group j, $\Delta \hat{F}_{k,j}^{hyd}(T)$, it is possible to determine the expected hydration effect of the k-type groups on structural changes studied as $\Delta F_k^{hyd}(T) = \sum_j \Delta ASA_{k,j} \times \Delta \hat{F}_{k,j}^{hyd}(T)$. The values of $\Delta \hat{F}_{k,j}^{hyd}(T)$ for the enthalpy and heat capacity effects of hydration are tabulated.[16–18]

[37] J. Gomez, V. J. Hilser, D. Xie, and E. Freire, *Proteins Struct. Funct. Genet.* **22,** 404 (1995).
[38] A. Ben-Naim, "Solvation Thermodynamics." Plenum, New York, 1987.

It should be noted that the values of the heat capacity effects of hydration for all aliphatic groups are positive and similar (see, e.g., Refs. 13, 21, 38, and 39). They are also positive for aromatic groups, but smaller in magnitude than for the aliphatic groups. It is absolutely incorrect therefore to assume that aliphatic and aromatic groups are identical in their hydration properties, as is often assumed in the literature; aromatic groups are not nonpolar, they are in fact slightly polar.[18,21] The heat capacity effects of hydration for the charged and polar groups, in contrast to the aliphatic groups, are negative and quite different for different groups. To a first approximation, however, an averaged value of the hydration heat capacity effect, which takes into account the typical distribution of various polar groups in globular proteins, may be used. For the standard temperature of 25°, the most appropriate approximation is[21]

$$\Delta_N^U C_p^{hyd} = -(2.14\Delta ASA_{np} + 1.55\Delta ASA_{arom} - 1.27\Delta ASA_{pol}) \text{ J K}^{-1} \text{ mol}^{-1}$$

However, the heat capacity effect of hydration depends on temperature and this dependence is different for each type of group. These dependencies for the aliphatic, aromatic, and polar groups are given in Ref. 21. It should be noted that there are other approximate expressions for calculating the hydration heat capacity effects (see, e.g., Refs. 13 and 39) but they do not distinguish between the nonpolar and aromatic group hydration effects and do not take into account the significant temperature dependencies for these groups.

Since the positive heat capacity effect of hydration of nonpolar and aromatic groups exceeds in magnitude the negative effect of hydration of polar groups, the overall heat capacity effect of hydration on protein unfolding is positive. If we know how large is the difference in heat capacities of the native and unfolded protein or nucleic acid, we can judge the surface area that is exposed on protein unfolding. The same is true for complexes of macromolecules: if we know the heat capacity value of the complex formation from the fully folded components, we can in principle determine the extent of the interface, or if we know the water-accessible surface areas of the complex and the components, we can determine the expected heat capacity change at complex formation.

A similar approach can be used to determine the enthalpy and entropy effects of hydration on protein unfolding or complex dissociation. The corresponding surface normalized hydration effects of various amino acid residues are given in Refs. 18 and 21. In contrast to the heat capacity effects, the enthalpy and entropy of hydration are negative for aliphatic, aromatic, and polar groups. By excluding the hydration effects from the calorimetri-

---

[39] R. S. Spolar, J. R. Livingstone, and M. T. Record, *Biochemistry* **32**, 3842 (1992).

cally measured enthalpy of protein unfolding (or of complex dissociation) we can determine the enthalpy of protein unfolding (or complex dissociation) without hydration effects, i.e., in a vacuum, which actually specifies the enthalpy of internal interactions stabilizing the protein (or complex) structure.[21]

Determination of Surface Area that Is Exposed to Water on Protein Unfolding or Complex Dissociation

The method of determining the water-accessible surface area (ASA) of proteins consists of rolling a probe of radius 1.4 Å over the considered surface. Several algorithms have been developed for this purpose,[40–43] and there are several commercial programs that are used for this purpose. But what can be done with an unfolded protein that does not have a definite structure? For determination of the ASA of the unfolded state, the polypeptide chain in a random coil conformation should certainly be used. Generation of this conformation, however, presents a problem for computer analysis. In existing reports, authors have used various model conformations approximating an unfolded polypeptide chain. In most of the earlier studies, the ASA of the unfolded state was estimated by calculating the surface area of each amino acid in an extended tripeptide, Gly-X-Gly or Ala-X-Ala, and then summing these contributions over the amino acid composition of the protein. Since glycine does not have a side chain and alanine has a small one, this method gives a net water-accessible surface area of the side chain X when not screened by the neighboring side chains. More recently, the ASA of the unfolded state of the polypeptide chain has been approximated by a fully extended conformation or β conformation. In the β conformation and especially in the fully extended conformation, the ASA values of polar, aromatic, and aliphatic groups are smaller than those calculated by simple summation of the surfaces of amino acid residues determined in Gly-X-Gly tripeptides because of the screening effects from neighboring side chains.[21] The use of the fully extended chain is thus unlikely to cause a significant overestimation of the ASA of the unfolded protein. In practice, the ASA of the polypeptide in the extended conformation appears to be the best that can be currently used to approximate the polypeptide in the random coil state.

[40] A. Shrake and J. A. Rupley, *J. Mol. Biol.* **79,** 351 (1973).
[41] A. A. Rashin, *Biopolymers* **23,** 1605 (1984).
[42] S. Miller, J. Janin, A. M. Lask, and C. Chothia, *J. Mol. Biol.* **196,** 641 (1987).
[43] C. Chothia, *J. Mol. Biol.* **105,** 1 (1976).

Approximating Interior of Proteins by Liquid Phase

We now know many protein structures at atomic resolution, and know that the interior of proteins cannot be approximated by the liquid phase. Nevertheless, in considering protein folding and discussing the forces responsible for stabilization of the native protein structure we use the concept of "hydrophobic force" which proceeds from studying the desolvation of liquid nonpolar solutes in water and was used to explain their low solubility.[44] The hydrophobic force is still regarded by many authors as a dominant factor responsible for the folding a polypeptide into a compact globular structure, although this entropic effect, which is not directed (i.e., is not a vector), cannot explain the crystal-like order in the protein interior or the cooperativity of the native structure. This is one of the reasons why the specific packing of groups with tight van der Waals contacts is attracting increasing attention in considering mechanisms of protein folding. Analysis of calorimetric data on protein folding, when corrected for dehydration effects, has shown that van der Waals interactions between tightly packed protein groups play a significant role in the stabilization of the native protein structure[21] and that the specific packing of such groups in the protein interior plays a definitive role in the cooperativity of the native structure.[45] It becomes more and more apparent that without a detailed analysis of such packing we cannot describe the energetics of protein structure and cannot efficiently use calorimetric data on protein unfolding or on dissociation of specific complexes. The packing density of the protein interior can be analyzed by the program Mole (www.mol3d.com), which is quite efficient and simple in operation.

Calorimetric Instruments for Studying Biological Macromolecules.

The heat effects of temperature-induced changes of macromolecules in dilute solutions are measured by differential scanning calorimeters (DSCs).[2,3,46] In calorimetric experiments two identical calorimetric cells are loaded first with solvent and the power required to maintain zero temperature difference between the heated/cooled cells is recorded as a function of temperature. This function is used as the baseline of the instrument. The solvent in one of the cells is then replaced with protein solution. It is important that the protein solution and the solvent, which is used as a reference, have identical concentrations of cosolutes (salts) and differ only in their protein content. Prior to the calorimetric experiment, therefore,

[44] W. Kauzmann, *Adv. Protein Chem.* **14,** 1 (1959).
[45] F. M. Richards and W. A. Lim, *Q. Rev. Biophys.* **26,** 423 (1994).
[46] P. L. Privalov and V. V. Plotnikov, *Thermochim. Acta* **139,** 257 (1989).

the protein solution must be exhaustively dialyzed against the solvent, which is used as a reference. Chromatography can also be used for this purpose, particularly when the molecules being studied are too small and can escape from the dialysis tubes. The other possibility, which permits the saving of time, is to wash the protein by centrifuging it in a Centricon membrane concentrator. (Amicon, Danvers, MA).

The difference in the power required to maintain identical temperatures in both cells in these two experiments on heating/cooling is converted into a difference heat capacity effect, $\Delta C_p^{app}(T)_{pr}^{sol/solv}$. This difference function is then converted into the partial specific heat capacity of the protein, $C_p(T)_{pr}$,

$$C_p(T)_{pr} = C_p(T)_{solv} \frac{V(T)_{pr}}{V(T)_{solv}} - \frac{\Delta C_p^{app}(T)_{pr}^{sol/solv}}{m(T)_{pr}}$$

Here $V(T)_{pr}$ is the partial specific volume of the protein at temperature T and $V(T)_{solv}$ is that of the solvent; $m(T)_{pr}$ is the mass of protein that is in the calorimetric cell at temperature T, and $C_p(T)_{solv}$ is the heat capacity of the solvent (see Ref. 2 for details).

The following defines the DSC requirements for quantitative studies of the thermal properties of biological macromolecules in dilute solutions:

1. The instrument should be not only supersensitive but also superprecise, i.e., it should have a stable reproducible baseline so as to determine with high accuracy the absolute partial heat capacity of macromolecules over a broad temperature range.

2. The operational temperature range of the instrument should be from -10 to $130°$. Below $-10°$ aqueous solutions cannot readily be studied in the liquid state because of the high probability of spontaneous freezing of the overcooled solution. Above $130°$, temperature-induced degradation of the polypeptide (or polynucleotide) chain is too fast and so there is little sense in heating proteins or nucleic acids above this temperature.

3. Measurements of the heat capacity should be made both on heating and on cooling.

Most of the commercially available sensitive scanning calorimeters are differential and adiabatic.[46] The adiabaticity means that the twin calorimetric cells of these instruments are completely isolated thermally from the surroundings. This is usually done by thermal screens that are maintained at temperatures close to the temperature of the cells, preventing heat exchange with the surroundings. The cells are heated by electric heaters, but they cannot be cooled. Correspondingly, adiabatic calorimeters measure the difference heat effect on heating but not cooling. Therefore, a new-generation DSC has been designed.[6] The main difference between the previous sensitive scanning microcalorimeters and this new instrument is that it is not adiabatic and the cells are heated or cooled by heat exchange with the surrounding block; the temperature of the block is regulated by computer-controlled Peltier elements. The advantage of this instrument is that it can scan up and down over a temperature range of −10 to 125° at any desired constant heating rate. This instrument is now produced by Calorimetry Science under the trademark Nano-DSC. The new model of Nano-DSC is equipped with cylindrical or capillary cells 0.3 ml in volume. The main advantage of capillary cells is that their washing and reloading is easier. The other advantage is that a calorimeter with such cells is not sensitive to changes in solution viscosity on heating, which frequently occurs with fibrillar macromolecules. The Nano-DSC is equipped with programs for the full analysis of calorimetric data: determination of the partial molar heat capacity and its deconvolution on the components in accordance with various models.

A modified isothermal titration microcalorimeter was first built by S. J. Gill[47,48] at Colorado University and by I. Wadso[49] at Lund University. The main principle of operation of this differential isothermal calorimeter is that the reagent is placed in one of the two cells and the reactant is introduced to it in small portions. The reactant and reagent are mechanically stirred and the heat effect of mixing is measured. At the present time several companies manufacture ITCs (in the United States: Microcal and Calorimetry Science Corporation, SCS). Both these instruments have a mechanical stirrer to mix the reagents. This somewhat limits their operational volume (about 1.5 ml) and also their sensitivity because of the considerable joule heat evolved by the stirrer. It was therefore suggested that the stirrer be eliminated and that for mixing the kinetic energy of the reagent injection or of the solution taken up in advance from the reactor be used.[50] The operation of the Jet-Titration Calorimeter is illustrated in Fig. 14. This instrument is now produced by Calorimetry Science.

Acknowledgment

Financial support from the NIH (GM48036) is gratefully acknowledged.

[47] R. B. Spokane and S. J. Gill, *Rev. Sci. Instrum.* **52,** 1728 (1981).
[48] I. R. McKinnon, L. Fall, A. Parody-Morreale, and S. J. Gill, *Anal. Biochem.* **139,** 134 (1984).
[49] I. Wadso, *Thermochim. Acta* **250,** 285 (1995).
[50] E. Freire, G. P. Prtivalov, P. L. Privalov, and V. V. Kavina, U.S. Patent 5,707,149 (1998).

[3] Probing Stability of Helical Transmembrane Proteins

By KAREN G. FLEMING

Introduction

Our understanding of the structural energetics of membrane proteins lags far behind that of soluble proteins. In part, this is because only a small number of membrane protein structures are known to atomic resolution. Equally responsible, however, is the lack of quantitative thermodynamic information on membrane protein folding and assembly.[1] Although several helical membrane proteins have been investigated by differential scanning calorimetry, their thermal unfolding transitions are either irreversible or are low in enthalpy.[2] Consistent with the low enthalpy changes characterizing the thermal denaturation of membrane proteins is the observation that, even at high temperatures, helical membrane proteins retain significant amounts of secondary structure. Thus, thermal denaturation of helical membrane proteins results in a transition from the native state of the protein not to a completely disordered, unfolded state, but rather to an intermediate denatured state that retains significant amounts of secondary structure and that is still associated with its hydrophobic environment.

It may be more instructive, then, to consider helical membrane protein folding by using a model for the folding pathway that includes this intermediate denatured state that is observed calorimetrically. Indeed, Popot and Engelman in 1990[3] used energetic considerations to outline a thermodynamic framework for helical membrane protein folding that explicitly considers both the fully unfolded state and the intermediate denatured state as well as the folded, native state. As summarized in Fig. 1, the two-stage model of Popot and Engelman proposes that the transition from a fully disordered, unfolded state to an intermediate denatured state (containing significant secondary structure and associated with membranes) occurs in a first stage where a polypeptide segment of hydrophobic amino acid residues is established across the phospholipid bilayer as a transmembrane α helix. Energetic considerations suggest that the insertion process can happen spontaneously and that it is driven by the hydrophobic effect. Further, the secondary structure of the transmembrane α helix is stabilized by the

[1] M. H. B. Stowell and D. C. Rees, Structure and stability of membrane proteins. *In* "Advances in Protein Chemistry" (F. M. Richards, ed.), pp. 279–311. Academic Press, New York, 1995.

[2] T. Haltia and E. Freire, *Biochim. Biophys. Acta* **1228**, 1 (1995).

[3] J.-L. Popot and D. M. Engelman, *Biochemistry* **29**, 4032 (1990).

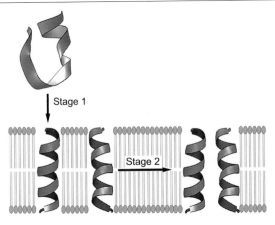

FIG. 1. A folding pathway for helical membrane proteins as proposed by Popot and Engelman.[3] A polypeptide segment of hydrophobic amino acid residues will spontaneously insert as a transmembrane α helix in a first thermodynamic stage. Side-to-side interactions of independently established transmembrane α helices then drive formation of the native, helical bundle in a second thermodyamic stage.

large enthalpic penalty that would occur if the intrahelical hydrogen bonds were to remain unsatisfied in the highly apolar environment. Thus, once a transmembrane α helix is established, it will neither unfold nor leave the bilayer. Although the secondary structure and membrane localization of a helical membrane protein are intact at this stage, tertiary contacts found in the native fold are yet to be established. The transmembrane α helix established in the first stage may therefore represent the experimentally observed, intermediate denatured state.

The transition from this intermediate denatured state to the folded, native state occurs in a second thermodynamic stage where the independently established transmembrane α helices are postulated to associate in the specific, side-to-side manner that gives rise to the bundle of helices defining the native fold of the helical membrane protein. This stage describes equally well helix–helix interactions occurring within a given monomeric membrane protein and those occurring between transmembrane α helices of different polypeptide chains. The forces driving this side-to-side association will minimally include an increase in helix–helix interactions, lipid–lipid interactions, and lipid entropy as well as a decrease in helix–lipid interactions and in helix entropy. The hydrophobic effect is not expected to play an important role in this second stage because water is already excluded from the apolar environment. The absence of the hydrophobic effect as a major driving force during this stage may explain why the experimental heat capacity changes are so low, compared with soluble proteins.[2]

Separation of helical membrane protein folding into two thermodynamically uncoupled stages may simplify approaches to investigating and understanding structural energetics of helical membrane proteins. While measurement of the free energy change that occurs during the stage 1 insertion process is not, for most sequences, experimentally feasible because of the insolubility of the transmembrane α helix when removed from the bilayer, the principal structure/stability determinant for this stage appears to be a requirement of a stretch of hydrophobic amino acid residues long enough to span the bilayer as a transmembrane α helix. Moreover, sequence analysis programs have good success rates at identifying stretches of amino acid residues that are hydrophobic enough and long enough to span the phospholipid bilayer as transmembrane α helices.[4] Thus, while a quantitative analysis of the insertion process awaits further experimentation, it may at least be possible in the meantime to identify most of the polypeptide segments that will be independently stable as transmembrane α helices. The membrane protein-folding questions then become more focused: Which of these transmembrane α helices will associate to form bundles? What are the forces driving bundle formation? How can stage 2 helix–helix associations be experimentally probed by quantitative techniques?

Sedimentation Equilibrium as Tool for Probing Thermodynamics of Helix–Helix Interactions

For helix–helix interactions occurring between transmembrane helices on separate polypeptide chains, transmembrane α-helix association will be accompanied by an increase in the membrane protein molecular weight. In these cases, sedimentation equilibrium is a well-suited technique for extraction of quantitative thermodynamic information. Sedimentation equilibrium has long been recognized as the technique of choice for solution interaction analysis of macromolecules.[5] In contrast to calorimetry, it provides a direct measure of a fundamental aspect of protein–protein association: a measurement of mass. In addition, relatively small amounts of sample are required.

Although the determination of the membrane protein molecular weight formally requires knowledge of the amount of detergent or lipid bound,

[4] D. M. Engelman, T. A. Steitz, and A. Goldman, *Annu. Rev. Biophys. Biophys. Chem.* **15,** 321 (1986).

[5] P. Hensley, *Structure* **4,** 367 (1996).

experimental approaches have demonstrated that, in practice, the contribution of the amphiphile can be minimized and the molecular weight of the protein alone can be obtained. Sedimentation equilibrium experiments can be carried out in a wide variety of detergent environments, over a range of pH values, temperatures, and ionic strengths. It is a nonperturbing technique, requires no standards for calibration, and does not necessitate labeling of the protein, as long as the detergent of choice does not absorb at the wavelength of interest. Thus, the molecular weight of a reconstituted membrane protein can be rigorously analyzed under a wide variety of solubilization conditions. Further, when conditions of reversibility are satisfied (as discussed below), sedimentation equilibrium can be used to obtain quantitatively rigorous estimates of the equilibrium constants describing the relationship between interacting species in solution. It follows that free energy changes (ΔG) can be calculated from the equilibrium constants. Other thermodynamic parameters such as the enthalpy (ΔH), entropy (ΔS), and heat capacity (ΔC_p) of an association reaction may also be estimated from the temperature dependence of the free energy change.

Glycophorin A Transmembrane Dimerization as Model System for Helix–Helix Interactions

A model system used extensively for testing the ideas embodied by the two-stage model is the dimerization mediated by the single transmembrane domain of the erythrocyte glycoprotein, glycophorin A. The transmembrane domain (GpATM) can drive dimerization of a chimeric staphylococcal nuclease (SN) fusion protein (SN/GpATM), and the dimer is stable even in solutions of sodium dodecyl sulfate (SDS). Because of its stability in SDS, Lemmon et al.[6,7] were able to use SDS–polyacrylamide gel electrophoresis (PAGE) in an extensive mutagenesis study in which the sequence dependence of dimerization was elucidated. A subset of the point mutants that were investigated is shown in Fig. 2, where it can be seen that even subtle mutations at specific positions along the helix profoundly affect dimerization. For example, an Ile → Ala mutation at position 76 completely abrogates dimerization, whereas the same mutation at position 77 has no effect. Because the mutagenesis information was so extensive, a dimerization motif could be identified. The motif was subsequently shown to be

[6] M. A. Lemmon, J. M. Flanagan, J. F. Hunt, B. D. Adair, B. J. Bormann, C. E. Dempsey, and D. M. Engelman, *J. Biol. Chem.* **267**, 7683 (1992).
[7] M. A. Lemmon, J. M. Flanagan, H. R. Treutlein, J. Zhang, and D. M. Engelman, *Biochemistry* **31**, 12719 (1992).

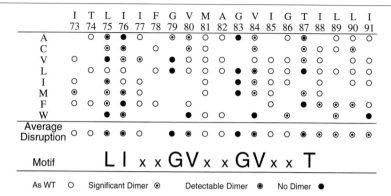

FIG. 2. The effect of point mutations on the apparent dimer stability of SN/GpATM as measured by SDS–PAGE. The subset of hydrophobic mutations is shown. [Adapted from M. A. Lemmon, J. M. Flanagan, H. R. Treutlein, J. Zhang, and D. M. Engelman, *Biochemistry* **31,** 12719 (1992). With permission.]

capable of driving dimerization of transmembrane α helices of diverse amino acid sequence backgrounds, including polyleucine and neu.[8,9]

The mutational sensitivity data were placed in a structural framework by the nuclear magnetic resonance (NMR) solution structure of the transmembrane dimer in dodecylphosphocholine micelles.[10] Not surprisingly, the dimer interface is composed of side chain and backbone atoms of those residues that had been found to be sensitive to mutation. The two glycine residues facilitate close approach of the helices to each other, and side-chain atoms of other motif residues are arranged in a detailed geometry of specific packing between the helices. No interhelical hydrogen bonds or salt bridges stabilize the dimer.

While the NMR structure provides a three-dimensional framework in which to consider the effects of point mutations, an understanding of the structural energetics of the glycophorin A dimer still lacked quantitative thermodynamic information. The development of a sedimentation equilibrium protocol to obtain estimates of the free energy of GpATM dimerization in micelles has made great strides toward this goal.[11] This chapter focuses on explaining the theoretical and experimental foundations for making these free energy measurements with the analytical ultracentrifuge.

[8] M. A. Lemmon, H. R. Treutlein, P. D. Adams, A. T. Brunger, and D. M. Engelman, *Nature Struct. Biol.* **1,** 157 (1994).

[9] C. L. Burke, M. A. Lemmon, B. A. Coren, D. M. Engelman, and D. F. Stern, *Oncogene* **14,** 687 (1997).

[10] K. R. MacKenzie, J. H. Prestegard, and D. M. Engelman, *Science* **276,** 131 (1997).

[11] K. G. Fleming, A. L. Ackerman, and D. M. Engelman, *J. Mol. Biol.* **272,** 266 (1997).

Methods

Theoretical Considerations on Interpretation of Experimentally Determined Buoyant Molecular Weight

The primary information obtained in a sedimentation equilibrium experiment is the buoyant molecular weight of the sedimenting particle. Because, by their nature, membrane proteins require detergent micelles or lipid vesicles in order to be soluble in solution, the buoyant molecular weight of the sedimenting particle in the ultracentrifuge, $M_{Pr}(1 - \phi'\rho)$, will contain contributions from both the protein as well as the detergent or lipid (where M_{Pr} is the molecular weight of the protein, ϕ' is the effective partial specific volume of the protein–detergent complex, and ρ is the solvent density). However, the molecular weight of the protein alone in a detergent or lipid solution can be determined unambiguously if the contribution of the detergent or lipid to the sedimenting particle is appropriately treated. To account for the separate contributions of the protein and detergent, the buoyant molecular weight of the complex can be rewritten in terms of its components[12,13]:

$$M_{Pr}(1 - \phi'\rho) = M_{Pr}[(1 - \bar{v}_{Pr}\rho) + \delta_{Det}(1 - \bar{v}_{Det}\rho)] \qquad (1)$$

where $\bar{v}_{Pr}$ is the partial specific volume of the protein, $\bar{v}_{Det}$ is the partial specific volume of the detergent when bound to the protein, δ_{Det} is the amount of detergent bound in grams per gram of protein, and other terms are as previously described. Typically, the $\bar{v}_{Det}$ value measured above the critical micelle concentration is used in Eq. (1). It is worth noting that use of this $\bar{v}_{Det}$ value, measured in the absence of protein, assumes that it does not significantly change on binding to protein.[13]

If the amount of bound detergent as well as its partial specific volume are known, then these values can be substituted into Eq. (1) in order to solve for the protein molecular weight. Often, however, the investigator is lacking one or both pieces of information. In this case, the contribution of the detergent can be experimentally minimized by matching the density of the detergent micelle with the solvent buffer such that $\rho = 1/\bar{v}_{Det}$. Under these conditions, the value calculated by the term $\delta_{Det}(1 - \bar{v}_{Det}\rho)$ in Eq. (1) approaches zero, and the analysis of this multicomponent system then becomes pseudo-two component, where the buoyant molecular weight of the protein–detergent complex, $M_{Pr}(1 - \phi'\rho)$, becomes essentially equal to the buoyant molecular weight of the protein alone, $M_{Pr}(1 - \bar{v}_{Pr}\rho)$. This density-matching strategy is similar in concept to contrast-matching ap-

[12] E. F. Cassassa and H. Eisenberg, *Adv. Protein Chem.* **19**, 287 (1964).

[13] J. A. Reynolds and C. Tanford, *Proc. Natl. Acad. Sci. U.S.A.* **73**, 4467 (1976).

proaches used in small-angle neutron and X-ray scattering.[14] Where an equilibrium between membrane protein oligomers is the desired information, use of the density-matching strategy is preferable over substitution and calculation of the molecular weight, because it is likely that different oligomeric states may bind different amounts of detergent. Technically, it is preferable to match the micelle density with heavy water as opposed to sucrose, because this should minimally perturb the water activity.[13,15] Thus, the amphiphiles that are the most experimentally favorable will have partial specific volumes in the range of 0.9 to 1.0 ml g^{-1}. Extensive compilations of data for detergent and phospholipid partial specific volumes can be found in Steele *et al.*,[16] and in Durshlag.[17]

Special Considerations for Sedimentation Equilibrium Analysis of Membrane Proteins

Aside from the considerations regarding interpretation of the experimental buoyant molecular weight (detailed above), the mechanics of an experimental setup for a membrane protein system in detergent micelles are identical to that for a soluble protein. For the experiments described in this chapter, standard six-sector cells equipped with quartz windows are used. To visualize the distribution of the protein alone (and not the detergent), the absorbance optics are used for detection of the equilibrium radial distributions at a wavelength where the detergent micelles have nominal absorbance.

Analysis of Transmembrane Helix–Helix Interactions in $C_{12}E_8$: Importance of Reversibility in Determining Equilibrium Constants

Initial attempts to measure the SN/GpATM dimerization dissociation constant were carried out in detergent solutions of $C_{12}E_8$. $C_{12}E_8$ is a "nondenaturing," nonionic polyoxyethylene ether detergent with a long history of use in the ultracentrifuge for determination of membrane protein molecular weights.[13,15,18] Technically, density matching of $C_{12}E_8$ micelles is feasible, because the micelle partial specific volume is reported to be 0.973 ml g^{-1}.[16] At 25° in buffer containing 20 mM sodium phosphate and 200 mM NaCl, the density can be matched with 18% (v/v) D_2O.

[14] C. Sardet, A. Tardieu, and V. Luzzati, *J. Mol. Biol.* **105,** 383 (1976).

[15] J. A. Reynolds and D. R. McCaslin, *Methods Enzymol.* **117,** 41 (1985).

[16] J. C. Steele, C. Tanford, and J. A. Reynolds, *Methods Enzymol.* **48,** 11 (1978).

[17] H. Durshlag, Specific volumes of biological macromolecules and some other molecules of biological interest. *In* "Thermodynamic Data for Biochemistry and Biotechnology" (H.-J. Hinz, ed.), pp. 45–128. Springer-Verlag, Berlin, 1986.

[18] A. Musatov and N. C. Robinson, *Biochemistry* **33,** 13005 (1994).

Sedimentation equilibrium data collected at different initial concentrations and speeds are globally analyzed by the NONLIN algorithm.[19] Use of this algorithm is advantageous because it uses the primary data for analysis and because it allows reversibility of an association reaction to be verified by fitting an apparent equilibrium constant as either a local or global parameter.[19,20] The simplest model that has been found to describe the data indicates that SN/GpATM distributes as monomeric, dimeric, and tetrameric forms in $C_{12}E_8$. Other models containing fewer parameters were first evaluated in accordance with standard protocols for fitting equilibrium data[20] (e.g., single species, monomer–dimer, dimer–tetramer); however, the minimal number of species required to describe the data was found to be three. The formal expression follows:

$$c_i = c_{ref} \exp[\sigma(\xi_i - \xi_{ref})] + c_{ref}^2 K_{1,2} \exp[2\sigma(\xi_i - \xi_{ref})]$$
$$+ c_{ref}^4 K_{1,4} \exp[4\sigma(\xi_i - \xi_{ref})] + base \qquad (2)$$

where c_i is the total absorbance at a radial position, r_i; c_{ref} is the monomer absorbance at a reference position r_{ref}; $\sigma = M(1 - \bar{v}_{Pr}\rho)\omega^2/RT$; M is the monomer molecular mass; $\bar{v}_{Pr}$ is the monomer partial specific volume; ρ is the solvent density; ω is the angular velocity (radians sec^{-1}); R is the universal gas constant; T is the absolute temperature; $\xi = r^2/2$; $K_{1,2}$ is the apparent monomer–dimer equilibrium constant; $K_{1,4}$ is the apparent monomer–tetramer equilibrium constant; and *base* is a baseline term for nonsedimenting material.

Obtaining an estimate for an equilibrium constant of an interacting system requires that the species in question reversibly associate with each other on the time scale of the experiment. In analysis of sedimentation equilibrium data, such reversibility will be reflected as a constant value for the equilibrium constant K over all initial concentrations and speeds. Thus, care must be taken to collect enough data at different initial concentrations (at least three) and at different rotor speeds (at least three). The concentrations are usually serial 1 : 3 dilutions of a stock protein solution into buffer. The speed choices will depend on the anticipated solution molecular weights.[21,22] Thus, nine data sets should typically be used for a sedimentation equilibrium analysis.

Reversibility of interacting species is verified during the global fitting procedure in NONLIN by treating K as a global parameter (i.e., a single

[19] M. L. Johnson, J. J. Correia, D. A. Yphantis, and H. R. Halvorson, *Biophys. J.* **36,** 575 (1981).
[20] T. M. Laue, *Methods Enzymol.* **259,** 427 (1995).
[21] G. Ralston, "Introduction to Analytical Ultracentrifugation," pp. 3–7. Beckman Instruments, Fullerton, California, 1993.
[22] D. K. McRorie and P. J. Voelker, "Self-Associating Systems in the Analytical Ultracentrifuge." Beckman Instruments, Fullerton, California, 1993.

value for K for all the data sets). When species are not reversibly associating with each other on the time scale of the experiment, the use of a global K in the fitting function will result in a poor fit to the data as evidenced by nonrandomness of the residuals and poor fit statistics. The contrast between the use of a local versus a global $K_{1,2}$ parameter is shown in Fig. 3[23,24] for two initial concentrations of SN/GpATM. While the data at each concentration are well described by Eq. (2) using local $K_{1,2}$ values (Fig. 3, top) forcing the fitting function to use a global $K_{1,2}$ results in a poor fit (Fig. 3, bottom). The fitting procedure thus suggests that the system is kinetically frozen on the time scale of the experiment, and it is not possible to obtain estimates of thermodynamic equilibrium constants by this technique. However, verification of the model (i.e., the oligomeric identities) is still possible by treating the K values as local parameters (i.e., a separate K for each data set) as long as such a fitting function describes the data with acceptable goodness of fit. Switching between models that use local and global K values is easily carried out in NONLIN.[19,20]

Attempts to globally fit SN/GpATM sedimentation equilibrium data collected in $C_{12}E_8$ by using Eq. (2) with global $K_{1,2}$ values have always been unsuccessful. In the past, considerations were made for the possibility of a slowly equilibrating process by collecting data for 7 days per speed (as opposed to the normal 16–24 hr). In addition, efforts were made to mobilize the reaction by raising the temperature to 37°. However, reversibility between SN/GpATM oligomers has never been observed during the time scale of sedimentation equilibrium when the protein is reconstituted in $C_{12}E_8$.

Analysis of Transmembrane Helix–Helix Interactions in C_8E_5:
SN/GpATM Dimerization with High Affinity

The distribution of SN/GpATM species in detergent solutions of C_8E_5 has also been investigated. This detergent is chemically similar to $C_{12}E_8$, and the partial specific volume of C_8E_5 has been estimated as 0.993 g ml^{-1}.[25] Thus, use of this detergent is technically advantageous, because its micelles are neutrally buoyant at 25° in buffer containing 20 mM sodium phosphate, 200 mM NaCl.

[23] E. J. Cohn and J. T. Edsall, Density and apparent specific volume of proteins. *In* "Proteins, Amino Acids, and Peptides" (E. J. Cohn and J. T. Edsall, eds.), pp. 370–381. Reinhold Publishing, New York, 1943.
[24] T. M. Laue, B. Shah, T. M. Ridgeway, and S. L. Pelletier, Computer-aided interpretation of analytical sedimentation data for proteins. *In* "Analytical Ultracentrifugation in Biochemistry and Polymer Science" (S. E. Harding, A. J. Rowe, and J. C. Horton, eds.), pp. 90–125. Royal Society of Chemistry, Cambridge, 1992.
[25] B. Ludwig, M. Grabo, I. Gregor, A. Lustig, M. Regenass, and J. P. Rosenbusch, *J. Biol. Chem.* **257** 5576 (1982).

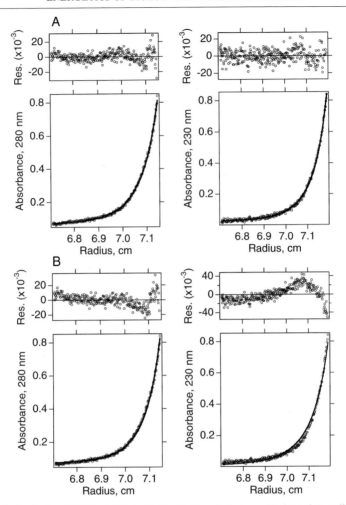

Fig. 3. Analysis of SN/GpATM in $C_{12}E_8$. Sedimentation equilibrium data collected at 24,500 rpm are shown for two initial concentrations (32 μM, left-hand side and 2 μM, right-hand side). (A) Results of fitting to Eq. (2), using local $K_{1,2}$ values; (B) results of fitting to Eq. (2), using a global $K_{1,2}$ value. The open circles are the data points and the solid line is the appropriate fit of Eq. (2) to the data. The residuals (Res.) of the fit for each data set are shown above each radial distribution. Sedimentation equilibrium experiments were conducted with an Optima XL-I ultracentrifuge, using the absorbance optics. The protein partial specific volume was calculated as 0.7476 ml g^{-1}, using the values of Cohn and Edsall.[23] The program Sednterp was used to calculate the solvent densities from the buffer components.[24] For experiments in $C_{12}E_8$, a solvent density of 1.027 g ml^{-1} was obtained by the addition of 18% (v/v) D_2O.

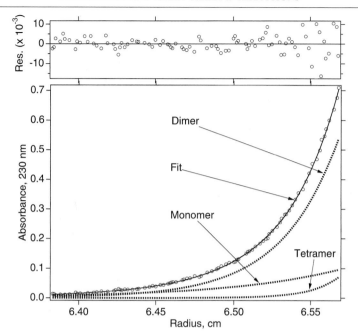

FIG. 4. Equilibrium distribution of wild-type SN/GpATM in C_8E_5. *Bottom:* A distribution of species at 30,000 rpm is shown. The open circles are the data points and the solid line is the fit of Eq. (2) to the data, using a global $K_{1,2}$ value. The dotted lines are the distribution of the monomer, dimer, and tetramer species whose sum gives rise to the observed fit. *Top:* The residuals of the fit are shown. For experiments in C_8E_5, no addition of D_2O was necessary, and the solvent density was calculated as 1.0075 g ml^{-1} at 25°.

Equation (2) has been found to be the simplest model that describes the oligomeric species distribution of SN/GpATM in C_8E_5. In contrast to the attempts to analyze data collected in $C_{12}E_8$, data fitting with a global $K_{1,2}$ value in the fitting function converged successfully, indicating reversibility of the monomeric and dimeric species in C_8E_5. A distribution of species at 30,000 rpm is shown in Fig. 4. At micromolar concentrations, it can be seen that SN/GpATM is predominantly dimeric in C_8E_5. The $K_{2,1}$ was estimated from the global fit to be 240 ($\pm$50) nM, corresponding to a dissociation free energy of 9.0 ($\pm$0.1) kcal mol^{-1} at 25°. Thus, SN/GpATM dimerizes with high affinity in the neutrally buoyant detergent C_8E_5.

Effect of Disruptive Mutations of Free Energy of Dimerization

With a protocol now in hand to rigorously measure interaction energies of transmembrane α helices, point mutants at two positions in the GpATM are analyzed for their ability to dimerize in C_8E_5. The mutations Leu75Ala and Ile76Ala had previously been shown to reduce dimerization significantly, as measured by SDS–PAGE (see Fig. 2). When assayed in the ultracentrifuge, both of these mutants showed reversible monomer–dimer associations in C_8E_5. Estimates of 1.4 (± 0.2) μM for Leu75Ala and of 4.2 (± 0.9) μM for Ile76Ala were obtained at 25°.[11] The differences in the association free energies between the wild-type sequence and the Leu75Ala and Ile76Ala mutants are 1.1 (± 0.1) and 1.7 (± 0.2) kcal mol^{-1}, respectively. The distribution of the dimeric species for these mutants compared with the wild-type sequence is shown in Fig. 5. This presentation also illustrates the large concentration range over which solution interaction analysis by analytical ultracentrifugation is carried out (the solid portions of the curves). Nearly 100-fold concentrations are actually assayed in the sedimentation equilibrium experiment, and each global fit is conducted with approximately 1000 primary data points. Thus, sedimentation equilibrium provides a powerful and quantitative method for analyzing the effect of mutations on the equilibrium of an associating membrane protein system.

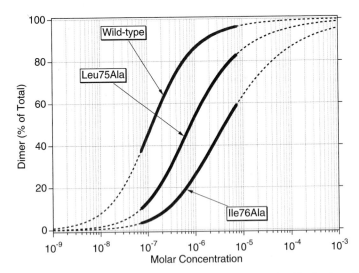

Fig. 5. Species distribution of SN/GpATM dimers in C_8E_5. The distribution of species was calculated from the monomer–dimer equilibrium constants determined from global fitting of sedimentation equilibrium data for each mutant. The solid lines are the concentration range over which the analysis was carried out. The dashed lines are extrapolations of those data. [Reprinted from K. G. Fleming, A. L. Ackerman, and D. M. Engelman, *J. Mol. Biol.* **272**, 266 (1997). With permission.]

Absolute Value of Free Energy

Values for the free energy of dimerization in this study have been calculated by estimating the equilibrium constant measured in 33 mM C_8E_5. As discussed in the introduction, the free energy change observed on transmembrane α-helix dimerization occurring in stage 2 of membrane protein folding will contain contributions not only from changes in helix–helix interactions, but also from changes in helix–lipid and lipid–lipid interactions, as well as from changes in helix and lipid entropy. Thus, the absolute value of the free energy cannot be simply interpreted as the free energy of protein–protein interaction, as is the case for soluble proteins. Further, this free energy value measured in C_8E_5 may not even directly translate to that which would be observed in biological membranes, if it could be measured. Still, by consideration of the effects of changes in the value of the free energy on perturbation of the system (compared with a standard state), these methods will allow useful studies on structural energetics of membrane proteins.

The two-stage model provides a framework for understanding the effects of point mutations on dimer stability. While mutations in the amino acid sequence might alter any of the three states available to the polypeptide sequence (shown in Fig. 1), the small point mutations used in this study minimally alter the hydrophobicity of the amino acid sequence and should therefore not significantly perturb the equilibrium of the stage 1 insertion process. Thus, the effect of the point mutations used in this study can be considered by using a thermodynamic cycle that includes only membrane-bound states[26] as shown in Fig. 6. The $\Delta\Delta G_{Mut}$ value is then given by the difference $(\Delta G_2 - \Delta G_1)$ or $(\Delta G_4 - \Delta G_3)$. The term $(\Delta G_2 - \Delta G_1)$ is experimentally measured, and the term $(\Delta G_4 - \Delta G_3)$ may be accessed computationally.[26]

Two computational approaches have been used to provide structure-based interpretations of the effect of point mutations on free energy changes. Despite completely ignoring all the energy terms associated with lipid interactions, both of these approaches show good agreement between the experimental free energy data and calculated parameters. Fleming *et al.* used molecular dynamics and energy minimization to generate structural models for mutant dimers.[11] The authors found good correlations between the experimental dimer stabilities and the interhelical van der Waals potential energies $(R^2 = 0.990)$ as well as with the buried occluded surface areas $(R^2 = 0.996)$ of the modeled structures. Subsequently, MacKenzie and Engelman derived structure-based empirical parameters for prediction of

---

[26] K. R. MacKenzie and D. M. Engelman, *Proc. Natl. Acad. Sci. U.S.A.* **95** 3583 (1998).

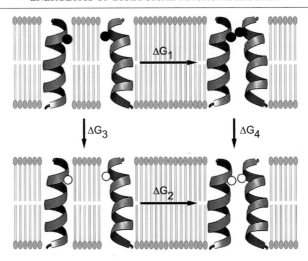

F‍IG. 6. Thermodynamic cycle depicting the effect of point mutations on the free energy of dimerization. The experimentally measured free energy is given by either ΔG_1 or ΔG_2. $\Delta\Delta G_{\text{Mut}}$ can be calculated from experimental differences in free energy changes, $(\Delta G_2 - \Delta G_1)$.

GpATM dimer stability.[26] Although determination of this parameter set was based on apparent dimer stabilities measured by SDS–PAGE,[6,7,26] a good correlation $(R^2 = 0.997)$ was found between their predicted dimerization propensities and the experimental free energies measured in C_8E_5. Free energy measurements of more mutants are clearly needed to evaluate the accuracy of these two approaches in describing the physical basis of transmembrane helix–helix interactions and in deriving a structure-based energy scale.

Summary

Sedimentation equilibrium in the analytical ultracentrifuge provides membrane biochemists with a tool to quantitatively probe thermodynamics of associating systems in detergent environments. As long as conditions of reversibility are met, the free energy of interaction can be measured in varied hydrophobic environments, pH values, ionic strengths, and temperatures. Although the absolute value of the interaction free energy of membrane protein subunits will no doubt depend on the hydrophobic environment, experiments in any one environment will allow subunit associations to be placed on a relative scale of interaction. The temperature dependence of the free energy change may provide more thorough information about the thermodynamics of helix–helix association in micelles.

Acknowledgments

The NONLIN algorithm is distributed freely from the RASMB Web site: *http://www.bbri. harvard.edu/RASMB/rasmb.html.* This work was supported by Grant GM16769 from the National Institutes of Health to K.G.F., and by grants from the NIH (GM54160) and by funds from the National Foundation for Cancer Research to Donald M. Engelman.

[4] Energetics of Vinca Alkaloid Interactions with Tubulin

By Sharon Lobert and John J. Correia

This chapter presents a detailed discussion of the energetics of interactions between *Vinca rosea L.* (vinca) alkaloids, antimitotic chemotherapeutic agents, and tubulin, the major protein constituent of microtubules and the mitotic spindle. Vinca alkaloids (Fig. 1) induce tubulin to form indefinite spirals and ordered paracrystals that compete with microtubule formation. Drug binding is known to be thermodynamically linked to spiral formation. *In vivo*, vinca alkaloids cause mitotic arrest at substoichiometric concentrations by acting at the ends of microtubules to diminish dynamic instability in mitotic spindles. We have obtained a full thermodynamic description of the energetics of vinca alkaloid-induced tubulin self-association over a range of temperature and buffer conditions for a group of vinca alkaloid congeners.[1-6] An important conclusion of these studies is that the self-association step, induced by a relatively constant drug-binding affinity, is the determining factor that differentiates weak drugs from strong drugs. The first part of this chapter briefly describes the use of sedimentation velocity, the quantitative fitting of weight average sedimentation coefficient data, and its applicability to extracting the energetics of vinca alkaloid–tubulin interactions. Sedimentation velocity has proven to be the best method for determining vinca alkaloid binding affinities and drug-induced tubulin spiraling potential.[1-7] The second part compares results obtained with four clinically useful vinca alkaloid congeners. The third part provides a quantitative description of the impact of allosteric effectors such as guanine nucleotides, pH, salt, and divalent cations on the system. The final part relates these

[1] S. Lobert, A. Frankfurter, and J. J. Correia, *Biochemistry* **34,** 8050 (1995).
[2] S. Lobert, B. Vulevic, and J. J. Correia, *Biochemistry* **35,** 6806 (1996).
[3] S. Lobert, C. A. Boyd, and J. J. Correia, *Biophys. J.* **72,** 416 (1997).
[4] B. Vulevic, S. Lobert, and J. J. Correia, *Biochemistry* **42,** 12828 (1997).
[5] S. Lobert, A. Frankfurter, and J. J. Correia, *Cell Motil. Cytoskel.* **39,** 107 (1998).
[6] S. Lobert, J. W. Ingram, B. T. Hill, and J. J. Correia, *Mol. Pharmacol.* **53,** 908 (1998).
[7] G. C. Na and S. N. Timasheff, *Methods Enzymol.* **117,** 459 (1985).

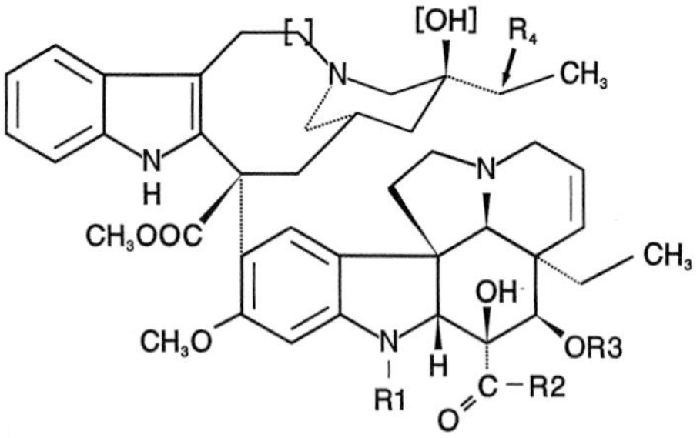

Vinblastine: R1 = CH₃; R2 = OCH₃; R3 = COCH₃
Vincristine : R1 = CHO; R2 = OCH₃; R3 = COCH₃
Vinorelbine: R1 = CH₃; R2 = OCH₃; R3 = COCH₃; [ ]= removed
Vinflunine : R1 = CH₃; R2 = OCH₃; R3 = COCH₃; [ ]= removed; R₄ = F₂

FIG. 1. Chemical structure of vinca alkaloids.

energetic findings to the structure of tubulin, the structure and dynamics of microtubules, and the implications for antimitotic and antineoplastic effectiveness with specific emphasis on tubulin isotype effects.[5]

I. Sedimentation Velocity as a Modern Tool to Study Ligand-Linked Association

The method of analyzing ligand-linked macromolecular self-association and the use of quantitative fitting of weight average sedimentation coefficient data to extract molecular energetics, particularly the interaction of vinblastine with tubulin, have been previously reviewed in these proceedings.[7,8] Here we mention only the features of our analysis that differ from these previous reports while clarifying the motivation for and advantages of these differences. In our studies we have exclusively used porcine brain tubulin free of microtubule-associated proteins (MAPs), obtained by two cycles of warm–cold polymerization–depolymerization followed by phosphocellulose chromatography to separate tubulin (PC-tubulin) from MAPs.[9,10] (Tubulin and microtubules have been extensively reviewed elsewhere.[11,12]) There is no evidence that the difference in source of tubulin, calf versus porcine, or the method of tubulin isolation[9] affects the results. One important difference between the extensive series of sedimentation

[8] G. C. Na and S. N. Timasheff, *Methods Enzymol.* **117,** 496 (1985).
[9] R. C. Williams, Jr. and J. C. Lee, *Methods Enzymol.* **85,** 376 (1982).
[10] J. J. Correia, L. T. Baty, and R. C. Williams, Jr., *J. Biol. Chem.* **262,** 17278 (1987).
[11] K. H. Downing and E. Nogales, *Curr. Opin. Cell Biol.* **10,** 16 (1998).
[12] J. S. Hyams and C. W. Lloyd, (eds.), "Microtubules." John Wiley & Sons, New York 1994.

studies conducted by Timasheff and co-workers[7,13,14] and our own is that they performed all their studies in a Beckman model E analytical ultracentrifuge equipped with schlieren optics, whereas we performed all our sedimentation studies in a Beckman (Fullerton, CA) Optima XLA analytical ultracentrifuge equipped with absorbance optics. Because of limitations of both optical systems, their studies were done at 50–200 μM tubulin, while all our studies are generally done at 1–4 μM PC-tubulin. Because of the high tubulin concentrations used in those previous studies it has been falsely assumed that vinca alkaloids weakly induce tubulin spirals and require both protein and drug concentrations that greatly exceed physiologic or clinical levels. This is clearly not true and in later sections of this chapter we discuss the functional consequences of this tightly coupled association.

In our binding studies the degree of self-association as a function of free drug concentration is monitored by an increase in the weight average sedimentation coefficient. Tubulin samples are equilibrated in spun Sephadex G-50 columns using appropriate buffers [10–100 mM piperazine-N,N'-bis(2-ethanesulfonic acid) (PIPES), pH 6.9] and drug concentrations ranging from 0.1 to 70 μM. The free drug concentration is obtained from the known drug concentration in the equilibration buffer. After equilibration, the protein is brought to the final concentration by dilution with the equilibration buffer, although it is typical to plan the equilibration step such that the protein concentration is as near as possible to the desired concentration (typically 2 μM). Because we use low tubulin concentrations in the absorbance optical system, the drug concentration is usually in significant excess, and thus during the sedimentation velocity run we neither observe ligand depletion nor the bimodal patterns that are typical of Cann–Goad systems and the typical data observed by Na and Timasheff.[7,13,14] In our system dilution with equilibration buffer does not significantly alter the free drug concentration, although it clearly causes a redistribution of polymer sizes by changing the protein concentration, an important variable in the analysis of these data. This technical difference necessitated by the use of different optical systems does not appear to affect the quantitative results gleaned from the analysis of otherwise similar data.[1] Nonetheless, there are implications and differences in interpretation, as already discussed above, that are clarified in later sections.

Temperature is calibrated by the new method of Liu and Stafford,[15] an absorbance technique that is highly appropriate for the new generation of Optima XLA/I analytical ultracentrifuges. The density and viscosity of

[13] G. C. Na and S. N. Timasheff, *Biochemistry* **19,** 1347 (1980).
[14] G. C. Na and S. N. Timasheff, *Biochemistry* **25,** 6222 (1986).
[15] S. Liu and W. F. Stafford III, *Anal. Biochem.* **224,** 199 (1995).

each buffer are measured as described.[1] Samples are spun at various temperatures and appropriate speeds that maximize the number of scans that can be used in the analysis. To achieve this, we currently collect velocity data at 278 nm, the absorption maxima for tubulin, and a radial spacing of 0.002 cm with 1 average in a continuous scan mode. Typically, 20 data scans are analyzed by DCDT, software that generates a distribution of sedimentation coefficients, $g(s)$.[16-19] These distributions are then converted to a weight average sedimentation coefficient ($\bar{s}_{20,w}$) by either appropriate manipulations in a spread sheet (Origin, Microcal Software, Inc., Northampton, MA) or directly by newer versions of DCDT (available from RASMB by anonymous FTP at bbri.harvard.edu). These newer versions of DCDT (DCDT_30z or _60z) also output the uncertainty of $\bar{s}_{20,w}$, which can be appropriately used in the fitting. The advantage of using DCDT is the ease of analysis and the improved signal averaging derived from using multiple scans. The experimental weight average sedimentation coefficient [$\int g(s)sds/\int g(s)ds$] is directly relatable to a molecular model corresponding to the weight average involving a distribution of discrete species ($\Sigma c_i s_i/\Sigma c_i$). Note, in this equivalence s is a (pseudo) continuous distribution that relates to the discrete distribution s_i, $g(s)$ is the derivative of concentration with respect to s, and $\int g(s)ds$ [$\int (dc/ds)ds$] and Σc_i correspond to the total macromolecule concentration in absorbance or weight concentration units, respectively. In the work of Timasheff and co-workers the second moment position was used and transformed into an $\bar{s}_{20,w}$ value. As discussed elsewhere, the second moment corresponds to the weight average polymer size and is linked through the energetics of the process to the experimental plateau concentration.[7,20] This is exactly equivalent to the weight average method described here.[21] Both methods require that a plateau exists and the boundary has cleared the meniscus; clearance of the meniscus is especially important for obtaining a useful $g(s)$ from DCDT.

A. Curve Fitting of Weight Average Sedimentation Coefficient Data

Na and Timasheff[7,8,13,14] established that vinca alkaloid binding is linked to spiral formation. The models that best describe the data are isodesmic

[16] W. F. Stafford III, *Anal. Biochem.* **203**, 295 (1992).

[17] W. F. Stafford III, *in* "Analytical Ultracentrifugation in Biochemistry and Polymer Science" (S. E. Harding, A. J. Rowe, and J. C. Horton, eds.), pp. 359–393. Royal Society of Chemistry, Cambridge 1992.

[18] W. F. Stafford III, *Methods Enzymol.* **240**, 478 (1994).

[19] W. F. Stafford III, *in* "Modern Analytical Ultracentrifugation" (T. M. Schuster and T. M. Laue, eds.), pp. 119–137. Birkhauser, Boston, 1994.

[20] R. J. Goldman, *J. Phys. Chem.* **57**, 194 (1953).

[21] W. F. Stafford III, *Methods Enzymol.* **323**, [13], 2000 (this volume).

$$2A + 2X \overset{K_1}{\rightleftharpoons} 2AX$$
$$K_4 \Big\updownarrow \qquad \Big\updownarrow K_2$$
$$A_2 \underset{K_3}{\overset{}{\rightleftharpoons}} A_2X_2$$

SCHEME 1

ligand mediated and isodesmic ligand mediated plus facilitated, or the combined model (Scheme 1), where A represents tubulin heterodimers and X represents the vinca alkaloids. In these models, K_1 is the affinity of drug for tubulin heterodimers, K_2 is the affinity of liganded heterodimers for spiral polymers, K_3 is the affinity of drug for polymers, and K_4 is the association constant for unliganded tubulin heterodimers.

Scheme 1 describes the ligand linkage that precedes the indefinite association. Liganded dimers further react to form liganded trimers, liganded tetramers, etc.[1,7,13] The isodesmic assumption is that each successive step, expressed as K_2^{app},[7] involves the same free energy. An alternative model, the isoenthalpic model (see discussion in Ref. 1), has constant enthalpy due to the formation of the same interface, but varying entropy for each step because of changes in polymer shape. To obtain binding parameters, sedimentation data are listed as $\bar{s}_{20,w}$ versus free drug (plotted in Fig. 2) and total protein concentration in the plateau, and binding constants are obtained by fitting these input data with the nonlinear least-squares program Fitall (MTR Software, Toronto, Canada), modified to include the appropriate fitting functions.[7,22] In general, Na and Timasheff performed experiments at a fixed drug concentration and varied the protein concentration, whereas we typically vary drug concentration at fixed protein concentrations. The protein concentration is required for both methods, and our fitting function allows for a global analysis over both drug and protein concentration.[1] We use a minimum of 15–21 different drug concentrations over the range of 0.1 to 70 μM. (Above 70 μM the signal-to-noise ratio is significantly reduced in the absorbance optical system.) Using the four-hole rotor in the XLA, this corresponds to five to seven sets of velocity runs. The absorbance optical system cannot collect data fast enough to optimally utilize the newer eight-hole rotor.

In this scheme for drug binding there are two independent variables, tubulin concentration and drug concentration, and therefore the data and fits are three-dimensional surfaces. In Fig. 2, the presentation of fits as a

[22] J. J. Correia, *Methods Enzymol.* **321,** 81 (2000).

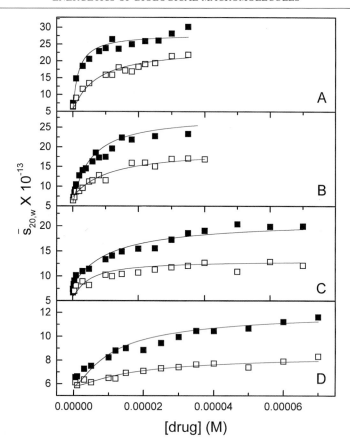

FIG. 2. Plots of $\bar{s}_{20,w}$ values versus free concentrations of vinca alkaloids at 25°. (□) 50 μM GTP; (■) 50 μM GDP. The solid lines represent fits using the ligand-mediated model. The equilibrium constants obtained from these fits are given in Table I. Tubulin in all experiments was 2 μM. (A). Vincristine; (B) vinblastine; (C) vinorelbine; (D) vinflunine. [The vincristine, vinblastine, and vinorelbine data are from Lobert et al.,[2] and the vinflunine data are from Lobert et al.[6]]

two-dimensional figure is shown at average protein concentrations that do not visually reflect goodness of fit. In the combined model there are only three independent constants since by microscopic reversibility $K_1 K_2 = K_3 K_4$. Na and Timasheff[7,8,13,14] demonstrated that K_4 values $<1 \times 10^4\ M^{-1}$ were indistinguishable, and thus in our work we also fix $K_4 = 1 \times 10^4\ M^{-1}$. This constrained model allows for more reproducible parameter estimation. The standard deviations of the constrained and unconstrained fits are indistinguishable from those of the two-parameter ligand-mediated fits.[1-6] Either the data lack sufficient information to differentiate between the two models or one branch of the thermodynamic loop is kinetically excluded, i.e., tubulin does not make spirals in the absence of drug (see Section IV). It is clear, however, that thermodynamic exclusion of the facilitated branch is not allowed. In fact, the binding of the drug to spiral polymers, K_3, is a direct conceptual indicator of thermodynamic linkage, and effectors in the system that alter K_2 necessarily alter K_3.[23] Therefore, if drugs induce assembly then drugs must bind to spiral polymers preferentially over tubulin heterodimers.[23,24]

Sedimentation is a hydrodynamic method and therefore analysis requires information about polymer shapes in addition to polymer mass. The simplest assumption, referred to as the spherical approximation,[1,7] is used in the fitting function for the sedimentation coefficient of i-mers ($s_i = i^{2/3}s_1$) to calculate the weight average ($\Sigma c_i s_i / \Sigma c_i$). It can be easily shown that for all polymers with constant axial ratio, not just spheres, the sedimentation coefficients will scale as $MW^{2/3}$. Because asymmetric molecules will have smaller sedimentation coefficients than spheres of the same molecular weight due to an enhanced shape factor (f/f_0), the equilibrium constants estimated with the spherical assumption are reduced in magnitude. In our initial attempts to determine the extent of this reduction we incorporated f/f_0 into the fitting function. We observed the expected increase in binding constants, but a poorer fit of the data,[1] an observation that remains unexplained and compromises the use of an isoenthalpic model. To resolve this uncertainty it is necessary to use direct measures of molecular mass such as sedimentation equilibrium or light scattering. In principle these are far more robust determinants of correct models and equilibrium constants. The major disadvantage of sedimentation equilibrium, and thus the strength and advantage of sedimentation velocity for this system, is that tubulin is unstable and, during the time required for equilibrium experiments, undergoes irreversible aggregation. This instability is exacerbated by the large size of vinca-induced spirals and the extreme times required to achieve equilibrium. Furthermore, evidence suggests that one vinca alkaloid, vincristine, induces tubulin denaturation.[5,25] Scattering techniques have been applied to a vinblastine–tubulin system,[26] but the higher protein concentrations required favor formation of paracrystals, or higher ordered association of spirals, which precludes studying the energet-

[23] J. Wyman, *Adv. Protein Chem.* **19**, 223 (1964).
[24] J. J. Correia, *Pharmacol. Ther.* **52**, 127 (1991).
[25] V. Prakash and S. N. Timasheff, *Biochemistry* **24**, 5004 (1985).
[26] E. Nogales, F. J. Medrano, G. P. Diakun, G. R. Mant, E. Towns-Andrews, and J. Bordas, *J. Mol. Biol.* **254**, 416 (1995).

ics of spiral formation. (As discussed in Section IV, spiral formation appears to be the relevant polymer form that accounts for the relationship between vinca alkaloid energetics and antimitotic activity.) Thus, we must conclude that sedimentation velocity is the preferred method for studying vinca alkaloid-induced tubulin spiral formation. We favor the ligand-mediated fits as the best quantitative indicators of overall association because fitting with this simple model gives standard deviations that are equivalent or better than when the combined model is used. While more work must be done to establish the appropriateness of the facilitated pathway, it is clear that K_3 values are important indicators of thermodynamic linkage and may ultimately prove to be essential to understanding the free energy differences among vinca alkaloids.

II. Comparison of Vincristine, Vinblastine, Vinorelbine, and Vinflunine Effects

Vinca alkaloids are natural products or semisynthetic derivatives of natural products that differ by subtle chemical modifications. The derivatives presented in Fig. 1 represent drugs that are clinically useful in the United States and Europe: vincristine, vinblastine, and vinorelbine, and a new drug, vinflunine,[6,27] that will soon enter clinical trials. Figure 2 presents a summary of $\bar{s}_{20,w}$ data plotted as a function of drug concentration for each vinca alkaloid depicted in Fig. 1. Each data set demonstrates a hyperbolic increase in $\bar{s}_{20,w}$ as a function of increasing drug concentration, consistent with an indefinite polymerization mechanism that does not exhibit cooperativity.[1,7] Tubulin is a heterodimeric guanine nucleotide-binding protein with one exchangeable nucleotide binding site on the β subunit that regulates microtubule assembly and the interaction of tubulin with antimitotic drugs.[10–12,24] (There is also a nonexchangeable nucleotide binding site on the α-subunit that appears to fulfill a structural role.) Thus, our experiments were performed in either 50 μM GDP- or 50 μM GTP-containing buffers. (*Note:* In the absorbance optical system guanine nucleotide concentrations must be kept at $<100 \ \mu M$.) If we compare results between (A)–(D) in Fig. 2 then it is apparent that vinca alkaloids have significantly different effects on tubulin spiral formation. For example, under GDP conditions the plateau levels vary from 30S for vincristine to 25S for vinblastine to 20S for vinorelbine to 12S for vinflunine. A similar trend exists under GTP conditions. The results of quantitative analysis of

[27] J. Fahy, A. Duflos, J. C. Jacqesy, M. P. Jouannetaud, C. Meheust, A. Kruczynski, B. Etievant, J. M. Barret, F. Colpaert, and B. T. Hill, *J. Am. Chem. Soc.* **119,** 8576 (1997).

TABLE I

EQUILIBRIUM CONSTANTS FOR INTERACTION OF VINCA ALKALOIDS WITH TUBULIN[a]

Drug	GXP	K_1 (M^{-1})	K_2 (M^{-1})	K_1K_2 (M^{-2})	SD[b]
Vincristine	GTP	1.4 ($\pm$0.4) $\times$ 10^5	1.7 ($\pm$0.4) $\times$ 10^7	2.3 $\times$ 10^{12}	1.2
Vinblastine	GTP	1.2 ($\pm$0.2) $\times$ 10^5	5.1 ($\pm$0.6) $\times$ 10^6	6.1 $\times$ 10^{11}	0.9
Vinorelbine	GTP	1.3 ($\pm$0.2) $\times$ 10^5	1.1 ($\pm$0.8) $\times$ 10^6	1.4 $\times$ 10^{11}	0.3
Vinflunine	GTP	8.8 ($\pm$2.0) $\times$ 10^4	3.0 ($\pm$0.4) $\times$ 10^5	2.6 $\times$ 10^{10}	0.2
Vincristine	GDP	2.1 ($\pm$0.4) $\times$ 10^5	3.9 ($\pm$0.5) $\times$ 10^7	7.9 $\times$ 10^{12}	1.7
Vinblastine	GDP	1.3 ($\pm$0.1) $\times$ 10^5	2.3 ($\pm$0.6) $\times$ 10^7	3.0 $\times$ 10^{12}	0.8
Vinorelbine	GDP	1.3 ($\pm$0.3) $\times$ 10^5	5.1 ($\pm$0.6) $\times$ 10^6	6.6 $\times$ 10^{11}	1.0
Vinflunine	GDP	1.0 ($\pm$0.3) $\times$ 10^5	1.2 ($\pm$0.2) $\times$ 10^6	1.2 $\times$ 10^{11}	0.5

[a] At 25°.
[b] Data were fit with the ligand-mediated model.

these data are presented in Table I. Two features are apparent. First, all the K_1 values are relatively constant for all drugs and conditions with a mean value of 1.31 ($\pm$0.34) $\times$ 10^5 M^{-1}. This implies that each drug binds to the tubulin heterodimer with nearly identical affinity. The parameter that accounts for the enhanced spiral potential is K_2, the indefinite polymerization constant. Within each nucleotide condition vincristine has the largest K_2 value while vinflunine has the smallest K_2 value and there is a difference of two orders of magnitude in K_2 and in the overall affinity K_1K_2. This corresponds to an enhancement of 2.6 kcal/mol in spiral formation for each step in the indefinite polymerization (Fig. 3), and accounts for the nearly threefold increase in the plateau values. The order observed in overall affinity for tubulin, K_1K_2 (Table I and Fig. 3), is always vincristine > vinblastine > vinorelbine > vinflunine regardless of conditions. When data are fit with the ligand-mediated plus ligand-facilitated model, this order is also found in K_3, the binding of the drug to unliganded polymers (data not shown).[2,6] These results mean that chemical modifications of these drugs alter the ability to induce and stabilize spiral polymers and that the activity of these drugs is linked to tubulin self-association.

Thermodynamic analysis of these data (Table II), derived from van't Hoff plots, demonstrates that the driving force for each of these drug-induced reactions is entropic[2,6] and therefore consistent with the release of water on the burial of hydrophobic surfaces.[28] In addition, the enthalpy is in general unfavorable; for the strongest drug, vincristine, ΔH is a small negative number or near zero and becomes more positive for each succes-

[28] B. Vulevic and J. J. Correia, *Biophys. J.* **72**, 1357 (1997).

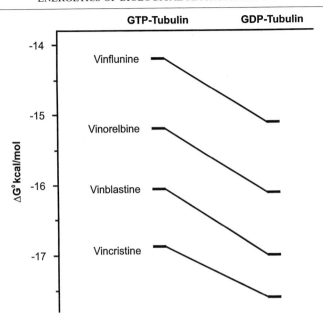

FIG. 3. Free energy of overall interaction K_1K_2 (derived from Table I) for vinca alkaloid-induced self-association of tubulin, showing the nearly constant enhancement by GDP for each vinca alkaloid.

sively weaker drug. The molecular origin of these unfavorable enthalpies is partially entropic but must also involve drug–protein and protein–protein interactions. For example, the enthalpy is always more unfavorable, more positive, for GTP rather than GDP conditions. Experimental attempts to understand more fully the molecular origins of these driving forces require

TABLE II
THERMODYNAMIC PARAMETERS FOR TUBULIN–VINCA ALKALOID INTERACTION[a]

Drug	Condition	K_1K_2 (M^{-2})	$\Delta G°$ (kcal/mol)	$\Delta S°$ (cal/mol-K)	$\Delta H°$ (kcal/mol)
Vincristine	GTP	2.3×10^{12}	-16.9	47	-2.9
	GDP	7.9×10^{12}	-17.6	60	-0.3
Vinblastine	GTP	6.1×10^{11}	-16.1	68	4.2
	GDP	3.0×10^{12}	-17.0	64	2.0
Vinorelbine	GTP	1.4×10^{11}	-15.2	76	7.4
	GDP	6.6×10^{11}	-16.1	69	4.4
Vinflunine	GTP	2.6×10^{10}	-14.2	98	15.1
	GDP	1.2×10^{11}	-15.1	87	10.9

[a] At 25°; data fit according to the ligand-mediated model.

both structural information and additional energetic data on drug derivatives that provide sufficient information to form complete thermodynamic cycles.[29]

We have also investigated the kinetics of spiral reequilibration by
stopped-flow light-scattering methods.[1–3,6] The data are best described by
a single average relaxation time where smaller polymers reequilibrate on
dilution more quickly than large polymers. This is consistent with a cas-
cade of dissociation events corresponding to the endwise loss of liganded
tubulin subunits from spiral polymers. Vincristine, which makes the
largest spirals, exhibits the longest relaxation times, while vinflunine,
which makes the smallest spirals, exhibits the fastest relaxation times.
These data indicate that the size of the spirals induced correlates with
the lifetime of drug retention (see Section IV). In addition, the relaxation
times are inversely proportional to protein concentration, consistent with
the occurrence of annealing or breaking of spiral polymers. This suggests
that direct addition of spirals to the ends of microtubules is feasible
(see Section IV).

III. Allosteric Effectors

A. Effects of Guanine Nucleotides on Vinca Alkaloid-Induced Tubulin Self-Association

In comparing the open and closed symbols within Fig. 2 A–D, it is clear
that there is a significant enhancement in spiral formation in the presence
of GDP for each vinca alkaloid studied. (*Note:* Vinca alkaloid-induced
spirals do not have GTPase activity, so these are equilibrium experiments.[24])
Quantitative analysis reveals that GDP enhances tubulin self-association
three- to fivefold in the presence of all four drugs. Relative to GTP, this
GDP enhancement for vincristine-, vinblastine-, vinorelbine-, and vinflun-
ine-induced spiral assembly is 0.94 ± 0.19 kcal/mol.[2,6] This corresponds to
0.94 kcal/mol for each step, K_2, in the indefinite polymerization process,
or, in the combined model, also to each occurrence of drug binding to the
spiral. Thus, nucleotide enhancement is linked to the self-association step
and appears to be an intrinsic property of the tubulin structure that is not
dependent on the specific vinca alkaloid used. A structural interpretation
of these data is now possible because the electron crystal structure of tubulin
has been determined[30] (see Section IV).

[29] Q. D. Dang, E. R. Guinto, and E. DiCera, *Nature Biotechnol.* **15,** 146 (1997).
[30] E. Nogales, S. G. Wolf, and K. H. Downing, *Nature* (*London*) **391,** 199 (1998).

B. Influence of Guanine Nucleotide Analogs

Because chemical modifications of the vinca alkaloid structure modulate spiral formation, it is likely that, by Wyman linkage, chemical modifications of guanine nucleotide allosteric effectors also modulate drug effects on tubulin. We compared the effects of the GTP and GDP α,β-methylene analogs, GMPCPP and GMPCP, on vinblastine-induced tubulin self-association.[4] When bound to tubulin, GMPCPP is a weakly hydrolyzable analog of GTP and strongly enhances microtubule assembly.[28] GMPCPP perfectly mimicks GTP in the energetics of vinca alkaloid-induced spiral assembly under all ionic strength and temperature conditions. However, GMPCP in 10 mM PIPES behaves as a GTP analog and not as a GDP analog. This was verified by demonstrating that GMPCP also acts like GTP in a Mg^{2+}-induced ring formation assay.[4,30,31] In 100 mM PIPES, the energetics of the GMPCP effect on tubulin self-association is intermediate between those of GDP and GTP. These results are consistent with previous observations that in a 2 M glycerol, 100 mM PIPES buffer, GMPCP, but not GDP, will induce microtubule formation.[28] The fact that an α,β-methylene suppresses cleavage at the β,γ-oxygen has important structural implications for understanding allosteric interactions that occur in the binding of guanine nucleotides to tubulin. We hypothesize that the α,β-methylene sterically or allosterically favors a salt-dependent GTP–tubulin conformation in various polymerized states (see Section IV).

C. Wyman Linkage Analysis

The thermodynamic linkage of other allosteric effectors (NaCl, divalent cations, pH) to vinca alkaloid-induced tubulin self-association is investigated by Wyman analysis.[23,24] The Wyman linkage relationship is

$$(\delta \ln K / \delta \ln a_3)_{T,P,m2} = \Delta \nu \qquad (1)$$

where K is the equilibrium constant for the reaction at constant temperature, pressure, and protein concentration, a_3 is the activity of ligand, and $\Delta \nu$ is the change in apparent additional binding of component 3 to the protein, component 2, during the reaction. According to this relationship, a change in the activity of any species affects the activity of all other species in solution. The equation reflects a change in binding and not a difference in the total amount bound, and is thus referred to as preferential interaction.[23,24] As discussed in Section IV, extra thermodynamic information, specifically kinetic and structural information, is required to interpret these data in terms of a molecular mechanism.

[31] J. J. Correia, A. H. Beth, and R. C. Williams, Jr., *J. Biol. Chem.* **263**, 10681 (1988).

D. Effect of NaCl on Vincristine- and Vinblastine-Induced Tubulin Spirals

The addition of NaCl to our standard buffer enhances overall vinblastine- or vincristine-induced tubulin self-association[3] (Fig. 4). In the presence of salt, the GDP enhancement of overall tubulin self-association is reduced. For example, in 150 mM NaCl GDP enhancement is 0.24 kcal/mol for vinblastine and 0.36 kcal/mol for vincristine versus an average enhancement of 0.87 ($\pm$0.34) kcal/mol for the same drugs in the absence of salt. Similar

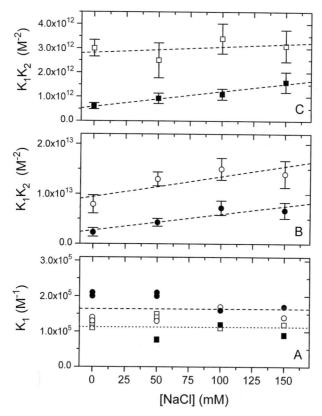

Fig. 4. Effect of NaCl on vincristine- and vinblastine-induced tubulin spiral formation in 10 mM PIPES, 1 mM MgSO$_4$, pH 6.9. Circles ($\bullet$, $\bigcirc$) indicate vincristine data and squares ($\blacksquare$, $\square$) indicate vinblastine data. Open and closed symbols represent the GDP and GTP condition, respectively. (B and C) Lines represent linear fits of the data; (A) lines are mean values. (A) NaCl effect on K_1. The average values of K_1 [1.13 ($\pm$0.2) $\times$ 10^5 M^{-1} for vinblastine and 1.64 ($\pm$0.28) $\times$ 10^5 M^{-1} for vincristine] are NaCl independent. (B) NaCl effect on $K_1 K_2$ for vincristine. (C) NaCl effect on $K_1 K_2$ for vinblastine. [Data are replotted from Lobert et al.[3]]

to the GDP enhancement, the NaCl enhancement is in K_2, the affinity of liganded heterodimers for spirals (Fig. 4B and C). Wyman linkage analysis of experiments with vinblastine or vincristine over a range of NaCl concentrations shows a twofold increase in $\Delta\nu_{NaCl}$, the change in NaCl bound to drug-induced spirals, in the presence of GTP compared with GDP. These data indicate that GDP enhancement of vinca alkaloid-induced tubulin self-association is in part due to electrostatic inhibition in the GTP–tubulin conformational state. As implied in the discussion of nucleotide effects, this state may be induced by polymerization. The locations of these repulsive interactions in the sequence and in the tertiary structure are unknown but partially reside in the highly variable and acidic carboxyl tail regions of both subunits (see the next section). Because the carboxyl regions are unordered in the electron crystal structure of tubulin,[30] it is likely they will influence GDP–tubulin and GTP–tubulin equally, suggesting that additional electrostatic interactions more localized to the nucleotide-binding site and the spiral polymer interface may account for these differential effects (see Section IV).

E. Effect of Divalent Cations and pH on Vinblastine-Induced Tubulin Spirals

Vinca alkaloids in the presence of 1 mM Mn^{2+} or Ca^{2+} cause tubulin precipitation and the formation of paracrystals.[3] We find that NaCl suppresses precipitation, probably by inhibiting lateral electrostatic interactions between spirals. Thus the presence of NaCl permits studies of spiral formation at high divalent cation concentrations. Studies at 0.1 to 1.0 mM Mg^{2+}, Mn^{2+}, and Ca^{2+} revealed a differential divalent cation influence on vinblastine- and vincristine-induced spiral formation. This suggested a potential antimitotic and antineoplastic role for divalent cations. To investigate this further we have performed more extensive titrations in the presence of 150 mM NaCl and $CaSO_4$ (0.1–10 mM) or $MgSO_4$ (0.1–15 mM). Results from a study with vinblastine-induced spiral formation are presented in Fig. 5. There are a number of features worth noting. First, as indicated by comparisons of different drugs and guanine nucleotides, divalent cation effects are primarily seen in K_2, the indefinite polymerization constant, and not K_1, the binding to tubulin heterodimers. The average values of K_1 are relatively independent of divalent cation concentration at 150 mM NaCl. The data in Fig. 5B indicate there is a divalent cation dependence for spiral formation, K_2. The data for GDP and Mg^{2+} appear to plateau by 5 mM cation, while the Ca^{2+} data appear to increase in a more gradual and linear manner. If the more robust and reliable product of the two parameters, $K_1 K_2$, is plotted (Fig. 5C), then a more consistent linear trend is apparent for both divalent

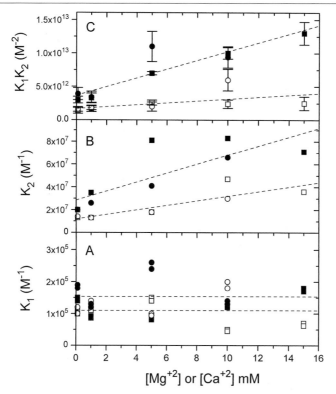

Fig. 5. Effect of Mg^{2+} or Ca^{2+} on vinblastine-induced tubulin spiral formation in 10 mM PIPES, 1 mM MgSO$_4$, 150 mM NaCl, pH 6.9. Circles (●, ○) indicate Ca^{2+} data and squares (□, ■) indicate Mg^{2+} data. Open and closed symbols represent the GTP and GDP condition, respectively. (B and C) Lines represent linear fits of the data; (A) lines are mean values. (A) Divalent cation effect on K_1. The average values of K_1 are 1.10 ($\pm$0.39) $\times$ $10^5 M^{-1}$ for Mg^{2+}, and 1.53 ($\pm$0.50) $\times$ $10^5 M^{-1}$ for Ca^{2+}. (B) Divalent cation effect on K_2. (C) Divalent cation effect on K_1K_2.

cations. The Mg^{2+} and Ca^{2+} data are in reasonable agreement and suggest the absence of a significant differential divalent cation effect at intracellular ionic strength. Furthermore, the trend is clearly nucleotide dependent, with both divalents having a larger effect on K_1K_2 in the presence of GDP. This is exactly the opposite of what we observed for NaCl. A transition from a single spiral to a double spiral has been reported to occur above 6 mM Mg^{2+}.[26] The early plateau in K_2 at 5 mM Mg^{2+} may represent this transition, although it is clearly not apparent in the K_1K_2 data.

A number of specific binding sites for divalent cations have been identified, although their exact structural location has not been confirmed because of the resolution level (3.7 Å) of the electron crystal structure.[30] There are two sites for either Mg^{2+} or Mn^{2+} that are tightly coupled to the binding of GTP at the exchangeable and the nonexchangeable nucleotide-binding site, located on the β and α subunits, respectively.[10,24,31] There is also a specific site for Zn^{2+} binding that induces the formation of sheets used to solve the electron crystal structure of tubulin.[30] Microtubule assembly is inhibited by high-affinity Ca^{2+}-binding sites located on the tubulin hetero-dimer,[32] while Mg^{2+} and Mn^{2+} stimulate microtubule assembly. This is in contrast to vinca alkaloid-induced spirals, which are equally stimulated by all of these divalent cations (Fig. 5).[3] As with NaCl effects, the sites involved in favorable spiral formation are in part low-affinity, nonspecific sites lo-cated on the acidic carboxyl tail region. This is suggested by the fact that the formation of microtubules and divalent cation-induced rings is enhanced by subtilisin digestion of tubulin, a procedure that removes the carboxyl tail region, thus removing repulsive electrostatic interactions.[33,34] Because subtilisin-digested tubulin responds to increasing divalent cation concentra-tions,[34] other domains must also be involved. In a microtubule the carboxyl regions are located on the outer surface[30] and thus cause the microtubule to be a highly charged polyelectrolyte. A polyelectrolyte character may ultimately prove to be an important determinant of the manner in which microtubules and other tubulin polymers interact with cations, MAPs, and molecular motors.

It is of interest that NaCl and divalent cations appear to have opposing effects on paracrystal formation. NaCl suppresses while divalent cations stimulate paracrystal formation. They both stimulate spiral assembly in a nucleotide-dependent manner (Figs. 4 and 5). Rai and Wolff[35] have re-ported that polyanions, including nucleoside triphosphates, inhibit paracrys-tal formation in a manner that depends on the charge of the anion: the more charge, the more inhibition. Spiral assembly appears to be insensitive to the nature of the polyanion. These data are consistent with the nucleotide effect observed for spiral assembly in our studies[1-6] being due to site-specific interactions, while the influences of salt,[3] cations,[3,14] and polyanions[35] on paracrystal formation are nonspecific effects that reflect the polyelectrolyte character of spirals and paracrystals. Rai and Wolff[35] also observed what they refer to as a nucleation event prior to paracrystal formation that must clearly be the formation of a critical concentration of spirals or spirals of the correct average length to nucleate paracrystals.[26]

[32] S. A. Berkowitz and J. Wolff, *J. Biol. Chem.* **256**, 11216 (1981).
[33] S. Lobert and J. J. Correia, *Arch. Biochem. Biophys.* **296**, 152 (1992).
[34] S. Lobert, B. S. Hennington, and J. J. Correia, *Cell Motil. Cytoskel.* **25**, 282 (1993).
[35] S. S. Rai and J. Wolff, *Eur. J. Biochem.* **250**, 425 (1997).

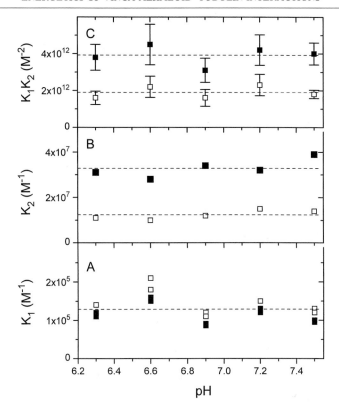

FIG. 6. Effect of pH on vinblastine-induced tubulin spiral formation in 10 mM PIPES, 1 mM MgSO$_4$, 150 mM NaCl. Open and closed squares represent the GTP and GDP condition, respectively. Lines represent mean values of the data. (A) pH effect on K_1. The average value of K_1 is 1.29 ($\pm$0.30) $\times$ 10^5 M^{-1}. (B) pH effect on K_2. (C) pH effect on K_1K_2.

Sedimentation velocity experiments carried out over a range of pH conditions (pH 6.3–7.5) indicate that pH has no effect on vinblastine-induced spiral formation in the presence of GDP or GTP (Fig. 6). This differs from microtubule formation, where the binding of a single H$^+$ is linked to polymerization and optimal assembly occurs between pH 6.6 and 6.9.[36] Thus H$^+$ ions appear to be involved in stabilizing lateral interactions of linear protofilaments not found in vinca alkaloid-induced tubulin spirals. Paracrystal formation, the lateral association of spirals,[26] exhibits a pH dependence similar to that of microtubules, partially dependent on the acidic carboxyl terminus.[37] The regions that make contact between microtu-

[36] J. C. Lee and S. N. Timasheff, *Biochemistry* **16**, 1754 (1977).
[37] S. S. Rai and J. Wolff, *Proc. Natl. Acad. Sci. U.S.A.* **95**, 4253 (1998).

bule protofilaments occur in both α- and β-tubulin in a loop located between B7 and H9 and an acidic region in H3 on the adjacent heterodimer.[30,38,39] These interactions may account for the pH dependence, specific Ca^{2+} inhibition, and Mg^{2+} activation of microtubule assembly.

IV. Structural Implications of Energetic Data

Because the electron crystal structure of tubulin has been reported,[30,38] we can offer a structural basis for many of our observations.[1-6] The exchangeable nucleotide- and vinca alkaloid-binding sites are both located on the β subunit. The guanine nucleotide site is located in the interface between the β subunit from one chain and the α subunit from the next heterodimer in a microtubule protofilament. This explains the nonexchangeability of this nucleotide in the microtubule lattice, and, because the β subunit is located at the positive ($+$) end of a microtubule, it also explains the exchangeability of this nucleotide at the ($+$) end only.[40] On the basis of the analogous structure of the bacterial protein FtsZ,[38,41] the catalytic group that cleaves GTP is actually located on the α subunit of the next heterodimer in the protofilament. Lethal mutants of FtsZ and homology with tubulin suggest the catalytic group is Glu254 in α-tubulin. This explains why hydrolysis is tightly linked to assembly of microtubules. It also seems to resolve an old uncertainty about the mechanism of colchicine induction of weak tubulin–GTPase activity. These data appear to require that colchicine binding must induce weak polymer formation to stimulate GTPase activity.[42] Only a dimer–dimer interface would provide the catalytic group.

Vinca alkaloid-induced spirals do not cleave GTP, and thus spiral assembly probably occurs with the catalytic group improperly positioned. The α,β-methylene of GMPCPP may have a similar influence on the positioning of an essential group. It has been proposed that in the presence of GDP longitudinal contacts between dimers are at an angle that favors closed rings or spiral structures.[43] By analogy the addition of a vinca alkaloid into or near the interface favors an open spiral that grows indefinitely. A vinblastine analog, Ant-Vin, is known to cross-link to a region of the β subunit, H5 through H6,[44] that is located near this dimer–dimer interface,

[38] E. Nogales, K. H. Downing, L. A. Amos, and J. Lowe, *Nature Struct. Biol.* **5**, 451 (1998).
[39] E. Nogales, K. H. Downing, M. Whittaker, and R. A. Milligan, in preparation (2000).
[40] T. J. Mitchison, *Science* **261**, 1044 (1993).
[41] J. Lowe and L. A. Amos, *Nature (London)* **391**, 203 (1998).
[42] C. Heusele and M.-F. Carlier, *B.B.R.C.* **103**, 332 (1981).
[43] J. F. Diaz, E. Pantos, J. Bordas, and J. M. Andreu, *J. Mol. Biol.* **238**, 214 (1994).
[44] S. S. Rai and J. Wolff, *J. Biol. Chem.* **271**, 14707 (1996).

toward the outside of the microtubule surface although buried beneath H11 and H12.[11,30,38] This region sits between the guanine nucleotide and a loop connecting H11 and H12 that is important for interactions with the next subunit in a zinc sheet.[30] We hypothesize that vinca alkaloid binding distorts the interface, possibly by a direct steric interaction or by interactions with the loop between H11 and H12. There are two reports that demonstrate vinblastine cross-links to both the α- and β-tubulin, with a preference for α-tubulin.[45,46] Although nonspecific weak binding sites may explain the cross-linking results, these cross-linking data support a model in which vinca alkaloids are located in the interface between two heterodimers and have a direct steric influence on the polymer form. For GTP–tubulin this steric effect distorts the straight orientation of heterodimers usually formed in the microtubule protofilament, while for GDP–tubulin it distorts the curved orientation usually found in Mg^{2+}-induced rings.[43,47,48] In both cases it creates a twisted orientation that forms an open indefinite spiral. From an energetic perspective GDP enhances vinca alkaloid-induced spiral formation because the influence of the two is additive and both favor a spiral, one closed and one open, while GTP disfavors a spiral.

The energetic difference between the strongest vinca alkaloid, vincristine, and the weakest vinca alkaloid, vinflunine, is 2.6 kcal/mol (Table 1 and Fig. 3). The origin of these differences can involve interactions with the solvent, interactions with tubulin, and allosteric coupling to conformational changes. The primary energetic difference is in the self-association step, K_2, and thus involves protein–protein interactions. Therefore, an explanation of these energetic differences must relate structural changes in the vinca alkaloids to structural changes in the spiral polymer interface. These can be direct steric influences, as suggested by the cross-linking data,[45,46] or allosteric effects, consistent with our hypothesis that vinca alkaloids might interact with a loop that is important for dimer–dimer contacts.[30] It is conceivable that different drugs induce spiral polymers with an altered helical pitch.[26] Because modifications of both the vindoline and catharanthine moieties of vinca alkaloids (Fig. 1) affect self-association, it is likely to be a complex set of linked interactions. For example, catharanthine alone induces weak self-association of tubulin, while vindoline does not.[49] Because K_1 values are not significantly affected by structural changes in the drugs (Table I), we must conclude modifications of vindoline indi-

[45] A. R. Safa, E. Hamel, and R. L. Felsted, *Biochemistry* **26,** 97 (1987).
[46] J. Wolff, L. Knipling, H. J. Cahnmann, and G. Palumbo, *Proc. Natl. Acad. Sci. U.S.A.* **88,** 2820 (1991).
[47] R. P. Frigon and S. N. Timasheff, *Biochemistry* **14,** 4559 (1975).
[48] R. P. Frigon and S. N. Timasheff, *Biochemistry* **14,** 4567 (1975).
[49] V. Prakash and S. N. Timasheff, *Biochemistry* **30,** 873 (1991).

rectly affect polymerization. A structural resolution of these questions must await a high-resolution structure of spirals in the presence of different drugs.

A. Reliability of Energetic Data

There has been considerable disagreement about the number of vinca sites, their affinity, and the implications for inhibition of microtubule assembly.[14,24,50–52] Our results and the results from Timasheff's laboratory are in reasonable agreement because we both analyze binding data and sedimentation velocity data with a coupled assembly mechanism.[1,7] The disagreement generally arises when investigators neither analyze the linkage between drug binding and self-association nor perform a rigorous thermodynamic analysis of the influence effectors such as divalent cation concentration have on the system.[3,13,14,24] As convincingly demonstrated by Na and Timasheff,[7,8,13,14] the K_{app} for vinca alkaloid binding must be a function of protein concentration. Therefore, the methods used to study energetics must be able to distinguish polymer forms under varying solution conditions. For example, one study[53] uses methods (primarily turbidity) that are sensitive only to paracrystal formation, and thus the authors incorrectly conclude that inhibition of microtubule assembly is separable from spiral formation. Thus while these authors appreciate the importance of protein concentration, they incorrectly equate spiral formation and paracrystal, or what they call aggregate, formation. As discussed above,[35] spiral formation precedes paracrystal formation.[26] In fact, this work[53] demonstrates that microtubule inhibition occurs in the absence of paracrystal formation, but in a regime where only spirals are formed. This agrees with our assertion that spiral formation is essential for the mode of action of these drugs and that sedimentation is the best method by which to study these interactions. Their study involved a novel fluorescent vinblastine analog, Ant-Vin, that was previously used to identify the location of the vinca alkaloid-binding site.[44,53] In this study they demonstrate that Ant-Vin inhibits microtubule assembly less well than vinblastine. This example is identical to the case of vinflunine, the weakest vinca alkaloid we have tested by sedimentation velocity methods.[6] Direct binding studies were unable to demonstrate interactions with tubulin even though vinflunine inhibited microtubule assembly and killed

[50] R. H. Himes, *Pharmacol. Ther.* **51,** 257 (1991).
[51] E. Hamel, *Med. Res. Rev.* **16,** 207 (1996).
[52] R. Bai, G. R. Pettit, and E. Hamel, *J. Biol. Chem.* **265,** 17141 (1990).
[53] S. S. Rai and J. Wolff, *FEBS Lett.* **416,** 251 (1997).

FIG. 7. A representation of the endwise-specific interaction of vinca alkaloids (V) with $\alpha\beta$-tubulin and microtubules. A single $\alpha\beta$-tubulin at the ends of the microtubule represents the polar character of the polymer and the simplest subunit scheme for elongation of microtubule protofilaments. At the (+) end of a microtubule V can bind to the exposed β subunit, creating a spiral interface binding site for the next subunit ($\alpha\beta$) or liganded subunit ($\alpha\beta_V$, not shown) or spiral polymer of any size (represented by a curved $\alpha\beta_V\alpha\beta_V\alpha\beta_V$) to add. At the (−) end of a microtubule the β subunit is buried and thus to create a spiral interface V must bind to free $\alpha\beta$-tubulin, making a liganded subunit ($\alpha\beta_V$). Either liganded subunits ($\alpha\beta_V$) or spiral polymers of any size (represented by a curved $\alpha\beta_V\alpha\beta_V\alpha\beta_V$) can then bind to the exposed α subunit. These reactions will compete with normal microtubule dynamics, but a spiral interface of any kind in one site will potentially disrupt lateral contacts as well. The distribution of spiral species and their relative participation in microtubule inhibition is regulated by the energetics for that particular drug, the free drug and tubulin concentration, and the solution conditions as discussed in the text.

cells in culture.[54] The only direct evidence of vinflunine binding to tubulin and inducing spiral formation came from our sedimentation velocity studies (Figs. 2 and 3, Table I).[6]

We can now suggest a structural explanation for our model preference in fitting binding data. In the discussion of the molecular model and fitting function, we stressed that the ligand-mediated model best described our data, and that the ligand-facilitated model was an unlikely pathway to spiral formation. Note that spirals such as microtubules are polar structures and polymer growth can occur in both directions. If vinca alkaloids are directly involved in stabilizing the spiral polymer interface such that both the β chain from one subunit and the α chain from the next subunit in the spiral can either make contact with the drug or are required for formation of the drug-binding site, then there is a simple structural and kinetic explanation for exclusion of the facilitated pathway. For growth in the plus (+) direction, drug must first bind to the β subunit to create the binding site for the next tubulin subunit in the spiral (Fig. 7). This corresponds to a mechanism involving an ordered addition of drug and tubulin heterodimer. For growth in the minus (−) direction, the drug must first bind to the heterodimer to create the binding site for attaching to the spiral. This would exclude any spiral association at either end that is not mediated by a drug molecule at

[54] A. Kruczynski, J. M. Barret, C. Etievant, F. Colpaert, J. Fahey, and B. T. Hill, *Biochem. Pharmacol.* **55**, 635 (1998).

the interface. Verification of this hypothesis may require high-resolution structural information. It is possible that kinetic data utilizing fluorescent drug analogs will provide verification, but it is anticipated that linkage to assembly is likely to cause average relaxation times similar to those observed with stopped-flow light scattering.[1–3,6]

Another example of the importance of selecting appropriate methods to study energetics is found in comparing binding affinities obtained by fluorescence quenching and sedimentation velocity. Timasheff and co-workers made attempts to measure K_1 independently of K_2 by reducing the protein concentration to 2.85 μM, omitting Mg^{2+} from the solution, and performing fluorescence quenching studies.[8,49,55,56] They obtained values of K_1 for vinblastine and vincristine of 0.4 and 0.35 $\times$ 10^5 M^{-1}. These values are in excellent agreement with our estimates of K_1 (0.52 $\times$ 10^5 M^{-1}) for vinblastine in 100 μM Mg^{2+} concentrations by sedimentation velocity. We have now independently confirmed the smaller values determined by fluorescence quenching studies for no Mg^{2+} conditions. However, fluorescence quenching experiments assume that only drug binding to tubulin heterodimer is occurring in the complete absence of self-association. This was originally based on the incorrect assumption that reducing the protein concentration from 50–200 to 2.85 μM would suppress the coupled polymerization. This assumption was untestable with schlieren optics but is easily tested with absorbance optics. Sedimentation velocity studies show a weak but measurable self-association at 2 μM tubulin (10 mM PIPES, 50 μM GTP, pH 6.9) in the absence of Mg^{2+} for vincristine and vinblastine (K_2 equals 8.3 $\times$ 10^5 and 1.1 $\times$ 10^5, respectively). As expected, we have in fact found the value of K_1 (strictly K_{app}) measured by fluorescence quenching experiments is tubulin concentration dependent for these drugs. The values obtained are in reasonable agreement with these previous estimates, but the contributions from binding and self-association are still inseparable. These results highlight two conclusions from our data. First, the parameters estimated by sedimentation velocity are reliable in that they agree with other methods and correctly attempt to parse the energetics into two linked thermodynamic steps. Second, coupled equilibria, as reflected by the overall energetics $K_1 K_2$, are powerful processes that can drive weak equilibria, even at low protein and drug concentrations. This has important consequences for the interaction of these drugs with microtubules and the suppression of microtubule dynamics.

[55] J. C. Lee, D. Harrison, and S. N. Timasheff, *J. Biol. Chem.* **250**, 9276 (1975).
[56] V. Prakash and S. N. Timasheff, *J. Biol. Chem,* **258**, 1689 (1983).

B. *Implications for Interactions with Microtubules*

We have proposed that the ability of vinca alkaloids to make spirals is a direct measure of their ability to inhibit microtubule assembly and promote microtubule disassembly.[1–6] This is an energetic argument that correlates spiraling potential, as measured by K_1K_2, with drug efficacy (see Fig. 3). The ability to disassemble microtubules by shifting the equilibrium from microtubule polymer to spirals clearly goes through a mechanism involving endwise disassembly and conversion of straight microtubule protofilaments to vinca alkaloid-induced spiral polymers. The microtubules fray from the ends inward toward the GDP core in a zipper-like fashion. These structures have been observed by electron microscopy (EM) and are consistent with a model in which spirals disrupt lateral contacts in the microtubule.[50,57] The nucleotide dependence of spiral formation, where GDP favors assembly, works to the advantage of vinca alkaloid-induced disassembly because the core of the microtubule is GDP–tubulin. Microtubules composed of nonhydrolyzable analogs are less sensitive to vinca alkaloids.[4,28] Thus the energetics of spiral formation are necessarily a factor and are entirely consistent with the phenomena observed at stoichiometric concentrations of drug.

Effects on microtubule dynamics at substoichiometric concentrations of drug are more complicated. Wilson, Jordan, and co-workers[58–61] have convincingly shown that mitotic arrest does not require microtubule disassembly. The suppression of microtubule dynamics alone is sufficient. They have argued that drug binding to the ends, not spiral formation of any kind, is the critical factor at low doses of drug. However, spirals attached to the ends of microtubules have been observed.[57] It is also true that drugs that do not induce alternate polymers also suppress microtubule dynamics.[62] The complexity is that microtubule dynamics are a cooperative process that is easily halted and placed into a so-called pause state.[63] This explains why drugs that disassemble microtubules initially stabilize them.[61,62] The argument for drug binding alone[58–61] is partially based on the low concentration of drug in the cell culture medium and the assumption that spirals are not involved at these concentrations. However, cells concentrate drugs, in

[57] L. Wilson, M. A. Jordan, A. Morse, and R. L. Margolis, *J. Mol. Biol.* **159,** 125 (1982).

[58] M. A. Jordan, R. H. Himes, and L. Wilson, *Cancer Res.* **45,** 2741 (1985).

[59] M. A. Jordan, D. Thrower, and L. Wilson, *Cancer Res.* **51,** 2212 (1991).

[60] M. A. Jordan, D. Thrower, and L. Wilson, *J. Cell Sci.* **102,** 401 (1992).

[61] R. Dhamadhanan, M. A. Jordan, D. Thrower, L. Wilson, and P. Wadsworth, *Mol. Biol. Cell* **6,** 1215 (1995).

[62] M. J. Schilstra, S. R. Martin, and P. M. Bayley, *J. Biol. Chem.* **264,** 8827 (1989).

[63] R. J. Toso, M. A. Jordan, K. W. Farrell, B. Matsumoto, and L. Wilson, *Biochemistry* **32,** 1285 (1993).

part by a mechanism involving binding to tubulin or in the case of taxol binding to microtubules, and the total cellular concentration may exceed the cell culture medium concentration by two orders of magnitude.[57,58] For vinca alkaloids the mechanism of concentrating drug is the coupled formation of tubulin spirals that, by linkage, draw drug into the cell. The free concentration of drugs in cells is not known. Our kinetic data demonstrate that the half-life of drug retention strongly correlates with the spiraling potential and the size of the spirals.[1-3,6,64] The removal of drugs by passive or active processes is thus limited by the release of drug from the spiral. Thus the relative cytotoxicity of vinca alkaloids, often correlated with lifetime of drug retention, is also explained by the energetics of spiral formation.

As discussed above for spiral polymers, microtubules are polar structures with the β subunit exposed on the (+) end and the α subunit exposed on the (−) end.[11,12,24,50,51] Thus, vinca alkaloids can bind directly to the (+) end, creating a site for another dimer [or the (−) end of a spiral] to bind and form a spiral interface (Fig. 7). At the (−) end drug cannot bind except as a liganded dimer [or the (+) end of a spiral polymer]. The addition of drug, liganded dimer, or spirals at the ends is presumed to be sufficient to disrupt successive microtubule growth, simply by lateral disruption of straight protofilament formation. A proposal that the two microtubule ends have intrinsically different responses to vinca alkaloids has been made.[65] The concentration regimen at which drug alone, or liganded dimer or spirals, contribute to these endwise-specific processes is dictated by the energetics of each vinca alkaloid derivative (Table I and Fig. 3) and the free concentration of drugs and tubulin. It is our interpretation of the energetic data,[1-3,6] the observation of microtubule disassembly at high drug concentrations, the presence of spirals attached to the ends of microtubules,[58] and the suppression of microtubule dynamics at extremely low drug concentrations[59] that all these interactions disrupt microtubule dynamics *in vivo* and *in vitro*.

C. Comparison of Vincristine, Vinblastine, and Vinorelbine Effects on Tubulin Isotypes

Generally, mammalian microtubules are formed from a mixture of α- and β-tubulin isotype classes.[66,67] It is possible that the tubulin isotype

[64] S. Lobert, *Crit. Care Nurse* **17**, 71 (1997).
[65] D. Panda, M. A. Jordan, K. C. Chu, and L. Wilson, *J. Biol. Chem.* **271**, 29807 (1996).
[66] S. A. Lewis, W. Gu, and N. J. Cowan, *Cell* **49**, 539 (1987).
[67] M. A. Lopata and N. J. Cleveland, *J. Cell Biol.* **105**, 1707 (1987).

composition in various tissues or tumors determines the clinically observed drug efficacy and toxicity. To test the possibility that the tubulin isotype composition is an important determinant in antineoplastic efficacy, we determined thermodynamic parameters for vinca alkaloid interactions with purified β-tubulin isotypes, $\alpha\beta$II or $\alpha\beta$III, as well as mixtures of $\alpha\beta$II and $\alpha\beta$III, $\alpha\beta$II and $\alpha\beta$I&IV, or $\alpha\beta$III and $\alpha\beta$I&IV (referred to as isotype-depleted tubulin), by quantitative sedimentation velocity.[5] We used immunoaffinity chromatography with single or two sequential columns to obtain pure isotypes or a mixture of isotypes. Vincristine-, vinblastine-, or vinorelbine-induced isotype self-association was studied at 25° in 10 mM PIPES (pH 6.9), 1 mM MgSO$_4$, and 2 mM EGTA in the presence of 50 μM GTP or GDP. For all three drugs, we observed no significant differences in the size of spirals or overall affinity, $K_1 K_2$, among isotypes or unfractionated tubulin. Furthermore, the GDP enhancement of purified isotypes is the same as for unfractionated tubulin. These data indicate that differential antitumor efficacy observed clinically for these vinca alkaloids is not determined by tissue isotype composition.

Small, but significant differences in the individual binding parameters, K_1 and K_2, are found in the vincristine data. In the presence of vincristine and GTP, K_1, the affinity of drug for tubulin heterodimers, tends to be two- to fivefold larger for purified $\alpha\beta$II- or $\alpha\beta$III-tubulin compared with unfractionated tubulin. The apparent dimerization constant, K_2^{app}, can be calculated from the K_1 and K_2 values for the ligand-mediated fits:

$$K_2^{app}(M^{-1}) = \frac{K_2}{[1 + (1/K_1[X])]^2} \qquad (2)$$

where X represents free drug.[8] It is a measure of the magnitude of spiral formation and overall drug binding. When K_2^{app} is plotted versus free vincristine concentration the differences between isotypes are most apparent at low and high drug concentrations (Fig. 8A). At physiologically significant drug concentrations (≤ 1 μM), K_2^{app} is larger for these purified isotypes. When $\alpha\beta$II- and $\alpha\beta$III-tubulin are combined, the cooperativity between drug binding and spiral formation approaches that of unfractionated PC-tubulin. Figure 8B is a plot of the slopes (ΔX) of the K_2^{app} plots versus free vincristine concentration. The error bars demonstrate that the difference found with pure $\alpha\beta$II- or $\alpha\beta$III-tubulin compared with multiple isotypes is significant. These differences are not observed in the presence of vinblastine or vinorelbine. The differences found with vincristine may be implicated in the dose-limiting neurotoxicity found with this drug, but not found with vinblastine or vinorelbine. Our current hypothesis is that vincristine induces tubulin denaturation, reflected in larger K_1 values, in addition to spiral

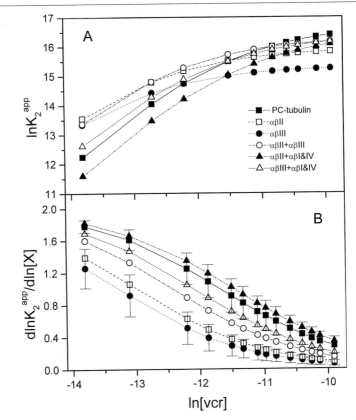

FIG. 8. Wyman analysis of vincristine binding to tubulin isotypes. K_2^{app} was calculated for 1.0 to 50 μM free vincristine concentrations. (A) Plots of ln K_2^{app} versus ln[vcr]. (B) Plots of the slopes of the lines in (A) (d ln K_2^{app}/d ln[X] = ΔX). The symbols and lines used for each isotype combination are PC-tubulin (■), $\alpha\beta$II (□), $\alpha\beta$III (●) $\alpha\beta$II + $\alpha\beta$III (○), $\alpha\beta$II + $\alpha\beta$I& IV (▲), and $\alpha\beta$III + $\alpha\beta$I&IV (△). ΔX is clearly smallest for the purified isotypes, indicating that there is a weaker linkage of drug binding to spiral formation compared with when multiple β isotypes are present. The uncertainty in ΔX (error bars), derived from the propagated errors in K_1, is indicated for PC-tubulin (■) and $\alpha\beta$III-tubulin (●) in order to demonstrate that these differences are significant. [From S. Lobert, A. Frankfurter, and J. J. Correia, *Cell Motil. Cytoskel.* **39,** 107 (1998).]

formation, and thus causes irreversible changes in tubulin structure and aggregate formation.[5,25,49,56]

D. Clinical Implications of Biophysical Data

Vinblastine and vincristine have become the mainstay of many chemo-therapy protocols since their discovery more than 30 years ago, yet little is understood regarding the origin of their differential antitumor and toxic effects. Our interest lies in how the energetics of drug action at the receptor tubulin correlate with *in vivo* effects. Both vincristine and vinblastine were first used in the early 1960s for the treatment of Hodgkin's lymphoma. Since that time they have been effective against a spectrum of hematologic and solid tumors. Vincristine is commonly used for leukemias, especially childhood leukemias.[64] Vinblastine may be more effective than vincristine in treating Hodgkin's lymphoma. Only three vinca alkaloids are currently available in the United States for cancer chemotherapy: vinblastine, vincristine, and vinorelbine. Vinorelbine is a new vinca alkaloid analog[68–70] used for the treatment of breast cancer and non-small cell lung carcinoma. A fourth drug, vindesine, a metabolite of vinblastine, is available in Europe; however, its antitumor efficacy is similar to vinblastine and it does not appear to have any demonstrable advantages in terms of toxicity.[71] The newest analog, vinflunine,[6,27] a difluorinated derivative of vinorelbine, will soon enter Phase I clinical trials.

The relative overall binding affinities of these drugs correlate with the weekly drug doses used clinically in cancer chemotherapy, where vincristine is used at the lowest dosages and vinorelbine at the highest.[64] Furthermore, dosage is in part determined by dose-limiting toxicities, which also vary among vinca congeners. Vincristine is highly neurotoxic, whereas vinblastine and vinorelbine doses are limited by bone marrow toxicity. Clinical data on vinflunine are not yet available. Surprisingly, K_1 values, the affinity of drug for tubulin heterodimers, are identical within error for all four drugs. When data are fit with the ligand-mediated model, the differences between drugs are found in K_2 values, the affinity of liganded heterodimers for spiral polymers, indicating that we can differentiate these drugs on the basis of the magnitude of the spirals formed. Thus we hypothesize that dosage and toxicity correlate with the magnitude of the spiral size induced by the drug. Weaker drugs potentially allow for a wider margin between effective and toxic doses or an improved therapeutic index.[6] This does not exclude the importance of other parameters that contribute to pharmacokinetics (e.g., lipid solubility and drug uptake into tissues and cells).

[68] R. L. Nelson, *Med. Pediatr. Oncol*. **10**, 115 (1982).

[69] R. Rahmani, F. Gueritte, M. Martin, S. Just, J. P. Cano, and J. Barbet, *Cancer Chemother. Pharmacol.* **16**, 223 (1986).

[70] A. Krikorian, R. Rahmani, M. Bromet, P. Bore, and J. P. Cano, *Semin. Oncol.* **16**, 21 (1989).

[71] X. J. Zhou and R. Rahmani, *Drugs* **44** (Suppl. 4), 1 (1992).

[5] Kinetics and Thermodynamics of Conformational Equilibria in Native Proteins by Hydrogen Exchange

By Cammon B. Arrington *and* Andrew D. Robertson

Introduction

Hydrogen exchange is a powerful and increasingly popular tool for the study of protein structure and dynamics. Proteins contain a number of functional groups bearing hydrogens that are labile to exchange with solvent hydrogens. The chemistry of these exchange or proton transfer events is well understood for small model compounds in which the functional groups are well exposed to solvent.[1,2] However, when the exchange-labile hydrogen participates in a hydrogen bond or is otherwise excluded from solvent, then exchange can be slowed significantly.[3,4] In proteins, the amide hydrogen (NH) of the peptide backbone has been the principal focus of exchange studies for two reasons. The first is practical: The kinetics of exchange at other functional groups are intrinsically more rapid and thus more difficult to measure even when exchange is slowed. The second reason reflects the structural properties of native proteins: Most of the hydrogen bonding interactions in proteins involve the peptide backbone.[5,6]

The growing use of NH exchange is attributable to advances in both technology and in our understanding of the mechanisms by which NH exchange is modulated in proteins. The major technical advances are multi-dimensional nuclear magnetic resonance (NMR) spectroscopy and mass spectrometry (MS). NMR spectroscopy can be used to monitor exchange at many individual NHs in small (<30 kDa) soluble proteins. MS can be used to measure NH exchange in much larger proteins and requires less protein than NMR. However, MS generally is used to measure the average exchange behavior of the protein as a whole or peptide fragments thereof. Many contributions to our understanding of the molecular mechanism of NH exchange have come from detailed analysis of NMR data. The execution and analysis of NMR experiments are the focus of this chapter.

[1] M. Eigen, *Angew. Chem. Int. Ed. Engl.* **3,** 1 (1964).
[2] Y. Bai, J. S. Milne, L. Mayne, and S. W. Englander, *Proteins Struct. Funct. Genet.* **17,** 75 (1993).
[3] S. W. Englander and N. R. Kallenbach, *Q. Rev. Biophys.* **16,** 521 (1984).
[4] F. Hibbert and J. Emsley, *Adv. Phys. Org. Chem.* **26,** 255 (1990).
[5] E. N. Baker and R. E. Hubbard, *Prog. Biophys. Mol. Biol.* **44,** 97 (1984).
[6] D. F. Stickle, L. G. Presta, K. A. Dill, and G. D. Rose, *J. Mol. Biol.* **226,** 1143 (1992).

The long-standing allure of NH exchange has been the opportunity to monitor the dynamic nature of native proteins.[7] What are the dynamic features of native proteins that lead to exchange of buried NHs? Studies have demonstrated that NH exchange, at many sites in native proteins, occurs as a result of conformational equilibria.[8] NMR permits measurement of residue-specific equilibria at sites distributed throughout the three-dimensional structure of a protein. More recent work suggests that the kinetics of these conformational interconversions can be measured as well.[9]

Several previous chapters in this series describe the use of NH exchange to study conformational changes in proteins. Bai *et al.*[10] explain how to derive thermodynamic parameters from chemical denaturant and temperature dependencies of NH exchange. The chapters by Roder[11] and Kim[12] detail the use of NH exchange to identify intermediates in protein folding. This chapter contains a detailed description of our laboratory's approach to the study of conformational dynamics in native proteins using NH exchange. Specifically, we will demonstrate how to obtain thermodynamic and kinetic parameters from the pH dependence of exchange. Included in this chapter are NH exchange protocols that describe sample preparation, data collection, and data analysis. The aim is to help those with little or no experience in using NH exchange to understand (1) the technical aspects of this experimental method and (2) the theoretical basis of NH exchange as a means to detect conformational fluctuations in native proteins.

Model for Slowed NH Exchange in Native Proteins

In native proteins, many NHs are not fully accessible to solvent and their rate of exchange can be slowed by many orders of magnitude relative to rates for model compounds. The principle of native-state NH exchange is that as conformational fluctuations occur, amide protons that are normally hydrogen bonded and buried within the interior of a protein are exposed to solvent. In the solvent-accessible state, amide protons are labile to exchange. This process is routinely measured by diluting protonated protein in deuterium oxide (D_2O) and monitoring the disappearance of NH resonances in the ^1H NMR spectrum. The NMR measurements are described in more detail later in the chapter.

[7] K. Linderström-Lang, *Spec. Publ. Chem. Soc.* **2**, 1 (1955).
[8] J. Clarke and L. S. Itzhaki, *Curr. Opin. Struct. Biol.* **8**, 112 (1998).
[9] C. B. Arrington and A. D. Robertson, *Biochemistry* **36**, 8686 (1997).
[10] Y. Bai, J. J. Englander, L. Mayne, J. S. Milne, and S. W. Englander, *Methods Enzymol.* **259**, 344 (1995).
[11] H. Roder, *Methods Enzymol.* **176**, 446 (1989).
[12] P. S. Kim, *Methods Enzymol.* **131**, 136 (1986).

A microscopic two-state model has been proposed to describe the exchange behavior of individual NHs in native proteins.[13]

$$\text{NH(closed)} \underset{k_{cl}}{\overset{k_{op}}{\rightleftharpoons}} \text{NH(open)} \overset{k_{ch}}{\rightharpoonup} \text{exchanged} \tag{1}$$

According to the model, a given slowly exchanging NH is in equilibrium between a solvent-inaccessible conformation [NH(closed)] and a solvent-accessible conformation [NH(open)]. Rate constants for opening and closing at a particular NH are denoted k_{op} and k_{cl}, respectively. While in the open state, exchange can occur with the chemical exchange rate constant, k_{ch}. This rate constant depends on sequence, pH, and temperature and can be predicted from model compound data, as described in the next section. When protein is placed in excess D_2O, NH to ND exchange occurs irreversibly.

From the microscopic two-state model shown in Eq. (1), the following quadratic equation has been derived[13] for the observed rate constant of exchange, k_{obs}:

$$k_{obs} = (k_{op} + k_{cl} + k_{ch}) \left\{ \frac{1 - \left[1 - \dfrac{4 k_{op} k_{ch}}{(k_{op} + k_{cl} + k_{ch})^2} \right]^{1/2}}{2} \right\} \tag{2}$$

Under conditions favoring the native state, the equilibrium constant for unfolding, k_{op}/k_{cl}, is much less than 1. This follows from the fact that the free energy of unfolding, ΔG_u°, for native proteins at 25° typically falls between 5 and 15 kcal/mol, which corresponds to equilibrium constants ranging from approximately 10^{-4} to 10^{-11}. Under these conditions $k_{op} \ll k_{cl}$ and Eq. (2) can be simplified as follows:

$$k_{obs} \cong \frac{k_{op} k_{ch}}{k_{cl} + k_{ch}} \tag{3}$$

Most studies use Eq. (3) as a point of departure for interpreting slowed NH exchange. In fact, most investigators further simplify Eq. (3) by assuming that $k_{cl} \gg k_{ch}$. This assumption leads to the following equation, which describes what is known as *EX2 exchange*.[14]

$$k_{obs} = \left(\frac{k_{op}}{k_{cl}} \right) k_{ch} \tag{4}$$

[13] A. Hvidt, *C.R. Trav. Lab. Carlsberg* **34**, 299 (1964).
[14] A. Hvidt and S. O. Nielsen, *Adv. Protein Chem.* **21**, 287 (1966).

The rationale for this assumption is based on knowledge from protein-folding studies. Protein folding can occur with rate constants as large as $10^3 \sec^{-1}$,[15] whereas k_{ch} averages about 10 $\sec^{-1}$ at 25° and pH 7. Interpretation of NH exchange data using Eq. (4) thus yields equilibrium constants for opening reactions at individual NHs.

Another possible simplification of Eq. (3) follows from conditions where $k_{ch} \gg k_{cl}$. Under these circumstances, the opening reaction is rate limiting and the observed rate constant for exchange is equal to the rate constant for opening.

$$k_{obs} = k_{op} \tag{5}$$

This is referred to as *EX1 exchange*.[14] EX1 exchange thus provides access to the kinetics of opening at individual NHs in native proteins.

Hydrogen Exchange in Unstructured Peptides

Quantitative interpretation of NH exchange in native proteins using Eqs. (3)–(5) is predicated on accurate values for k_{ch}. Values of k_{ch} depend on pH because exchange is subject to specific acid and base catalysis, with a minor contribution from pH-independent water catalysis. The following equation pertains to NH exchange in D_2O:

$$k_{ch} = k_A[D^+] + k_B[OD^-] + k_W \tag{6}$$

where k_A and k_B are the second-order rate constants for acid- and base-catalyzed exchange, respectively, and k_W is the pseudo-first-order rate constant for water-catalyzed exchange. $[D^+]$ is the concentration of acid and equals 10^{-pD}, where pD = pH* + 0.4.[16] pH* is the pH meter reading from a standard glass electrode that has not been corrected for isotope effects. The deuteroxide concentration, $[OD^-]$, equals $10^{-(pK_D - pD)}$, where K_D is the dissociation constant for D_2O. Values of k_B are approximately 8 orders of magnitude greater than values of k_A. Consequently, k_{ch} is dominated by the base-catalyzed reaction above pH 3 and increases by a factor of 10 for every unit increase in pH.

Calculation of k_{ch}

Bai et al.[2] developed a method to calculate k_{ch} for any amino acid in a protein. The method accounts for pH, temperature, and steric and inductive

[15] W. A. Eaton, V. Muñoz, P. A. Thompson, C. K. Chan, and J. Hofrichter, *Curr. Opin. Struct. Biol.* **7**, 10 (1996).

[16] P. F. Glasoe and F. A. Long, *J. Phys. Chem.* **64**, 188 (1960).

effects due to neighboring amino acids. Exchange data for denatured proteins suggest that the predicted values of k_{ch} are accurate to within a factor of 2 or 3. The following paragraphs illustrate how to use the method of Bai et al.[2] to calculate k_{ch} for an individual amide. The method is easily implemented in a spreadsheet program.

The general procedure first accounts for the effects of sequence and pH and then makes corrections for temperature if it differs from the reference temperature of 20°. Sequence effects are incorporated into the method by multiplying a reference rate constant by "side-chain factors." For proteins, reference rate constants for acid, base, and water catalysis, $k_{A,ref}$, $k_{B,ref}$, and $k_{W,ref}$, respectively, are derived from data for poly-DL-alanine (PDLA[2]). Acid and base side-chain factors, A and B, respectively, are empirical corrections derived from studies of many model compounds.[2] Values of A and B are used to correct $k_{A,ref}$ and $k_{B,ref}$, respectively. There are A and B values for each of the two side chains flanking the NH of interest.

As an example, the value of k_{ch} will be determined for the backbone NH of phenylalanine in the tripeptide sequence –Val–(NH)Phe–Ile–. This tripeptide is considered to reside within a protein because parameters used in the calculation of k_{ch} for proteins are different from those used for small peptides.[2] In the present example, the NH of phenylalanine lies between the side chains of valine (residue $n - 1$) and phenylalanine (residue n). In this method, the identity of the $n + 1$ residue, isoleucine in the present example, is irrelevant. In the nomenclature of Bai et al.,[2] the letter R refers to residue $n - 1$ and L refers to residue n. Hence, for residue $n - 1$, side-chain correction factors for acid and base rate constants are denoted A_R and B_R, respectively, while those for residue n are called A_L and B_L.

The effect of adjacent side chains on the rate constants in Eq. (6) is accounted for using the following equations:

$$k_A = k_{A,ref}(A_R A_L) \tag{7a}$$
$$k_B = k_{B,ref}(B_R B_L) \tag{7b}$$
$$k_W = k_{W,ref}(B_R B_L) \tag{7c}$$

In NH exchange, water appears to act as a base. For this reason, the side-chain factors for base-catalyzed exchange, B_R and B_L, are used in Eq. (7c). Equations (8a)–(8c) incorporate the effects of pH on the pseudo-first-order rate constants for acid-, base-, and water-catalyzed exchange, k(acid), k(base), and k(water), respectively.[2]

$$k(\text{acid}) = k_{A,ref}(A_R A_L)[D^+] \tag{8a}$$
$$k(\text{base}) = k_{B,ref}(B_R B_L)[OD^-] \tag{8b}$$
$$k(\text{water}) = k_{W,ref}(B_R B_L) \tag{8c}$$

The reference rate constants and side-chain correction factors are reported in logarithmic form by Bai et al.,[2] so Eqs. (8a)–(8c) are usually modified to facilitate direct utilization of the empirical parameters.

$$k(\text{acid}) = \text{antilog}(\log k_{A,\text{ref}} + \log A_R + \log A_L - \text{pD}) \qquad (9a)$$
$$k(\text{base}) = \text{antilog}(\log k_{B,\text{ref}} + \log B_R + \log B_L - \text{pOD}) \qquad (9b)$$
$$k(\text{water}) = \text{antilog}(\log k_{W,\text{ref}} + \log B_R + \log B_L) \qquad (9c)$$

The calculated rates are thus far specific for 20°. They can be corrected to other reaction temperatures (absolute temperature, T) by using the integrated form of the Arrhenius equation:

$$k(x)_T = k(x)_{293}\exp(-E_a(x)[(1/T) - (1/293)]/R) \qquad (10)$$

where $k(x)$ refers to $k(\text{acid})$, $k(\text{base})$, or $k(\text{water})$. The activation energies (E_a) for these rate constants are 14, 17, and 19 kcal/mol, respectively, and R is the gas constant, 1.987 cal/(mol · K).

The following is a protocol to calculate k_{ch} for the NH of phenylalanine in –Val–Phe–Ile–. The reaction conditions are as follows: 30°, pH* 9 (measured at 30°), and the "low-salt" conditions of Bai et al.[2]

1. Look up logarithmic reference rate constants. The values for PDLA under low-salt conditions apply to the present example: $\log k_{A,\text{ref}} = 1.62\ M^{-1}\ \text{min}^{-1}$, $\log k_{B,\text{ref}} = 10.05\ M^{-1}\ \text{min}^{-1}$, and $\log k_{W,\text{ref}} = -1.5\ \text{min}^{-1}$.[2]
2. Look up logarithmic side chain correction factors for the $n - 1$ or R residue.[2] The R residue in this example is valine and the correction factors are as follows: $\log A_R = -0.30$ and $\log B_R = -0.14$.
3. Look up values for the n or L residue, phenylalanine in this example: $\log A_L = -0.52$ and $\log B_L = -0.24$.[2]
4. Calculate pD and pOD. The pH* for this example is 9.0, so pD = 9.0 + 0.4 = 9.4. The value of pOD = pK_D – pD = 15.05 – 9.4 = 5.65. pK_D, the molar ionization constant for D_2O, is 15.05 at 20°.[17] The temperature corrections for pK_D are included in the temperature corrections for $k(\text{base})$, described below, and are not explicitly included in any further calculations.
5. Substitute the parameters from steps 1–4 into Eqs. (9a)–(9c). For this sample calculation, the resulting pseudo-first-order rate constants for acid-, base-, and water-catalyzed exchange are as follows: $k(\text{acid})$ = antilog (1.62 – 0.30 – 0.52 – 9.4) = $2.51 \times 10^{-9}\ \text{min}^{-1}$; $k(\text{base})$ = antilog(10.05 – 0.14 – 0.24 – 5.65) = $1.05 \times 10^4\ \text{min}^{-1}$; and $k(\text{water})$ = antilog (–1.5 – 0.14 – 0.24) = $1.32 \times 10^{-2}\ \text{min}^{-1}$.
6. Use Eq. (10) to adjust the rate constants from step 5 to the desired temperature, which is 30° in this example. The following rate constants are

[17] A. K. Covington, R. A. Robinson, and R. G. Bates, J. Phys. Chem. 70, 3820 (1966).

obtained: $k(\text{acid})_{303} = 5.55 \times 10^{-9}$ min^{-1}, $k(\text{base})_{303} = 2.74 \times 10^4$ min^{-1}, and $k(\text{water})_{303} = 3.87 \times 10^{-2}$ min^{-1}.

7. Sum the three temperature-corrected rate constants. The predicted value of k_{ch} for the NH of Phe in $-$Val$-$Phe$-$Ile$-$at pH* 9, 30° is 2.74×10^4 min^{-1}.

Use of ^{1}H Nuclear Magnetic Resonance to Detect Hydrogen Exchange

Before initiating a hydrogen exchange study, NMR resonance assignments should be determined for the NHs of interest in a protein. For macromolecules, Evans[18] provides a detailed description of strategies that can be employed to make resonance assignments. In studies of NH exchange, useful information is often obtained through systematic variation of perturbants such as pH, temperature, and chemical denaturants. However, resonance identification can be complicated in such studies because chemical shifts can change by as much as 1 ppm or more as solution conditions are changed.[19] In such cases NH chemical shifts can be tracked by systematically varying the perturbant and obtaining NMR spectra of protein dissolved in H_2O over the range of solution conditions in which NH exchange is to be investigated.

A number of different NMR experiments can be used to monitor NH to ND exchange in proteins. The most commonly used experiments are simple one dimensional (1D) ^{1}H NMR (e.g., Arrington and Robertson[9]), ^{1}H$-^1$H correlation spectroscopy (COSY) (e.g., Swint-Kruse and Robertson[20]), and ^{1}H$-^{15}$N HSQC if one has ^{15}N-labeled protein (e.g., DeLorimier et al.[21]).

In this chapter we focus on the use of pH as the perturbant and present two ^{1}H NMR protocols that can be used to monitor NH exchange as a function of pH. In addition, we demonstrate how thermodynamic and kinetic parameters can be obtained for individual NHs in a protein from analysis of exchange data collected over a range of pH.

Measurement of Real-Time Hydrogen Exchange

A general procedure used in our laboratory begins with preparation of the protein in H_2O by preadjusting the pH to the experimental value. The

[18] J. N. S. Evans, "Biomolecular NMR Spectroscopy." Oxford University Press, New York, 1995.
[19] W. Schaller and A. D. Robertson, *Biochemistry* **34,** 4714 (1995).
[20] L. Swint-Kruse and A. D. Robertson, *Biochemistry* **35,** 171 (1996).
[21] R. DeLorimier, H. W. Hellinga, and L. D. Spicer, *Protein Sci.* **5,** 2552 (1996).

protein is lyophilized and then dissolved in buffered D_2O at the desired pH to initiate exchange. A typical experiment is performed as follows.

1. Dissolve 1–2 μmol of protein into a minimal volume (e.g., 1 ml) of H_2O.
2. Using NaOH or HCl, adjust the solution to the desired experimental pH and lyophilize the protein to a constant weight.
3. Prior to initiating exchange, spectrometer settings should be adjusted to minimize dead time (i.e., the time between initiation of exchange and the start of data acquisition).
   a. Calibrate the variable temperature controller of the spectrometer to the desired reaction temperature, using a standard such as methanol.[22]
   b. Tune the NMR probe and adjust shim settings with a mock sample that is similar in composition to the experimental sample.
   c. Set predetermined values for all data acquisition parameters: pulse widths, spectral widths, number of transients, number of data points, number of increments (for 2D experiments), relaxation delays, etc. The residual solvent signal is usually suppressed by presaturation with a low-power pulse during the relaxation delay.[23] With 2 mM protein solutions on a 500-MHz spectrometer, satisfactory 1D ^{1}H spectra can be obtained with 16 transients consisting of 8000 time-domain data points. With the spectral width set to 6000 Hz and the recycle delay set to 1.5 sec, the time required to collect a single FID is 38 sec.
   d. The total acquisition time for each NMR experiment should be confirmed with a mock acquisition prior to initiating exchange; estimated acquisition times reported by the spectrometer may not properly account for all of the delays in an NMR experiment.
4. Initiate exchange by dissolving the dried protein into 800 μl of buffered D_2O that has been preadjusted to the desired experimental pH* with DCl or NaOD. The protein concentration will be approximately 1–2 mM.
5. Rapidly transfer the dissolved protein solution into a 5-mm NMR tube and insert it into the spectrometer.
6. Monitor the deuterium lock value on the spectrometer until it levels off. A steady lock value signifies that the sample has reached the reaction temperature. To minimize the time required for the sample to reach the reaction temperature, the NMR tube and the buffered

[22] A. L. Van Geet, *Anal. Chem.* **40**, 2227 (1968).
[23] P. J. Hore, *Methods Enzymol.* **176**, 64 (1989).

D$_2$O solution used in step 4 can be preequilibrated at the reaction temperature prior to initiating exchange.

7. With the lock value steady, rapidly recheck the shim settings and initiate data acquisition. Spectra can be acquired in as little as seconds with 1D NMR, a few minutes with HSQC, and $\geq$30 min with COSY. In this protocol, 300-sec dead times are common; therefore, the maximum value of k_{obs} that can be measured in real-time experiments is $\sim$0.005 sec^{-1}. This maximum is realizable only when data acquisition times are brief, such as with 1D NMR and HSQC experiments.

8. For NH exchange that is complete in a matter of minutes to hours, most modern spectrometers can be set up to collect an array of spectra at specific time points. When exchange takes weeks or months to go to completion, samples can be removed from the spectrometer and stored in a thermostatted water bath between measurements. Because NH exchange is a pseudo-first-order process, the zero time point can be arbitrarily defined. It is often easiest to consider the start of data acquisition as zero time.

Ideally, 500–800 μl of a 1–2 mM protein solution is needed to achieve a reasonable ratio of signal to noise while minimizing spectral acquisition time. The volume needed depends on the technical specifications of the spectrometer. When the half-time for exchange is on the order of hours to days, less protein is needed. More signal averaging and consequently longer acquisition times can be used to generate adequate sensitivity. In general, NMR spectra should be acquired in as little time as possible because substantial exchange during the acquisition time can compromise spectral quality.

Steps 1 and 2 are designed to ensure that the ionization state of the protein is roughly what it will be during the exchange experiment. This is important because the protein acts as a buffer and affects the final reaction pH. Steps 1 and 2 will facilitate better prediction of the final reaction pH. However, the actual sample pH must be checked immediately after NMR analysis. Alternatively, a mock sample can be prepared and used to check pH. pH measurements should be made at the experimental temperature.

For step 3c, instrument settings for the ^{1}H–^{15}N HSQC experiment might include 64 or 128 increments in the ^{15}N dimension. If the spectrometer is equipped with gradient capabilities, then each increment can consist of as little as one transient with concentrated protein solutions.[21] The resulting acquisition time can be as brief as 2 min. COSY experiments require a minimum of about 128 increments in the second dimension, each consisting of 4 or more transients. An optimization procedure for measuring NH exchange with COSY experiments has been described previously.[24]

Choice of buffer in step 4 depends on the experimental pH. At alkaline pH we use glycine and glycylglycine, whereas monobasic and dibasic potassium phosphate are used at neutral and acidic pH. Buffers with proton resonances between 7 and 10 ppm should be avoided because they will obscure the amide region of the spectrum. Alternatively, deuterated buffers can be used to minimize spectral interference. Suppliers of deuterated buffers include Cambridge Isotope Laboratories (Andover, MA) and Isotec (Miamisburg, OH).

In step 4 of this protocol, lyophilized protein is directly dissolved into a buffered D_2O solution. Although this is the simplest way to initiate exchange, in some instances NH exchange may occur more rapidly at some sites than would normally happen in native protein. This issue is of greatest importance at high pH, where exchange occurs as fast or faster than the refolding rate (i.e., EX1 conditions). It is probably less important under conditions in which refolding is rapid compared with the rate of exchange (i.e., EX2 conditions). Another issue that can pose problems is that some proteins do not tolerate lyophilization well.

To avoid possible artifactual exchange with lyophilized protein, it can first be dissolved in D_2O at pH 3, 0°, conditions under which the half-life for exchange at fully solvent-exposed NHs is on the order of tens of minutes to hours. This allows the protein to adopt a native fold under conditions of minimal NH exchange. Exchange can then be initiated by rapidly mixing the low-pH protein solution with a predetermined amount of base designed to achieve the desired reaction pH. For protein that cannot be lyophilized, exchange can be initiated by passing a solution of protein in H_2O (e.g., 0.5 ml) over a small gel-filtration column (e.g., 1 × 4 cm) equilibrated in buffered D_2O.[25]

Quenched Hydrogen Exchange Measurements

When NH exchange is fast or the protein concentration is too low for timely spectral acquisition, a quench protocol can be used to analyze exchange. This protocol consists of allowing a sample to exchange for a specific amount of time under a given set of reaction conditions, then slowing or stopping the exchange process by rapid adjustment of sample

[24] J. M. Scholtz and A. D. Robertson, in "Methods in Molecular Biology," Vol. 40: "Protein Stability and Folding: Theory and Practice" (B. A. Shirley, ed.), p. 291. Humana Press, Totowa, New Jersey, 1995.
[25] S. W. Englander and J. J. Englander, Methods Enzymol. 26, 406 (1972).

pH and temperature. The progress of exchange is followed by NMR analysis of samples corresponding to different reaction times. Sample preparation is as follows.

1. Dissolve lyophilized protein into ice-cold, acidic D_2O (e.g., pH 3). The amount of protein required is dependent on the number of time points to be analyzed and the protein concentration required for adequate spectral analysis. In general, the final protein concentration of each sample for NMR analysis should be greater than 0.1 mM with a sample volume of 500–800 μl.

2. After complete dissolution of the protein, heat the protein solution in a water bath to the desired reaction temperature for a brief period of time (e.g., 1–2 min).

3. Use a vortex mixer to rapidly combine the protein solution with enough deuterated base to raise the pH to the desired reaction pH. Prior to the mixing event, the alkaline solution should also be incubated at the reaction temperature. Mixture of the low-pH protein solution with the high-pH solution is zero time.

4. Remove aliquots from the reaction solution at regular intervals and rapidly quench the exchange process by rapid dilution of the aliquots into ice-cold, acidic D_2O (e.g., pH 3 or the most acidic pH tolerated by the protein). Although exchange is slowed significantly under these conditions, it will still occur. Consequently, all postquench samples must be handled identically to ensure that subsequent exchange is uniform across all samples.

5. Transfer each sample to a 5-mm NMR tube for spectral analysis or store the samples at −20° for analysis at a later time.

The protocol described above calls for rapid manual mixing, where the dead times are likely to be about 5 sec. The use of a quench-flow apparatus in an otherwise similar experiment permits measurement of rate constants as rapid as 1000 sec^{-1}.[11]

Interpretation of NH exchange data depends on precise knowledge of the reaction conditions, particularly the reaction pH. For this reason, it is best to ensure that a sufficient amount of unquenched reaction solution remains at the end of an experiment to confirm the reaction pH. Two other pH-related issues may need to be considered. First, the protein solutions may need to be buffered to facilitate accurate pH adjustments in this protocol. Zwitterionic buffers such as glycine and glycylglycine are ideal because they have buffering capacity at both low and high pH. Second, if protein solubility is limited or if protein is denatured at pH 3, then the lowest possible pH that preserves solubility and native structure will need to be used for protein dissolution (step 1) and quenching of exchange (step 4). If this pH is significantly higher than 3, then care must be taken to minimize exchange by the judicious use of temperature. In addition, all samples should be treated identically with regard to any further manipulations prior to NMR analysis. Under these circumstances, any further exchange during sample handling is the same for all samples and *relative* peak intensities in the NMR spectra are preserved.

One drawback of the quench protocol is that each quenched sample represents a single time point. Consequently, to determine residue-specific values of k_{obs}, numerous quenched samples must be analyzed. In general, quench-type protocols require at least five times the protein needed for real-time measurements.

Analysis of Nuclear Magnetic Resonance-Derived Exchange Data

Determination of k_{obs}

NH signal intensities can be quantified by measuring peak height or peak area in 1D NMR spectra and peak height, area, or volume in 2D NMR spectra. Most NMR software packages contain routines to make these measurements. Theoretically, peak heights, areas, and volumes should yield identical results but the accuracy of these measurements depends on proper phasing and baseline adjustments.[26] In 1D NMR spectra, the accuracy of peak area measurements is more sensitive to the choice of baselines and phasing than is peak height determination. Resonance intensity is therefore most often monitored by peak height measurements. For 2D NMR spectra, peak volumes are most commonly used.

To control for fluctuations in resonance intensity that can result from spectrometer instability (i.e., magnetic field disturbances, temperature fluctuations, etc.), NH resonance intensities should be normalized to a nonexchanging peak. This is especially important when using the quench protocol, wherein multiple samples must be analyzed to determine exchange rate constants. Nonexchanging peaks with strong signals, such as aromatic or methyl resonances, act as good controls. Because the aromatic region lies adjacent to the amide region in ^1H spectra, some overlap may exist between the two regions. Therefore, it is necessary to ensure that aromatic peaks chosen as controls do not overlap with exchanging amide resonances.

A variety of commercially available packages are available for nonlinear least-squares analysis of experimental exchange data. Through the use of

[26] J. A. Ferretti and G. H. Weiss, *Methods Enzymol.* **176,** 3 (1989).

such programs, residue-specific values of k_{obs} can be obtained by fitting normalized peak height versus time to Eq. (11):

$$h = a \exp(-k_{obs}t) + c \tag{11}$$

where h is the normalized peak height, a is the amplitude of the exchange curve at zero time, k_{obs} is the observed hydrogen exchange rate constant, t is the time in minutes, and c is the intensity value at infinite time of exchange. The constant c should be included in the equation because the baseline in NMR experiments often deviates from zero. To maximize the precision and accuracy of fitted k_{obs} values, at least four measurable intensities should be observed for each NH and exchange should be at least 75% complete at the longest time point.

Thermodynamic Parameters from Hydrogen Exchange

The majority of exchange studies to date have focused on EX2 exchange. According to the microscopic two-state model [Eq. (1)], when exchange is monitored under EX2 conditions, Eq. (4) can be used to determine the equilibrium constant for the opening reaction at a particular NH, $k_{op}/k_{cl} = k_{obs}/k_{ch}$. In turn, residue-specific free energies of opening, ΔG_{HX}, can be calculated by the following relation:

$$\Delta G_{HX} = -RT \ln \left(\frac{k_{obs}}{k_{ch}} \right) \tag{12}$$

How is ΔG_{HX} related to known conformational changes in proteins? Results from a number of studies show that ΔG_{HX} for the most slowly exchanging residues in a protein are similar or identical to the free energy of unfolding, ΔG_u°, measured independently of NH exchange.[8] For example, Fig. 1A shows ΔG_{HX} versus ΔG_u° for two of the most slowly exchanging NHs in turkey ovomucoid third domain (OMTKY3), Leu-23 and Asn-28.[20] In Fig. 1, the dashed line represents the case of perfect agreement between the two free energy measurements. The excellent agreement between the two independent measures of global stability underscores the growing consensus that EX2 exchange can be used to monitor and quantify the energetics of conformational changes in native proteins.

If the most slowly exchanging NHs report on the complete unfolding of a protein, then what type of reaction is responsible for more rapid exchange at other slowly exchanging sites? For these NHs, exchange probably occurs as a result of more localized fluctuations in the native conformation.[27,28] The simplest mathematical models treat exchange from such sites

[27] A. Rosenberg and K Chakravarti, *J. Biol. Chem.* **243**, 5193 (1968).
[28] K. Y. Kim, J. A. Fuchs, and C. K. Woodward, *Biochemistry* **32**, 9600 (1993).

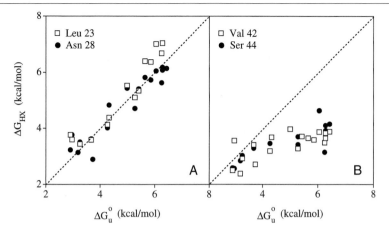

FIG. 1. Comparison of ΔG_{HX} with ΔG_u° for Leu-23, Asn-28, Val-42, and Ser-44 in OMTKY3. The dashed line represents perfect agreement between ΔG_{HX} and ΔG_u°. Using Eq. (12), values of ΔG_{HX} were calculated from exchange data obtained over a wide range of pH and temperature [L. Swint-Kruse and A. D. Robertson, *Biochemistry* **35**, 171 (1996)]. Values of ΔG_u° were extrapolated from thermal denaturation experiments [L. Swint-Kruse and A. D. Robertson, *Biochemistry* **34**, 4724 (1995)]. Results for Leu-23 and Asn-28 (A) reveal near-perfect agreement between the two independent free energy measurements, suggesting that complete unfolding is required for exchange to occur at these sites. Results for Val-42 and Ser-44 (B) show that two separate processes can lead to NH exchange from these sites. When ΔG_u° is <3.5 kcal/mol, then global unfolding is responsible for exchange whereas a local fluctuation leads to exchange when ΔG_u° is >3.5 kcal/mol.

as the sum of two processes: a local fluctuation in structure and the overall unfolding reaction.[29,30]

$$\text{NH(closed)} \rightleftharpoons \text{NH(open)}_G \rightharpoonup \text{exchanged}$$
$$\updownarrow$$
$$\text{NH(open)}_L \qquad\qquad\qquad (13)$$
$$\downarrow$$
$$\text{exchanged}$$

The kinetics of exchange from the two-process model are described as follows:

$$k_{obs} = k_{op,L}\frac{k_{ch}}{k_{cl,L} + k_{ch}} + k_{op,G}\frac{k_{ch}}{k_{cl,G} + k_{ch}} \qquad (14)$$

[29] H. Qian, S. L. Mayo, and A. Morton, *Biochemistry* **33**, 8167 (1994).
[30] S. N. Loh, C. A. Rohl, T. Kiefhaber, and R. L. Baldwin, *Proc. Natl. Acad. Sci. U.S.A.* **93**, 1982 (1996).

where the subscripts L and G refer to local fluctuations and global unfolding, respectively. Under most circumstances, one of the two processes will dominate the observed rate of NH exchange. For example, when protein is stable then $k_{op,G}/k_{cl,G} \ll k_{op,L}/k_{cl,L}$ and local fluctuations will dominate the observed rate of exchange. Destabilization of protein by mutation or changing solution conditions leads to increases in $k_{op,G}/k_{cl,G}$ and global unfolding can become the dominant contributor to exchange.

These trends are illustrated in Fig. 1B with data for two NHs in OMTKY3, Val-42 and Ser-44.[20] When $\Delta G_u^\circ > \sim 3.5$ kcal/mol, then exchange at Val-42 and Ser-44 is more rapid than expected if global unfolding alone were responsible for slowed exchange. Hence, exchange from these NHs is dominated by a local fluctuation in the native structure. According to the model, the free energy of this local opening reaction is ~ 3.5 kcal/mol (Fig. 1B). When the overall stability of OMTKY3, ΔG_u°, falls below 3.5 kcal/mol, then global unfolding dominates exchange and $\Delta G_{HX} = \Delta G_u^\circ$. Thus, NH exchange measured by NMR may provide access to many types of conformational equilibria in native proteins.

Most studies to date show that local fluctuations appear to be much less sensitive than global unfolding to changes in protein stability. Thus, identification of NHs that exchange through local fluctuations or global unfolding is achieved by monitoring exchange as a function of protein stability. Protein stability can be perturbed in a variety of ways, including mutagenesis and varying temperature, pH, and denaturant concentration.

Kinetic Parameters from Hydrogen Exchange

The potential for measuring k_{op} directly under EX1 conditions, Eq. (5), has been recognized for some time. However, few investigators have pursued such studies partially because of concerns regarding the accuracy of the microscopic two-state model for slow NH exchange [Eq. (1)]. As described above, equilibrium studies demonstrate that the model is fairly accurate under EX2 conditions. Consequently, there is reason to believe that EX1 exchange can indeed provide access to the kinetics of opening reactions at individual residues in native proteins.

A switch from EX2 [Eq. (4)] to EX1 [Eq. (5)] exchange occurs when k_{ch} becomes larger than k_{cl}. In principle, one of the simplest ways to detect this switch is through the use of pH.[9,31,32] To illustrate this point, let us assume that k_{cl} at a particular NH is 1000 sec^{-1} and k_{cl} is independent of pH. At pH 5 and 25° the average value of k_{ch} for a backbone NH is 0.1 sec^{-1}. Consequently, EX2 exchange predominates under these con-

[31] C. K. Woodward and B. D. Hilton, *Annu. Rev. Biophys. Bioeng.* **8,** 99 (1979).
[32] H. Roder, G. Wagner, and K. Wüthrich, *Biochemistry* **24,** 7396 (1985).

ditions. A switch to EX1 exchange occurs above pH 9, where k_{ch} is 1000 sec^{-1} and equals k_{cl}.

One intriguing aspect of combined analysis of EX2 and EX1 exchange is the possibility of determining residue-specific closing rate constants, k_{cl}, in native proteins. From EX2 exchange, k_{op}/k_{cl} can be determined while k_{op} can be measured under EX1 conditions. Therefore, a mathematical consequence of monitoring the switch from EX2 to EX1 exchange is that values of k_{cl} can be calculated.[9]

The major focus of this section is on the use of pH to increase k_{ch} relative to k_{cl} and drive NH exchange from EX2 to EX1 conditions. Alternatively, the EX1 rate limit can be reached by decreasing k_{cl} relative to k_{ch}. Thus, conditions that decrease k_{cl}, such as protein destabilization, are likely to lead to EX1 exchange.[30,33]

Any investigation of a switch from EX2 to EX1 exchange, using pH, must consider the possible pH dependence of k_{op} and k_{cl}. These values are assumed to be constant in the following discussion. This assumption is reasonable if protein stability is constant over the range of pH to be studied. However, the stability of many proteins is sensitive to pH. The possible effects of pH on k_{op} and k_{cl} are discussed later in the chapter.

To illustrate the appearance of a switch from EX2 to EX1 exchange, simulated exchange data are presented in Fig. 2. In Fig. 2 the logarithm of k_{obs} is plotted versus pH*. The straight line near the top represents exchange from a solvent-exposed NH. The value of k_{obs} is equal to k_{ch} and increases by a factor of 10, or 1 unit on the logarithmic scale, per unit increase in pH above 3. The curves near the bottom of Fig. 2 represent simulations of exchange from slowly exchanging NHs in native protein. For each of these three curves, the equilibrium constant for the opening reaction (i.e., k_{op}/k_{cl}) is held constant at 10^{-5} over the pH range shown, while the magnitudes of k_{op} and k_{cl} are systematically varied from curve to curve. Values used for k_{cl} are shown to the right of each simulated curve.

The slopes for all the curves at low pH are equal to 1. The difference between the top curve and the bottom three for native protein is that NH exchange in native protein is slowed by k_{op}/k_{cl} or the equilibrium fraction of that NH in the open state. Differences between the bottom curves at higher pH result from the different values of k_{op} and k_{cl} used in the simulations. The switch from EX2 to EX1 exchange is evidenced by the plateaus, where the value of k_{obs} at the plateau is approximately k_{op}. The switch occurs at pH values where k_{ch} overtakes k_{cl}; the smaller the value of k_{cl}, the lower the pH at which k_{ch} surpasses it.

[33] A. Miranker, C. V. Robinson, S. E. Radford, R. T. Aplin, and C. M. Dobson, *Science* **262**, 896 (1993).

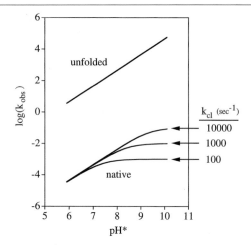

FIG. 2. Simulations of NH exchange plotted as $\log(k_{obs})$ versus pH*. The top line represents NH exchange in unfolded protein (i.e., fully solvent-exposed NHs) where $k_{obs} = k_{ch}$. Because of the pH dependence of k_{ch} [see Eq. (6)], k_{obs} increases by a factor of 10 per unit increase in pH above pH 3. The bottom three curves represent simulations of slowed exchange in native protein, using Eq. (3). In these curves, the magnitudes of k_{op} and k_{cl} were systematically varied while k_{op}/k_{cl} was held constant at 10^{-5}. Values used for k_{cl} are shown to the right of the plot. The simulated curves illustrate a switch from pH-dependent EX2 exchange to pH-independent EX1 exchange.

Similar trends are seen in experimental data for the slowly exchanging NHs of Gly-25 and Lys-29 in native OMTKY3, shown in Fig. 3.[9] Between pH 6 and 8, $\log(k_{obs})$ increases by a factor of 10 per unit increase in pH, which is consistent with EX2 exchange. Above pH 8, the pH dependence of k_{obs} decreases and even disappears for Gly-25. This behavior is indicative of a switch in the rate-limiting step for NH exchange to the opening reaction, where k_{obs} approaches the opening rate, k_{op}.

In the case of Lys-29, the plateau appears to be defined by few data points (Fig. 3). In reality, curvature in the data begins well below the pH where the plateau is evident. This is illustrated by superimposing a dotted line with a unit slope on the exchange data for Lys-29. If EX2 exchange had persisted above pH 8, values of $\log(k_{obs})$ should have continued to follow the dotted line. Instead, systematic deviation from a unit slope beginning at pH 8 is observed.

The data have been fit to a modified form of Eq. (3) by nonlinear least-squares analysis.[34] Values of k_{ch} are dominated by base-catalyzed exchange at pH > 3, so $k_{ch} \cong k_B [OH^-]$. Substitution of this relationship and the

[34] M. L. Johnson and L. M. Faunt, *Methods Enzymol.* **210,** 1 (1992).

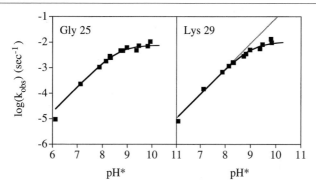

FIG. 3. Semilogarithmic plots of the observed exchange rate constant, k_{obs}, versus pH* for two of the most slowly exchanging residues in native OMTKY3, Gly-25 and Lys-29. The solid lines represent nonlinear least-squares fits of the data using Eq. (15). From the nonlinear regression analysis, values of k_{op} and k_{cl} were determined for each NH. For Gly-25, $k_{op} = 0.008$ sec^{-1} and $k_{cl} = 2500$ sec^{-1}; for Lys-29, $k_{op} = 0.01$ sec^{-1} and $k_{cl} = 2400$ sec^{-1} [C. B. Arrington and A. D. Robertson, *Biochemistry* **36**, 8686 (1997)]. A dotted line with a slope of 1 is superimposed on the data for Lys-29. This line represents pure EX2 exchange and facilitates visualization of the switch from EX2 to EX1 exchange.

temperature dependence for base-catalyzed exchange, Eq. (10), into Eq. (3) results in the equation used to fit the data of Fig. 3.

$$k_{obs} = \frac{k_{op}k_B 10^{-(pK_D - pH^* - 0.4)}\exp(-17,000[(1/293)]/R)}{k_{cl} + k_B 10^{-(pK_D - pH^* - 0.4)}\exp(-17,000[(1/293)]/R)} \qquad (15)$$

From nonlinear least-squares analysis, residue-specific values of k_{op} and k_{cl} can be determined for both Gly-25 and Lys-29. For Gly-25, $k_{op} = 0.008$ sec^{-1} and $k_{cl} = 2500$ sec^{-1}; for Lys-29, $k_{op} = 0.01$ sec^{-1} and $k_{cl} = 2400$ sec^{-1}. As stated previously, one assumption in this analysis is that k_{op} and k_{cl} do not change over the range of pH studied. This assumption is probably valid in the case of OMTKY3 because its global stability is constant from pH 5 to 10. Overall, NMR-derived NH exchange data may provide access to rapid conformational fluctuations at the level of individual residues in native proteins.

Least-Squares Analysis of pH Dependence of Exchange

Two assumptions are made with least-squares analysis in general.[35] First, all error is in the dependent variable, k_{obs} in Eq. (15). In practice, this condition is satisfied when the precision of the independent variable, pH*

[35] M. L. Johnson and S. G. Frasier, *Methods Enzymol.* **117**, 301 (1985).

in Eq. (15), is much greater than the precision of the dependent variable. The second assumption is that the error distribution is Gaussian. Therefore, while the logarithmic form of Eq. (15) is useful for diagrammatic purposes (e.g., Figs. 2 and 3), it is not suitable for least-squares analysis because the logarithmic transformation distorts the error distribution. Nonlinear regression analysis of the pH dependence of exchange should be performed with Eq. (15) as written.

Proper use of least-squares analysis may require weighting of the data. Different data points within a set may have significantly different relative errors and the data should be weighted to account for these differences.[35] The relative errors in k_{obs} values for Fig. 3 are similar, and so no such weighting scheme is used to account for this source of bias during fitting to Eq. (15). However, k_{obs} values vary by several orders of magnitude (Fig. 3). An unweighted fit to Eq. (15) would be biased toward larger values of k_{obs} in minimizing the residuals. To remove this bias, k_{obs} values are weighted by the reciprocals of their absolute values.

How well must the data describe the plateau region of Figs. 2 and 3 in order to obtain reasonable fits to Eq. (15)? This question can be addressed by truncating the exchange data for Gly-25 point by point, starting with the data at the highest pH. The truncated data sets are then subjected to nonlinear least-squares analysis. Analysis of the resulting variances suggests that data must be collected up to a pH value at which k_{ch} is at least two to three times the fitted value of k_{cl}. Graphically, this corresponds to a pH at which the instantaneous slope of log (k_{obs}) versus pH is slightly less than 0.5.

Stability and pH Dependence of Exchange

Ideally, protein stability remains constant over the range of conditions in which the switch from EX2 to EX1 exchange occurs. In this case, kinetic analysis of NMR-derived NH exchange data is relatively straightforward. However, the stability of many proteins varies with pH, which means that k_{op}, k_{cl}, or both will vary with pH. Under these circumstances, thermodynamic and kinetic analysis of NH exchange data is more complicated.

The focus of discussion has been on increasing pH into the alkaline region, conditions under which proteins tend to become less stable. For these proteins, interpretation of NH exchange is greatly facilitated by measurements of protein stability, which can be made in a number of ways.[36-38] In the following discussion we illustrate how knowledge of stability might

[36] M. R. Eftink, *Methods Enzymol.* **259,** 487 (1995).
[37] R. Lumry, *Methods Enzymol.* **259,** 628 (1995).
[38] E. Freire, *Methods Enzymol.* **259,** 144 (1995).

be used to interpret exchange data for the most slowly exchanging NHs in native proteins.

As stated previously, the most slowly exchanging NHs tend to exchange only when the protein unfolds completely; thus, ΔG_{HX} for these sites is approximately equal to ΔG_u°. By replacing ΔG_{HX} with ΔG_u° in Eq. (12), the pH dependence of k_{op}/k_{cl} can be calculated for these NHs and used in Eq. (4) to generate the predicted pH dependence for EX2 exchange. This curve serves as the "EX2 hypothesis" for evaluation of experimental exchange data.

What happens to the pH dependence of slow NH exchange as protein stability decreases at alkaline pH? To illustrate what might be seen, a decrease in ΔG_u° with increasing pH has been simulated, as shown in Fig. 4A.

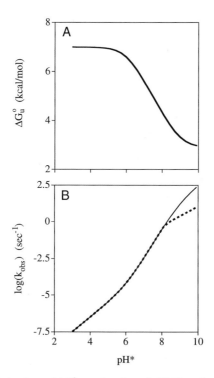

FIG. 4. (A) Simulated values of ΔG_u° as a function of pH. Over the neutral to alkaline pH range, protein stability drops approximately 4 kcal/mol. (B) Simulations of $\log(k_{obs})$ versus pH*. The solid line represents the pH dependence of EX2 exchange, or the EX2 hypothesis. The dashed line represents simulated exchange data, in which a switch from EX2 to EX1 exchange occurs at approximately pH* 8. Deviation from the EX2 hypothesis indicates that k_{cl} decreases with increasing pH. From the EX1 region of the curve, the pH dependence of k_{op} is also discernible.

This pH dependence and Eq. (4) are used to generate the "EX2 hypothesis" shown as the solid line in Fig. 4B. The pH dependence of ΔG_u° results in deviations from a unit slope at pH* > 6.

How might a switch from EX2 to EX1 exchange be manifested in cases of decreasing ΔG_u° at alkaline pH? If the pH-induced decrease in ΔG_u° is due solely to increases in k_{op}, then a switch from EX2 to EX1 exchange is indistinguishable from the EX2 hypothesis. Alternatively, if k_{cl} decreases with increasing pH, then $\log(k_{obs})$ must deviate from the EX2 hypothesis, whether or not k_{op} changes with pH. The dashed line in Fig. 4B illustrates an example of such deviation, in which the switch to EX1 exchange occurs above pH* 8.

[6] Mathematical Modeling of Cooperative Interactions in Hemoglobin

By MICHAEL L. JOHNSON

The quintessential example of cooperativity in macromolecular systems is the binding of oxygen to hemoglobin. For this specific case, the binding of the first oxygens to hemoglobin changes the affinity of subsequent oxygens. In hemoglobin this cooperativity is usually positive, i.e., the binding of the first oxygens increases the affinity for subsequent oxygens. However, cases exist in which cooperativity is negative, i.e., the binding of some ligands lowers the affinity for subsequent ligands.

The cooperative interactions in a macromolecule such as hemoglobin are not limited to ligands of the same type. The *Bohr effect* is a well-known example of the binding of protons by hemoglobin changing the affinity for oxygen.[1] When the binding of protons alters the affinity for oxygens, it must also be true that the binding of oxygens will also change the affinity for protons.

As is seen below, cooperative interactions within macromolecules are not limited to the interactions of one, or more ligands (e.g., oxygen and protons). The range of possible cooperative interactions within macromolecules includes, but is not limited to, ligand binding, subunit association or assembly, and conformational changes.

Cooperativity is an inherent thermodynamic property of macromolecules. It is observed as changes in the free energy of thermodynamic interac-

---

[1] C. Bohr, K. A. Hasselbalch, and A. Krogh, *Skand. Arch. Physiol.* **16,** 401 (1904).

tions that take place within a macromolecule. These thermodynamic interactions must obviously be related to structural changes within the macromolecule. However, the theoretical basis required to predict thermodynamic relationships based exclusively on structural data does not exist. Cooperativity cannot currently be predicted solely on structural data (e.g., X-ray crystal structures).

The simplest models of ligand binding are those that are based only on specific numbers of binding sites. For example, hemoglobin tetramers have four binding sites. A stoichiometry-based model for tetrameric hemoglobin is the four-site Adair model [see Eq. (9) below]. The parameters of this model provide macroscopic average information about the binding process. In the present example, the first binding constant (i.e., Adair constant) is the average affinity to bind the first oxygen. However, tetrameric human hemoglobin contains two α and two β subunits. Thus, the first Adair constant is a macroscopic average of the microscopic affinities of the two α and two β subunits in deoxygenated hemoglobin tetramers.

The advantage of these stoichiometry-based thermodynamic models is that the parameters of these models can be used to constrain microscopic mechanistic models. For a microscopic mechanistic model to describe a set of experimental data correctly the model must predict the correct macroscopic average Adair constants. This requirement is independent of any of the assumptions of the microscopic model. Simply put, if a microscopic mechanistic model cannot predict the correct values for observed macroscopic equilibrium constants then that microscopic mechanistic model is incorrect.

This chapter reviews some experimental observations of hemoglobin that all realistic models of hemoglobin must be able to describe. This chapter also reviews some classic and modern models of ligand binding, with particular emphasis on hemoglobin cooperativity. The literature on models of hemoglobin oxygen binding and cooperativity spans more than a century. Thus, by necessity, this chapter does not include all of the literature on the subject. An appendix is included that discusses some mathematical methodology required to test models of cooperativity against actual experimental observations.

Simplest Binding Model

The first conceptual model for the binding of oxygen by hemoglobin[2,3] was proposed by Hüfer in 1890. In this model, it was assumed that one

---

[2] J. T. Edsall, *J. Hist. Biol.* **5,** 205 (1972).
[3] G. Hüfer, *Arch. Anat. Physiol.* (*Physiol. Abt.*), 1 (1890).

molecule of oxygen is bound per iron atom within hemoglobin; it was 35 years later before it was known that hemoglobin is a distinct macromolecule of approximately 70,000 Da.[4,5] The Hüfer model is the simplest mass action relationship:

$$HB + O_2 \rightleftharpoons HbO_2 \tag{1}$$

This relationship predicts that the fractional saturation of hemoglobin with oxygen is a hyperbolic relationship of the form

$$\overline{Y} = \frac{k[O_2]}{1 + k[O_2]} \tag{2}$$

where $[O_2]$ is the free, or unbound, concentration of oxygen and k is the binding affinity (i.e., the ligand association constant).

Sigmoid-Shaped Oxygen Binding and Bohr Effect

It was 15 years after Hüfer proposed his mechanistic and mathematical model for the binding of oxygen to hemoglobin before the complete oxygen-binding curves were first measured by Bohr and co-workers.[1,2,6] The actual experimental observations are fundamentally different from the hyperbola predicted by Hüfer. Bohr and co-workers observed the sigmoid-shaped curve that is characteristic of cooperative oxygen binding by hemoglobin. An example of such sigmoid-shaped hemoglobin oxygen-binding data[7] is shown in Fig. 1.

Also shown in Fig. 1 is the least-squares calculated curve for these data based on the Hüfer model [Eq. (2)]. It is clear from the comparison of the data and calculated curve shown in Fig. 1 that the Hüfer model [Eq. (2)] does not provide a good description of the actual experimental data. At the time this discrepancy could not be explained. At the time, it was not known that hemoglobin was a distinct tetrameric molecule with four oxygen-binding sites.

It is interesting that the observation of cooperativity (i.e., the sigmoid-shaped oxygen-binding curves) is not what Bohr is known for discovering. It is the cooperative relationship between the binding of oxygen and pH (i.e., proton binding) that is known as the *Bohr effect*.

---

[4] G. S. Adair, *Proc. R. Soc.* **109A**, 292 (1925).
[5] T. Svedberg and J. B. Nichols, *J. Am. Chem. Soc.* **49**, 2920 (1927).
[6] J. Barcroft, "The Respiratory Function of Blood," 1st Ed., p. 21. Cambridge University Press, Cambridge, 1914.
[7] F. C. Mills, M. L. Johnson, and G. K. Ackers, *Biochemistry* **15**, 5350 (1976).

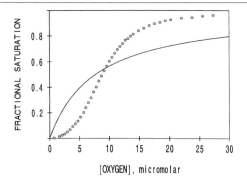

FIG. 1. Sigmoid-shaped hemoglobin oxygen-binding curve. These data were measured at pH 7.4, 21.5°, 0.1 M NaCl, 0.1 M Tris, 1.0 mM Na$_2$EDTA, and a hemoglobin A concentration of 382.5 μM (heme). Also shown is the least-squares estimated hyperbola based on the model of Hüfer, i.e., Eq. (2) with $k = 1.3 \times 10^5$. [Data points were redrawn from Mills *et al.*[7]]

Hill Equation

A. V. Hill[8] was the first to propose a mechanistic hypothesis and mathematical equation that could approximate the sigmoid shape of the observed oxygen-binding curves. Hill assumed that hemoglobin was a unique-sized aggregate that reversibly binds n oxygens. Given this assumption, the fractional saturation function can be written as

$$\overline{Y} = \frac{k[O_2]^n}{1 + k[O_2]^n} \qquad (3)$$

Figure 2 presents an analysis of the data shown in Fig. 1 according to the Hill equation [Eq. (3)]. Clearly, this formulation provides a much better description of these experimental data.

The value of n is known as the *Hill coefficient.* The value of n is not always an integer. For the present example, $n = 3.064$ (see Table I). This was rationalized by hypothesizing that hemoglobin existed as a mixed aggregate instead of a discrete aggregate. Thus, the observed noninteger values of n were simply averages over aggregates of variable size. The qualitative explanation of the Hill coefficient is that it is a measure of the number of binding sites that interact cooperatively.

It is also interesting that the values of the Hill equation parameters depend on the hemoglobin concentration. Mills *et al.*[7] published data at both high and low hemoglobin concentrations. Table I shows that at 5 μM hemoglobin the apparent Hill coefficient, n, decreases and the binding affinity, k, increases.

[8] A. V. Hill, *J. Physiol.* **40,** iv (1910).

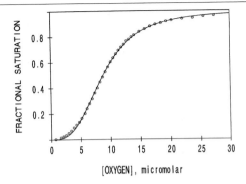

FIG. 2. Sigmoid-shaped hemoglobin oxygen-binding curve and analysis according to the Hill equation. These data were measured at pH 7.4, 21.5°, 0.1 M NaCl, 0.1 M Tris, 1.0 mM Na$_2$EDTA, and a hemoglobin concentration of 382.5 μM. Also shown is the least-squares estimation of the model of Hill, i.e., Eq. (3) with $k = 1.14 \times 10^5$ and $n = 3.06$. [Data points are the same as shown in Fig. 1 and were redrawn from Mills *et al.*[7]]

TABLE I
EFFECT OF HEMOGLOBIN CONCENTRATION ON
HILL EQUATION PARAMETERS

Concentration	n	k
382.5 μM [heme]	3.064 ± 0.025	1.139(±0.003) × 10^5
5.36 μM [heme]	2.885 ± 0.035	1.325(±0.006) × 10^5

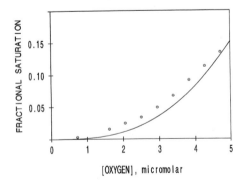

FIG. 3. An expansion of the lower section of Fig. 2.

Hill reported[9] in 1913 that the Hill hypothesis and mathematical model [Eq. (3)] predicts that if $n > 1$ then the limiting slope of the binding curve is zero as the concentration of oxygen approached zero.

$$\lim_{O_2 \to 0} \frac{d\overline{Y}}{d[O_2]} = nk[O_2]^{n-1} = 0 \quad \text{for } n > 1 \tag{4}$$

Hill noted that if the actual experimental data do not exhibit a limiting slope of zero then using the Hill equation is inappropriate. At the time it was difficult to measure the saturation curves accurately at these low oxygen concentrations and so it was, at the time, an unresolved question.

Figure 3 is an expansion of the lower section of Fig. 2. It is clear from this expanded view that the limiting slope of the Hill equation does not agree with the limiting slope of the data. Hence, by Hill's criteria,[9] the Hill equation is not appropriate for these data! It is also interesting that the Hill equation is still commonly used for the analysis of this type of experimental data.

Adair Equations

By the mid 1920s it was obvious from osmotic pressure[4] and sedimentation equilibrium[5] experiments that human hemoglobin was a discrete tetramer with four oxygen-binding sites. Obviously, the all-or-nothing binding mechanism of Hill does not accurately describe the experimental data. Therefore, G. S. Adair[4] formulated a four binding-site model that allowed for the intermediate partially oxygenated species that the Hill equation specifically excluded.

$$\begin{aligned}
Hb_4 + O_2 &\rightleftharpoons Hb_4O_2 \\
Hb_4O_2 + O_2 &\rightleftharpoons Hb_4(O_2)_2 \\
Hb_4(O_2)_2 + O_2 &\rightleftharpoons Hb_4(O_2)_3 \\
Hb_4(O_2)_3 + O_2 &\rightleftharpoons Hb_4(O_2)_4
\end{aligned} \tag{5}$$

This four binding-site model has four separate average binding affinities, one for each of the reaction steps shown in Eq. (5). These stepwise average binding constants are

$$k_{4i} = \frac{[Hb_4(O_2)_i]}{[Hb_4(O_2)_{i-1}][O_2]} \quad i = 1, 2, 3, 4 \tag{6}$$

These are stepwise because they refer to sequential steps for the addition of oxygens. Adair constants are also commonly written as product constants

[9] A. V. Hill, Biochem. J., 481 (1913).

that refer to the formation of the intermediate states from unligated hemoglobin and unbound oxygen:

$$K_{4i} = \frac{[Hb_4(O_2)_i]}{[Hb_4][O_2]^i} \quad i = 1, 2, 3, 4 \tag{7}$$

The stepwise Adair constants, k_{4i}, and the product Adair constants, K_{4i},
are related by

$$K_{4i} = \prod_{j=1}^{i} k_{4j} \tag{8}$$

The four-site Adair fractional saturation equation in terms of product
Adair constants is

$$\overline{Y}_4 = \left(\frac{1}{4}\right) \frac{K_{41}[O_2] + 2K_{42}[O_2]^2 + 3K_{43}[O_2]^3 + 4K_{44}[O_2]^4}{1 + K_{41}[O_2] + K_{42}[O_2]^2 + K_{43}[O_2]^3 + K_{44}[O_2]^4} \tag{9}$$

Or, in terms of the stepwise Adair constants:

$$\overline{Y}_4 = \left(\frac{1}{4}\right) \frac{k_{41}[O_2] + 2k_{41}k_{42}[O_2]^2 + 3k_{41}k_{42}k_{43}[O_2]^3 + 4k_{41}k_{42}k_{43}k_{44}[O_2]^4}{1 + k_{41}[O_2] + k_{41}k_{42}[O_2]^2 + k_{41}k_{42}k_{43}[O_2]^3 + k_{41}k_{42}k_{43}k_{44}[O_2]^4} \tag{10}$$

The major difference between the n-site Adair formulation and the
n-site Hill formulation is in the magnitude of the cooperativity being modeled. The Adair formulation allows any amount of cooperativity, positive
or negative. Conversely, the Hill formulation assumes that the n sites have
such a high positive cooperativity that the intermediate ligated states have
an extremely low concentration and thus do not contribute significantly to
the data.

The derivation of these Adair equations is presented in [7] in this
volume.[10]

Figure 4 presents the analysis of the same set of experimental data[7]
according to the four-site Adair product equation [Eq. (9)]. Clearly, the
inclusion of the intermediate species provides a much better description
of these experimental data than either the Hüfer or Hill model. Table II[11]
presents the stepwise and product Adair constants that correspond to the
analysis in Fig. 4.

The Adair constants presented in Table II are statistical macroscopic
averages over each of the microscopic configuration of the tetrameric molecule. For example, there are four ways to create a singly ligated hemoglobin

---

[10] M. L. Johnson and M. Straume, *Methods Enzymol.* **323**, [7], 2000 (this volume).
[11] B. Efron and R. J. Tibshirani, "An Introduction to the Bootstrap." Chapman & Hall, New
York, 1993.

TABLE II
ADAIR CONSTANTS FOR ANALYSIS PRESENTED IN FIGURE 4

Constant	Value	Confidence interval[a]	Statistical factor
K_{41}	4.076×10^4	3.444×10^4 to 4.578×10^4	4
K_{42}	2.169×10^9	1.177×10^9 to 3.094×10^9	6
K_{43}	2.583×10^{14}	1.909×10^{14} to 3.341×10^{14}	4
K_{44}	1.804×10^{20}	1.777×10^{20} to 1.816×10^{20}	1
k_{41}	4.076×10^4	3.444×10^4 to 4.578×10^4	4
k_{42}	5.322×10^4	2.608×10^4 to 8.968×10^4	3/2
k_{43}	1.191×10^5	0.619×10^5 to 2.857×10^5	2/3
k_{44}	6.984×10^5	5.331×10^5 to 9.453×10^5	1/4

[a] Plus or minus 1 standard deviation confidence intervals as evaluated by a bootstrap procedure.[11]

tetramer corresponding to an oxygen on each of the four binding sites. Thus, if the assumption is made that all the binding sites are identical, then the values of K_{41} and k_{41} presented in Table II are each four times the expected microscopic values (i.e., the binding constant to the individual subunits). There are six ways to create a doubly ligated tetramer from unligated tetramers and thus the statistical factor for K_{42} is six. The statistical factors for the other Adair constants are given in Table II.

If we assume that the hemoglobin tetramers consist of four identical subunits then the microscopic (i.e., corrected for statistical factors) stepwise affinities for the present example are $k'_{41} = 1.019 \times 10^4$, $k'_{42} = 3.548 \times 10^4$,

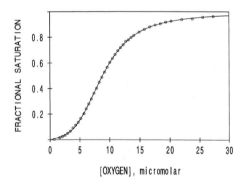

FIG. 4. Sigmoid-shaped hemoglobin oxygen-binding curve and analysis according to the four-site Adair equation. These data were measured at pH 7.4, 21.5°, 0.1 M NaCl, 0.1 M Tris, 1.0 mM Na$_2$EDTA, and a hemoglobin concentration of 382.5 μM. Also shown is the least-squares estimation of the model of the four-site Adair equation [i.e., Eq. (9)] with $K_{41} = 4.076 \times 10^4$, $K_{42} = 2.169 \times 10^9$, $K_{43} = 2.583 \times 10^{14}$, and $K_{41} = 1.804 \times 10^{20}$. [Data points are the same as shown in Figs. 1 and 2 and were redrawn from Mills et al.[7]]

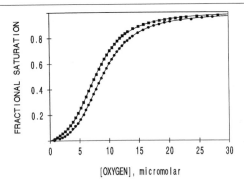

Fig. 5. Sigmoid-shaped hemoglobin oxygen-binding curves at different heme concentrations. These data were measured at pH 7.4, 21.5°, 0.1 M NaCl, 0.1 M Tris, 1.0 mM Na$_2$EDTA. The heme concentrations are 382.5 μM for the circles (*right*) and 5.36 μM for the squares (*left*). [Data are from Mills *et al.*[7]]

$k'_{43} = 1.787 \times 10^5$, and $k'_{44} = 4.794 \times 10^6$. This sequence clearly shows the apparent positive cooperativity present in tetrameric hemoglobin A.

There is however, a serious problem with the analysis as presented here. The oxygen-binding data depend on the concentration of the hemoglobin used for the experiment. Figure 5 presents two different oxygen-binding curves from Mills *et al.*,[7] one at 382 μM (heme) and the other at 5 μM (heme). Because the data depend on hemoglobin concentration, and the four-site Adair formulation assumes no concentration dependence, the four-site Adair formulation is less than correct and cannot be used under these conditions.

Linkage between Oxygenation and Subunit Dissociation

Human hemoglobin tetramers consist of two identical dimers, each containing one α and one β subunit. Tetrameric hemoglobin readily dissociates into $\alpha\beta$ dimers, but the dimers do not readily dissociate into monomers. It is known that the major transformation in quaternary structure concomitant with oxygen binding occurs in the contact region between the two $\alpha\beta$ dimers.[12–15] Ackers and Halvorson[14] analyzed the linkage between the binding of oxygen by human hemoglobin A and the dissociation of hemoglo-

[12] M. F. Perutz, *Nature (London)* **228,** 726 (1970).
[13] M. F. Perutz and L. F. Ten Eyck, *Cold Spring Harbor Symp. Quant. Biol.* **36,** 295 (1971).
[14] G. K. Ackers and H. R. Halvorson, *Proc. Natl. Acad. Sci. U.S.A.* **71,** 4312 (1974).
[15] G. K. Ackers, M. L. Doyle, D. Myers, and M. A. Daugherty, *Science* **255,** 54 (1992).

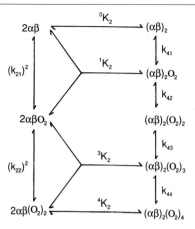

FIG. 6. The oxygenation-linked dimer–tetramer assembly scheme for human hemoglobin A. The k_{ij} values are the stepwise macroscopic (i.e., not corrected for statistical factors) Adair oxygen-binding constants: $i = 2, 4$ for dimer or tetramer; and $j = 1, 2, 3, 4$, indicating the number of the oxygen bound; and $^xK_2$ is the subunit assembly equilibrium constant to form a hemoglobin tetramer with x oxygens bound.

bin A tetramers into dimers. This linkage is shown pictorially in Fig. 6. The k_{4i} values are the stepwise Adair constants (not corrected for statistical factors) to bind the ith oxygen to tetrameric hemoglobin, the k_{2i} values are the stepwise Adair constants (not corrected for statistical factors) to bind the ith oxygen to dimeric hemoglobin, and the $^xK_2$ values are the dimer–tetramer association constants to form tetramers with x oxygens bound. $^2K_2$ is ambiguous, and not shown, because the doubly oxygenated tetramer can be formed from two singly ligated dimers or it can be formed from a doubly ligated and an unligated dimer.

Ackers and Halvorson[14] derived the mathematical form for the saturation function as a function of oxygen concentration, $[O_2]$, and heme concentration, $[Pt]$, as

$$\overline{Y} = \frac{Z'_2 + Z'_4 \dfrac{\sqrt{Z_2^2 + 4{}^0K_2Z_4[Pt]} - Z_2}{4Z_4}}{Z_2 + \sqrt{Z_2^2 + 4{}^0K_2Z_4[Pt]}} \tag{11}$$

where

$$\begin{aligned}
Z_2 &= 1 + K_{21}[O_2] + K_{22}[O_2]^2 \\
Z'_2 &= K_{21}[O_2] + 2K_{22}[O_2]^2 \\
Z_4 &= 1 + K_{41}[O_2] + K_{42}[O_2]^2 + K_{43}[O_2]^3 + K_{44}[O_2]^4 \\
Z'_4 &= K_{41}[O_2] + 2K_{42}[O_2]^2 + 3K_{43}[O_2]^3 + 4K_{44}[O_2]^4
\end{aligned} \tag{12}$$

Z_2 and Z_4 are the "binding polynomials" (i.e., macroscopic analogs of the partition functions) for the binding of oxygen to hemoglobin dimers and tetramers, respectively. Z_2' and Z_4' are the corresponding derivatives with respect to the logarithm of the oxygen concentration. This fractional saturation function has seven independent parameters: K_{21}, K_{22}, K_{41}, K_{42}, K_{43}, K_{44}, and $^0K_2$.

Ackers and Halvorson[14] also derived the mathematical form for the subunit association constants, $^xK_2$, as a function of oxygen concentration.

$$^xK_2 = {}^0K_2 \frac{Z_4}{(Z_2)^2} \tag{13}$$

These subunit assembly constants also have seven independent parameters: K_{21}, K_{22}, K_{41}, K_{42}, K_{43}, K_{44}, and $^0K_2$.

It is obvious from the outside circular pathway shown in Fig. 6 that the fully oxygenated association constant, $^4K_2$, depends on K_{22}, K_{44}, and $^0K_2$.

$$^4K_2 = {}^0K_2 \frac{K_{44}}{(K_{22})^2} \tag{14}$$

Thus, a measurement of any three of K_{22}, K_{44}, $^0K_2$, or $^4K_2$ will determine the fourth.

A precise determination of either the fractional saturation as a function of hemoglobin and oxygen concentration or the subunit assembly constant as a function of oxygen concentration should be sufficient to determine all seven linkage parameters. The approach taken by Ackers and co-workers has been to simultaneously determine both the subunit assembly constant at various oxygen concentrations, and the fractional saturation as a function of hemoglobin and oxygen concentrations. Specifically, Ackers and co-workers measured the equilibrium constant for unligated dimers to form unligated tetramers ($^0K_2$), and the equilibrium constant for fully oxygenated dimers to form fully oxygenated tetramers ($^4K_2$). Ackers and co-workers also measured oxygen-binding data as a function of oxygen and hemoglobin concentration. Equation 14 and the values of $^0K_2$ and $^4K_2$ were used to specify two of the seven parameters of the oxygenation subunit assembly linkage scheme. The remaining five parameters were evaluated by a least-squares analysis[7,16–21] of the oxygen-binding data. These experiments have

[16] M. L. Johnson and S. G. Frasier, *Methods Enzymol.* **117**, 301 (1985).

[17] M. L. Johnson and L. M. Faunt, *Methods Enzymol.* **210**, 1 (1992).

[18] M. L. Johnson, *Methods Enzymol.* **240**, 1 (1994).

[19] M. Straume and M. L. Johnson, *Methods Enzymol.* **210**, 87 (1992).

[20] D. M. Bates and D. G. Watts, "Nonlinear Regression Analysis and Its Applications." John Wiley & Sons, New York, 1988.

[21] D. G. Watts, *Methods Enzymol.* **240**, 23 (1994).

TABLE III
THERMODYNAMIC PROPERTIES OF LINKAGE SYSTEM IN
HEMOGLOBINS A AND KANSAS

Free energy change[a]	Hemoglobin A[b]	Hemoglobin Kansas[c]
$-RT \ln(k'_{21})$	-8.38 ± 0.2	-8.40
$-RT \ln(k'_{22})$	-8.38 ± 0.2	-7.34
$-RT \ln(k'_{41})$	-5.45 ± 0.2	-5.81 ± 0.1
$-RT \ln(k'_{42})$	-5.28 ± 0.5	-5.41 ± 0.4
$-RT \ln(k'_{43})$	-7.80 ± 0.6	-5.59 ± 0.5
$-RT \ln(k'_{44})$	-8.65 ± 0.4	-6.36 ± 0.4
$-RT \ln(^{0}K'_2)$	-14.38 ± 0.2	-13.63 ± 0.2
$-RT \ln(^{1}K'_2)$	-11.46 ± 0.3	-11.41 ± 0.2
$-RT \ln(^{3}K'_2)$	-7.78 ± 0.2	-7.48 ± 0.2
$-RT \ln(^{4}K'_2)$	-8.05 ± 0.2	-5.29 ± 0.2

[a] In kilocalories per mole; the prime (') indicates that the values
have been corrected for statistical factors.
[b] For the Mills et al. (1976) data.[7]
[c] For the Atha et al. (1979) data.[28]

been done on a variety of hemoglobins and under various conditions (e.g.,
temperature and pH).

Studying the subunit assembly properties and the concentration dependence of the oxygen-binding properties of the human hemoglobin system
provides a second dimension to the study of the system. The major transformation in quaternary structure concomitant with oxygen binding occurs in
the contact region between the two $\alpha\beta$ dimers.[12–15] Thus, the simultaneous
analysis (i.e., what was subsequently named "global analysis") of both types
of experimental data, obtained over a range of oxygen and hemoglobin
concentrations, provides additional resolving power to the analysis. This
improved resolution obtained by the simultaneous analysis is analogous
to the differences between one- and two-dimensional nuclear magnetic
resonance (NMR) or electrophoresis analysis.

This thermodynamic approach has been used to resolve all the energetic
relationships for hemoglobins under many different experimental conditions.[7,15,22–28] The first results for hemoglobins A and Kansas are shown in
Table III.

[22] F. C. Mills and G. K. Ackers, Proc. Natl. Acad. Sci. U.S.A. **76,** 273 (1979).
[23] F. C. Mills and G. K. Ackers, J. Biol. Chem. **254,** 2881 (1979).
[24] A. H. Chu, B. W. Turner, and G. K. Ackers, Biochemistry **23,** 604 (1984).
[25] M. L. Doyle, J. M. Holt, and G. K. Ackers, Biophys. Chem. **64,** 271 (1997).
[26] Y. Huang, M. Doyle, and G. K. Ackers, Biophys. J. **71,** 2094 (1996).
[27] Y. Huang and G. K. Ackers, Biochemistry **35,** 704 (1996).
[28] D. H. Atha, M. L. Johnson, and A. F. Riggs, J. Biol. Chem. **254,** 12390 (1979).

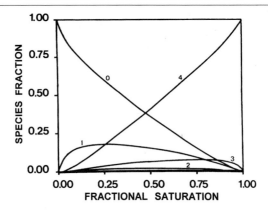

FIG. 7. Relative abundance of various oxygenation states based on the free energy values for human hemoglobin A tetramers[7] presented in Table III.

Figure 7 presents the distribution of the various ligation states of tetrameric hemoglobin A as calculated from the Mill *et al.*[7] free energy changes presented in Table III. The small fractions of doubly and triply oxygenated tetramers are a consequence of the highly cooperative nature of hemoglobin A. The sequentially increasing stepwise oxygen-binding affinities for tetrameric hemoglobin A require that oxygens be preferentially bound to doubly oxygenated and triply oxygenated hemoglobin species. Thus, concentrations of these species are depleted compared with the other species. For comparison, the Hill coefficient for tetrameric hemoglobin Kansas is approximately 1.25 at 50% saturation while that for hemoglobin A is approximately 3. The calculated maximal fraction of doubly oxygenated hemoglobin Kansas tetramers is approximately 30%, as is expected for the less cooperative hemoglobin Kansas.[7]

Presence and Influence of Dimers in Experiments

Consider how many dimers are present under the specific conditions for the experiments. Is it possible to do the experiments at a sufficiently high hemoglobin concentration that the presence of dimers will have little influence? What are the errors in the Adair binding constants introduced by neglecting the presence of dimeric species?

Figure 8 presents the calculated fraction of dimeric species as a function of the concentration of fully oxygenated hemoglobins A and Kansas. Oxygen-binding experiments are commonly done at 60 μM heme. From Fig. 8 it is obvious that oxygenated hemoglobin A is ~10% dimers and hemoglobin Kansas is ~60% dimers, at 60 μM heme. The percentage of dimers present

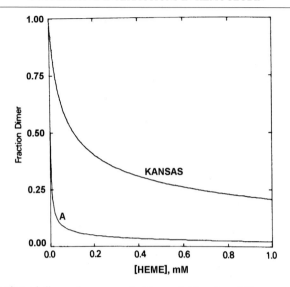

Fɪɢ. 8. Fraction of dimers for oxygenated hemoglobins A and Kansas as a function of heme concentration. These are calculated from the free energy values presented in Table III. [Redrawn from Atha et al.[28]]

in both deoxygenated hemoglobins is close to zero. It should also be noted that at the hemoglobin concentrations found within red blood cells, 7% of hemoglobin Kansas will be dissociated dimers.[28] Clearly, it is impossible to conduct experiments on this hemoglobin at a high enough concentration that dimers are not present in significant amounts.

Nevertheless, how much influence will the presence of dimers have on the evaluation of the Adair oxygen-binding constants? This was answered by a series of computer simulations.[29] Oxygen-binding data were simulated according to the seven-parameter oxygenation-linked oxygen-binding isotherm [Eq. (11)] and then the data were least-squares fit according to the tetramer-only oxygen-binding isotherm [Eq. (9)]. For these simulations the values for the linkage scheme parameters [Eq. (11)] for hemoglobin A were taken from Chu et al.[24] and those for hemoglobin Kansas are taken from Atha et al.[28] Any differences between the least squares-fitted values using the tetramer-only isotherm and the values used for the simulations represent the induced systematic errors that are introduced by neglecting the presence of dimers.

[29] M. L. Johnson and A. E. Lassiter, Biophys. Chem. 37, 231 (1990).

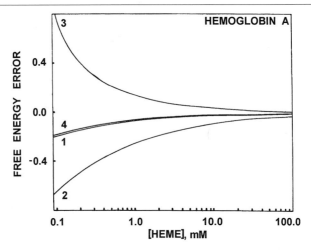

Fig. 9. The induced systematic error in free energies of oxygen binding caused by the presence of dimers during the measurement of the free energy changes for oxygen binding to hemoglobin A. The induced systematic error in free energy corresponds to a change in the product Adair oxygen-binding constants. This simulated result is based on the experimental results of Chu et al.[24] [Redrawn from Johnson and Lassiter.[29]]

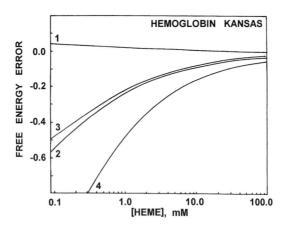

Fig. 10. The induced systematic error in free energy caused by the presence of dimers during the measurement of the free energy changes for oxygen binding to hemoglobin Kansas. The induced systematic error in free energy corresponds to a change in the product Adair oxygen-binding constants. This simulated result is based on the experimental results of Atha et al.[28] [Redrawn from Johnson and Lassiter.[29]]

Figures 9 and 10 present the predicted induced systematic errors expected for hemoglobins A and Kansas. These induced systematic errors correspond to errors in the product Adair constants introduced when the analysis methodology neglected the presence of the dimers of hemoglobin contained in the solution being investigated. Clearly, the evaluation of the free energies of oxygen binding is significantly affected by neglecting the presence of dimeric species. This is true even when the heme concentration is 3 mM or above, as shown in Table IV. A heme concentration of 3 mM is greater than can be accommodated by the experimental techniques commonly used to study hemoglobin oxygen binding. Thus, for these two hemoglobins it is impossible to design an experimental protocol using standard methodologies that will allow a high enough hemoglobin concentration to negate the systematic errors introduced by neglecting the presence of dimers.

It is also important to note that the induced systematic errors for hemoglobins A and Kansas have different patterns. This means that predicting the direction and size of the expected systematic errors for other variants of hemoglobin and other experimental conditions is impossible. The directions and magnitudes of these induced systematic errors cannot be predicted a priori without having previously solved the complete linkage scheme, for the particular hemoglobin and experimental conditions, and then doing the preceding simulations.

TABLE IV
INDUCED SYSTEMATIC ERRORS[a] IN OXYGEN-BINDING
FREE ENERGIES CAUSED BY NEGLECTING DIMERS
IN HEMOGLOBINS A AND KANSAS

Error	Hemoglobin A		Hemoglobin Kansas	
	1.0 mM	3.0 mM	1.0 mM	3.0 mM
$\delta \Delta G_{41}$	−0.07	−0.04	0.02	0.01
$\delta \Delta G_{42}$	−0.26	−0.16	−0.24	−0.15
$\delta \Delta G_{43}$	0.14	0.07	−0.22	−0.14
$\delta \Delta G_{44}$	−0.06	−0.03	−0.48	−0.29
$\delta \Delta g_{42}$	−0.19	−0.12	0.26	0.16
$\delta \Delta g_{43}$	0.40	0.23	−0.02	−0.01
$\delta \Delta g_{44}$	−0.20	−0.10	0.26	0.15

[a] $\delta \Delta G_{4i}$ is the error in the free energy change corresponding to the product Adair binding constants and $\delta \Delta g_{4i}$ is the error in the free energy change corresponding to the stepwise Adair binding constants.

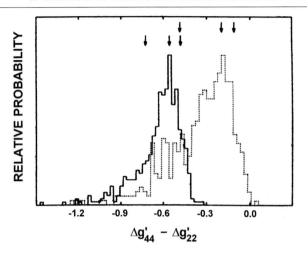

FIG. 11. A Monte Carlo analysis of the statistical significance of the quaternary enhancement effect for the experimental data of Mills and Ackers[23] (dotted line) and Chu et al.[24] (solid line). These complete probability distributions were determined by a Monte Carlo analysis.[30] The irregular shape of these distributions is simply a consequence of using a finite number of Monte Carlo cycles. In the limit of an infinite number of cycles these distributions will be smooth curves. The arrows correspond to the most probable value, the lower and the upper ±1 SEM confidence intervals from the Monte Carlo analysis. [Redrawn from Straume and Johnson.[30]]

Quaternary Enhancement

Quaternary enhancement is simply the observation that the fourth oxygen in tetrameric hemoglobin is bound with a higher affinity than the oxygens bound to either isolated α and β subunits or dissociated $\alpha\beta$ dimers. This can be seen from the linkage scheme values for hemoglobin A shown in Table III, where the free energy change to bind the fourth oxygen to a triply oxygenated hemoglobin (corrected for statistical factors), $-RT \ln(k'_{44})$, is -8.65 kcal while the corresponding values for the dimers, $-RT \ln(k'_{21})$ and $-RT \ln(k'_{22})$, are both -8.38 kcal. While this difference is small, it is significant in this data set.[30] Furthermore, the experimental results have been repeated[22–24] and quaternary enhancement is observed in those, and other,[24] experiments as well.

Figure 11 presents a detailed analysis of the experimental data of Mills and Ackers[23] (dotted line) and Chu et al.[24] (solid line) specifically to evaluate the complete probability distribution for the magnitude of the quaternary

[30] M. Straume and M. L. Johnson, *Biophys. J.* **56,** 15 (1989).

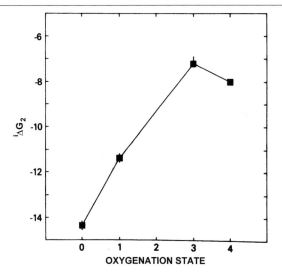

FIG. 12. The intrinsic free energy of association as a function of oxygenation for hemoglobin A. [Redrawn from Atha *et al.*[28]]

enhancement effect. These complete probability distributions were determined by a Monte Carlo analysis.[30] Figure 11 shows that for the Chu *et al.*[24] data the quaternary enhancement effect is −0.6 kcal/mol and the distribution does not include zero. As noted above, for the Mills and Ackers[23] data the quaternary enhancement effect is −0.3 kcal/mol, and only an insignificant fraction of the distribution is above zero. Thus, the quaternary enhancement effect is significant in these data.

From the lower thermodynamic cycle in Fig. 6 it can be seen that

$$k_{44}\,^3K_2 = k_{22}\,^4K_2 \tag{15}$$

And thus

$$\Delta g_{44} - \Delta g_{22} = {}^4\Delta G_2 - {}^3\Delta G_2 \tag{16}$$

The quaternary enhancement effect is observed as the last oxygen-binding step of tetrameric hemoglobin having a higher affinity than the last oxygen-binding step of the dimeric hemoglobin. From Eqs. (15) and (16) it is clear that the quaternary enhancement effect can also be observed as an increase in the subunit association constant (i.e., more negative free energy change) for the formation of fully oxygenated hemoglobin as compared with triply oxygenated hemoglobin. Figure 12 presents the observed subunit association free energy changes for the Mills and Ackers data.[23]

Note that the free energy change is more negative for the quad-oxygenated as compared with the triply oxygenated hemoglobin, thus demonstrating the quaternary enhancement effect. The analogous results for the Mill et al.[7] data are given in Table III.

For a complete discussion of the quaternary enhancement effect the reader is referred to Ackers and Johnson.[31]

MWC Allosteric Model

In 1965 Monod, Wyman, and Changeux[32] proposed a model for the cooperativity of hemoglobin based on a two-state allosteric transition. For this model, the hemoglobin tetramer is assumed to exist in either of two configurations that can rapidly interchange, as in Fig. 13: a low-affinity tense configuration, T, and a high-affinity relaxed configuration, R. This is, of course, analogous to the known oxygenated and deoxygenated X-ray crystallographic structures.[12,13] The intrinsic stepwise oxygen-binding constant within the T configuration tetramers is K_T, and in the R configuration tetramers it is K_R with $K_T < K_R$. Transitions between the T and R configurations are "concerted" because all the binding site affinities within a particular molecule change simultaneously. Unligated (i.e., deoxygenated) hemoglobin tetramer is predominantly in the lower affinity T configuration. Thus, the initial oxygen is predominantly bound to the low-affinity T configuration. However, as the number of bound oxygens increases the mass action equilibrium shifts to where the higher affinity R configuration is predominant. Consequently, according to this model the oxygen-binding affinity shifts from a low affinity to higher affinity as oxygenation increases. This is, by definition, an example of positive cooperativity. This classic MWC model cannot predict any form of negative cooperativity.

MWC is a model because it makes specific assumptions about the molecular mechanism by which the average oxygen-binding affinity changes during the ligation process. In contrast, the Adair binding mechanisms [e.g., Eqs. (9)–(14)] make no specific assumptions about the relationships between the binding constants.

The MWC model has three parameters; L, c, and K_R or K_T. L is the isomerization equilibrium constant for the unligated species, i. e., the ratio of the concentrations of the unligated T configuration tetramer to the unligated R configuration tetramers.

$$L = T_0/R_0 \qquad (17)$$

[31] G. K. Ackers and M. L. Johnson, *Biophys. Chem.* **37**, 265 (1990).
[32] J. Monod, J. Wyman, and J.-P. Changeux, *J. Mol. Biol.* **12**, 88 (1965).

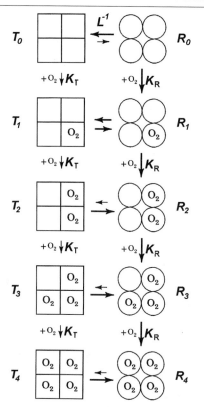

FIG. 13. Schematic of the MWC model with two configurations, R and T, with four identical binding sites each; i.e., the MWC model for human hemoglobin tetramers.

The constant ratio of the T and R configuration intrinsic oxygen-binding affinities is commonly used as a parameter of the MWC model:

$$c = K_T/K_R \tag{18}$$

The MWC model requires that the values of K_T, K_R, L, and c remain constant during the ligation cycle. Models can be imagined in which any, or all, of these vary as a function of oxygen and/or hemoglobin concentration, but these models are not the model proposed by Monod, Wyman, and Changeux.[32]

For tetrameric hemoglobin the four-site MWC model contains one less parameter than the Adair formulation [Eq. (9)]. Specifically, the four-site MWC binding isotherm in terms of L, K_T, and K_R is

$$\overline{Y}_4 = \frac{LK_T[O_2](1 + K_T[O_2])^3 + K_R[O_2](1 + K_R[O_2])^3}{L(1 + K_T[O_2])^4 + (1 + K_R[O_2])^4} \qquad (19)$$

Or, in terms of L, c, and K_R:

$$\overline{Y}_4 = \frac{LcK_R[O_2](1 + cK_R[O_2])^3 + K_R[O_2](1 + K_R[O_2])^3}{L(1 + cK_R[O_2])^4 + (1 + K_R[O_2])^4} \qquad (20)$$

This classic MWC model is commonly used in hemoglobin research even though it has several limitations. The classic model cannot predict any form of negative cooperativity. It assumes that the subunits are identical so it cannot be used for hemoglobins such as Kansas, where the α and β subunit intrinsic affinities differ by 0.84 kcal/mol. It does not include the possibility for tetrameric hemoglobin to dissociate into $\alpha\beta$ dimers.

The classic MWC model was extended by Ackers and Johnson[33] to include the possibility of nonidentical α and β subunits, and for the dissociation into $\alpha\beta$ dimers. This extended MWC model was tested by analyzing the Mills et al.[7] data for hemoglobin A and the Atha et al.[28] data for hemoglobin Kansas.

The extended MWC model with dimers assumed to be identical to R configuration tetramers (extended two-state MWC) does not provide a reasonable description of the Mills et al.[7] data for hemoglobin A. The distribution of the residuals from this analysis is presented in the lower panel of Fig. 14. Residuals are simply the differences between the data points and the best least-squares estimated curve. For comparison, the residuals from the analysis based on the subunit association-linked Adair formulation [Eq. (11)] are presented in the upper panel of Fig. 14. Clearly these residuals from the extended two-state MWC model have significant systematic trends. This shows that the extended MWC model with dimers assumed to be identical to R configuration tetramers is not consistent with the Mills et al.[7] data for hemoglobin A. The variance of fit for this extended MWC model is 7.4×10^{-5} as compared with 2.6×10^{-5} when the data were analyzed by the subunit association-linked Adair formulation [Eq. (11)]. This nearly threefold increase in the variance of fit also shows that the extended two-state MWC model is not an adequate description of the Mills et al.[7] data. The extended two-state MWC model parameters are shown in Table V.

The cause of the failure of the extended two-state MWC model's to

[33] G. K. Ackers and M. L. Johnson, J. Mol. Biol. **147**, 559 (1981).

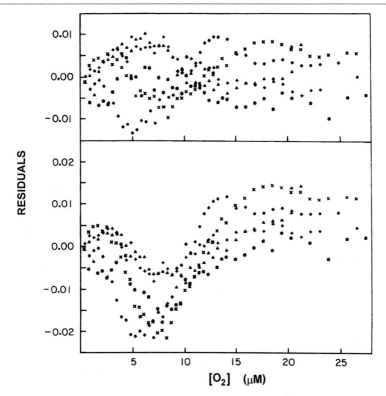

FIG. 14. Residuals for the least-squares analysis of the Mills *et al.*[7] data. *Top:* Using the model-independent ligand-linked subunit association [Eq. (11)]. *Bottom:* Using an extended two-state MWC model. [Redrawn from Ackers and Johnson.[33]]

TABLE V

EXTENDED MWC MODEL PARAMETERS FOR MILLS *et al.*[7]
HEMOGLOBIN A DATA[a]

Parameter	Two-state MWC	Three-state MWC
L	5.2×10^4	5.6×10^5
c	6.1×10^{-3}	3.5×10^{-3}
K_R	1.73×10^6	3.1×10^6
δ_{04}	0	1.39 kcal

[a] As presented by Ackers and Johnson.[33]

adequately describe the Mills *et al.*[7] data is its inability to describe the quaternary enhancement effect. For this model the intrinsic subunit assembly constant is given by[33]

$$^{i}K_2' = K_{2R}'(1 + Lc^{i}) \tag{21}$$

where K_{2R}' refers to the intrinsic (i.e., corrected for statistical factors) association constant of R configuration dimers to form R configuration tetramers, K_2' is the intrinsic association constant to form a mixture of R and T configuration tetramers with i oxygens bound, and L and c are MWC model parameters. It is clear from Eq. (21) that $^{i}K_2'$ is a monotonic function of the degree of oxygenation, i. However, as shown in Fig. 12, the experimentally determined values of $^{i}K_2'$ are not a monotonic function of the degree of oxygenation, i, because of the quaternary enhancement effect.

Ackers and Johnson[33] also showed that an extended three-state MWC model can describe these experimental data to the same precision as Eq. (11). This analysis, however, required that the dimers had an affinity that was 1.39 kcal lower than the R configuration tetramers.

Analogous results and conclusions were presented by Ackers and Johnson[33] for hemoglobin Kansas and data from a series of different temperatures.[22,23]

Asymmetric Doubly Ligated Hemoglobin

It is known that the major transformation in quaternary structure concomitant with oxygen binding occurs in the contact region between the two $\alpha\beta$ dimers.[12–15] Therefore, as noted previously, oxygen-binding cooperativity can be studied by an analysis of the energetics of subunit assembly. In 1985 Smith and Ackers[34] developed a method to measure the subunit assembly (dimer to tetramer) free energies for each of the 10 nonredundant microstates of the human hemoglobin system. The microstates are determined by all the possible ways to bind from zero to four ligands to a tetrameric molecule consisting of dimers of $\alpha\beta$ dimers. These microstates are shown in Fig. 15.

The Smith and Ackers[34] strategy used tightly bound ligands (e.g., CO and cyanomet) in combination with metal iron substitutions (e.g., Mn and Co) to circumvent the problem of ligand lability. The experimental techniques for resolving the assembly energetics of the various species in-

[34] F. R. Smith and G. K. Ackers, *Proc. Natl. Acad. Sci. U.S.A.* **82**, 5347 (1985).

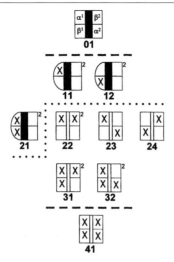

Fig. 15. The 16 ligation states (microstates) of tetrameric human hemoglobin. Only the 10 unique (i.e., nonredundant) states are shown. The two-digit names of the microstates are shown below them. The first digit refers to the number of bound ligands. The second digit is an index. The X's represent the ligand. The 2 at the upper right of some states indicates that the particular state has a statistical redundancy of 2. The locations of the α and β subunits are shown in the 01 microstate. The contact region between the two $\alpha\beta$ dimers is shown in black for the T configuration tetramers and white for the R configuration tetramers.

cluded kinetic methods,[34,35] analytical gel chromatography,[35,36] and low-temperature electrophoresis techniques.[37–40]

On the basis of these microstate dimer–tetramer assembly energies, Ackers and co-workers[41] proposed the *molecular code mechanism*. In the molecular code mechanism, unligated hemoglobin tetramers (i.e., microstate 01) are in a T quaternary configuration. When the first ligand is bound to either an α or β subunit within the tetramer (i.e., microstates 11 and 12), the tetramer remains in the T configuration, but the presence of the ligand imposes a stress on the tetramer. This stress is observed as a lower assembly free energy for microstates 11 and 12. When the second ligand

[35] P. C. Speros, V. J. LaCata, T. Yonetani, and G. K. Ackers, *Biochemistry* **30,** 7254 (1991).

[36] G. K. Ackers, *Adv. Protein Chem.* **24,** 323 (1970).

[37] M. Perrella and L. Rossi-Bernardi, *Methods Enzymol.* **76,** 133 (1981).

[38] M. Perrella, L. Benazzi, M. A. Shea, and G. K. Ackers, *Biophys. Chem.* **35,** 97 (1990).

[39] M. Perrella, A. Colosimo, L. Benazzi, M. Ripamonti, and L. Rossi-Bernardi, *Biophys. Chem.* **37,** 211 (1990).

[40] M. Samaja, E. Rovida, M. Niggeler, M. Perrella, and L. Rossi-Bernardi, *J. Biol. Chem.* **262,** 4528 (1987).

[41] G. K. Ackers, M. L. Doyle, D. Myers, and M. A. Daugherty, *Science* **255,** 54 (1992).

is bound to the same $\alpha\beta$ dimer within the tetramer as the previous ligand (i.e., microstate 21), the tetramer remains in the T configuration, but may have some additional stress and thus a lower assembly free energy. When the second ligand is bound to a site on the other $\alpha\beta$ dimer of the tetramer (i.e., microstates 22, 23, and 24), the tetramer shifts predominantly to the R configuration that is strained by the presence of unligated subunits. The subunit assembly free energy changes of the triply ligated hemoglobins (i.e., microstates 31 and 32) are similar to the doubly ligated R configuration tetramers (i.e., microstates 22, 23, and 24). When the last ligand is bound (i.e., microstates 41), the strain within the R configuration tetramer is released. Thus, the model predicts two quaternary structures: T configurations for microstates 01, 11, 12, and 22; R configurations for microstates 22, 23, 24, 31, 32, and 41. The model also predicts at least four energetic levels: one for microstate 01; a second for microstates 11, 12, and 21; a third for microstates 22, 23, 24, 31, and 32; and another for microstate 41. An example of these different energy levels can be seen in the right-hand column of Table VI.

In 1992 Ackers et al.[41] used this molecular code mechanism to predict the values of the microstate free energies of association for native hemoglobin with oxygen as a ligand. That prediction is shown in the first column of Table VI. The right-hand column of Table VI is the actual experimental determination of the association free energy changes in native hemoglobin with oxygen as a ligand[26] that became available in 1996. The agreement is remarkable.

These results are typical of the results obtained for other ligands (e.g.,

TABLE VI

EXPERIMENTAL ASSEMBLY FREE ENERGIES $^{ij}\Delta G_2$ FOR
HEMOGLOBIN A MICROSTATES

Microstate, ij	Predicted 1992[a]	Resolved 1996[b]
01	−14.4	−14.4
11	−11.5	−11.6
12	−11.5	−11.6
21	−9.2	−9.4
22	−7.2	−7.7
23	−7.2	−7.6
24	−7.2	−7.8
31	−7.2	−7.5
32	−7.2	−7.5
41	−8.0	−8.1

[a] From Ref. 41.
[b] From Ref. 26.

CO and cyanomet) in combination with metal iron substitutions (e.g., Mn and Co).[26,42–44]

Microstate 41 will be different from microstates 31 and 32 by a small amount because of the quaternary enhancement effect. When the distribution of microstates was first published[34] the small differences between microstate 41 and microstates 22, 23, 24, 31, and 32 were not obvious and the direct experimental observations of the microstates were not available for oxygen as a ligand. Consequently Straume and Johnson[45] tested all 6561 possible three-state combinatorial switch models for consistency with the Mills *et al.*[7] and the Chu *et al.*[24] oxygen-binding data. In these models the free energy changes for association to microstates 01 and 41 were assumed to have two distinct values. The remaining eight unique microstates were assumed to be (1) like microstate 01; (2) like microstate 41, or (3) have some other free energy changes for association. Thus, 3^8, or 6561, different models were tested. The conclusion was that none of the three-state combinatorial switch models could describe the actual oxygen-binding data. However, models that included quaternary enhancement effect as a fourth state were consistent with the experimental hemoglobin concentration-dependent oxygen-binding data.

The experimentally observed microstate free energies of association and the molecular code mechanism are inconsistent with the MWC model. The MWC model predicts that the microstate free energies of association must be the same for each of the ligation stages. However, the experimental observations are that the free energies of association are different within ligation step 2 (i.e., microstate 21 is different from microstates 22, 23, and 24). Thus, these experimental observations provide a direct experimental disproof of the classic MWC model.

For a discussion of the experimental basis of the molecular code mechanism the reader is referred to Ackers *et al.*[41]

Discussion

The generation of realistic models to describe cooperativity in human hemoglobin requires three general categories of information. First and foremost is a series of high-precision experimental data. Second is a conceptual idea, or theory, about the molecular mechanism. Third is the availability

[42] G. K. Ackers, M. Perrella, J. M. Holt, I. Denisov, and Y. Huang, *Biochemistry* **36,** 10822 (1997).

[43] Y. W. Huang, M. L. Koestner, and G. K. Ackers, *Biophys. J.* **71,** 2106 (1996).

[44] Y. W. Huang, T. Yonetani, A. Tsuneshige, B. M. Hoffman, and G. K. Ackers, *Proc. Natl. Acad. Sci. U.S.A.* **93,** 4425 (1996).

[45] M. Straume and M. L. Johnson, *Biochemistry* **27,** 1302 (1988).

of numerical analysis tools to evaluate the consistency of the model with the data. Some of these numerical analysis tools are discussed in the appendix.

It is known that human hemoglobin tetramers dissociate to form identical $\alpha\beta$ dimers. These $\alpha\beta$ dimers do not readily dissociate into monomers. It is also known that the major transformation in quaternary structure concomitant with oxygen binding occurs within the contact region between the two $\alpha\beta$ dimers.[12–15] Thus, the cooperative nature of oxygen binding to hemoglobin can be studied experimentally by studying (1) the oxygen-binding properties as a function of hemoglobin concentration, or (2) the dimer–tetramer association properties as a function of oxygen concentration, or (3) both simultaneously. While either of these should be sufficient, using both simultaneously adds significantly to the resolving power in a manner analogous to two-dimensional NMR or two-dimensional electrophoresis analysis.

One experimental observation is that the $\alpha\beta$ dimer of human hemoglobin A binds oxygen noncooperatively.[7] Furthermore, the oxygen-binding properties (i.e., binding affinities) of the $\alpha\beta$ dimers are essentially the same as those of the isolated α and β subunits.[7] The binding affinities of the $\alpha\beta$ dimers and isolated α and β subunits of human hemoglobin A are much higher than the average affinity of the tetrameric form.[7]

Another interesting experimental observation that all models of cooperativity in human hemoglobin must be consistent with is the quaternary enhancement effect.[31] The quaternary enhancement effect[31] is the experimental observation that the last oxygen bound to tetrameric human hemoglobin A is bound with an affinity that is higher than the affinities of the isolated α and β subunits and higher than the affinities of the $\alpha\beta$ dimers. This experimental observation is inconsistent with the classic two-state MWC model of hemoglobin cooperativity.

The molecular code mechanism is based on another important experimental observation. That observation is that the double-ligated microstates in tetrameric hemoglobin do not have equal association free energy changes (i.e., they have different dimer–tetramer association constants). Microstate 21, in which a double-ligated tetrameric hemoglobin has both ligands on the same $\alpha\beta$ dimer within the tetramer, is in a quaternary T configuration. The remaining doubly ligated microstates (i.e., 22, 23, and 24) are in a quaternary R configuration. This experimental observation is also inconsistent with the classic two-state MWC model of hemoglobin cooperativity.

Experimental observations of the assembly free energies for the microstates of tetrameric human hemoglobin show that the microstates group into at least four different energy levels based on the free energies of association. These levels are shown in Ackers et al.[41] in red, yellow, green, and blue. These are delineated in Fig. 15 by the dashed and dotted lines.

The microstates above the dotted line in Fig. 15 are in the T quaternary conformation while those below are in an R quaternary configuration. The dashed lines further divide the microstates into four individual energy levels. All realistic models of hemoglobin cooperativity must be capable of describing these four distinct free energy of association levels.

The most commonly utilized mechanistic model of hemoglobin cooperativity is the classic two-state allosteric model of Monod, Wyman, and Changeux.[32] This classic model is inconsistent with the experimentally observed quaternary enhancement effect. The classic model assumes that the individual subunits of hemoglobin exist in one of two affinity states that are in rapid equilibrium. Ackers and Johnson[33] extended the classic two-state model to include dimer–tetramer association and a third affinity state. This extended three-state MWC model[33] can describe the quaternary enhancement effect.

Both the classic two-state MWC model[32] and the extended three-state MWC model[33] are inconsistent with the experimental observation that the doubly ligated tetrameric microstate with both ligands on the same $\alpha\beta$ dimer (i.e., microstate 21) is in the T configuration while the remaining doubly ligated microstates (i.e., 22, 23, and 24) are in an R configuration. The doubly ligated microstate with both ligands on the same $\alpha\beta$ dimer (i.e., microstate 21) has a different free energy for dimer–tetramer association than the remaining doubly ligated microstates (i.e., 22, 23, and 24). These MWC models assume that the subunits within tetrameric hemoglobin are functionally identical. Consequently, these MWC models cannot predict that particular doubly ligated microstates exist in different configurations, and free energies of association, depending on which subunits are ligated. Ackers and Johnson[33] also extended the MWC model to include functionally nonidentical α and β subunits. However, this extension does not consider the placement of the ligands on individual α and β subunits within the specific $\alpha\beta$ dimers of tetrameric hemoglobin. Thus, this extension of the MWC model is also inconsistent with the experimentally observed differences between the doubly ligated microstates.

The molecular code mechanism of Ackers et al.[41] was specifically formulated to consider the experimentally observed doubly ligated microstate heterogeneity (as shown in Fig. 15) and the quaternary enhancement effect. The molecular code mechanism predicts that initially hemoglobin tetramers are predominantly in the T quaternary configuration (i.e., microstate 01 and the first energy level). When the first ligand is bound (i.e., microstates 11 and 12), the tetramer remains predominantly in the T quaternary configuration, but the presence of the ligand introduces a strain on the tetramer. This strain is experimentally observed as a lower dimer–tetramer association free energy (i.e., the second energy level). When the second ligand is bound to the same $\alpha\beta$ dimer within the tetramer as the first ligand (i.e., microstate 21), the tetramer remains in the T quaternary configuration with approximately the same dimer–tetramer association free energy as the singly ligated tetramers. However, when the second ligand is bound to the other $\alpha\beta$ dimer within the tetramer (i.e., microstates 22, 23, and 24), then the tetramer shifts to being predominantly in the R quaternary configuration with a different dimer–tetramer association free energy change. Triply ligated microstates (i.e., 31 and 32) are predominantly in the R configuration with a dimer–tetramer association free energy close to that of microstates 22, 23, and 24. Fully ligated tetramers are also predominantly in the R quaternary configuration. The presence of an unligated subunit within an R configuration tetramer introduces a tertiary constraint with these tetramers. The tertiary constraint is observed as microstate 41 having a different dimer–tetramer association free energy of association than microstates 22, 23, 24, 31, and 32. These different energetic levels are separated in Fig. 15 by dotted and dashed lines.

The molecular code mechanism of Ackers et al.[41] is consistent with the quaternary enhancement effect. It is also consistent with the heterogeneity of the doubly ligated microstate energy levels. Furthermore, in 1992 Ackers and co-workers used the molecular code mechanism to predict (shown in Table VI) the microstate energy levels for oxygen binding to native human hemoglobin.[41] The microstate energy levels for oxygen binding were not experimentally available until 1996 (also shown in Table VI).[26] These predicted and observed energy levels are consistent, showing that the molecular code mechanism is a good description of the mechanism of cooperativity in human hemoglobin.

Appendix: Numerical Analysis Tools for Comparing Models and Experimental Data

The general numerical analysis procedure for comparing models and experimental data is first to "fit" the experimental data to the mathematical form dictated by the conceptual or mechanistic model being tested. Once the "fit" has been accomplished, the validity of the model is evaluated by testing how well the fitted model describes the actual experimental observations. This procedure should address four distinct components:

Evaluation of the optimal model parameter values
Evaluation of realistic estimates of the precision of the estimated parameter values
Evaluation of the uniqueness of the parameter values
Evaluation of the goodness of fit

The most common procedure for estimating the optimal model parameter values is one of the weighted nonlinear least-squares parameter estimation algorithms.[16–21] These are successive approximation algorithms that require the user to specify an initial estimate, or guess, of the parameter values. The algorithm is then applied to obtain a better approximation of the parameter values. Next, this better approximation of the parameter values is used as a new guess and the algorithm is applied again. This process is repeated until the resulting better approximation at the end of a cycle is the same as the parameter estimates that were used to begin the cycle.

The validity of using any least-squares technique is based on a series of assumptions about the nature of the experimental uncertainties contained within the data. For hemoglobin data it is generally assumed that the experimental uncertainties follow a Gaussian, or bell-shaped, distribution. The validity of this assumption and the nature of the weighting functions are discussed elsewhere.[46] Given this, and several other reasonable assumptions about the nature of the experimental uncertainties concomitant within the data, least-squares algorithms will provide the parameter values with the highest probability (i.e., maximum likelihood) of being correct.

Once the parameter values have been estimated, an investigator should question how accurately the parameters values have been determined. This is important because it is the confidence limits of the estimated model parameter values that allow tests of hypotheses about the specifics of the models. For example, hemoglobin was initially determined to be tetrameric on the basis of light scattering[4] and sedimentation equilibrium[5] measurements that showed that the molecular mass was approximately 67,000 Da. It was also known that the molecular mass per iron was approximately 17,000 Da. The ratio of 4 indicated the stoichiometry. But what if the numbers were $67,000 \pm 15,000$ and $17,000 \pm 4,000$ Da? With these uncertainties the ratio would be 4 ± 1.3 and we would know that there is a 68.2% chance (i.e., ± 1 SEM) that hemoglobin was somewhere between a trimer and a pentamer.

The problem is that there is no exact algorithm for the evaluation of confidence intervals for nonlinear parameter estimation procedures. Asymptotic standard errors are by far the most commonly used, and abused, estimates of the precision of parameters determined by nonlinear least-squares procedures. Asymptotic standard errors are also by far the least accurate and usually provide a gross underestimation of the true uncertainties. Mills et al.[7] used a search algorithm[47] that is far more accurate, but

[46] M. L. Johnson and A. E. Lassiter, *Biophys. Chem.* **38,** 159 (1990).
[47] M. L. Johnson, *Biophys. J.* **44,** 101 (1983).

sometimes overestimates the true uncertainties. Overestimating the uncertainties is preferable to underestimating them. When Straume and Johnson[30] needed a precise estimate of parameter precision, a Monte Carlo approach was used. Given that computers are rapidly becoming faster the reader is encouraged to examine Monte Carlo[30] and bootstrap[48] techniques for the analysis of the precision of model parameters estimated by least-squares procedures.

There is no known method to guarantee that the parameters estimated by any nonlinear procedure (i.e., least squares or other) are unique. This is a classic problem in numerical analysis with nonlinear equations. For nonlinear fitting equations the possibility always exists that there are completely different sets of parameter values that will provide equivalent descriptions of the data (i.e., least squares or minimum variance of fit). These are known as relative minima. Being sure that all the relative minima for a nonlinear fitting problem have been found and tested is virtually impossible.

A grid search method was used by Johnson and Ackers[49] to find several relative minima for the Szabo–Karplus model of hemoglobin. In a grid search procedure, the entire physically meaningful region of the multiple parameter space is divided by a coarse multidimensional grid. The grid is searched for minima in the apparent variance of fit by evaluating the variance of fit at every intersection of the grid. Every apparent minimum found by the grid search is then used as an initial starting point (i.e., initial guess of the parameter values) for the nonlinear least-squares procedure. If the nonlinear least-squares procedure converges to the same set of parameter values when started at all these apparent relative minima, then assuming that relative minima do not exist is reasonable. Conversely, if the nonlinear least-squares procedure converges to different sets of parameter values then relative minima exist.

When testing models it is critically important to address the question, "Does the least-squares fit adequately describe the experimental data?" Clearly, if the fitted model cannot describe the data then either the model is wrong, or the data are wrong, or both. Numerical analysis cannot distinguish these alternatives. Conversely, if the least-squares fit appears to characterize the experimental data adequately then the model can be considered consistent with the data.

It is also important to note that if the least-squares estimated model cannot describe the experimental data then the values of the model parameters have no physical meaning. Under these conditions there is no need to

[48] B. Efron and R. J. Tibshirani, "An Introduction to the Bootstrap." Chapman & Hall, New York, 1993.
[49] M. L. Johnson and G. K. Ackers, *Biochemistry* **21,** 201 (1982).

evaluate the precision of the estimated parameters. However, it is important to test for other relative minima that may provide a better description of the experimental data.

The statistical methods for assigning a probability to how well a model describes an experimental data set are known as goodness-of-fit criteria. Conceptually these are simple. The application of the least-squares procedures assumes, among other things, that (1) the experimental uncertainties concomitant on the experimental data follow a Gaussian, or bell-shaped, distribution, and (2) the mathematical model is correct. This implies that the differences between the data points and the fitted curve (i.e., the residuals) resulting from a valid analysis of the experimental data should also follow a Gaussian distribution. Thus, if the residuals are not consistent with a Gaussian distribution then either (1) the mathematical model is not correct, or (2) the experimental uncertainties concomitant on the experimental data did not follow a Gaussian distribution, or (3) both. Consequently, goodness-of-fit criteria are usually tests of the Gaussian nature of the residuals. Commonly used and recommended procedures include (1) runs test, (2) autocorrelations of the residuals, and (3) the Kolmogorov–Smirnov test.[19]

Acknowledgments

The author acknowledges the support of the National Science Foundation Science and Technology Center for Biological Timing at the University of Virginia (NSF DIR-8920162), the General Clinical Research Center at the University of Virginia (NIH RR-00847), and the University of Maryland at Baltimore Center for Fluorescence Spectroscopy (NIH RR-08119).

[7] Deriving Complex Ligand-Binding Formulas

By MICHAEL L. JOHNSON and MARTIN STRAUME

Introduction

Understanding the mechanisms by which small molecules (i.e., ligands) bind to macromolecules is important to many fields of biomedical research. Examples are the binding of drugs to receptors and the binding of oxygen to hemoglobin. The study of ligand-binding problems usually involves the collection of a set of data reflecting the quantity of ligand bound as a function of ligand and possibly macromolecular concentration. The binding constants are then determined by least-squares fitting[1-4] an assumed equation to these data. The quality of the fit (i.e., goodness of fit[5]) of the assumed equation is then used to test if the assumed functional form of the equation is consistent with the experimental data (i.e., hypothesis testing). If the fit does not provide a good description of the experimental data then either the function form is wrong, the fitting procedure was in error, or the data contained systematic errors.

Fundamental to the process of testing ligand-binding hypotheses is the derivation of the mathematical functional form of the binding equations that correspond to particular biochemical hypotheses about the ligand-binding process. This chapter outlines a general approach that can be used to derive mathematical equations for even the most complex ligand-binding hypotheses.

Examples

In 1925 Adair[6] presented a macroscopic formulation to describe the fractional saturation of oxygen to the four binding sites of human hemoglobin:

$$\overline{Y} = \frac{1}{4} \frac{K_1[X] + 2K_2[X]^2 + 3K_3[X]^3 + 4K_4[X]^4}{1 + K_1[X] + K_2[X]^2 + K_3[X]^3 + K_4[X]^4} \qquad (1)$$

In Eq. (1) the fractional saturation is expressed as a function of the free (i.e., unbound) ligand concentration, X. The K values are the macroscopic product Adair binding constants. K_n corresponds to the average binding constant for n ligands simultaneously bound to an unliganded macromolecule with four binding sites. Equation (1) can also be written with stepwise macroscopic binding constants where k_n corresponds to the binding of the nth ligand when $n - 1$ ligands are already bound:

$$\overline{Y} = \frac{1}{4} \frac{k_1[X] + 2k_1k_2[X]^2 + 3k_1k_2k_3[X]^3 + 4k_1k_2k_3k_4[X]^4}{1 + k_1[X] + k_1k_2[X]^2 + k_1k_2k_3[X]^3 + k_1k_2k_3k_4[X]^4} \qquad (2)$$

Adair equations [e.g., Eqs. (1) and (2)] provide an exact formulation for the binding with macroscopic, or average, binding constants. Neither product nor stepwise macroscopic binding constants contain explicit terms

[1] M. L. Johnson and S. G. Fraser, *Methods Enzymol.* **117**, 301 (1985).
[2] M. L. Johnson and L. M. Faunt, *Methods Enzymol.* **210**, 1 (1992).
[3] M. L. Johnson, *Methods Enzymol.* **240**, 1 (1994).
[4] D. G. Watts, *Methods Enzymol.* **240**, 23 (1994).
[5] M. Straume and M. L. Johnson, *Methods Enzymol.* **210**, 87 (1992).
[6] G. S. Adair, *Proc. R. Soc.* **109A**, 292 (1925).

to describe specific microscopic interactions that take place during the binding process. Examples of specific microscopic interactions that may be altered on ligand binding are the formation of salt bridges, proton binding, phosphate binding, conformational changes, electrostatic interactions, etc. Any detailed study of molecular ligand-binding mechanisms clearly needs specifically to include a more detailed description of the binding process than is afforded by the macroscopic Adair formulation.

The Adair formulas can easily be derived by simple algebraic methods. However, these methods become difficult for ligand-binding problems with multiple microscopic steps. The simple approach is to use the partition function for the macromolecular binding problem, Ξ. In this context, the partition function is simply a statement of the total probability of all the microscopic states of the macromolecule. Noting that the total probability need not be normalized to unity is important; it may simply be expressed in terms of relative statistical weights. Once the partition function is defined then the mean number of ligands bound can be evaluated on the basis of Hill[7]:

$$\overline{N} = \frac{\partial \ln \Xi}{\partial \ln X} = \frac{X}{\Xi} \frac{\partial \Xi}{\partial X} \tag{3}$$

The mean number of bound ligands is simply the fractional saturation multiplied by the total number of binding sites.

The formulation of any ligand-binding problem by statistical thermodynamics is a four-step process. First, define all the possible states, or configurations, of the macromolecule. Second, define the energy levels of each accessible state of the macromolecules. These energy levels define the thermodynamic "rules" that the macromolecule must follow to change states. Third, use the information defined in the first two steps to create the partition function. The final step is to apply Eq. (3) to the partition function. As is seen below, the third and final steps of this process are simply algebra that can be easily done by a computer. Thus, deriving a complex ligand-binding equation simply involves creating a statement of all the possible configurations of the macromolecule and the thermodynamic rules for the interconversion of the configurations.

Example 1

Example 1 is quite simple: a noncooperative two-binding site macromolecule. It is used to introduce the use of statistical thermodynamics in

---

[7] T. L. Hill, "An Introduction to Statistical Thermodynamics." Addison-Wesley, London, 1960.

the derivation of ligand-binding equations. It also provides a simple example of the statistical factors that exist in all complex ligand-binding problems.

Consider a simple ligand-binding example in which the macromolecule contains two identical noninteracting binding sites, α and β. The Adair formulation for the mean number of ligands bound to a two-site macromolecule is

$$\overline{N} = \frac{K_1[X] + 2K_2[X]^2}{1 + K_1[X] + K_2[X]^2} \tag{4}$$

The question is, how do K_1 and K_2 relate to microscopic binding constants of the individual binding sites? The answer to this question is seen by reformulating Eq. (4) as a function of the individual binding to each subunit.

The macromolecule in example 1 can exist in four configurations: unliganded; with only the α site occupied; with only the β site occupied; and with both sites occupied. The choice of thermodynamic reference state is arbitrary. For convenience the reference state is taken to be the unliganded species. Thus, the relative free energy levels for the four states are zero, ΔG_α, ΔG_β, and $\Delta G_{\alpha\beta}$. The probability of each of the i states is evaluated from its free energy level by a Boltzmann distribution:

$$\text{Probability}_i \propto e^{-\Delta G_i/RT}[X]^{m_i} \tag{5}$$

where m_i is the number of ligands bound to the ith form. For the unliganded species the free energy level and number of ligands bound are both zero and thus the corresponding statistical weight (i.e., the unnormalized probability) is unity. The corresponding statistical weight of the α only species is

$$e^{-\Delta G_\alpha/RT}[X] = K_\alpha[X] \tag{6}$$

By analogy the statistical weights of the other macromolecular species are given by $K_\beta[X]$ and $K_{\alpha\beta}[X]^2$. The partition function is simply the sum of all the statistical weights:

$$\Xi = 1 + K_\alpha[X] + K_\beta[X] + K_{\alpha\beta}[X]^2 \tag{7}$$

Partition functions written as Eq. (7) are sometimes called *binding polynomials*. The application of Eq. (3) to the partition function given in Eq. (7) provides the binding isotherm in terms of the individual site binding constants.

$$\overline{N} = \frac{(K_\alpha + K_\beta)[X] + 2K_{\alpha\beta}[X]^2}{1 + (K_\alpha + K_\beta)[X] + K_{\alpha\beta}[X]^2} \tag{8}$$

A simple comparison of Eqs. (4) and (8) is informative. Obviously, for this example $K_2 = K_{\alpha\beta}$ and $K_1 = K_\alpha + K_\beta$. If the microscopic binding constants to the two binding sites are identical, $K_\alpha = K_\beta$, then it is clear that the macroscopic Adair binding constant for the first binding site is twice the microscopic site binding constant. This factor of two is a statistical factor that arises from the two different ways that the macromolecule can bind a ligand: to the α site or to the β site. These statistical factors are obviously important for the microscopic interpretation of the macroscopic Adair binding constants. The corresponding statistical factors for the stepwise Adair constants are 2 and 1/2.

The statistical factors for a four-binding site model, such as oxygen binding to hemoglobin tetramers, are 4, 6, 4, and 1 for the product Adair constants. The corresponding statistical factors for the stepwise Adair constants are 4, 3/2, 2/3, and 1/4. The product statistical factors are simply the products of the stepwise statistical factors, analogous to the product and stepwise Adair constants. Note the progressions that occur in the numerator and denominator of the statistical factors for the stepwise Adair binding constants.

What is the average energy level for the singly liganded macromolecule? It is not simply the arithmetic average of the free energy changes for binding to the α and β sites! Consider the comparison of Eqs. (4) and (8) above. Clearly, the corrected macroscopic binding constant $K_1' = (K_\alpha + K_\beta)/2$, where the 2 is to remove the statistical factor. The macroscopic equilibrium constant for the binding of a single ligand is the average of the equilibrium constants for the binding to each of the binding sites, not the average of the free energy changes. Or, expressed as free energy changes:

$$e^{-\Delta G_1'/RT} = \frac{e^{-\Delta G_\alpha/RT} + e^{-\Delta G_\beta/RT}}{2} \tag{9}$$

Example 2

The previous example did not consider any type of interaction between the binding sites. Example 2 introduces one possible mechanism to describe cooperative binding. Consider example 1 with an additional thermodynamic interaction, ΔG_f, between the binding sites that is sensitive to the state of ligation of the sites. Furthermore, define this thermodynamic interaction to be of first order[8] if it is eliminated by a single ligand binding to either binding site. An analogous thermodynamic interaction, ΔG_s, is of second

---

[8] G. Weber, *Proc. Natl. Acad. Sci. U.S.A.* **81**, 7098 (1984).

TABLE I
STATES AND RULES FOR EXAMPLE

| State | Thermodynamic constraints | |
	First-order model	Second-order model
Unliganded	ΔG_f	ΔG_s
α Site liganded	ΔG_α	$\Delta G_\alpha + \Delta G_s$
β Site liganded	ΔG_β	$\Delta G_\beta + \Delta G_s$
Both liganded	$\Delta G_\alpha + \Delta G_\beta$	$\Delta G_\alpha + \Delta G_\beta$

order[8] if it is eliminated only by ligands binding simultaneously to both binding sites. Table I lists the "states" and "rules" for these models.

The two-site Adair equation [i.e., Eq. (4)] can easily be generated with Eq. (3) and a partition function of the form

$$\Xi = 1 + K_1[X] + K_2[X]^2 \tag{10}$$

where the K values are the macroscopic binding constants uncorrected for statistical factors. This partition function can be generalized for a binding model with any number of sites in terms of the microscopic interaction energies of each species:

$$\Xi = \sum_i e^{-\Delta G_i/RT}[X]^{m_i} \tag{11}$$

where m_i is the number of ligands bound to the ith form and ΔG_i is the ith free energy difference between the ith state and the reference (i.e., unliganded) state. Equation 11 can also be grouped by the number of ligands bound:

$$\Xi = \sum_j \xi_j[X]^j = \xi_0 + \xi_1[X] + \xi_2[X]^2 + \cdots \tag{12}$$

where ξ_j is the subsystem partition function for states with j ligands bound. These subsystem partition functions are summations over all the macromolecular states with the particular number of ligands bound. A comparison of Eq. (10) and (12) shows that the uncorrected macroscopic Adair binding constants can be evaluated from the subsystem partition functions as

$$K_j = \xi_j/\xi_0 \tag{13}$$

Thus, any microscopic thermodynamic binding model can be formulated as a macroscopic Adair equation with Adair constants evaluated by Eq. (13).

For the first-order case of the present example, the subsystem partition functions are

$$
\begin{aligned}
\xi_0 &= e^{-\Delta G_f/RT} \\
\xi_1 &= e^{-\Delta G_\alpha/RT} + e^{-\Delta G_\beta/RT} \\
\xi_2 &= e^{-(\Delta G_\alpha + \Delta G_\beta)/RT}
\end{aligned}
\tag{14}
$$

For the second-order case of the current example the subsystem partition functions are

$$
\begin{aligned}
\xi_0 &= e^{-\Delta G_s/RT} \\
\xi_1 &= e^{-(\Delta G_\alpha + \Delta G_s)/RT} + e^{-(\Delta G_\beta + \Delta G_s)/RT} \\
\xi_2 &= e^{-(\Delta G_\alpha + \Delta G_\beta)/RT}
\end{aligned}
\tag{15}
$$

An alternative method to specify a model is to define a configuration matrix. In this matrix each row corresponds to a particular state, or configuration, of the macromolecule. It is critical that all the states be represented in configuration matrices. Each column refers to constraints of a particular type. The individual elements of this matrix are the numbers of the constraints of the particular type corresponding to the distinct state of the macromolecule. For example, the configuration matrix for the present two-site second-order coupling model is presented in Table II. This matrix has a column for each of the thermodynamic constraints: ΔG_s, ΔG_α, and ΔG_β. The free energy level of an individual state is simply the sum of the number of each constraint times the free energy value of the particular constraint.

Two other columns are added to the configuration matrix for convenience: one containing the total number of ligands bound and one for a statistical (i.e., redundancy) factor. For the present model all the states are unique. However, for complex models it is common for several states to be indistinguishable. These indistinguishable states can be included either

TABLE II
TWO-SITE SECOND-ORDER CONFIGURATION MATRIX

State	Number of ligands bound	Statistical factor	Thermodynamic constraints		
			ΔG_s	ΔG_α	ΔG_β
Unliganded	0	1	1	0	0
α site liganded	1	1	1	1	0
β site liganded	1	1	1	0	1
Both liganded	2	1	0	1	1

TABLE III

TETRAMERIC MWC CONFIGURATION MATRIX

State	Number of ligands bound	Statistical factor	Thermodynamic constraints		
			L	K_R	C
T_0	0	1	1	0	0
T_1	1	4	1	1	1
T_2	2	6	1	2	2
T_3	3	4	1	3	3
T_4	4	1	1	4	4
R_0	0	1	0	0	0
R_1	1	4	0	1	0
R_2	2	6	0	2	0
R_3	3	4	0	3	0
R_4	4	1	0	4	0

as multiple identical rows in the matrix or as a statistical factor corresponding to the specific number of duplicates.

The derivation of the functional form for ligand binding to a macromolecule is simply to define the "states" (i.e., configurations) of the macromolecule and the "rules" (i.e., the thermodynamic constraints) that define each state. Once this has been accomplished then the functional form is totally defined as an Adair equation with uncorrected macroscopic Adair constants as defined in Eq. (13).

Example 3

As another example of the statistical thermodynamic approach Table III presents the configuration matrix for the allosteric model of Monod, Wyman, and Changeux[9,10] (MWC) as applied to a four-site macromolecule such as human hemoglobin. In this model hemoglobin is assumed to exist in two conformational states, R and T. Within the R-state tetramers each of the four binding sites is assumed to be identical, with an oxygen (ligand) affinity of K_R. For the T-state macromolecule each of the binding sites is assumed to be identical, with an oxygen affinity of K_T. In the MWC model K_T is expressed as C times K_R. Thus, a K_T constraint is incorporated into Table III as simultaneous C and K_R constraints. The isomerization constant to convert from the R state to the T state is L. Thus, the MWC model

[9] J. Monod, J. P. Wyman, and J. J. Changeux, *J. Mol. Biol.* **12,** 88 (1965).
[10] G. K. Ackers and M. L. Johnson, *J. Mol. Biol.* **147,** 559 (1981).

contains three adjustable parameters (L, K_R, and C), which correspond to the statistical thermodynamic constraints of the model.

The MWC model has a total of 32 thermodynamic states. There is one unliganded T state and one fully liganded T state, T_0 and T_4, respectively. There are four ways (e.g., one for each binding site) to form a T-state tetramer with only a single oxygen bound, T_1. There are also four ways (e.g., one for each unoccupied binding site) to form a T-state tetramer with three oxygens bound, T_3. Similarly, there are six ways to distribute two oxygens within the four binding sites of doubly oxygenated T-state tetramers, T_2. The 16 R-state configurations are analogous to the T-state configurations. Only 10 of the 32 states are distinguishable. Thus, Table III contains only 10 states, some of which have nonunity statistical factors. These statistical factors literally specify how many times to count the particular state when generating the partition function.

Example 4

Many biological macromolecules can bind multiple different ligands to the same binding site. This example is a macromolecule with one binding site that can bind either of two ligands, X or Y.

Here the partition function has three terms: one for the unliganded macromolecule, one for X bound to the macromolecule, and one for Y bound to the macromolecule.

$$\Xi = 1 + K_X[X] + K_Y[Y] \tag{16}$$

In Eq. (16) K_X and K_Y are the binding constants for X and Y. Using Eq. (3) this can be transformed to express the mean number of ligands bound.

$$\overline{N_X} = \frac{K_X[X]}{1 + K_X[X] + K_Y[Y]} \tag{17}$$

$$\overline{N_Y} = \frac{K_Y[Y]}{1 + K_X[X] + K_Y[Y]}$$

Example 5

A more realistic multiple-ligand problem is when there are two ligands, X and Y, that can bind to either or both sites on a two-site macromolecule. In this example each of the two sites can exist in three possible states: unliganded, liganded with X, or liganded with Y. This gives a total of nine possible configurations. These configurations are denoted as $\alpha\{\#\}\beta\{\#\}$, where $\#$ can be – for no ligand, and X or Y if it is liganded.

TABLE IV
CONFIGURATION MATRIX FOR TWO LIGANDS AND TWO IDENTICAL
BINDING SITES

State	Number of ligands bound	Statistical factor	Thermodynamic constraints	
			ΔG_X	ΔG_Y
$\alpha\{-\}\beta\{-\}$	0	1	0	0
$\alpha\{-\}\beta\{X\}$	1	1	1	0
$\alpha\{-\}\beta\{Y\}$	1	1	0	1
$\alpha\{X\}\beta\{-\}$	1	1	1	0
$\alpha\{X\}\beta\{X\}$	2	1	2	0
$\alpha\{X\}\beta\{Y\}$	2	1	1	1
$\alpha\{Y\}\beta\{-\}$	1	1	0	1
$\alpha\{Y\}\beta\{X\}$	2	1	1	1
$\alpha\{Y\}\beta\{Y\}$	2	1	0	2

Tables IV and V present the configuration matrix where the binding sites are identical but the ligands have different binding affinities for X and Y. Tables IV and V are functionally identical. In Table V the three redundant configurations are replaced by setting the corresponding redundancy factors to two. The partition function for this example is presented as Eq. (18). The terms from left to right correspond to the configurations in Table V from top to bottom.

$$\Xi = e^0 + 2e^{-\Delta G_X/RT}[X] + 2e^{-\Delta G_Y/RT}[Y] + e^{-2\Delta G_X/RT}[X]^2$$
$$+ 2e^{-(\Delta G_X+\Delta G_Y)/RT}[X][Y] + e^{-2\Delta G_Y/RT}[Y]^2$$
$$\Xi = 1 + 2K_X[X] + 2K_Y[Y] + (K_X)^2[X]^2$$
$$+ 2K_XK_Y[X][Y] + (K_Y)^2[Y]^2 \tag{18}$$

TABLE V
CONFIGURATION MATRIX FOR TWO LIGANDS AND TWO IDENTICAL BINDING SITES

State	Number of ligands bound	Statistical factor	Thermodynamic constraints	
			ΔG_X	ΔG_Y
$\alpha\{-\}\beta\{-\}$	0	1	0	0
$\alpha\{-\}\beta\{X\}$	1	2	1	0
$\alpha\{-\}\beta\{Y\}$	1	2	0	1
$\alpha\{X\}\beta\{X\}$	2	1	2	0
$\alpha\{X\}\beta\{Y\}$	2	2	1	1
$\alpha\{Y\}\beta\{Y-\}$	2	1	0	2

The fractional saturations of X and Y are shown in Eq. (19).

$$\overline{N_X} \frac{2K_X[X] + 2(K_X)^2[X]^2 + 2K_XK_Y[X][Y]}{1 + 2K_X[X] + 2K_Y[Y] + (K_X)^2[X]^2 + 2K_XK_Y[X][Y] + (K_Y)^2[Y]^2}$$

$$\overline{N_Y} = \frac{2K_Y[Y] + 2(K_Y)^2[Y]^2 + 2K_XK_Y[X][Y]}{1 + 2K_X[X] + 2K_Y[Y] + (K_X)^2[X]^2 + 2K_XK_Y[X][Y] + (K_Y)^2[Y]^2}$$

$$(19)$$

It is clear from this that the fractional saturation of X depends on Y and thus Y depends on X.

Example 6

It is known that human hemoglobin undergoes a dimer-to-tetramer self-association in the concentration range where oxygen-binding experiments are usually performed.[11] The dimers contain one α and one β subunit, each of which binds oxygen. The tetramers contain two of each subunit and bind four oxygens. The binding properties of a subunit within a dimer are different from the binding properties of a subunit within a tetramer.[12] Furthermore, the α chains may bind differently than the β chains.[13]

One method to account for simultaneous ligand binding and self-association was used by Ackers and Halvorson.[11] They derived an explicit binding equation that contains two Adair binding constants for the oxygen binding to the dimeric species, four Adair binding constants for the oxygen binding to the tetrameric species, and a dimer–tetramer association constant for the unliganded species for a total of seven macroscopic equilibrium constants.

For more complex microscopic models one approach that we have used is to consider the partition functions for the dimeric and tetrameric species separately.[14] A configuration matrix is generated for each and the microscopic thermodynamic constraints are translated into the dimer and tetramer Adair constants. These Adair constants and the unliganded dimer–tetramer association constant are used with the explicit expression of Ackers and Halvorson[11] to evaluate the fractional saturation at any oxygen and hemoglobin concentration.

A more general approach is to write the fractional saturation as the

[11] G. K. Ackers and H. R. Halvorson, *Proc. Natl. Acad. Sci. U.S.A.* **71**, 4312 (1974).
[12] F. C. Mills, M. L. Johnson, and G. K. Ackers, *Biochemistry* **15**, 5350 (1976).
[13] D. H. Atha, M. L. Johnson, and A. F. Riggs, *J. Biol. Chem.* **254**, 12390 (1979).
[14] M. L. Johnson, *Adv. Biophys. Chem.* **5**, 179 (1995).

weight average of all of the individual oligomers' fractional saturations, as
in Eq. (20) for the present example.

$$\overline{Y} = f_{dimer}\overline{Y}_2 + f_{tetramer}\overline{Y}_4$$

$$f_{dimer} = \frac{[dimer]\Xi_2}{[dimer]\Xi_2 + 2^0K_2[dimer]^2\Xi_4} \tag{20}$$

$$f_{tetramer} = \frac{2^0K_2[dimer]^2\Xi_4}{[dimer]\Xi_2 + 2^0K_2[dimer]^2\Xi_4}$$

The factor of two in the second term on the right of Eq. (20) is because
the fractions are in mass units and the dimer–tetramer association constant
is in molar units.

Equation (20) can be used to evaluate the fractional saturation if, and
only if, the concentration of unliganded dimer, [dimer], is known. The
concentration of unliganded dimer, [dimer], depends on both the ligand
(e.g., oxygen) and the macromolecule (e.g., hemoglobin) concentrations.
It also depends on the Adair binding constants and the values of the
thermodynamic constraints of the model. Thus, the value of [dimer] is not
a constant!

The total mass of the macromolecule, for this example, is shown in
Eq. (21):

$$\text{Total mass} = [dimer] + 2^0K_2[dimer]^2 \tag{21}$$

For any specific set of thermodynamic constraints, Adair constants, $^0K_2$,
and oxygen and hemoglobin concentrations Eq. (21) has only one unknown,
[dimer]. For the present example Eq. (21) is a quadratic equation and thus
has an explicit solution. It is this solution that is the basis of the Ackers
and Halvorson[11] equation. However, for the general case of any arbitrary
self-association reaction the analogous analytical solution does not exist.
However, a numerical solution can always be found. The [dimer] can be
evaluated as a numerical root of Eq. (21) for any specific set of thermodynamic constraints, Adair constants, $^0K_2$, and oxygen and hemoglobin concentrations. This numerical value of [dimer] can subsequently be used in
Eq. (20) to evaluate fractional saturation of oxygens that corresponds to
the specific set of thermodynamic constraints, Adair constants, $^0K_2$, and
oxygen and hemoglobin concentrations.

Conclusion

The basic goal of this chapter is to present a simple statistical thermodynamic approach to the derivation of equations to describe ligand-binding

experiments. The advantage of this approach is that it can be used to translate virtually any conceptual mechanistic model into a mathematical formula. The statistical thermodynamic approach requires the investigator to define the conceptual model in terms of "states" and "rules." The states are simply all the microscopic configurations in which the macromolecule can exist. The rules are simply statements of the free energy levels corresponding to each accessible state of the macromolecule. Once these have been defined then a partition function can be written and Eq. (3) applied to obtain an equation to describe the binding of ligand.

Acknowledgments

The authors acknowledge the support of the National Science Foundation Science and Technology Center for Biological Timing at the University of Virginia (NSF DIR-8920162), the Clinical Research Center at the University of Virginia (NIH RR-00847), the University of Maryland at the Baltimore Center for Fluorescence Spectroscopy (NIH RR-08119), and the National Institutes of Health (Grant GM-35154).

[8] Calculation of Entropy Changes in Biological Processes: Folding, Binding, and Oligomerization

By L. Mario Amzel

Introduction

In contrast to processes involving the formation of covalent bonds, the free energy changes in protein folding and binding are small, usually in the range of 5–15 kcal/mol. This free energy change is the difference between the two larger quantities ΔH and $T\Delta S$. The change in enthalpy ΔH results from the differences in the energies of the bonds present in the initial and the final state (protein–protein, solvent–solvent, protein–solvent, ligand–solvent, ligand–protein, etc.). The change in entropy ΔS reflects changes in the configurational entropy of the protein and the ligand, as well as changes in the organization of the solvent. In binding and oligomerization, ΔS includes, in addition, changes in translational and rotational degrees of freedom.

The changes in configurational, rotational, and translational entropy are amenable to treatments using molecular mechanics calculations.[1–10] In

[1] L. M. Amzel, *Proteins Struct. Funct. Genet.* **28,** 144 (1997).
[2] J. A. D'Aquino, J. Gomez, V. J. Hilser, K. H. Lee, L. M. Amzel, and E. Freire, *Proteins Struct. Funct. Genet.* **25,** 143 (1996).

contrast, solvent effects are difficult to evaluate quantitatively by calculations involving pairwise interactions and are better estimated by empirically parameterized equations.[5,11] In this chapter we discuss methods that can be used to evaluate changes in entropy in protein folding, binding, and oligomerization. Only methods to calculate changes in configurational entropy are presented. The empirical calculation of solvent effects is outside the scope of this presentation.

General Considerations and Formulas

The unfolded state of a protein consists of an ensemble of different structures, each having a different set of values for the dihedral angles that define the conformation of the backbone and side chain of every residue. Folding of the protein involves a large reduction in the number of conformations accessible to the side chains and the backbone. This reduction results in a large negative change in the entropy, ΔS_{conf}, that is the main unfavorable effect that needs to be overcome by the folding protein. Precise estimation of ΔS_{conf} is critical for accurate predictions of protein stability because small variations in entropy (for each residue) result in large changes in the predicted overall stability.

Changes of entropy in binding and oligomerization involve mainly changes in the configurational entropy of side chains, even though immobilization of flexible loops is sometimes also observed. Binding of flexible compounds involves, in addition, changes in the configurational entropy of the ligand. When the ligand is, for example, a short peptide (3–15 residues), the unbound peptide does not have a well-defined structure: It adopts a large number of conformations similar to those observed in the unfolded state of proteins. The degrees of freedom that are lost on binding include

[3] S. J. Leach, G. Nemethy, and H. A. Scheraga, *Biopolymers* **4,** 369 (1966).

[4] K. H. Lee, D. Xie, E. Freire, and L. M. Amzel, *Proteins Struct. Funct. Genet.* **20,** 68 (1994).

[5] K. P. Murphy, D. Xie, K. C. Garcia, L. M. Amzel, and E. Freire, *Proteins Struct. Funct. Genet.* **15,** 113 (1993).

[6] K. P. Murphy, D. Xie, K. S. Thompson, L. M. Amzel, and E. Freire, *Proteins Struct. Funct. Genet.* **18,** 63 (1994).

[7] M. S. Searle and D. H. Williams, *J. Am. Chem. Soc.* **114,** 10690 (1992).

[8] I. Z. Steinberg and H. A. Scheraga, *J. Biol. Chem.* **238,** 172 (1963).

[9] J. Janin, *Proteins Struct. Funct. Genet.* **21,** 30 (1995).

[10] G. Nemethy and H. A. Scheraga, *Biopolymers* **3,** 155 (1965).

[11] I. Luque, J. Gomez, N. Semo, and E. Freire, *Proteins Struct. Funct. Genet.* **30,** 74 (1998).

side-chain as well as main-chain dihedrals. The change in entropy includes the following terms:

$$\Delta S = \Delta S_{conf,\,mc} + \Delta S_{conf,\,sc} + \Delta S_{rt}$$

where $\Delta S_{conf,\,mc}$ and $\Delta S_{conf,\,sc}$ are the changes in the configurational entropies of the main chain and the side chain, respectively. ΔS_{rt} is the change in rotational and translational entropy. The detailed estimation of these terms is discussed in the following sections.

The loss of backbone and side-chain configurational entropy in all these processes has two main components: (1) restriction in the number of conformers accessible to each dihedral; and (2) restriction in the amplitude of the movements within the energy well corresponding to each conformer. This partition of ΔS_{conf} results in two general cases: one in which the process restricts only the number of conformers (case 1), and one in which the movement within the energy wells is also restricted (case 2). Restriction of accessible conformers can also result in enthalpic effects. In case 1 above, the resulting conformer could be in a strained configuration and result in a net unfavorable ΔH. In case 2 a tighter energy well results in an oscillator with higher average energy. This change, however, is in general small because the energy ("zero point") of the oscillator is dependent on the square root of the force constant, and burying a residue produces only small changes in the tightness of the energy well.

Estimation of ΔS_{conf} requires the calculation of the full energy profile of the system: the value of the energy for all possible coordinates of all atoms. The entropy loss can then be calculated as

$$\Delta S_{conf} = S_{folded} - S_{unfolded}$$

where

$$S_X = \sum_{\text{all states}} p_i \ln p_i \tag{1}$$

with

$$p_i = \frac{\exp(-E_i/kT)}{\sum \exp(-E_i/kT)} \tag{2}$$

In most cases, carrying out these calculations over all possible states is computationally too onerous. On the other hand, it is clear from the preceding expressions that low-energy states make larger contributions to the value of the entropy. One good approximation that can be used to make the computation tractable consists of calculating the energy profile over all values of the dihedral angles only; such a procedure would use a manageable number of states while sampling the largest number of low-energy conformations. However, simply calculating the energy as a function of the dihedrals does not result in a realistic energy profile. In particular, it exaggerates the number of high-energy conformations because by simply rotating around a dihedral angle, numerous short distance contacts are encountered that result in large positive repulsion terms. In many cases, conformations that show unfavorable interactions at a given dihedral value can be relaxed locally by minimizing the energy with respect to the other conformational variables, in particular the valence bond angles.[4] Because small changes in the valence bond angles produce only small changes in the corresponding energy terms [$\Delta E = k_\theta(\theta - \theta_o)^2$, with k_θ typically equal to 0.03–0.05 kcal/mol · degree], the total energy at an apparently unfavorable dihedral can actually be low.*

Calculations

General Methods

The methods described in this section follow closely those proposed by Lee et al.[4] and D'Aquino et al.[2] Estimation of entropies requires the evaluation of the energy of the system for all relevant conformations. It is clear that if accurate energies for the complete system could be calculated in a simple manner it would also be possible to estimate the total energy of the system and therefore changes in enthalpy [because $\Delta(PV)$ is small]. This is not the case and, therefore, the changes in energy are calculated as one or more dihedral angles are scanned. The energy of the system as a function of a limited set of dihedrals is what we call an *energy profile*.

Estimation of Energy Profiles. For each dihedral (either backbone or side chain) a full energy profile is calculated every 1° to 10° for all combinations of dihedrals. Energies are calculated with the potential functions (AMBER/OPLS) proposed by Weiner et al.[12] and by Jorgensen and Tirado-Rives.[13] Electrostatic terms—calculated with a dielectric constant of 78—include partial charges. For each combination of dihedrals the structures were allowed to relax (taken to the equilibrium configuration) by minimizing the energy as a function of the valence bond angles associated with the relevant dihedrals. Initially, minimizations were performed with a specially written code using conjugate gradients.[4] More recently,

---

* The repulsion terms are proportional to r^{-12} while the valence bond angle terms are proportional to $(\Delta r)^2$ (for small distortions).

[12] S. J. Weiner, P. A. Kollman, D. A. Case, V. C. Singh, C. Ghio, G. Alagona, S. Profeta, and P. Weiner, *J. Am. Chem. Soc.* **106**, 765 (1984).

[13] W. L. Jorgensen and J. Tirado-Rives, *J. Am. Chem. Soc.* **110**, 1657 (1988).

calculations have been carried out with the program X-PLOR.[14] With this program, the energy is minimized with respect to the coordinates of all atoms defining the relevant dihedrals. During minimization the dihedrals are constrained to their selected values by the CONSTRAIN DIHEDRAL[†] option of X-PLOR, with a constant of 400 kcal/mol · rad.[2] After convergence of the minimization the energy is recalculated without the additional term.

Estimation of Changes in Configurational Entropy. For the purpose of calculating entropy differences, the configurational entropy of the native and denatured states can be estimated by summing all terms in Eq. (1) over the entire energy profile, using the probabilities of each state estimated from Eq. (2). For profiles calculated with fine enough intervals, this integration can provide reliable estimates. Fortunately, we are interested only in entropy differences and therefore it is not necessary to evaluate accurately the configurational entropy of each state: We need only evaluate the contributions of those effects that change during the transition.

Therefore, it is sometimes convenient to think about the configurational entropy as having two contributions: one coming from the population of different conformers, the other from the population of oscillation (libration) states within each conformer.

$$S_{\text{configurational}} = S_{\text{conformations}} + S_{\text{oscillations}} \tag{3}$$

This separation is particularly useful if, for example, there is no change in the shape of the well during the process (case 1). For unfolding, for example,

$$\Delta S_{\text{configurational}} = S_{\text{configurational, unfolded}} - S_{\text{configurational, folded}}$$

Equation (3) can be used to obtain

$$\Delta S_{\text{configurational}} = \Delta S_{\text{conformations}} + \Delta S_{\text{oscillations}} \tag{4}$$

For the calculation of $\Delta S_{\text{conformations}}$, the sum in Eq. (1) is carried out over a small number of conformers for each dihedral. The conformers are defined as the main wells in the energy profile diagrams. For example, for a valine side chain the conformers are *gauche*[+], *trans,* and *gauche*[−]. The value of the probability p_i for each conformer is estimated by adding the normalized probabilities [calculated with Eq. (2)] of all the values of the dihedral angles corresponding to the given conformer. This calculation provides the value of $\Delta S_{\text{conformations}}$ in Eq. (4). The definition of the conformers for backbone and side-chain dihedrals is discussed below.

[14] A. T. Brunger, "X-PLOR: A System for Crystallography and NMR," version 3.1. Yale University Press, New Haven, Connecticut, 1992.
[†] This option adds a quadratic constraint of the form $E = (1/2) k(\chi - \chi_0)^2$ to the total energy.

If, on folding (or binding), there is a significant change in the shape of the energy well (case 2), it is necessary to include the term $\Delta S_{\text{oscillations}}$, which reflects this change. To estimate this term, the energy at the bottom of each potential well can be approximated by a quadratic equation of the form

$$E = E_{\min} + (1/2)k(\chi - \chi_{\min})^2 \tag{5}$$

After finding the value of the constant k by adjusting Eq. (5) to the energy in the energy profile within 10–20° from $\chi_{\min}$, this value of k can be used to calculate the frequency ω of the rotational oscillator and with it, θ_ω, the Einstein temperature of the oscillator.[‡] S_i, the contribution of the kinetic energy states to the entropy of the ith well, can then be calculated with the expression

$$S_i = R\{[(\theta_i/T)\exp(\theta_i/T) - 1] - \ln[1 - \exp(-\theta_i/T)]\}$$

The vibrational entropy due to the oscillations is calculated as

$$S_{\text{vib}} = \Sigma\, w_i S_i$$

where $w_i = \exp(-E_{\min,\,i}/RT)/\Sigma\,\exp(-E_{\min,\,i}/RT)$ is the Boltzmann factor for each well. The contributions to the entropy calculated in this manner are added to the conformational entropy calculated as described above to provide the complete configurational entropy.

Alternatively, $\Delta S_{\text{configurational}}$ can be calculated by summing over all the sampled points in the energy profile. For the differences to be meaningful the energy profiles must be calculated with identical sampling. That is, both profiles must contain the same number of points. If that is not convenient, the value of ΔS must be corrected for the difference in the number of states.

$$\Delta S = S_2 - S_1 + RT\ln(n_2/n_1)$$

where n_1 and n_2 are the number of points in the energy profiles of states 1 and 2.

Calculation of Main-Chain Entropy Changes

One way of calculating the backbone entropy of the unfolded chain was presented by D'Aquino *et al.*[2] For these calculations, dipeptides of the form N-acetyl-(aminoacyl)-N'-methylamide were used. Calculations were carried out for dipeptides containing glycine, alanine, valine, leucine, and

---

[‡] The moment of inertia $I = \Sigma\, m_i r_i^2$ is calculated only for the rotation around the axis of the dihedral angle; therefore, the r_i values are distances to the axis.

lysine. Side chains were set to all-*trans* conformations. Energies were calculated for all the dipeptides for all combinations of the main-chain dihedrals ϕ and ψ every 10°, between 0° and 360°. The energy of every ϕ and ψ combination was minimized as described above, using the program X-PLOR. The CONSTRAIN DIHEDRAL option with a force constant of 400 kcal/mol was used to maintain the dihedrals at their specified values. Energies of the final conformations were calculated without including this additional term.

The backbone entropy of the folded chain was estimated by calculating the energy profile for a nine-residue helix.[2] The calculations were carried out for two helices, $(Ala)_9$ and $(Gly)_9$. Structures corresponding to all combinations of ϕ and ψ within a 20° interval from the helical conformation were generated and their energies minimized as for the dipeptides. The entropy per residue was obtained with the equations described above.

The entropy of long unfolded peptides cannot be obtained simply by summing the entropy per residue over all residues in the peptide: It must be corrected to account for the restriction in the number of conformations that results from the clashes that occur from the peptide folding back on itself. One way proposed to quantitate this excluded volume effect is to consider the fraction of nonoverlapping conformations in a cubic lattice. The total number of conformations for the case in which a residue is not allowed to occupy the same volume as the previous one (nearest-neighbor excluded volume) is $6 \times 5^{n-1}$. The total number of allowed conformations excluding all possible clashes is $n^\gamma \times \mu^n$, where $\gamma = 1/6$ and $\mu = 4.68$.[2] The fraction of accessible conformations for a chain of n residues, $F_{acc}(n)$, is

$$F_{acc}(n) = (n^\gamma \times \mu^n)/(6 \times 5^{n-1})$$

Results of Calculations. Calculations were carried out for selected residues in both dipeptides and in helical conformations.[2] For both alanine and glycine the calculations in the helical conformation give similar results: alanine, 6.68 e.u. (1 e.u. = 1 cal/K · mol); glycine, 6.66 e.u. The calculations for the dipeptide, representing the unfolded state, give 13.19 e.u. for glycine and 11.23 e.u. for alanine. Thus, the $\Delta S_{conf, helix \rightarrow unfolded}$ for glycine and alanine are −6.51 and −4.57 e.u., respectively. The difference in stability between a helix containing an alanine residue and one containing a glycine is then 1.97 e.u. The increased stability of the alanine-containing helix is due entirely to the entropic stabilization of the unfolded state in the glycine-containing peptide. This result is in agreement with the experimental data on the stability of variants of the coiled-coil helix of GCN4.[2] In this system, a calorimetric study showed that the added stability of the alanine-containing helix was all due to entropic effects and the magnitude of the stabilization was found to be 2 e.u.

Calculation of Side-Chain Entropy Changes

The proposed methods can be used to estimate the configurational entropy of a side chain in the specific structure in which the side chain occurs.[4] During the calculation of the energy profiles of a given side chain, all other atoms in the structure are kept in their original positions. Starting with the side chain in the observed conformation, all side-chain dihedrals are set to their *trans* conformation. Starting at this conformation, the χ angles are varied at constant intervals ($1–10°$) and the resulting structures are energy minimized as described above, allowing only the atoms of the side chain to move. To reduce the computational requirements, for surface side chains only χ_1 and χ_2 are scanned in this manner; for longer side chains, $R \ln 3$ per χ angle is added to the final entropy. Using the energy profiles, the entropy is calculated by Eq. (2). For calculations involving only conformational entropy differences, the sum is carried out over the major conformations of the side chain. Populations for dihedrals around a single bond between two sp^3 carbons are estimated for three conformers: *trans* ($120°$ to $240°$), and *gauche*$^+$ ($0°$ to $-120°$), and *gauche*$^-$ ($0°$ to $120°$). For dihedrals around single bonds between an sp^3 and an sp^2 atom, there are six conformers to be considered ($30°$, $90°$, $150°$, $210°$, $270°$, and $330°$). If the planar group containing the sp^2 atom is symmetrical (i.e., phenylalanine, tyrosine, etc.) a symmetry number of two is used resulting in a reduction of the conformational entropy by $R \ln 2$.

Results of Calculations. By this method Lee *et al.*[4] calculated the conformational entropy of all side chains in a helical environment. The calculations were carried out for nonapeptides of sequence $(Ala)_4–AA–(Ala)_4$, where AA represents one of the amino acids. The results of these calculations agree with the entropies estimated from the distribution of side-chain conformers in the Protein Data Bank.[15] For the estimation of differences in entropy between helix and unfolded chain, the entropies of the unfolded conformations of Creamer and Rose[16] were used. These changes in entropy, corrected for solvent effects, can be compared with several experimental measurements. The results of these calculations show good correlation with experimental estimates of the contribution of the side-chain entropy to the free energy of helix unfolding.[4]

[15] S. D. Pickett and M. J. E. Sternberg, *J. Mol. Biol.* **231**, 825 (1993).
[16] T. P. Creamer and G. D. Rose, *Proc. Natl. Acad. Sci. U.S.A.* **89**, 5937 (1992).

Translational Entropy

Several methods have been proposed for the estimation of the loss of translational entropy in these processes.[1,6,8,17,18] They are all based on estimating the translational entropy, using the partition function with energies based on the "particle in a box" plus the interaction energy.

$$Q = \frac{1}{N}\frac{1}{h^{3N}}\int_V \cdots \int_V \exp\left(-\frac{H(p,q)}{kT}\right) dp \, dq$$

where

$$H(p,q) = \sum \frac{1}{2m_i}(p_x^2 + p_y^2 + p_z^2) + U(q)$$

Because the moments and the coordinates are independent, the integral over the moments can be evaluated first, leading to

$$Q = \frac{1}{N}\frac{1}{\Lambda^{3N}}\int_V \cdots \int_V \exp\left(-\frac{U(q)}{kT}\right) dq$$

where Λ equals $h/(2\pi m k T)^{1/2}$. The integral in the equation is often called the configurational integral Z. If the particles do not interact and $U(q) = 0$, the configurational integral is equal to the volume and

$$Q = \frac{1}{N!}\left(\frac{V}{\Lambda^3}\right)^N$$

The resulting entropy is given by (Sakur–Tetrode equation)

$$S_{\text{trans}} = Nk \ln\left(\frac{v}{\Lambda^3} + \frac{5}{2}Nk\right)$$

This equation was used to estimate the loss of translational entropy in binding and oligomerization. There is no doubt that this equation provides the correct values for processes in gas phase. However, in aqueous solutions the configurational integral must account for the fact that most of the volume is physically occupied by water molecules. Thermodynamic cycles that estimate the entropy of association in the gas phase must account for the differences in entropy of the solutes in the gas phase and in solution.[17] This value cannot be estimated by assuming that the entropy of a solute in solution is identical to that of the same molecule in gas phase at a density of 1 mol/liter. In condensed phases, the configurational is equal to v_f, the

[17] G. P. Brady and K. A. Sharp, *Curr. Opin. Struct. Biol.* **7**, 215 (1997).
[18] A. V. Finkelstein and J. Janin, *Proteins Struct. Funct. Genet.* **15**, 1 (1989).

free volume of the liquid.[1] If the molecules of the liquid are considered rigid spheres, the free volume can be estimated from the equation

$$v_f = (2v^{1/3} - 2d)^3 = 8(v^{1/3} - d)^3$$

Using the free volume the entropy becomes

$$S_{trans} = Nk \ln \left(\frac{v_f}{\Lambda^3} + \frac{5}{2} Nk \right)$$

In aqueous solutions, the free volume accessible to a solute at 1 M concentration (the standard state) is equal to 55 times the free volume accessible to a water molecule (where 55 is the molarity of water).[1] Using this value for the free volume of the solutes, the difference in entropy between the complex and the two molecules can be calculated. The translational entropy of the ligand in the binding site of the protein requires two modifications: (1) The free volume is the volume in the binding site of the protein ($v_{f,bs}$); (2) the translational entropy of the ligand in the complex must be reduced by Nk, to account for the fact that in the complex the ligand cannot move independently over the whole volume of the container. The final expression is

$$\Delta S_{trans} \approx - \left[Nk \ln \left(55 \frac{v_{f,w}}{v_{f,bs}} + Nk \right) \right]$$

where $v_{f,w}$ is the free volume of water, and $v_{f,bs}$ is the free volume in the binding site of the protein. The quotient of free volumes can be estimated with the partial molar volume of the protein (0.735 cm^3/g) and the molar volume of water (1 cm^3/g).

Most of the assumptions involved in the derivation of this estimate of the change in translational entropy have been tested and confirmed by experimental data. The results of these analyses indicate that the use of the Sakur–Tetrode equation overestimates the value of the change in translational entropy. In addition, in one experimental study, the value of the loss of translational entropy was measured by comparing the thermodynamics of folding of a dimeric protein in a covalent and in a noncovalent dimer.[19] The experimental entropy difference between the two cases is approximately 2 e.u. This value is much lower than the value estimated by the Sakur–Tetrode equation but also lower than that estimated by the equation based in free volumes. However, the difference between the experimental value and that calculated with the equations reported here can be accounted for by the additional bond conformational entropy that exists in the disulfide-linked dimer.

[19] A. Tamura and P. L. Privalov, *J. Mol. Biol.* **273**, 1048 (1997).

Summary and Conclusions

Changes in configurational entropy represent one of the major contributions to the thermodynamics of folding, binding, and oligomerization. Methods have been developed to estimate changes in the entropy of the backbone and side chains, and for the loss of translational entropy. These methods have been used in combination with empirical methods that provide estimates of the changes in entropy of solvation as well as estimates of the changes of enthalpy. The results of such calculations are in excellent agreement with experimentally observed values.

Acknowledgment

Supported by NIGMS Grant GM 51362.

[9] Evaluating Energetics of Erythropoietin Ligand Binding to Homodimerized Receptor Extracellular Domains

By Preston Hensley, Michael L. Doyle, David G. Myszka,
Robert W. Woody, Michael R. Brigham-Burke,
Connie L. Erickson-Miller, Charles A. Griffin,
Christopher S. Jones, Dean E. McNulty, Shawn P. O'Brien,
Bernard Y. Amegadzie, Laurie MacKenzie, M. Dominic Ryan,
and Peter R. Young

Introduction

In contrast to the structures of macromolecules, which can in principle be defined by application of either X-ray crystallography or nuclear magnetic resonance (NMR), a description of their function requires the application of multiple technologies. When function is defined in terms of reversible molecular association, the mechanism, thermodynamics and kinetics of the interaction must all be elucidated (Fig. 1[1]; see color insert). Consequently, approaches such as analytical ultracentrifugation (AUC), isothermal titration calorimetry (ITC), and surface plasmon resonance (SPR) need to be coordinately employed (Fig. 2).

Analytical ultracentrifugation is a robust mass-determining tool that

---

[1] J. A. Wells, *Proc. Natl. Acad. Sci. U.S.A.* **93,** 1 (1996).

Analytical Ultracentrifuge	Titration Calorimeter	SPR Optical Biosensor

Solution Mass Stoichiometry	Thermodynamics Affinity	Kinetics Mechanism

FIG. 2. Three instruments used to define the kinetics and thermodynamics of the interactions of macromolecules in solution. Analytical ultracentrifugation is the most robust solution mass-determining tool and is therefore ideal for determining the stoichiometry of macromolecular interactions.[3] Isothermal titration calorimetry is ideal for thermodynamic characterizations, providing affinities and rich thermodynamic data,[8] and surface plasmon resonance optical biosensors have made the kinetics of molecular associations nearly routinely accessible and provide insight into the mechanisms of macromolecular assembly processes.[5]

provides strong validation of the stoichiometry of the assembly process in solution.[1a–3] Since its introduction in 1990, surface plasmon resonance biosensors have made the kinetics of macromolecular interactions routinely accessible.[4–6] Isothermal titration calorimetry is the method of choice for defining reaction thermodynamics.[7–9] Integration of these technologies in

[1a] S. E. Harding, A. J. Rowe, and J. C. Horton (eds.), "Analytical Ultracentrifugation in Biochemistry and Polymer Science." The Royal Society of Chemistry, Cambridge, 1992.

[2] T. M. Schuster and T. M. Laue (eds.), "Modern Analytical Ultracentrifugation: Emerging Biochemical and Biophysical Techniques" (T. M. Schuster, ed.). Birkhauser, Boston, 1994.

[3] P. Hensley, *Structure* **4**, 367 (1996).

[4] D. G. Myszka, X. He, M. Dembo, T. A. Morton, and B. Goldstein, *Biophys. J.* **75**, 583 (1998).

[5] D. G. Myszka, *Curr. Opin. Biotechnol.* **8**, 50 (1997).

[6] D. G. Myszka, M. D. Jonsen, and B. J. Graves, *Anal. Biochem.* **265**, 326 (1998).

[7] M. L. Doyle G. Louie, P. R. Dal Monte, and T. D. Sokoloski, *Methods Enzymol.* **259**, 183 (1995).

[8] M. L. Doyle, *Curr. Opin. Biotechnol.* **8**, 31 (1997).

[9] M. L. Doyle and P. Hensley, *Methods Enzymol.* **295**, 88 (1998).

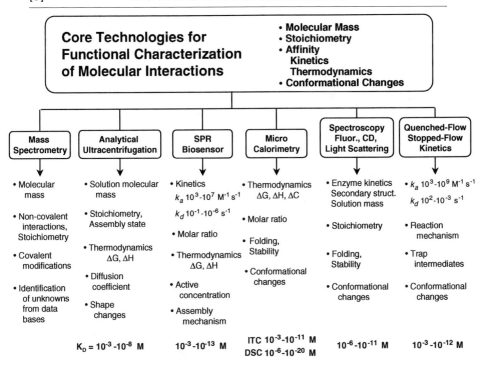

FIG. 3. Complementary technologies used in a modern molecular interactions core facility. [Modified from Hensley.[3]]

a molecular interaction core facility is outlined in Fig. 3.[10] A number of spectroscopic tools are included with, and extend, the preceding technologies, including fluorescence, light scattering, and circular dichroism (CD).[11,12] In this chapter, the power of combining these tools to characterize macromolecular interactions is demonstrated using the erythropoietin (EPO) ligand–receptor system.

EPO ligand is a protein hormone that binds to type I transmembrane receptors on target erythroid progenitor cells (Fig. 4A), resulting in their

[10] M. L. Doyle and P. Hensley, Experimental dissection of protein–protein interactions in solution. In "Advances in Molecular and Cell Biology," pp. 280–337. JAI Press, Stamford, CT, 1997.

[11] R. W. Woody, Methods Enzymol. 246, 34 (1995).

[12] R. W. Woody and A. K. Dunker, Aromatic and disulfide side-chain circular dichroism in proteins. In "Circular Dichroism: Conformational Analysis of Biopolymers" (G. D. Fasman, ed.), pp. 109–157. Plenum Press, New York, 1996.

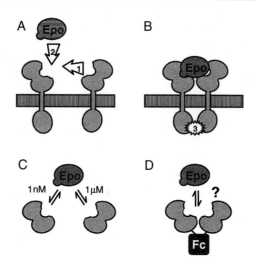

FIG. 4. Structure of the EPO receptor on a cell surface and the dimerized receptor expressed as an Fc chimera used in these studies. (A) EPO binds to cell surface receptors, which may exist as monomers or as a preformed dimer.[22] (B) Whichever model is correct, on EPO binding, there is likely to be a conformational change in the receptor, promoting signal transduction.[20] (C) Studies by Philo *et al.* have shown that there are low- and high-affinity sites on EPO for the monomeric receptors.[21] (D) In these studies, we construct an EPOR-Fc to use as a model for a soluble form of the dimerized receptor. The dimerized receptor is a chimera of the EPO receptor extracellular domain (residues 1–226) and the Fc domain of an IgG$_1$.

homodimerization and subsequent signal transduction (Fig. 4B). In turn, this results in cellular proliferation and differentiation into mature red blood cells.[13–15] A large amount of information has been forthcoming on the structures of free ligand, receptor, and the ligand–receptor complex,[16–18] as well as receptor complexed with agonist and antagonist peptides.[17,19]

[13] S. B. Krantz, *Blood* **77**, 419 (1991).

[14] H. Youssoufian, G. Longmore, D. Neumann, and A. Yoshimura, *Blood* **81**, 2223 (1993).

[15] H. F. Lodish, D. J. Hilton, U. Klingmuller, S. S. Watowich, and H. Wu, *Cold Spring Harbor Symp. Quant. Biol.* **60**, 93 (1995).

[16] J. C. Cheetham, D. M. Smith, K. H. Aoki, J. L. Stevenson, T. J. Hoeffel, and R. S. Syed, *Nature Struct. Biol.* **5**, 861 (1998).

[17] O. Livnah, D. L. Johnson, E. A. Stura, F. X. Farrell, F. P. Barbone, Y. You, and K. D. Liu, *Nature Struc. Biol.* **5**, 993 (1998).

[18] O. Livnah, E. A. Stura, S. A. Middleton, D. L. Johnson, L. K. Jolliffe, and I. A. Wilson, *Science* **283**, 987 (1999).

[19] O. Livnah, E. A. Stura, D. L. Johnson, S. A. Middleton, L. S. Mulcahy, N. C. Wrighton, W. J. Dower, L. K. Joliffe, and I. A. Wilson, *Science* **273**, 464 (1996).

Summarizing these structural data, Ballinger and Wells noted that dimerization is necessary, but not sufficient, for the initiation of signal transduction.[20] Less is known about the dynamics of these ligand–receptor interactions.

Previous studies using monomeric soluble receptor (EPOR) demonstrate that EPO will dimerize the free subunits, exhibiting low (micromolar) and high (nanomolar) affinity sites[21] (Fig. 4C). However, data suggest that the EPOR may exist in a predimerized form on the cell surface.[22] As performing biophysical studies on cell surface receptors is difficult, our work employs a soluble, covalently homodimerized form of the receptor, in which the extracellular domains of the receptor were genetically fused to the Fc domain of a human IgG$_1$ (Fig. 4D). The work described here addresses the question of whether this chimeric construct, which is amenable to biophysical analysis, provides a reasonable model of the dimerized receptor.

Protein Reagents

Cloning and Expression of Erythropoietin ArgHis$_6$

The cDNA for EPO is constructed by a combination of oligonucleotide synthesis and polymerase chain reaction (PCR). Initially, four overlapping synthetic oligonucleotides are hybridized and amplified by PCR, and the DNA products used as a template for a second round of PCR with two oligonucleotide primers derived from the expected ends of the synthetic gene product. The final cDNA products are cloned into PCR2000 (InVitrogen, San Diego, CA), and one is confirmed to encode EPO through DNA sequencing. The synthetic DNA insert is engineered with a carboxy-terminal ArgHis$_6$ peptide sequence, and inserted into pCDN, a mammalian expression vector that places the gene under the control of the cytomegalovirus (CMV) promoter. This vector is then transfected into chinese hamster ovary (CHO) cells, and stably transfected cells are selected by G418 resistance. Conditioned media containing recombinant carboxy-terminal ArgHis$_6$-tagged EPO are collected from the stable cells grown at 37° for 7 days. EPOArgHis$_6$ is purified by Ni-NTA chelate chromatography followed by size-exclusion chromatography.

Cloning and Expression of EPOR-Fc

The EPOR-Fc chimera (Fig. 4D) is constructed by fusing the extracellular domain of the EPOR with the Fc domain of an IgG$_1$. The EPOR is

[20] M. D. Ballinger and J. A. Wells, *Nature Struc. Biol.* **5**, 938 (1998).
[21] J. S. Philo, K. H. Aoki, T. Arakawa, L. O. Narhi, and J. Wen, *Biochemistry* **35**, 1681 (1996).
[22] I. Remy, I. A. Wilson, and S. W. Michnick, *Science* **283**, 990 (1999).

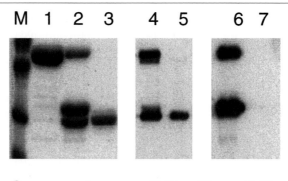

Coomassie   anti-EpoR   anti-Fc

FIG. 5. Expression and purification of EPOR-Fc: SDS–PAGE of factor Xa digestion of purified EPOR-Fc. Lane M, markers in decreasing order: 69, 46, 30, and 21 kDa; lane 1, EPOR-Fc; lane 2, factor Xa partial digest of EPOR-Fc; lane 3, supernatant of digest of lane 2 incubated with protein A beads and separated by centrifugation. Lanes M–3 were stained with Coomassie blue. Lanes 4 and 5 are Western blots stained with anti-EPOR. Lanes 6 and 7 are Western blots stained with anti-Fc.

amplified via PCR from a human fetal liver cDNA library purchased from Clontech (Palo Alto CA) using primers containing the published nucleotide sequence,[23] cloned into the Invitrogen vector PCR2000, and sequenced. This DNA fragment encodes amino acids 1–225 of the extracellular domain of the human EPOR. An *SpeI/XbaI* fragment containing the EPOR insert is then cloned into the *Drosophila* vector mtalell, which places the gene insert under the control of the *Drosophila* copper metallothionein promoter.[24] The construction of the vector is completed by replacing a *Bss*H2–*Xba*I C-terminal fragment with a synthetic *Bss*H2–*Kpn*I linker, containing the C terminus of the EPOR extracellular domain linked in frame to the four-amino acid recognition sequence for factor Xa cleavage (IEGR), and a *Kpn*I–*Xba*I fragment containing the human IgG$_1$Fc region. Transfection and expression in *Drosophila* S2 cells are as previously described.[24] EPOR-Fc is purified from *Drosophila* media by passage over a protein A-Sepharose column. To generate EPOR, the EPOR-Fc fusion is cleaved with factor Xa and the EPOR recovered by recovery of the flowthrough from a protein A-Sepharose column.

The purified EPOR-Fc is shown in Fig. 5. Lane 1 (Fig. 5) is the intact chimeric receptor. Lane 2 (Fig. 5) shows a partial factor Xa digest stained

[23] S. S. Jones, A. D. D'Andrea, L. L. Haines, and G. G. Wong, *Blood* **76,** 31 (1990).
[24] H. Johannsen, W. Jelkmann, G. Wiedemann, M. Otte, and T. Wagner, *Eur. J. Haematol.* **43,** 201 (1989).

with Coomassie blue. Lane 3 (Fig. 5) shows the mixture after incubation with protein A to remove species containing an Fc domain. The remaining band is the EPOR. In Fig. 5 Lanes 4 and 5 are lanes 2 and 3 stained with anti-EPOR and lanes 6 and 7 show lanes 2 and 3 stained with anti-Fc. These gels demonstrate that the EPOR-Fc is uniquely composed of Fc and EPOR domains.

Mass Spectrometry

Prior to an analysis of molecular function, the molecular masses of these proteins are established by mass spectrometry (MS). These data also provide a basis for analyzing some of the AUC data (see below). Matrix-assisted laser desorption ionization mass spectrometry MALDI-MS data are obtained with a PerSeptive Biosystems (Framingham, MA) Voyager RP laser desorption time-of-flight mass spectrometer. Samples are diluted 1:5 with 3,5-dimethoxy-4-hydroxy cinnamic acid [10 mg/ml in 0.1% (w/v) trifluoroacetic acid–acetonitrile (2:1, v/v)] for a final protein concentration of 1–10 pmol/μl. Bovine β-lactoglobulin A (MH$^+$ 18,364 Da) and rabbit muscle phosphorylase b (MH$^+$ 97,219.5 Da) (Sigma, St. Louis, MO) are included as internal calibrants. Desorption/ionization is accomplished by photon irradiation from a 337-nm pulsed nitrogen laser and 25-keV accelerating energy. Spectra are averaged over ~100 laser scans. EPO and EPOR-Fc are shown by MALDI-MS to have molecular masses of 28.0 and 107.1 kDa, respectively (Fig. 6). EPO exhibits a broad peak (half-height peak width, ~5 kDa), as expected for a highly glycosylated protein.

Reaction Stoichiometry

Solution Assembly States of EPO, EPOR-Fc, and EPO–EPOR-Fc Complex by Analytical Ultracentrifugation

Prior to an analysis of the kinetics and thermodynamics of the interaction of EPO with its dimerized extracellular domain, the assembly model must be defined. To this end, analytical ultracentrifuge data are collected in a Beckman (Fullerton, CA) XL-I analytical ultracentrifuge, using double sector cells with sapphire windows.[3] For analysis of data corresponding to a single macromolecular species, the primary data, absorbance at 280 nm versus radius, are fit to Eq. (1). For multiple species, data are fit to Eq. (2) or Eq. (3).

$$A_{tot} = A_m \exp\left[\frac{M_m(1 - \bar{v}_m\rho)\omega^2(r^2 - r_m^2)}{2RT}\right] + base \qquad (1)$$

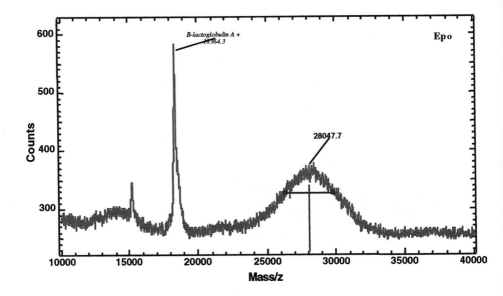

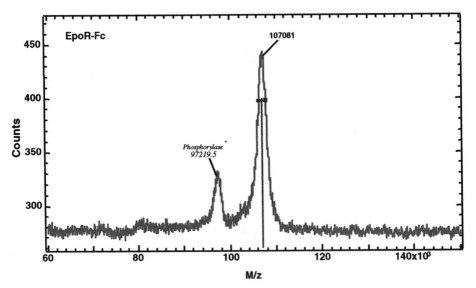

Fig. 6. MS of EPOR-Fc. *Top:* Mass spectrum of EPOArgHis$_6$ expressed from CHO cells as described in Cloning and Expression of Erythropoietin ArgHis$_6$. Extreme carbohydrate heterogeneity is evident with a half-height peak width corresponding to ~5000 Da. The protein molecular mass of EPO is predicted from the sequence to be 19,221 Da with a total mass of 28,047.7 Da, yielding a protein of 32% carbohydrate, as expressed here. *Bottom:* The mass spectrum of recombinant EPOR-Fc expressed from *Drosophila* S2 cells. This spectrum exhibits much less carbohydrate heterogeneity. The protein molecular mass of EPOR-Fc is predicted from the sequence to be 102,828 Da with a total mass of 107,081 Da, yielding a protein of 4% carbohydrate, as expressed here.

$$A_{tot} = A_m \exp\left[\frac{M_m(1 - \bar{v}_m \rho)\omega^2(r^2 - r_m^2)}{2RT}\right]$$

$$+ A_t \exp\left[\frac{4M_m(1 - \bar{v}_t \rho)\omega^2(r^2 - r_m^2)}{2RT}\right] + \text{base} \qquad (2)$$

$$A_{tot} = A_A \exp\left[\frac{M_A(1 - \bar{v}_A \rho)\omega^2(r_m^2)}{2RT}\right]$$

$$+ A_C \exp\left[\frac{M_C(1 - \bar{v}_C \rho)\omega^2(r^2 - r_m^2)}{2RT}\right] + \text{base} \qquad (3)$$

Here, A_{tot} is the total absorbance, A_i is the absorbance of various forms (monomer, tetramer, A, B, etc.) at a reference position, M_i and $\bar{v}_i$ are the molecular mass and partial specific volume of the forms, ρ is the solvent density, ω is the angular velocity, and r and r_m are the measured and reference radial positions. R and T are the gas constant and the absolute temperature, respectively; "base" is a term that corresponds to absorbing material that is nonsedimenting. The partial specific volume is estimated by the method of Cohn and Edsall[25] and Laue et al.[26] for EPO, EPOR-Fc, and the 1:1 complex of EPO and EPOR-Fc to be 0.707, 0.731, and 0.726 ml/g, and the solvent density is estimated to be 1.006 ml/g.[26] The molecular masses of EPO and EPOR-Fc are determined by MALDI-MS (Fig. 6) to be 28,124 Da (32% carbohydrate) and 107,098 Da (4% carbohydrate). Carbohydrate is incorporated into the determination of $\bar{v}_i$, as suggested by Laue et al.[26] Equilibrium is determined by demonstrating no change in the protein distributions from data sets taken 4 hr apart.

The test for self-assembly is to include in Eq. (1) terms for higher order aggregates, as in the progression from Eq. (1) to Eq. (2). Equation (3) is a "species analysis,"[27] which is used to determine the stoichiometry of macromolecules in a high-affinity complex. For instance, if molecules A (in a 50% molar excess) and B are mixed together to form a complex, C, then the mixture will have excess molecules, A, and the complex, C. Equation (3) can then be used to determine the mass of C and, hence, the stoichiometry

[25] E. J. Cohn and J. T. Edsall, "Proteins, Amino Acids, and Peptides," pp. 370–381. American Chemical Society, New York, 1943.
[26] T. M. Laue, B. D. Shah, T. M. Ridgeway, and S. L. Pelletier, Computer-aided interpretation of analytical sedimentation data for proteins. In "Analytical Ultracentrifugation in Biochemistry and Polymer Science" (S. E. Harding, A. J. Rowe, and J. C. Horton, eds.), pp. 90–125. The Royal Society of Chemistry, Cambridge, 1992.
[27] W. Chan, L. R. Helms, I. Brooks, G. Lee, S. Ngola, D. McNulty, B. Maleeff, P. Hensley, and R. Wetzel, Folding Design 1, 77 (1996).

of molecules in the complex. Hybrids of Eqs. (2) and (3) can then be used to determine if the complex self-associates. In the early stages of AUC data analysis, such species analyses are preferred to analyzing the data in terms of a model in which the preexponential terms contain equilibrium constants.[3,27] In this situation, the analysis suffers from significant numerical correlation, decreasing its robustness. In general, it is best to start with a species analysis, until the model is defined. Then a transition to a thermodynamic model is possible, using the results from the species analysis as starting guesses.

Figure 7 shows equilibrium sedimentation (ES) AUC data for EPO purified at 20,000 rpm. Under these conditions, it weakly associates into a tetramer, with 1% association occurring at 10^{-5} M and 50% association occurring at 10^{-4} M, when fit to Eq. (2). This is consistent with the NMR results of Cheetham et al.,[16] who observed a long overall correlation time, 16 nsec, for the deglycosylated form of EPO, suggesting that the protein self-associated. In contrast, the EPOR-Fc fits well to Eq. (1) and is a monomer within the limits of detection, up to 10^{-5} M (see inset to Fig. 7). These data also yield a molecular mass of 102.4 ± 0.2 kDa, which is in good agreement with the value of 107.1 kDa, determined by MS (above).

Figure 8 shows ES AUC data for a mixture of free EPO (20% excess on a molar basis) and EPOR-Fc. The data are analyzed according to Eq. (3). The fits converge only when terms for the free EPO monomer and the 1 : 1 complex of free EPO and EPOR-Fc are included. No other species can accommodate the data. In the analyses the molecular mass of the complex is determined to be 132.9 ± 0.9 kDa, in good agreement with the value of 135.5 predicted from MS for the 1 : 1 complex. These data establish the stoichiometry of the complex and also demonstrate that the complex itself does not self-associate, at least within these detection limits. A summary of the AUC results is given in Fig. 9.

Biological Function

Differential Inhibition of Erythropoietin Biological Activity with EPOR and EPOR-Fc

Prior to further biophysical analysis of the function of the cloned molecules, their function in a biological context is determined. It is also of interest to confirm that the EPO ligand is complexed between the two EPORs in the Fc-chimera. Here, the affinity of EPO for cell surface receptors on UT-7/EPO cells is determined in the presence and

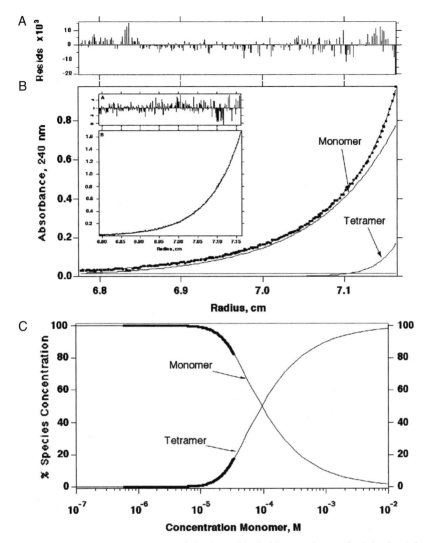

FIG. 7. ES of EPO and EPOR-Fc. (A) The residuals (theory–observed) of the fit of the primary data shown in (B) to Eq. (2) (see Reaction Stoichiometry). (B) Fit of absorbance equilibrium sedimentation data to a monomer–tetramer equilibrium. The total absorbance is shown as the sum of components for the monomer and tetramer, as indicated. (C) The fraction of monomer and tetramer as a function of total monomer concentration as predicted from the fit of the data to Eq. (2). The thick portions of the lines indicate the concentration range over which the data were actually analyzed. The thin portions are extrapolations based on the fitted parameters. This analysis shows that the tetramer is <1% of the total at a total EPO concentration of 10^{-5} M and is 50% of the total at 10^{-4} M. These data were obtained at 20,000 rpm at 25° in 20 mM phosphate buffer, 150 mM NaCl, pH 7.4, with an initial protein concentration of 1.8×10^{-5} M. *Inset* (B): Primary equilibrium sedimentation data (B) and residuals (A) for the EPOR-Fc fit to Eq. (1). The analysis yields a molecular mass of 102.4 ± 0.2 kDa. The experimental conditions were the same as above, with a rotor speed of 12,000 rpm and a starting concentration of 2.6×10^{-6} M. The sample was at equilibrium after 16 hr.

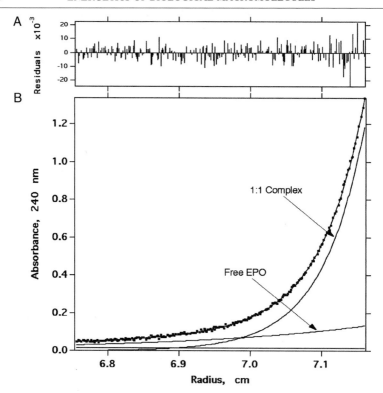

FIG. 8. Complex formation by ES. (A) Residuals (theory–observed) of the fit of the primary equilibrium sedimentation data (B) to Eq. (3). (B) Equilibrium sedimentation data for a 1.2:1 mixture of EPO and EPOR-Fc. The fit to the data was broken down into exponentials corresponding to free EPO and the 1:1 complex. The data could not be fit to any other combination of species. The fitted mass of the complex is 132.9 ± 0.9 kDa, which is in good agreement with a mass of 135.5 kDa predicted from the sums of the masses of EPO and EPOR-Fc, as determined by MALDI-MS. These data were obtained at 12,000 rpm under the same solvent conditions as in Fig. 7.

absence of competing receptor constructs, using cell proliferation as the readout.

The effects of EPOR and EPOR-Fc binding to EPO are initially examined with the UT-7/EPO cell line, which is known to respond to EPO by incorporating thymidine.[28] Dilutions of cleaved EPOR or EPOR-Fc fusion samples are in Iscove's modified Dulbecco's medium–10% (v/v) fetal

---

[28] N. Komatsu, M. Yamamoto, H. Fujita, A. Miwa, K. Hatake, T. Endo, H. Okano, T. Katsube, Y. Fukumaki and S. Sassa, *Blood* **82,** 456 (1993).

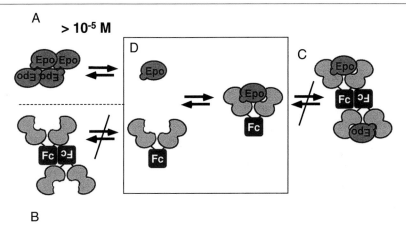

FIG. 9. Stoichiometry of assembly of EPO and the EPO-Fc to form a 1:1 complex. (A) AUC experiments described in Fig. 7 demonstrate that EPO self-assembles to a tetramer above 10^{-4} M. (B) Data in Fig. 7 (*inset*) demonstrate that the EPOR-Fc does not self-assemble below 10^{-5} M. (C) Data in Fig. 8 demonstrate that the EPOR-Fc does not associate under the concentration ranges examined. (D) EPO and the EPOR-Fc form a 1:1 complex and no other assembly processes occur below 10^{-5} M. protein. These data establish the assembly model that may be used in subsequent experiments.

calf serum (FCS). The samples are added to wells of a 96-well microtiter plate in quadruplicate along with EPO at 0.2 U/ml. UT-7/EPO cells are washed twice in IMDM–10% FCS, diluted to 1×10^5 cells/ml and plated at 100 μl/well. After the plates are incubated at 37°, 5% CO_2 for 3 days, [^{3}H]thymidine is added to a final concentration of 10 μCi/ml and the plates are incubated for 4 hr. The 96-well plates are harvested onto glass fiber filters with a Tomtec (Hamden, CT) plate harvester and cold 10% (w/v) trichloroacetic acid (TCA) and 95% (v/v) ethanol. Solid scintillant is melted onto the filters and the samples are counted. The mean and standard error of quadruplicate samples are determined. The data are reported as the percentage of positive (0.2 U/ml) EPO control (specific activity, 1.2×10^5 U/mg; therefore 0.2 U/ml = 5.9 pM).

Figure 10 shows that the median inhibitory concentration (IC$_{50}$) of EPOR-Fc i.e., the concentration of EPOR-Fc that inhibits UT-7/EPO cell proliferation by 50% under these conditions, is in the range of 2 ng/ml, while that of sEPOR is 1500 ng/ml. On the basis of these data, the EPOR-Fc is 750-fold more potent than sEPOR on a molar basis. This argues strongly for an avidity effect in which EPO is bound by both receptors on a single EPOR-Fc chimera.

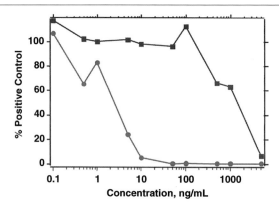

FIG. 10. Effects of EPOR-Fc. (●) and sEPOR (■) on EPO stimulation of UT-7/EPO cells. Dilutions of EPOR-Fc and sEPOR were added simultaneously with EPO at 0.2 U/ml. It was found that 750-fold more sEPOR than EPOR-Fc was necessary to inhibit the EPO-induced proliferation of UT-7/EPO cells. These data suggest that EPO is bound between the receptor subunits on the EPOR-Fc.

Reaction Kinetics

Surface Plasmon Resonance Analysis of EPO–EPOR-Fc Interaction

With the assembly model established, it is now possible to undertake an analysis of the kinetics of the EPO–EPOR interaction, using a BIACORE biosensor. A detailed discussion of the analysis is available from Morton *et al.*[29,30] In these studies, a BIACORE 2000 biosensor, CM5 sensor chips (research grade), NHS, EDC, ethanolamine hydrochloride, and P20 are obtained from Biacore AB (Uppsala, Sweden). Protein A and bovine serum albumin (BSA) are from Sigma. To produce high-quality biosensor data suitable for detailed kinetic analysis, EPOR-Fc is captured on a protein A surface, creating an oriented surface. Protein A is immobilized onto a research-grade carboxymethyldextran chip (CM5) by amine-coupling procedures described previously.[31] Flow cell 1 is activated with NHS–EDC [N-hydroxysuccinimide and N-ethyl-N'-(3-dimethylaminopropyl) carbodiimide · HCl] for 5 min. Protein A is injected at a concentration of 1 μg/ml in sodium acetate buffer (10 mM, pH 5.0) until 1000 resonance units (RUs) of protein is coupled. The remaining activated groups are blocked by a 7-min injection of 1 M ethanolamine. A control surface is created by repeating the coupling procedure in flow cell 2 without incorporating protein A.

[29] T. A. Morton, D. G. Myszka, and I. M. Chaiken, *Anal. Biochem.* **227,** 176 (1995).
[30] T. A. Morton and D. G. Myszka, *Methods Enzymol.* **295,** 268 (1998).
[31] U. Jonsson, L. Fagerstam, B. Ivarsson, B. Johnsson, R. Karlsson, K. Lundh, S. Lofas, B. Persson, H. Roos, and I. Ronnberg, *BioTechniques* **11,** 620 (1991).

All kinetic binding experiments are done at a flow rate of 100 μl/min with a buffer containing 10 mM HEPES (pH 7.4), 150 mM NaCl, 0.005% (w/v) P20 (Tween 20), and BSA (0.1 mg/ml). The flow path is set to include flow cells 1 and 2. Two different surface densities of EPOR-Fc are created on the protein A surface (flow cell 1) by injecting either 15 or 45 μl of receptor at a concentration of 70 nM. No binding of EPOR-Fc to the surface without protein A (flow cell 2) is detected. Samples of EPO (150 μl) are sequentially injected through both flow cells at a concentration of 2, 0.66, 0.22, 0.074, 0.025, and 0 nM. EPO dissociation is monitored for 250 sec before regenerating the protein A surface with a 10-μl injection of 10 mM HCl. The binding experiments for each EPO concentration are repeated five times and the samples are randomized.

Figure 11A shows the capture step as well as an overlay of the binding profiles for subsequent EPO injections. The capture step is highly reproducible and the EPOR-Fc–protein A complex is stable. In fact, EPOR-Fc does not dissociate from protein A during the time frame in which we have monitored the interaction with EPO, simplifying data analysis. The effect of flow rate is determined by monitoring EPO binding (0.22 nM) at 100, 33, and 11 μl/min (Fig. 11B).

Sensor data are prepared for kinetic analysis by zeroing the time and response before each EPO injection. Data from a control surface are used to correct for refractive index changes, nonspecific binding, and instrument drift.[32] Overlays of the normalized EPO binding profiles from each surface are shown by the black lines in Fig. 12A and B. The binding responses are highly reproducible even on a low-capacity surface (<15 RU). However, the binding rates on the lower density surface (Fig. 12A versus B; see color insert) appear faster than the higher density surface. For example, compare the binding profiles from the third highest EPO concentration (0.22 nM) in Fig. 12 (see color insert). This indicates that the reaction may be partially limited by mass transport. Lowering the surface density minimizes the effects of mass transport and provides better information about the intrinsic rate constants.[4,5,33]

Association and dissociation phase data for each concentration of EPO across both surface densities of EPOR-Fc are simultaneously fitted by numerical integration and nonlinear least-squares analysis procedures described previously.[29,30] Binding responses are tested against two reaction

---

[32] D. G. Myszka, *Methods Enzymol.* **323**, [14], 2000 (this volume).

[33] R. Karlsson, H. Roos, L. Fagerstam, and B. Persson, Kinetic and concentration analysis using BIA technology. *In* "Methods: A Companion to *Methods in Enzymology*," Vol. **6**, pp. 99–110. Academic Press, San Diego, California, 1994.

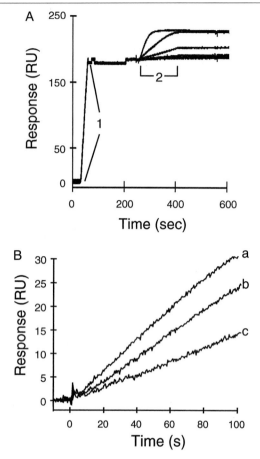

FIG. 11. SPR primary data. (A) Receptor construct capture step and subsequent EPO injections. Region 1 marks the association phase for the EPOR-Fc chimera. Samples (45 μl) of a 70 nM receptor solution were injected repeatedly over the protein A surface. Region 2 marks the association phase for the EPO samples. Samples (150 μl) of EPO were injected at concentrations of 2, 0.66, 0.22, 0.074, 0.025, and 0.0 nM. EPO binding experiments were done at a flow rate of 100 μl/min with a buffer containing 10 mM HEPES (pH 7.4), 150 mM NaCl, 0.005% (w/v) P20 surfactant, and BSA (0.1mg/ml). (B) Flow rate dependence of SPR signal. The effect of flow rate was determined by monitoring EPO binding (0.22 nM) at 100 μl/min (a), 33 μl/min (b), and 11 μl/min (c) over the 181-RU receptor surface.

Stoichiometry - Thermodynamics - Kinetics
Affinity

$$\Delta G = \Delta H - T\Delta S = RT \ln (K_D = k_d/k_a)$$

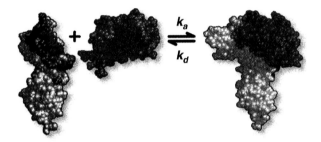

FIG. 1. The association of growth hormone with one of its two receptor extracellular domains. To define the dynamics of macromolecular assembly, the stoichiometry, kinetics, and thermodynamics of association need to be understood. Obtaining these parameters by independent experimental approaches acts both to provide a complete description of the assembly process and as internal controls. Here, the structures of these interaction partners have been defined by X-ray crystallography. Solution studies were then used to define the energetics of assembly and to define residues critical to the interaction. Hot spots are shown in red.[1]

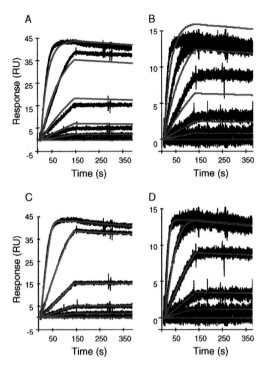

FIG. 12. Global analysis of response data for the EPO–EPOR-Fc interaction. Experimental data (black lines) represent 5 repeated injections of each EPO concentration (2, 0.66, 0.22, 0.074, 0.025 and 0 nM) over 181-RU (A and C) and 58-RU (B and D) EPOR-Fc surfaces. All binding experiments were done at a flow rate of 100 μl/min with a buffer containing 10 mM HEPES (pH 7.4), 150 mM NaCl, 0.005% P20, and BSA (0.1 mg/ml). The association and dissociation phase data from each concentration of EPO and across both surface densities of EPOR-Fc were fitted simultaneously. The red lines show the best fit of the binding data to a simple bimolecular reaction [Eq. (4)] (A and B) and a two-step mass transport limited reaction [Eq. (5)] (C and D).

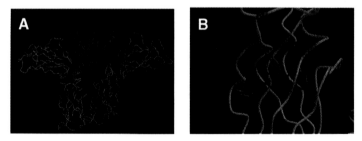

FIG. 16. A molecular model of the EPOR complexed with EPO. This structure was determined by Syed *et al.* (1998)[42] and is available from the PDB (*www.rcsb.org/pdp/*), citation 1EER. The aromatic residues discussed in the text are indicated.

models: a simple bimolecular reaction [see Eq. (4)] and a two-step mass transport limited reaction [see Eq. (5)].[4,33,34]

$$EPO + EPOR\text{-}Fc \underset{k_d}{\overset{k_a}{\rightleftharpoons}} EPO^*EPOR\text{-}Fc \qquad (4)$$

$$EPO_1 \underset{k_m}{\overset{k_m}{\rightleftharpoons}} EPO + EPOR\text{-}Fc \underset{k_d}{\overset{k_a}{\rightleftharpoons}} EPO^*EPOR\text{-}Fc \qquad (5)$$

The best fit of the data to a simple bimolecular reaction [see Eq. (4)] is shown by the red lines in Fig. 12A and B. This model fails to accurately describe the data and gives an unacceptable residual standard deviation of 1.77 RU. This is almost four times the replication standard deviation of 0.46 RU, which is a model-independent assessment of total experimental noise. A two-step mass transport limited reaction[4,5,33,35] describes very well the data from both surfaces as shown by the red lines in Fig. 12C and D. The residual standard deviation is 0.51 RU, which is much closer to the replication standard deviation. This reaction, which has only one additional parameter over the simple bimolecular reactions, accounts for transport of EPO to and from the sensor surface.

A total of five parameters are required to describe the data for EPO binding to the two different surface densities of EPOR-Fc. Table I shows the values returned for these parameters along with their standard errors and correlation coefficients. An apparent molar binding ratio is determined by comparing the maximum capacity for EPO versus the amount of EPOR-Fc captured. On the basis of the molecular weight of each protein, the high- and low-density surfaces show similar binding stoichiometries of 0.91 : 1 and 0.89 : 1, respectively. This indicates that the EPOR-Fc captured on the sensor surface binds one molecule of EPO per EPOR-Fc regardless of the surface density.

The resulting association rate constant (k_a) is fast at 8.09 (± 0.06) $\times$ $10^7 \ M^{-1} \ sec^{-1}$ while the dissociation rate constant (k_d) is slow at 2.44 (± 0.07) $\times 10^{-4} \ sec^{-1}$. Together these rate constants predict an equilibrium dissociation constant, K_D, of 3.01 (± 0.06) pM. The mass transport coefficient (k_m) is found to be 5.80 (± 0.20) $\times 10^8 \ sec^{-1}$. Binding responses are shown to be highly dependent on flow rate (Fig. 11B), further supporting the mass transport limited model.[4,34] To minimize the effects of mass trans-

[34] D. G. Myszka, T. A. Morton, M. L. Doyle, and I. M. Chaiken, *Biophys. Chem.* **64**, 127 (1997).
[35] D. G. Myszka, P. R. Arulanantham, T. Sana, Z. Wu, T. A. Morton, and T. L. Ciardelli, *Protein Sci.* **5**, 2468 (1996).

TABLE I
ANALYSIS OF SURFACE PLASMON RESONANCE DATA[a]

| Parameter | Parameter value | Correlation coefficient | | | |
		R_{max_1}	R_{max_2}	k_a	k_d
R_{max_1}	13.52 ± 0.02	0.38			
R_{max_2}	43.22 ± 0.04	−0.24	−0.08		
k_a, M^{-1} sec^{-1}	8.09 (±0.06) × 10^7	0.43	0.71	0.15	
k_d, sec^{-1}	2.44 (±0.07) × 10^{-4}	0.21	−0.01	−0.90	−0.09
k_m, sec^{-1}	5.80 (±0.20) × 10^8				
K_D, M	3.01 (±0.06) × 10^{-12}				

	High-density surface	Low-density surface
Fc immobilized, RU	181	58
Molar ratio, apparent	0.91:1	0.89:1
Replicate SD, RU	0.46	
Residual SD, RU	0.51	

[a] At 25°.

port we monitor binding to low surface densities of EPOR-Fc and at a high flow rate (100 μl/min).

An independent validation of the mass transport model used here comes from an analysis of the mass transport coefficient (k_m) in terms of the translational diffusion constant. This relation is defined explicitly as

$$k_m = 0.98(D/h)^{2/3}(f/bx)^{1/3}$$

where D is the translational diffusion constant, h, b, and x are machine constants describing the height, width, and length of the flow cell, respectively, and f is the flow rate.[33,36] With the known values of these parameters, the translational diffusion constant ($D_{20,w}$) may be estimated from k_m to be 6.28 × 10^{-7} cm^2/sec, or 6.28 F. The value for $D_{20,w}$ may also be directly determined by time derivative velocity sedimentation.[37] From this analysis (data not shown), a value of 6.87 F is obtained. These two values are remarkably close, given that the model used in the SPR data analysis ignores the presence of the dextran layer, which may slow the transport rate.

[36] S. Sjolander and C. Urbaniczky, *Anal. Chem.* **63**, 2338 (1991).
[37] W. F. Stafford III, *Curr. Opin. Biotechnol.* **8**, 14 (1997).

Reaction Thermodynamics

Titration Calorimetry of Erythropoietin Binding to sEPOR and EPOR-Fc

Direct equilibrium binding experiments for EPO binding to monomeric and dimeric forms of the receptor are carried out on a Microcal (Northhampton, MA) MCS isothermal titration calorimeter. Protein samples are filtered through a 0.22-μm pore size filter prior to measuring concentrations by absorbance at 280 nm. Extinction coefficients at 280 nm are calculated from the amino acid sequence.[38] Molar extinction coefficients are calculated as 22,700, 152,100, and 41,100 M^{-1} cm^{-1} for EPO, EPOR-Fc, and sEPOR, respectively. Data are analyzed with Microcal Origin software according to a single-site binding model in each case.[39] The model includes an equilibrium constant, K_D, a molar binding enthalpy change, ΔH, and a molar binding ratio.

Figure 13A shows that the binding interaction with EPOR-Fc is stoichiometric, i.e., there are no points in the transition region. Hence, the affinity is too tight to measure by calorimetry. The molar binding ratio for the EPOR-Fc is determined by ITC as 1.05 ($\pm$0.10) EPO per EPOR-Fc for the average of five experiments. In contrast, Fig. 13B shows calorimetry data for EPO binding to sEPOR, which demonstrates that the affinity is much weaker than for the EPOR-Fc form, i.e., there are several points in the transition region. The apparent K_D from the data in Fig. 13B is 100 nM, but interpretation of this apparent K_D value at a molecular level is complicated by the presence of multiple equilibria. In particular, Philo *et al.*[21] have shown that EPO can bind two sEPOR molecules, one with nanomolar affinity and the other with micromolar affinity. Thus, titration calorimetry data measured at micromolar concentrations require inclusion of both equilibria in the binding model. However, the data in Fig. 13B are not restrictive enough to allow deconvolution of a two-site binding model because of the limited concentration range studied. Nevertheless, Fig. 13 shows that the EPO affinity for sEPOR is much weaker than for EPOR-Fc. These data are consistent with the cell-based data shown in Fig. 9A and further demonstrate that EPO is likely to be bound between the two subunits of EPOR-Fc, the result of an avidity effect.

Figure 14 shows the EPO-binding enthalpy change for EPOR-Fc and monomeric sEPOR versus temperature. The slope of the data in Fig. 14 yields a binding heat capacity change of -1.23 kcal/mol per degree for EPO bound to EPOR-Fc. This value is significantly larger than is typical

[38] C. N. Pace, F. Vajdos, L. Fee, G. Grimsley, and T. Gray, *Protein Sci.* **4**, 2411 (1995).
[39] T. Wiseman, S. Williston, J. F. Brandts, and L. N. Lin, *Anal. Biochem.* **179**, 131 (1989).

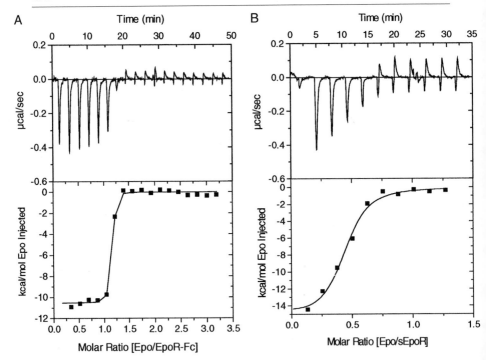

Fig. 13. Isothermal titration calorimetry data for EPO binding to EPOR-Fc and sEPOR. Top portion of each graph shows the raw data tracings of microcalories per second versus time. Lower portions show integrated heats per mole of EPO injected on each injection. (A) Injections (4 μl) of 284 μM EPO were made into 4.8 μM EPOR-Fc at 25°. (B) Injections (4 μl) of 99 μM EPO were made into 6 μM sEPOR at 37°. Curves in both cases represent best fits to a single-site binding model. Conditions: pH 7.4, 10 mM NaHPO$_4$, 150 mM NaCl, and 2 mM EDTA.

for protein–protein interactions.[40] The observed enthalpy and heat capacity changes for EPOR-Fc and sEPOR are distinct at all temperatures. Although it is difficult to interpret these parameters at a molecular level for sEPOR because of the expected multiple equilibria,[21] the differences between sEPOR and EPOR-Fc demonstrate that the natures of their EPO-binding reactions are different.

Interpretation of Binding Thermodynamics

The heat capacity change of -1.2 kcal/mol per degree that accompanies the EPO–EPOR-Fc reaction is large and suggests that a large amount of

[40] R. S. Spolar and M. T. Record, Jr., *Science* **263**, 777 (1994).

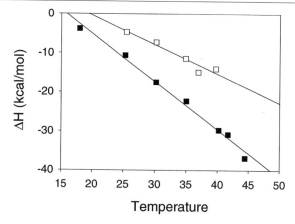

FIG. 14. Observed EPO binding enthalpy changes for sEPOR (□) and EPOR-Fc (■) versus temperature. Linear regression lines are shown through both data sets. The slope for EPOR-Fc yields the binding heat capacity change of −1.23 kcal/mol per degree. The slope for the sEPOR data yields only an apparent heat capacity change (−0.75 kcal/mol per degree) because of expected complications from multiple equilibria.[21] Titrations were done in phosphate buffer so that artifactual heats of buffer ionization are of minimal concern.

water-accessible surface area is buried during binding. It has been known for some time that the heat capacity change for reactions in aqueous solvent correlates with burial of water-accessible surface area. For example, a correlation has been reported that is based on a large database of protein-folding and model compound dissolution thermodynamics.[41] Application of this relationship to the EPO–EPOR-Fc case predicts the burial of 7500 Å^2 of water-accessible surface area when EPO binds EPOR-Fc (2990 Å^2 as polar and 4510 Å^2 as apolar). This is an unusually large value and is consistent with the expectation that one EPO is chelated by both EPOR domains of the EPOR-Fc. The crystal structure of the EPO–sEPOR complex shows 3160 Å^2 is buried at the EPO–EPOR interfaces (sites 1 and 2)[42] and none is buried between the receptor molecules, per se. The thermodynamically predicted value for the buried surface area is thus more than twice that observed in the complex, which raises the possibility that binding is coupled to significant conformational rearrangement.

The structure of the uncomplexed receptor has also been solved.[18] The two subunits associate in a way distinct from the receptors in either the complex with EMP-1[19] or with EPO.[42] For the receptor–receptor complex,

[41] K. P. Murphy and E. Freire, *Adv. Protein Chem.* **43**, 313 (1992).
[42] R. S. Syed, S. W. Reid, C. Li, J. C. Cheetham, K. H. Aoki, and B. Liu, *Nature (London)* **395**, 511 (1998).

TABLE II
ERYTHROPOIETIN BINDING THERMODYNAMICS OF EPOR-Fc

$K_D{}^b$	$\Delta G°$ (kcal/mol)	$\Delta H°$ (kcal/mol)	$-T\Delta S°$ (kcal/mol)	$\Delta C°$ (kcal/mol·deg)
10 pM	-15.6 ± 0.4	-26.6 ± 0.5	11.0 ± 0.6	-1.23 ± 0.10

[a] At 37°, determined by titration calorimetry and surface plasmon resonance. Conditions: pH 7.4, 10 mM NaHPO$_4$, 150 mM NaCl, and 2 mM EDTA.
[b] K_D at 37° for EPOR-Fc was measured by SPR at 25° and corrected for temperature by the van't Hoff equation, using $\Delta H°$ and $\Delta C°$ measured by calorimetry.

the buried surface area is only 1430 Å². Again, this is substantially less than the 7500 Å² value predicted from thermodynamics. In comparison, the total amount of surface area buried at the interfaces of the growth hormone–receptor crystal structure complex is 5060 Å², which includes interfaces between growth hormone and two receptor molecules, plus an interface between the two receptor molecules themselves.[43]

The large predicted amount of buried surface area for the EPO–EPOR-Fc reaction may reflect conformational changes that bury surface area remote from the binding interfaces on binding. An analysis of the thermodynamics in Table II with an independent empirical method that evaluates the binding entropy change[40] also raises the possibility that EPO binding to EPOR-Fc may be accompanied by conformational change. The results in Table II lead to a predicted ~70 amino acid residues folding when EPO binds EPOR-Fc. This value is large in comparison with other protein–protein interactions reported.[40] Although part of this effect may originate from loss of degrees of freedom of the EPOR domains on the Fc fusion on binding to EPO.

Spectroscopic Analysis

Circular Dichroism Spectroscopy

Narhi *et al.* have suggested that the binding of EPO to the free monomeric receptor (sEPOR) promotes a conformational change, based on observations of changes in the near- and far-UV CD spectra.[44] To determine to what extent such changes might be evident in the homodimerized receptor, similar spectral studies were undertaken. CD spectra are recorded with a Jasco (Easton, MD) J710 spectropolarimeter, using 1-mm path length cells

[43] J. A. Wells and A. M. de Vos, *Annu. Rev. Biophys. Biomol. Struct.* **22,** 329 (1993).
[44] L. O. Narhi, K. H. Aoki, J. S. Philo, and T. Arakawa, *J. Protein Chem.* **16,** 213 (1997).

at 25°. The proteins are in 10 mM sodium phosphate, 50 mM NaCl, pH 7.3. The CD spectrum of the EPO complex with its receptor differs significantly from the sum of the CD spectra of the separate components (Fig. 15A). The difference spectrum, shown in the inset of Fig. 15A, exhibits a positive band near 230 nm, and is negative below 214 nm. The shape of this difference spectrum is not consistent with that expected for changes in secondary structure, such as α helix converting to β sheets or random coil, or vice versa.[11]

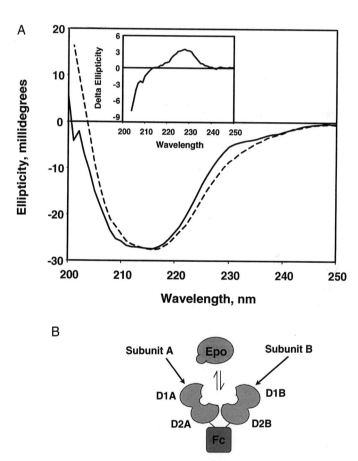

FIG. 15. (A) CD spectra of a stoichiometric mixture of EPO and EPOR-Fc. The solid line is the spectrum of the mixture and the dashed line is the mathematical sum of the individual spectra of EPO and EPOR-Fc. *Inset:* The difference spectrum, with a peak at ~228 nm. (B) Schematic structure of EPOR-Fc (from Fig. 4D). Attached to the lgG$_1$-Fc domain are two receptor subunits (A and B). Each subunit has an N-terminal domain (D1A, D1B) and a C-terminal, membrane-proximal domain (D2A, D2B) (nomenclature of Livnah et al.[19]).

Methods for Prediction of Circular Dichroism Spectra

CD spectra for domains D1 and D2 (for nomenclature, see Fig. 15B and Ref. 19) of EPOR are calculated using methods and parameters previously described by Grishina and Woody.[45] The calculations are based on the crystal structure of the EPOR complex with the EMP-1 peptide agonist,[19] which contains two D1–D2 fragments and two decapeptides in the asymmetric unit. Calculations are performed for both receptor fragments, labeled A and B, and for each of the separate domains, i.e., D1A, D2A, etc. In addition, each of the tyrosine and tryptophan side chains is omitted from the calculation, one at a time, to obtain an estimate of the individual contributions. A few test calculations indicate that omission of phenylalanine residues has no significant effect on the predicted CD above 220 nm, and so systematic calculations were not performed for phenylalanine side chains. The $\Delta[\Theta]_{230}$ values reported here are differences between subunits, assuming that these differences would be zero in the unliganded form, i.e., the unliganded receptor subunits are symmetric.

As in earlier calculations on proteins,[45,46] the calculated absolute spectra show poor agreement with experiment. However, as shown in these studies, difference spectra obtained by subtracting the CD calculated for a mutant, in which one aromatic residue has been replaced by a residue with a weaker chromophore, from the calculated CD of the wild type gives, in most cases, results that agree qualitatively with experimental CD difference spectra. This result, together with those from a study of bovine pancreatic ribonuclease,[47] indicate that current theoretical methods can provide reasonable predictions for the aromatic contributions, but are much less satisfactory for the peptide contributions. The best estimate of the aromatic CD contributions is obtained by subtracting the calculated peptide CD from that of the whole protein, including both peptide and aromatic contributions.

Magnitudes of Circular Dichroism Changes at 230 nm

The positive peak near 230 nm strongly suggests that one or more tryptophan or tyrosine side chains are perturbed when EPO binds to its receptor. The change in ellipticity at the maximum is 3×10^{-3} deg, which corresponds to $\Delta[\Theta]_{230} = 6 \times 10^5$ deg $\cdot$ cm^2/dmol of complex. Although this seems like a large change in CD, especially if it is attributed to the perturbation of a small number of aromatic side chains (perhaps only one),

[45] I. B. Grishina and R. W. Woody, *Faraday Disc.* **99,** 245 (1994).
[46] T. M. Thompson, B. L. Mark, C. W. Gray, T. C. Terwilliger, N. Sreerama, and R. W. Woody, *Biochemistry* **37,** 7463 (1998).
[47] G. Kurapkat, P. Kruger, A. Wollmer, J. Fleischhauer, B. Kramer, E. Zobel, A. Koslowski, H. Botterweck, and R. W. Woody, *Biopolymers* **41,** 267 (1997).

there are precedents.[12] For example, the conversion of chymotrypsinogen to chymotrypsin produces $\Delta[\Theta]_{230} = 5 \times 10^5$ deg·cm²/dmol,[48] which is attributable to the perturbation of a pair of tryptophan side chains.[45] Mutation of Trp-74 to leucine in *Escherichia coli* dihydrofolate reductase leads to $\Delta[\Theta]_{230} = 2 \times 10^5$ deg·cm²/dmol.[49] The oxidation of a single tryptophan in the coat protein of bacteriophage fd leads to $\Delta[\Theta]_{223} = 4 \times 10^5$ deg·cm²/dmol.[50] In the first two cases, the 230-nm band in the difference spectrum is accompanied by a band near 220 nm of opposite sign and this pattern is attributable to the exciton coupling of the strong B_b transitions of two tryptophan residues in close proximity. The difference spectrum induced by the EPO–EPOR interaction is not of this type. It resembles more closely, except for sign, that accompanying the chemical modification of the single tryptophan of the fd coat protein, for which the difference spectrum has a negative band near 223 nm and a positive band near 212 nm. This cannot be due to coupling of B_b transitions in two nearby tryptophan side chains because the cross-over wavelength is at 217 nm rather than 225 nm. The CD difference spectrum for the EPO–EPOR complex does not show a discrete negative band near 210 nm, but this may be obscured by a stronger negative band at shorter wavelengths.

CD spectra have been reported for domains from several members of the cytokine receptor family: domains D1 and D2 of EPOR,[44] the BC domain of granulocyte-colony stimulating factor receptor (G-CSFR[51]), the D3 domain of gp 130,[52] and the extracellular domain of the interleukin 6 receptor (IL-6R) (A. Wollmer, personal communication, 1999). The data are summarized in Table III. All these receptor domains exhibit a qualitatively similar pattern in the 200- to 240-nm region of the CD spectrum, with a positive band near 230 nm and a negative band at wavelengths ranging from 208 to 219 nm. Anaguchi *et al.*[51] have mutated both tryptophan residues of the WSXWS motif in G-CSFR to alanine, but have been able to measure the CD spectra of the mutant only at the first tryptophan residue in the WSXWS motif. Both bands decrease in magnitude, but the qualitative features of the spectrum persist.

[48] C. R. Cantor and S. N. Timasheff, Optical spectroscopy of proteins. *In* "The Proteins" (H. A. H. Neurath, ed.), pp. 145–306. Academic Press, NY, 1982.

[49] K. Kuwajima, E. P. Garvey, B. E. Finn, C. R. Matthews, and S. Sugai, *Biochemistry* **30**, 7693 (1991).

[50] G. E. Arnold, L. A. Day, and A. K. Dunker, *Biochemistry* **31**, 7948 (1992).

[51] H. Anaguchi, O. Hiraoka, K. Yamasaki, S. Naito, and Y. Ota, *J. Biol. Chem.* **270**, 27845 (1995).

[52] G. Müller-Newen, S. Pflanz, U. Hassiepen, J. Stahl, A. Wollmer, P. C. Heinrich, and J. Grötzinger, *Eur. J. Biochem.* **247**, 425 (1997).

TABLE III
FAR-UV BANDS OF CYTOKINE RECEPTOR DOMAINS[a]

Protein	λ_1	$[\Theta]_1$	λ_2	$[\Theta]_2$	λ_3	$[\Theta]_3$
EPOR[b]	231	+2,500	214	−3,000	—	—
G-CSFRwt[c]	230	+3,000	214	−6,000	197	+5,000
W294A	230	+2,300	213	−4,000	197	+6,200
C242/285A	232	+1,000	212	−4,500	197	+4,000
gp130[d]	230	+4,200	218	−1,500	196	−10,400
IL-6R[e]	230	+2,500	208	−2,500	192	−6,800

[a] Wavelengths in nanometers, mean residue ellipticities in deg/cm^2/dmol residue.
[b] D3 domain.[44]
[c] D3 domain.[51]
[d] D3 domain.[52]
[e] A. Wollmer (private communication, 1999), extracellular domain.

Calculation of Circular Dichroism Spectra for Erythropoietin Receptor

Theoretical calculations of the aromatic CD contributions based on the crystal structure of the EPOR–EMP-1 complex[19] can determine whether the large change in far-UV CD observed on complex formation in the present study can be accounted for, at least qualitatively, by conformational changes (inferred from the ITC data) that might affect aromatic side chains. Further, these calculations may help to define which of the domains, and within each domain, which residues contribute most strongly to the CD spectrum and, finally, whether the WSXWS motif common to this family of cytokine receptors gives rise to a significant and characteristic CD signal.

Table IV shows both the total CD and the aromatic CD contribution calculated at 230 nm for the two monomers, the four domains, and for the dimer. In all cases, the total CD at 230 nm is predicted to be negative, reflecting a relatively strong, negative peptide contribution. In contrast,

TABLE IV
THEORETICAL CIRCULAR DICHROISM FOR ERYTHROPOIETIN
RECEPTOR DOMAINS[a]

Fragment	Molecule A		Molecule B	
	$[\Theta]_{tot}$	$[\Theta]_{aro}$	$[\Theta]_{tot}$	$[\Theta]_{aro}$
D1D2	−3565	+221	−1969	+1702
D1	−4495	−446	−2996	+1542
D2	−3786	+695	−958	+1708

[a] At 230 nm. Mean residue ellipticities in deg · cm^2/dmol residue.
tot, Total calculated ellipticity; aro, aromatic contribution.

TABLE V
AROMATIC CIRCULAR DICHROISM CONTRIBUTIONS AT 230 nm FOR SINGLE RESIDUES

	Domain 1				Domain 2		
Residue	Molecule A	Molecule B	Difference	Residue	Molecule A	Molecule B	Difference
W40	296	1156	860	W142	−64	−155	−91
Y53	−370	87	457	Y156	343	−241	−554
Y57	379	6	−373	Y192	79	132	53
W64	−637	−97	540	W209[a]	223	968	745
W82	−281	308	589	W212[a]	576	1442	866
Y109	−1192	−730	462				

[a] Residues indicated in boldface are part of the WSXWS motif.

except for domain D1A, the aromatic contributions are all positive at 230 nm. The most striking feature of the results in Table IV is the difference in the calculated CD of molecules A and B and of the domains that comprise them. Domain D1 shows the largest difference, with opposite signs for the aromatic CD contributions of D1A and D1B. Domains D2A and D2B both have positive aromatic CD contributions, but that of D2B is substantially larger than that of D2A. The total aromatic contribution for molecule B is 1500 deg · cm^2/dmol residue more positive than for molecule A, corresponding to $\Delta[\Theta]_{230} = 3 \times 10^5$ deg · cm^2/dmol complex, a value comparable to the observed change in CD associated with EPO binding to the EPOR-Fc chimera, i.e., $\Delta[\Theta]_{230} = 6 \times 10^5$ deg · cm^2/dmol complex (from above).

Origin of Circular Dichroisn Changes in Erythropoietin Receptor

Livnah *et al.*[19] state that the complex of the EMP-1 dimer with the EPOR dimer exhibits an almost perfect twofold symmetrical arrangement. They note that superposition of the D1 and D2 domains from the two halves of the EPOR dimer gives root mean square (rms) deviations of 0.53 and 0.47Å, respectively, presumably because of an alignment of C$_\alpha$ atoms. However, there are significant differences in the aromatic side chains between the two halves of the EPOR dimer. If the aromatic side chains in D1A and D1B are aligned, the rms deviation is 0.93 Å, and for D2A and D2B, the rms deviation is 0.80 Å. Visual inspection of the superposed domains from molecules A and B, aligned with respect to the C$_\alpha$ atoms, shows a reorientation of a number of aromatic side chains in domains D1A versus D1B. This also occurs, but to a lesser extent, in domains D2A versus D2B. Table V shows the aromatic contributions for individual residues to the 230-nm CD. In domain D1A, the contributions of all six tyrosine and tryptophan residues are significant, ranging from −1192 for Y109 to +379 for Y57. In contrast, the aromatic CD of domain D1B has a large positive contribution from W40, a large negative contribution from Y109, and smaller contributions of both signs from the other aromatic side chains. All these residues, except Y57, give rise to much more positive values in molecule A than in molecule B. The 230-nm CD of Y57 has a sizable positive value in molecule A and a negligible value in molecule B.

In domain 2, the principal contributors to the aromatic CD are Y156, W209, and W212, and this is true for the domain in both molecules A and B. The tryptophans of the WSXWS motif (W209 and W212) dominate the aromatic CD of D2B, with a smaller contribution of opposite sign from Y156, but in D2A, the contributions of W209, Y156, and W212 are all positive, increasing in the order given. W209 and W212 make large positive contributions to the difference in 230-nm CD between molecules A and B, whereas Y156 makes a negative contribution to this difference. The WSXWS motif therefore is predicted to make a significant but not dominant positive contribution to the CD of the EPOR. This is consistent with the site-directed mutagenesis results of Anaguchi et al.[51] for the homologous G-CSFR. The D1 domain is predicted to be more important than the WSXWS-containing D2 domain in the CD differences between molecules A and B.

Nonaromatic Side-Chain Circular Dichroism Chromophores

Side-chain chromophores other than those of tryptophan, tyrosine, and phenylalanine have been omitted in these calculations. The ones most likely to have an impact in the 230-nm region are the disulfide and guanidinium groups. The D1 domain has two disulfides and the D2 domain has none. The site-directed mutagenesis study of G-CSFR,[51] a cytokine receptor in the same family as EPOR with homologous disulfides, shows that replacement of the cysteine residues forming one of the disulfides by alanine leads to a larger decrease in the positive 230-nm band than is achieved by replacing the first tryptophan of the WSXWS motif. It seems likely that this effect is not directly due to the disulfide group, because unstrained disulfides are not expected to have a strong CD band near 230 nm.[11] However, electronic interactions between the disulfide group and one or more aromatic side chains may account for this observation. Alternatively, the disulfide deletion might lead to a conformational change, at least locally, that perturbs the aromatic side chains.

An arginine side chain (R197) is sandwiched between the indole groups of the two tryptophan residues in the WSXWS motif and another (R199) is located above the indole ring of W209.[19] The guanidinium chromophore has two strong transitions near 190 nm[53] that could couple electronically

[53] R. S. McDiarmid, thesis, 1965, Harvard University.

with the tryptophan transitions. In addition, the positive charge of the guanidinium groups could perturb the transitions of the tryptophan groups. Further theoretical studies are underway to assess the significance of these neglected interactions. Although their inclusion may alter the quantitative details, it is unlikely that the results will be changed qualitatively.

In summary, the positive change in ellipticity at 230 nm may be interpreted in terms of small changes in the relative geometries of tryptophans and tyrosines in the separate domains of the EPOR on EPO binding (Fig. 16; see color insert). Here, the difference in ellipticity calculated for the formation of the EPOR–agonist peptide (EMP-1) complex, $\Delta[\Theta]_{230} = 3 \times 10^5$ deg $\cdot$ cm^2/dmol of complex, is comparable in magnitude to that observed for the binding of EPO to the EPOR-Fc, $\Delta[\Theta] = 6 \times 10^5$ deg $\cdot$ cm^2/dmol of complex. Because the binding of EMP-1 results in a complex with receptors in a nearly identical overall structure, these predicted CD changes must result from a relatively subtle change in conformation. Thus the observed CD changes on formation of the EPO–EPOR-Fc complex do not require conformational changes of the magnitude suggested from the calorimetry.

Summary

A number of techniques have been employed to characterize the energetics of EPO–EPOR-Fc interactions. AUC studies have shown that EPO and EPOR-Fc exist as monomers at concentrations less that 10 μM. Under these conditions, EPO and the EPOR-Fc associate to form a 1 : 1 complex and this complex does not undergo any further assembly processes. Studies in which the biological activity of EPO at a cell surface is competed by free and dimerized receptor show that the dimerized receptor is 750-fold more potent. This suggests that EPO is bound by both receptor subunits on the Fc chimera, as shown in Fig. 9D. This assembly model provides a foundation for interpretation of the kinetic, thermodynamic, and spectral results.

SPR kinetic analyses of the EPO–EPOR-Fc interaction yields association and dissociation rate constants of 8.0×10^7 M^{-1} sec^{-1} and 2.4×10^{-4} sec^{-1}, respectively, for an overall affinity of 3 pM (see Fig. 12). The half-maximal response in a cellular proliferation assay is evoked at an EPO concentration of 10 pM,[54] which is similar to the affinity kinetically determined for the EPOR-Fc. This value suggests that the EPOR-Fc chimera may be a reasonable model for the receptor on a cell surface (see Fig. 17). The use of this reagent is also supported by the studies of Remy et al.,[22]

[54] D. L. Johnson, F. X. Farrell, F. P. Barbone, F. J. McMahon, J. Tullai, and K. Hoey, *Biochemistry* **37**, 3699 (1998).

A   B

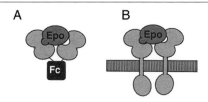

FIG. 17. These studies demonstrate that the EPOR-Fc (A) forms a 1 : 1 complex with EPO, that EPO interacts with both subunits in the Fc chimera, and that the affinity for EPO is high (3 pM). This result suggests that the EPOR-Fc is a valid model for the dimerized receptor on the cell surface (B) and, as such, will be a useful tool for probing the differences in the interactions of the receptor dimer with other EPO agonists and antagonists.

who demonstrate that the EPOR is likely to exist as a dimer on the cell surface, in the absence of ligand.

Titration calorimetry confirms the 1 : 1 stoichiometry, observed by AUC and SPR approaches. Further, the temperature dependence of the enthalpy yields a heat capacity that can be interpreted in terms of a large conformational change in the EPOR on EPO binding. Comparing the structures of the free and complexed receptor, some conformational changes are noted in loops L3 and L6.[18] However, these changes are small compared with the conformational changes predicted from an analysis of the calorimetric data reported here, i.e., equivalent to the folding of ~70 amino acids. The change in buried surface area between the free and complexed EPOR, determined from structural data, is also quite small when compared with the predicted value of 7500 Å^2 from calorimetry. Further studies need to be done to rationalize these observations. These may include an attempt to determine if conformational changes are communicated to the Fc domain and the extent to which EPOR extracellular domains are oriented on the Fc domain in a manner that faithfully reflects their orientation on a cell surface.

Finally, while changes in the CD spectra are observed on binding of EPO to the EPOR-Fc, and the monomeric receptor, these changes may be due to subtle changes in the microenvironments of tryptophans and tyrosines and do not require conformational changes of the magnitude suggested from the calorimetry results.

In summary, to define macromolecular interactions in solution, the stoichiometry, thermodynamics, and kinetics of assembly need to be understood. This task requires that a multitechnology approach be implemented. Here, AUC established an assembly model and provided a foundation on which SPR, ITC, and CD studies could be based and from which interpretation of these data could be extended. SPR established that the affinity of the dimerized receptor was high and ITC suggested that there may be a significant conformational change on binding. CD suggested that observed spectral changes may be due to these presumed conformational changes, but would also be consistent with more subtle changes. These studies further demonstrate that the EPOR-Fc is a valid model for the dimerized receptor on the cell surface and, as such, will be a useful tool for probing the differences in the interactions of the receptor dimer with EPO agonists and antogonists.

Acknowledgments

Clare Lynn is gratefully acknowledged for experimental assistance. R.W.W. acknowledges the assistance of Narasimha Sreerama in comparisons of the domains of EPOR and grant support from the NIH (GM22994). The authors gratefully acknowledge Frank DiCapua for constructing Figure 16.

[10] Measurement of Protein Interaction Bioenergetics: Application to Structural Variants of Anti-sCD4 Antibody

By Michael L. Doyle, Michael Brigham-Burke, Michael N. Blackburn, Ian S. Brooks, Thomas M. Smith, Roland Newman, Mitchell Reff, Walter F. Stafford III, Raymond W. Sweet, Alemseged Truneh, Preston Hensley, and Daniel J. O'Shannessy

Introduction

Biothermodynamic methods are capable of rigorously characterizing the functional chemistry of protein–protein and protein–ligand interactions. Analysis of the energetics for a binding reaction reveals insight into the molecular binding mechanism. Furthermore, a full bioenergetic analysis is a more stringent test of the fidelity of protein-engineering procedures than is possible from affinity measurements alone. For example, previous studies have shown that a series of structurally related monoclonal antibodies (MAbs) can have similar antigen-binding affinities, but analysis of the binding thermodynamics revealed otherwise hidden perturbations in binding mechanism.[1–3] Kinetic studies have also shown that structurally

[1] D. A. Brummell, V. P. Sharma, N. N. Anand, D. Bilous, G. Dubuc, J. Michniewicz, C. R. MacKensie, J. Sadowska, B. W. Sigurskjold, B. Sinnott, N. M. Young, D. R. Bundle, and S. A. Narang, *Biochemistry* **32**, 1180 (1993).

related peptide antigens can have similar affinities but different binding kinetics.[4]

However, commensurate with the more rigorous potential afforded by biothermodynamic methods comes a requisite higher standard of experimental rigor. To rigorously interpret biothermodynamic data, it is imperative to obtain experimental information about the reactants and products from several biophysical methods. Information is needed to define the basic chemistry of the reaction, such as the binding stoichiometry and the self-association and thermal stabilities of the reactants and products. Otherwise the observed bioenergetics and kinetics may contain systematic error contributions from linked aggregation or folding processes. It is also generally advantageous to utilize more than one method for measuring the affinity of an interaction, so that possible systematic instrumentation errors inherent in each methodology may be recognized if present.

In the present chapter we outline a strategy for the quantitative analysis of protein–protein interactions, using as an example the interaction between a protein antigen [human extracellular soluble CD4 (sCD4) receptor] and three structural variants of a monoclonal antibody. We demonstrate how multiple biophysical methods can be used in combination to provide a stringent examination of the functional chemistry of protein–protein interactions, and use this information to evaluate the consequences of protein-engineering procedures. Particular emphasis is placed on characterizing the reactants and products, in addition to the binding reaction per se. Finally, we discuss ways in which binding energetics and kinetics relate to the binding molecular mechanism.

Description of Protein Interaction Studied

Antibodies are composed of two identical light chains linked to two identical heavy chains (Fig. 1). These multifunctional proteins bind specific antigen through their N-terminal V regions and mediate effector functions through the interaction of their Fc domain with Fc receptors and transport and complement proteins. The two identical antigen-binding sites are composed primarily of six hypervariable loops, three each contributed by the V domains of the light chain (V_L) and the heavy chain. During antibody

[2] R. F. Kelley, M. P. O'Connell, P. Carter, L. Presta, C. Eigenbrot, M. Covarrubias, B. Snedecor, J. H. Bourell, and D. Vetterlein, *Biochemistry* **31**, 5434 (1992).

[3] K. Tsumoto, Y. Ueda, K. Maenaka, K. Watanabe, K. Ogasahara, K. Yutani, and I. Kumagai, *J. Biol. Chem.* **269**, 28777 (1994).

[4] L. Leder, C. Berger, S. Bornhauser, H. Wendt, F. Ackermann, I. Jelesarov, and H. R. Bosshard, *Biochemistry* **34**, 16509 (1995).

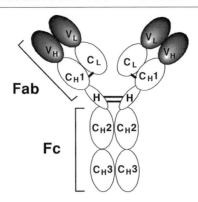

FIG. 1. Schematic of an IgG antibody. The molecule has two identical heavy chains (H subscript) and two identical light chains (L subscript). Antigen binding occurs in the hypervariable regions of each chain (designated by V) in the Fab segment. Each chain also has constant regions (C_H and C_L for constant domain in heavy and light chain, respectively) that form part of the Fab segment and all of the Fc segment. A hinge region (large H) in each heavy chain connects Fab and Fc segments. The four chains are covalently cross-linked by interchain disulfide bonds (shown as thick lines between C_H1 and C_L domains and also between the two hinge regions). The quaternary structure is also stabilized by noncovalent interchain interactions. All MAbs in the present study are composed of identical variable domains (shaded).

maturation, the same V_H can become associated with different heavy constant domains. One isotype is immunoglobulin G (IgG) and in humans there are four different IgG isotypes, 1–4, each having different effector properties. Thus, the biological activity of a specific antibody can be modulated by altering the nature of the heavy chain constant domain. A premise of this engineering is that these different Fc domains do not alter its antigen-binding properties.

Here we characterize the antigen-binding energetics of structural variants of the anti-human sCD4 monoclonal antibody CE9.1 (Keliximab; IDEC, San Diego, CA). The MAb CE9.1 is a PRIMATIZED MAb derived from the parent cynomolgus macaque monkey MAb E9.1.[5] It combines the antigen-binding V_H and V_L variable domains from MAb E9.1 with constant domains from a human IgG_1 MAb. Two IgG_4 isotype derivatives of CE9.1 were genetically engineered. The MAbs CE9γ4E and CE9γ4PE (Clenoliximab; IDEC) have the same V_H and V_L domains and light chain as CE9.1, but have been isotype switched by replacing the constant human γ1 heavy chain with a human γ4 heavy chain. MAb CE9γ4E contains a

[5] R. A. Newman, J. Alberts, D. Anderson, C. Heard, F. Norton, R. Raab, S. Shuey, and N. Hanna, *Bio/Technology* **10,** 1455 (1992).

L236E mutation near the hinge, which markedly reduces binding to FcRγI.[6] MAb CE9γ4PE contains in addition an S229P mutation in the hinge region, which enhances disulfide bond formation in this region and thus improves stability.[7]

The antigen sCD4 is a recombinant, soluble form of the extracellular human T cell coreceptor CD4. Full-length CD4 is a 55-kDa surface glycoprotein composed of a short intracellular C-terminal domain, a transmembrane domain, and the four extracellular immunoglobulin-like domains, D1–D4.[8] CD4 plays a central role in T cell ontogeny and activation, and is the attachment/trigger receptor for initiation of infection by the human immunodeficiency virus (HIV). The sCD4 protein used in our studies consists of the four extracellular domains.[9]

Description of Biophysical Methods

Here we present a brief overview of the biophysical methods used. More comprehensive descriptions can be found elsewhere.[10,11]

Analytical Ultracentrifugation

Sedimentation Equilibrium. Experiments are performed with a Beckman (Fullerton, CA) XL-A analytical ultracentrifuge. The absorbance at 280 nm at equilibrium is related to the distribution of a homogeneous species as

$$A_{280,r} = A_{280,m} \exp \left[\frac{M(1 - \bar{v}\rho)\omega^2(r^2 - r_m^2)}{2RT} \right] + \text{offset} \tag{1}$$

$A_{280,m}$ is absorbance at the meniscus, M is protein molecular mass, ω is angular velocity, r is distance in centimeters from the center of rotation, r_m is the radial position of the reference position (the meniscus) in centimeters, $\bar{v}$ is partial specific volume, ρ is solvent density, offset is an offset absorbance

---

[6] M. L. Alegre, A. M. Collins, V. L. Pulito, R. A. Brosius, W. C. Olson, R. A. Zivin, R. Knowles, J. R. Thistlethwaite, L. K. Jolliffe, and J. A. Bluestone, *J. Immunol.* **148**, 3461 (1992).

[7] S. Angal, D. J. King, M. W. Bodmer, A. Turner, G. Lawson, G. Roberts, B. Pedley, and J. R. Adair, *Mol. Immunol.* **30**, 105 (1993).

[8] P. J. Maddon, D. R. Littman, M. Godfrey, D. E. Maddon, L. Chess, and R. Axel, *Cell* **42**, 93 (1985).

[9] K. C. Deen, J. S. McDougal, R. Inacker, G. Folena-Wasserman, J. Arthos, J. Rosenberg, P. J. Maddon, R. Axel, and R. Sweet, *Nature (London)* **331**, 82 (1988).

[10] M. L. Doyle and P. Hensley, *Methods Enzymol.* **295**, 88 (1998).

[11] T. A. Morton and D. G. Myszka, *Methods Enzymol.* **295**, 268 (1998).

that is constant along radial position, R is the gas constant, and T is temperature. Data sets are acquired after attainment of equilibrium, as judged by an unchanging absorbance versus radial position signal over a 4-hr period. Partial specific volumes of 0.738 and 0.735 ml g^{-1} are calculated from sequence for sCD4 and the MAbs, respectively.[12]

Time-Derivative Sedimentation Velocity. The $g(s^*)$ data are acquired and analyzed by the time derivative method as described elsewhere.[13,14] The $g(s^*)$ data are fit to a Gaussian distribution, wherein the peak gives the sedimentation coefficient, s, and the standard deviation, σ, is related to the translational diffusion constant, D.

$$D = \frac{(\sigma \langle \omega^2 t \rangle r_{\mathrm{m}})^2}{2 \langle t \rangle} \tag{2}$$

Here, $\langle \omega^2 t \rangle$ is the harmonic mean of $\omega^2 t$ values for one $g(s^*)$ analysis, and t is the time for a particular scan, in seconds, from the start of the run. $\langle t \rangle$ is the harmonic mean of the times of the scans used in the $g(s^*)$ analysis. From s and D the molecular mass is determined from the Svedberg equation as[15]

$$M = \frac{RTs}{D(1 - \bar{v}\rho)} \tag{3}$$

Microcalorimetry

Isothermal Titration Calorimetry. Titrations are performed with a Microcal (Amherst, MA) Omega isothermal titration calorimeter (ITC).[16] sCD4 and MAbs are dialyzed and degassed prior to filling the calorimeter syringe and cell. Nonlinear least-squares analysis of the titration data is carried out with Microcal Origin software, using a single-site model, which is mathematically equivalent to an independent and identical two-site binding model.[17]

Differential Scanning Calorimetry. Thermal unfolding stabilities of sCD4 and MAbs CE9.1, CE9γ4E, and CE9γ4PE are measured with a differential

[12] E. T. Cohn and J. T. Edsall, *in* "Proteins, Amino Acids, and Peptides," p. 370. Reinhold, New York, 1943.
[13] W. F. Stafford III, *Anal. Biochem.* **203**, 295 (1992).
[14] W. F. Stafford III, *Methods Enzymol.* **240**, 478 (1994).
[15] W. F. Stafford III, *Curr. Opin. Biotechnol.* **8**, 14 (1997).
[16] T. Wiseman, S. Williston, J. F. Brandts, and L.-N. Lin, *Anal. Biochem.* **179**, 131 (1989).
[17] J. Wyman and S. J. Gill, *in* "Binding and Linkage: Functional Chemistry of Biological Macromolecules." University Science Books, Mill Valley, California, 1992.

scanning calorimeter (NanoDSC; Calorimetry Sciences, Provo, UT) with a scan rate of 1 deg/min.

Surface Plasmon Resonance

The surface plasmon resonance (SPR) data are measured with a BIAcore 1000 (Pharmacia Biosensor AB, Uppsala, Sweden) in 20 mM sodium phosphate, 150 mM sodium chloride, 0.05% (v/v) Tween 20, pH 7.4, and 25°. Immobilization of the MAbs to CM5 sensor chips is performed as previously described by reaction of the ligand-associated amines with an N-hydroxysuccinimide ester-activated surface.[18] sCD4 samples are passed over the surface at 60 μl/min and data are collected at 1 Hz throughout the analysis. The association and dissociation data from the sensorgrams are selected to minimize mass transport and rebinding effects, using BIA-evaluation 2.0. Nonlinear regression analysis is done via the Levenberg–Marquardt algorithm to a simple Langmuir reaction model to determine association and dissociation rate constants.[18,19]

Experimental Determination of Binding Energetics

Determination of Protein Concentrations

An accurate measurement of binding parameters for protein–protein interactions requires accurate determination of protein concentration. Protein concentration is critical for accurate determination of the binding molar ratio, which normally equates to stoichiometry, and for defining the binding physical parameters such as the affinity and kinetic constants, and binding thermodynamics. For ITC and SPR binding measurements the most critical parameters are the concentration of the reactant in the syringe and in the solution phase, respectively.

Several methods are reported for determination of protein concentrations, including absorbance at 280 nm,[20] absorbance in the far UV,[21] refractive index,[22] and amino acid analysis. We find that absorbance at 280 nm based on molar extinction coefficients calculated from amino acid sequence[20] is normally of sufficient accuracy for characterizing protein–protein interactions ($\pm$5–10%).

[18] D. J. O'Shannessy, M. Brigham-Burke, K. K. Soneson, P. Hensley, and I. S. Brooks, *Anal. Biochem.* **212,** 457 (1993).

[19] D. J. O'Shannessy, M. Brigham-Burke, K. K. Soneson, P. Hensley, and I. Brooks, *Methods Enzymol.* **240,** 323 (1994).

[20] C. N. Pace, F. Vajdos, L. Fee, G. Grimsley, and T. Gray, *Protein Sci.* **4,** 2411 (1995).

[21] C. M. Stoscheck, *Methods Enzymol.* **182,** 50 (1990).

[22] J. Babul and E. Stellwagon, *Anal. Biochem.* **28,** 216 (1969).

In the present study the concentration of sCD4 is most important because it is the reactant in the syringe for ITC and is in the solution phase for SPR. We thus determine the concentration of sCD4 by two independent methods and use the average value for our investigation. The theoretical extinction coefficient at 280 nm calculated from amino acid sequence is 60,200 M^{-1} cm^{-1} and the value determined from measurement of refractive index and the relation 4.1 fringes · ml/mg is 56,200 M^{-1} cm^{-1}. The average of these is 58,200 M^{-1} cm^{-1}. The MAb concentrations are determined from absorbance at 280 nm, using molar extinction coefficients of 260,000 and 280,000 M^{-1} cm^{-1}, respectively, for CE9.1 and the isotype 4 derivatives, as calculated from their sequences.

Determination of Self-Association Status of Reactants and Products

Analysis of the self-association status of reactants and products is necessary to define the binding model. In the absence of this knowledge, the physical meaning of observed binding parameters at a molecular level is unknown and may be obscured by coupled events between binding and self-association/dissociation processes. We therefore have characterized the self-association status of sCD4, MAbs, and the sCD4–MAb complex by analytical ultracentrifugation under the same conditions to be used in the kinetic and thermodynamic measurements.

Figure 2 shows sedimentation equilibrium analysis of sCD4. The data are well described by a single species as judged by the randomness of the residuals. The best-fit molecular mass is 44 ± 1 kDa and is in good agreement with the matrix-assisted laser desorption ionization-mass spectrometry (MALDI-MS) value of 44.62 kDa. The sedimentation equilibrium data demonstrate that sCD4 does not self-associate at concentrations up to about 1 mg/ml (20 μM). Similar results are found by sedimentation equilibrium analysis of MAbs CE9.1, CE9γ4PE, and CE9γ4PE. Their solution masses are in good agreement with values determined by MALDI-MS (Table I). The slightly lower mass values (3%) of the MAbs found by sedimentation equilibrium are within the expected error of calculating their partial specific volumes (which assumes additivity of individual amino acid partial specific volumes).

Since ligand-binding reactions of proteins may also cause an increase in self-association, it is also important to evaluate the self-association status of the product of the binding reaction. Again, any change in self-association will alter the observed bioenergetics and kinetics of the binding interaction. In the present study we have analyzed the self-association status of the complex between sCD4 and MAb CE9γ4PE by the time-derivative sedimentation velocity method as described below (see Stoichiometry). No self-

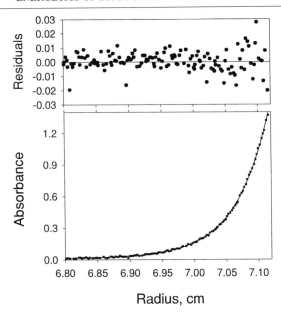

FIG. 2. Sedimentation equilibrium data for sCD4 (D1–D4) shown as absorbance at 281 nm versus radial position. *Bottom:* Best-fit curve [Eq. (1)] is for a single species of solution mass equal to 44 ± 1 kDa. *Top:* Residuals for best-fit curve in absorbance units. Conditions: 23,000 rpm, 20 mM sodium phosphate, 200 mM NaCl, pH 7.4, 0.5 mM EDTA, and 25°.

TABLE I
MOLECULAR MASS OF sCD4 AND ANTI-sCD4 MAbs[a]

Protein	Molecular mass (kDa)		
	Mass spectrometry	Sedimentation equilibrium	$g(s^*)$
sCD4	44.62	44	ND
CE9.1	148.4	143	ND
CE9γ4E	146.5	143	ND
CE9γ4PE	146.1	142	144

Abbreviation: ND, Not determined.

[a] Errors are approximately 0.2% for mass spectrometry, and 3% for sedimentation equilibrium and velocity values. Matrix-assisted laser desorption mass spectrometry data were measured with a Vestec LaserTec Research time-of-flight instrument.

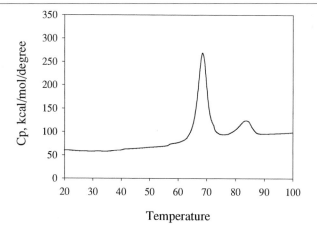

FIG. 3. DSC data for the temperature dependence of the heat capacity of MAb CE9.1 (2 mg/ml). The MAb C_p does not reveal any thermally induced transitions over the experimental temperature range used for the present ITC studies, suggesting the MAb exists in a single, native conformation. Unfolding begins to occur at about 60°. Multiple transitions are indicative of multiple domains and are typical of antibody unfolding. Conditions similar to those in Fig. 2.

association of the sCD4–mAb complex is detected under the conditions of the present study at 1-mg/ml concentrations.

Determination of Thermal Stabilities

The thermal stabilities of the reactants are important to characterize in order to define the temperature range over which the proteins can be studied without altering the protein structure through thermal unfolding. This in turn may relate to changes in apparent stoichiometry at elevated temperatures, or coupling of binding reactions to thermal folding processes. If present, such coupled processes would alter the observed bioenergetics and kinetics of the reaction.[23] We therefore have measured the thermal unfolding properties of sCD4 and MAbs CE9.1, CE9γ4E, and CE9γ4PE by DSC (Fig. 3). Thermal unfolding is irreversible under the present conditions, so that the bioenergetics of the unfolding processes are not determined; however, the data have allowed us to establish the maximum temperature at which binding experiments with the native forms of the reactants should be done. The half-unfolded temperatures have been found to be 70 (±2)° for all three MAbs and 62 (±2)° for sCD4. Thus, the highest temperature of the ITC measurements is 47° in order to avoid coupling of

[23] J. Thomson, G. S. Ratnaparkhi, R. Varadarajan, J. Sturtevant, and F. M. Richards, *Biochemistry* **33**, 8587 (1994).

the binding reactions to thermal unfolding of sCD4. Temperatures higher than 47° have been found to alter the apparent stoichiometry by ITC, indicating thermal loss of native sCD4.

Determination of Binding Stoichiometry and Molar Ratio

The sCD4-binding stoichiometry of the MAbs can be evaluated by several techniques, including time-derivative sedimentation velocity, titration sedimentation velocity, ITC, and SPR. However, of these measurements, only the time-derivative sedimentation velocity method provides a direct measurement of the molecular mass of the bound complex. It is thus the only method used in the present study that measures a true stoichiometry. The other methods we have used yield only a molar ratio, which is the concentration of one reactant that reacts with another. The accuracy of the molar ratio thus depends directly on the accuracy with which the reactant concentrations are determined initially and also assumes the reactants are 100% active. Nevertheless, it is valuable to compare the molar ratios and stoichiometries determined by several methods. The comparison serves as a useful control to firmly establish the stoichiometry and to assess the extent to which the reactants are active in each of the measurements.

Stoichiometry. The time-derivative,[13] and related sedimentation velocity methods, are especially useful for establishing stoichiometry because they are capable of measuring the molecular mass of the bound complex directly. In the time-derivative method, transformed sedimentation velocity data produce a single Gaussian peak for a single sedimenting species. Time-derivative patterns for MAb CE9γ4PE in the presence and absence of saturating sCD4 are shown in Fig. 4. The sedimentation coefficients ($s_{20,w}$) and translational diffusion constants ($D_{20,w}$) determined from this analysis are 6.85 and 7.31 Svedbergs, and 4.33 and 2.71 Ficks, respectively, for the MAb and sCD4–mAb complex. These values and Eq. (3) yield a molecular mass of 146 kDa for the MAb alone and 245 kDa for the complex (compared with 235.3 kDa from the sum of the MALDI-MS masses of the MAb and two sCD4). These results therefore provide a direct assessment of the stoichiometry, which is calculated to be 2.1 ± 0.1 sCD4 per MAb. These data also demonstrate that the MAb and the sCD4–MAb complex do not self-associate to form higher aggregates at micromolar concentration.

Molar Ratio. One means for determining the binding molar ratio for sCD4 binding to MAb CE9.1 is by titration sedimentation velocity. Figure 5 (*top*) presents a composite of three separate sedimentation experiments. The boundaries for sCD4 and MAb CE9.1 are clearly resolved. When a mixture of the two samples is sedimented the two boundaries

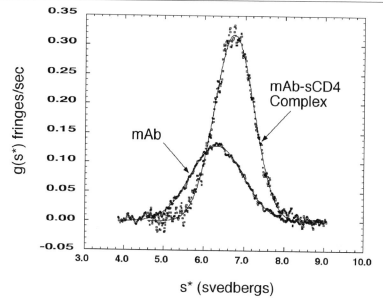

FIG. 4. Time-derivative sedimentation velocity analysis of MAb CE9γ4PE and the complex of MAb CE9γ4PE with sCD4. Conditions: 56,000 rpm, 150 mM NaCl, 10 mM phosphate, pH 7.4, and 20°. The MAb concentration in each case was 50 μg/ml. CD4 was added to the MAb in a 2:1 ratio to create the fully bound complex. Stoichiometry determined from this analysis is 2.07 ± 0.1 sCD4 per MAb.

observed correspond to unbound sCD4 and the CD4–MAb CE9.1, and the height of each boundary plateau is proportional to the concentration of each species; the free sCD4 concentration in the mixture (6.0 μM sCD4 plus 1.5 μM MAb CE9.1) is depleted by 50% relative to sCD4 alone. In Fig. 5 (*bottom*) increasing amounts of the MAb are titrated against samples of sCD4 that are present at a fixed concentration of 6.0 μM. The relative concentration of unbound sCD4 in each experiment is measured as the height of the plateau of the sCD4 boundary. The concentration of free sCD4 decreases linearly with increasing concentration of MAb. Extrapolation of the data to zero free sCD4 gives the equivalence point as 3 μM MAb for titration of 6 μM sCD4. The binding molar ratio was thus determined as two sCD4 per MAb. This method is also preferred for demonstrating that the majority (>98%) of the sCD4 material is active and capable of binding MAb, as judged by the absence of a sCD4 plateau when MAb is in excess.

The other methods we have used for determining molar ratio (to be described below) are ITC, which relies strongly on the accuracy of determin-

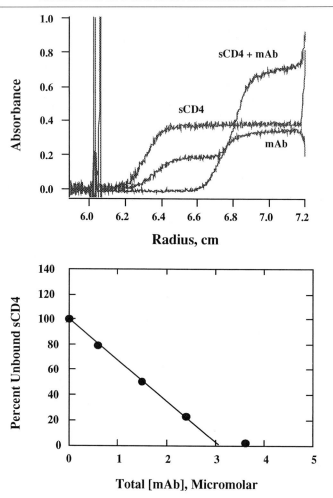

FIG. 5. Titration sedimentation velocity data of sCD4 and MAb CE9.1. *Top:* Data are shown as absorbance at 280 nm versus radial position in the centrifuge cell. Samples, as labeled: 6.0 μM CD4; 1.5 μM MAb CE9.1; a mixture of 6.0 μM CD4 plus 1.5 μM MAb CE9.1. Conditions: pH 7.4, 150 mM NaCl, 10 mM sodium phosphate, 25°, and 60,000 rpm. *Bottom:* Plot of percent unbound sCD4 (proportional to plateau height) versus total concentration of CE9.1 as determined by titration sedimentation velocity. Total sCD4 concentration was constant at 6.0 μM. The unbound sCD4 was depleted at a total concentration of 3.0 μM CE9.1, yielding a binding stoichiometry of 2 sCD4 per MAb CE9.1.

ing both sCD4 and MAb concentrations, and SPR, which is independent of concentration determinations but assumes all the immobilized MAb is active. Comparing the true stoichiometry measured by the time-derivative sedimentation velocity method (2.1 ± 0.1) to the molar ratios determined by titration sedimentation velocity (1.9 ± 0.1), ITC (1.90 ± 0.15), and SPR (1.64 ± 0.10) demonstrates that the majority of sCD4 and MAb reactants are active during all these measurements, in addition to defining the molecular stoichiometry as two sCD4 per MAb.

Determination of Binding Affinity and Thermodynamics by Isothermal Titration Calorimetry

We have measured the sCD4 affinities of the MAbs by two independent methods: an equilibrium method (ITC) and a kinetic method (SPR). Both procedures involve some degree of modeling during analysis and assumptions. For example, the ITC method we use is indirect, relying on measuring the affinity at high temperature followed by extrapolation to 25° by the integrated van't Hoff relation. The SPR measurements rely on modeling the kinetics with single on- and off-rate constants and then calculating the affinity K_d as their ratio. Thus, one reason for characterizing protein–protein binding interactions by more than one method is to provide a cross-check between the methods and ascertain whether systematic error is present.

Affinity from Isothermal Titration Calorimetry Measurements over Range of Temperature. Representative ITC data for titrating MAb CE9γ4PE with sCD4 are shown in Fig. 6. The data in the lower portion are analyzed by nonlinear least squares to a two-site independent and equivalent binding model, as is expected and widely applied to soluble antigen binding to symmetric MAb molecules. In addition, the present study provides further support for the simple model, in that we found that a single enthalpy change and binding equilibrium constant are sufficient to describe individual isotherms measured over a wide range of temperature and pH.[24,25] The independent and identical binding model is also supported by the SPR kinetics data (below), which are adequately described individually by single on- and off-rate constants.

The calorimetry results for all three MAbs are summarized in Table II. The three MAbs have best-fit molar binding ratios of 1.90 ± 0.15 sCD4 per MAb, which is indicative of a 2 : 1 stoichiometry and also demonstrates the high percentage of active antigen-binding sites in the MAb preparations.

[24] M. L. Doyle and P. Hensley, *Adv. Mol. Cell Biol.* **22A,** 279 (1996).
[25] M. L. Doyle, G. Louie, P. R. Dal Monte, and T. Sokoloski, *Methods Enzymol.* **259,** 183 (1995).

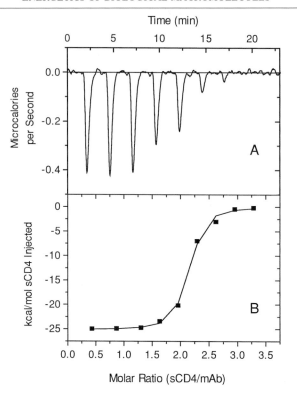

FIG. 6. Binding of sCD4 to MAb CE9γ4PE by titration microcalorimetry. (A) Raw data for consecutive injections of 130 μM sCD4 into 1.4 ml of 0.8 μM MAb CE9γ4PE. Three 4-μl injections were followed by six 3-μl injections. At each injection the reaction heat was observed as a deflection in microcalories per second, followed by a recovery to baseline. (B) Heats of reaction in kilocalories per mole of sCD4 injected versus cumulative molar ratio of sCD4 to MAb. Heats were determined by integration of the areas under the injection peaks in (A). The curve is the best fit to a single-site equilibrium binding model, with a K_d of 6 nM and a ΔH of -24.6 kcal/mol sCD4, and the molar ratio of sCD4 binding per MAb CE9γ4PE was 1.97 at pH 7.4, 20 mM potassium phosphate, 150 mM NaCl, and 45°.

The sCD4 affinities of all three MAbs are, within error, equivalent. Although a slightly higher affinity for CE9.1 may be suggested by its lower K_d value, this difference may be within experimental error, given the difficulty of measuring tight binding constants.

The parameters listed in Table II are the intrinsic values for the individual binding sites (corrected for the statistical factors for a two-site macromolecule[17]). The binding affinities at 25° are calculated from measurements of the equilibrium binding constant at 45° in combination with the experi-

TABLE II
INTRINSIC THERMODYNAMIC PARAMETERS FOR sCD4 BINDING TO MAbs[a]

MAb	$K_d{}^b$ (nM)	ΔG (kcal/mol sCD4)	ΔH (kcal/mol sCD4)	$-T\Delta S$ (kcal/mol sCD4)	ΔC_p (kcal/mol per sCD4 per degree)
CD9.1	0.2	-13.2 ± 0.5	-16.8 ± 1.0	3.6 ± 1.1	-0.48 ± 0.07
CD9γ4E	0.6	-12.6 ± 0.5	-15.8 ± 0.2	3.2 ± 0.6	-0.42 ± 0.01
CE9γ4PE	0.4	-12.8 ± 0.5	-15.5 ± 0.7	2.7 ± 0.9	-0.44 ± 0.05

[a] At 25°, pH 7.4, 20 mM potassium phosphate, 150 mM NaCl.
[b] Uncertainty in K_d is a factor of 2–3.

mentally determined enthalpy and heat capacity changes over a range of temperature (Fig. 7) according to[24]

$$K^T = K^0 e^{\frac{-\Delta H^0}{R}\left(\frac{1}{T}-\frac{1}{T_0}\right)} e^{\frac{\Delta C T_0}{R}\left(\frac{1}{T}-\frac{1}{T_0}\right)} \left(\frac{T}{T_0}\right)^{\frac{\Delta C}{R}} \tag{4}$$

Here K^T and K^0 are the binding equilibrium constants (inverse of K_d) at temperature T (degrees Kelvin) and reference temperature T_0, respectively. ΔH^0 is the enthalpy change at T_0, ΔC is the binding heat capacity change, and R is the gas constant.

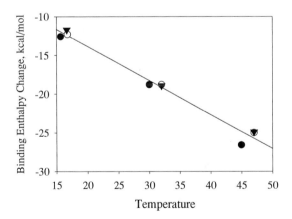

FIG. 7. Enthalpy change for sCD4 binding to MAbs versus temperature: (●) MAb CE9.1; (▼) MAb CE9γ4E; (○) MAb CE9γ4PE. Global analysis of all the data yielded the best-fit line. The slope of the plotted line is −0.450 kcal/mol of sCD4 per degree and represents the heat capacity change of the binding reactions. Analysis of individual data sets is summarized in Table II.

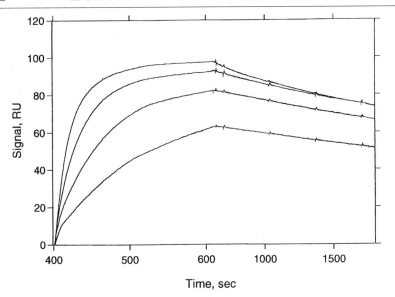

Fɪɢ. 8. SPR kinetics for sCD4 binding to immobilized MAb CE9.1 shown as response units versus time. The left-hand portion shows association progress curves, and the right-hand portion shows dissociation progress curves. Concentrations of sCD4 were 7, 14, 29, and 55 nM from bottom to top. Conditions: pH 7.4, 10 mM HEPES, 150 mM NaCl, 0.05% surfactant P20, and 25°.

Molecular Binding ΔH. The intrinsic sCD4-binding enthalpy change for MAb CE9.1 is determined by ITC by measuring sCD4-binding heats in the presence of two buffers that have distinct ionization heats (i.e., phosphate and Tricine buffers have ionization enthalpies of 1.0 and 7.7 kcal/mol at 30°, respectively[26,27]). Any difference in observed binding heat would thus indicate the existence of coupled protons, and a correction for the contribution of the buffer ionization heat would be necessary.[10] However, no difference in enthalpy change could be detected between the two buffer systems at pH 7.4, indicating the absence of linked protons at pH 7.4. Thus, the enthalpy changes reported in Table II are equal to the intrinsic heats of sCD4 binding to MAb CE9.1. The lack of coupled protons for the sCD4–MAb CE9.1 interaction at neutral pH has been reported previously.[25]

Binding Heat Capacity Changes (ΔC_p). Heat capacity changes for sCD4 binding to the MAbs are determined from ITC measurements of the binding

[26] J. J. Christensen, L. D. Hansen, and R. M. Izatt, *in* "Handbook of Proton Ionization Heats and Related Thermodynamic Quantities." John Wiley & Sons, New York, 1976.
[27] K. P. Murphy, D. Xie, K. S. Thompson, L. M. Amzel, and E. Freire, *Proteins* **15**, 113 (1993).

TABLE III
SURFACE PLASMON RESONANCE RATE CONSTANTS AND EQUILIBRIUM
DISSOCIATION CONSTANTS FOR REACTION OF MAbs WITH sCD4[a]

MAb	k_a ($\times 10^{-6}\ M^{-1}\ sec^{-1}$)	k_d ($\times 10^4\ sec^{-1}$)	K_d (nM)
CE9.1	1.1 ± 0.2	2.1 ± 0.5	0.19 ± 0.05
CE9γ4PE	2.1 ± 0.7	1.5 ± 0.8	0.07 ± 0.05

[a] At pH 7.4 and 25°.

enthalpy as a function of temperature. As seen in Fig. 7, the temperature dependence of the binding enthalpy changes are linear from 16 to 46°. The linearity indicates that temperature-induced side reactions (e.g., self-association of reactants or products, thermal unfolding) are not detected over this temperature range. The data in Fig. 7 also indicate that the binding enthalpy changes and heat capacity changes for sCD4 binding to MAbs CE9.1, CE9γ4E, and CE9γPE are, within error, equal.

Determination of Binding Affinity and Kinetics

SPR is used to measure the sCD4-binding affinity and kinetics of MAbs CE9.1 and CE9γ4PE. The SPR data are analyzed to obtain association and dissoction rate constants k_a and k_d as described elsewhere.[18,19] The data are acquired by methods that minimize the contribution from mass transport limitations.[11,28] Thus, for a simple binding mechanism the progress curves should correspond to the chemical rates of MAb–antigen binding, and the ratio of the rate constants is equal to the equilibrium dissociation constant, K_d (i.e., $K_d = k_d/k_a$). Raw data for sCD4 binding and dissociation phases of MAb CE9.1 are shown in Fig. 8. The kinetic rate and equilibrium binding constants determined by SPR for the two MAbs are presented in Table III. Within experimental error the sCD4-binding kinetics of the MAbs are equivalent. The association rate constants are small enough that mass transport effects are not likely to be severe.[29]

The molar ratios for sCD4 binding to the MAbs are also determined by SPR. Table IV lists the mass of each MAb immobilized and the mass of sCD4 bound to the immobilized MAb. These molar binding ratios indicate that two sCD4 bind per MAb, and also demonstrate that the majority of antigen-binding sites on the MAbs are active.

[28] D. J. O'Shannessy and D. J. Winsor, *Anal. Biochem.* **236,** 275 (1996).
[29] D. R. Hall, J. R. Cann, and D. J. Winzor, *Anal. Biochem.* **235,** 175 (1996).

TABLE IV
BINDING MOLAR RATIO DETERMINATIONS BY SURFACE PLASMON RESONANCE

MAb	MAb immobilized (pg)	sCD4 bound (pg)	Molar binding ratio
CE9.1	2500	1170	1.56
CE9γ4PE	2600	1370	1.70

Interpretation of Binding Thermodynamics

Binding Thermodynamics for Antibody–Protein Antigen Reactions

The binding thermodynamics of an interaction portray the underlying molecular forces of the reaction, such as bonding energies, hydrophobic effects, coupled conformational changes, and release of linked ions and water molecules. The enthalpic and entropic components of antigen–MAb reactions vary significantly in the literature and depend on the specific chemistry of each interaction. For example, in one instance[30] two different MAbs bound the same protein antigen with similar binding free energies (-7 and -10 kcal/mol) but with extremely different enthalpic contributions ($+2$ and -67 kcal/mol, respectively). Nevertheless, the thermodynamics that have been reported for MAb–protein antigen reactions tend to be enthalpically driven (-10 to -25 kcal/mol), involve heat capacity changes in the range of -200 to -700 cal/mol per degree, and have equilibrium dissociation constants in the range of 10^{-6} to 10^{-10} M.[2,27,31–37] The binding thermodynamics in Table II are thus normal for protein antigen–MAb reactions.

[30] D. Tello, E. Eisenstein, F. P. Schwarz, F. A. Goldbaum, B. A. Fields, R. A. Mariuzza, and R. J. Poljak, *J. Mol. Recognition* **7,** 57 (1994).

[31] T. N. Bhat, G. A. Bentley, G. Boulot, M. I. Greene, D. Tello, W. Dall' Acqua, H. Souchon, F. P. Schwarz, R. A. Mariuzza, and R. J. Poljak, *Proc. Natl. Acad. Sci. U.S.A.* **91,** 1089 (1994).

[32] K. A. Hibbits, D. S. Gill, and R. C. Willson, *Biochemistry* **33,** 3584 (1994).

[33] R. F. Kelley and M. P. O'Connell, *Biochemistry* **32,** 6828 (1993).

[34] K. P. Murphy, E. Freire, and Y. Paterson, *Proteins* **21,** 83 (1995).

[35] C. S. Raman, M. J. Allen, and B. T. Nall, *Biochemistry* **34,** 5831 (1995).

[36] F. P. Schwarz, D. Tello, F. A. Goldbaum, R. A. Mariuzza, and R. J. Poljak, *Eur. J. Biochem.* **228,** 388 (1995).

[37] K. Tsumoto, K. Ogasahara, Y. Ueda, K. Watanabe, K. Yutani and I. Kumagai, *J. Biol. Chem.* **270,** 18551 (1995).

TABLE V
PREDICTED CHANGES IN ACCESSIBLE SURFACE AREA OF
APOLAR AND POLAR GROUPS FOR sCD4–MAb
INTERACTIONS FROM BINDING THERMODYNAMICS[a]

MAb	ΔASA_{apolar}	ΔASA_{polar}	ΔASA_{total}
CE9.1	1990	1600	3590
CE9γ4E	1700	1450	3220
CE9γ4PE	1820	1470	3290

[a] Per sCD4 bound. Based on empirical correlations.[44] Propagated error in the ΔASA values is approximately 10%. ASA ($\mathring{A}^2$).

Binding Thermodynamics and Structure

Several correlations between binding thermodynamics and protein structure have been reported.[38–41] These correlations have been used to characterize structural aspects of several MAb–protein antigen interactions.[3,27,32–35] Here we present an example of how protein-binding thermodynamics may relate to protein structure. We also point out that correlations like these are expected to become refined with time as new structural and thermodynamic information becomes available. It is also important to keep in mind that deviations from the general correlations may occur, particularly when protein-binding interactions are coupled to binding of ions.[42,43]

Xie and Freire[44] have outlined a correlation between the amount of accessible surface area buried in protein–ligand or protein-folding reactions and the corresponding heat capacity and enthalpy changes:

$$\Delta C_p = 0.45\Delta ASA_{apol} - 0.26\Delta ASA_{pol} \tag{5}$$

$$\Delta H^{60} = -8.44\Delta ASA_{apol} + 31.4\Delta ASA_{pol} \tag{6}$$

Here ΔASA_{apol} and ΔASA_{pol} are the changes in accessible surface area of apolar and polar residues, ΔC_p is the binding heat capacity change, and ΔH^{60} is the binding enthalpy change at 60°. From experimental measurements of the binding heat capacity and enthalpy changes (Table II), Eqs. (5) and (6) were solved for ΔASA_{apol} and ΔASA_{pol} (Table V). Not surprisingly,

[38] J. R. Livingston, R. S. Spolar, and M. T. Record, *Biochemistry* **30**, 4237 (1991).
[39] K. P. Murphy and E. Freire, *Adv. Protein Chem.* **43**, 313 (1992).
[40] R. S. Spolar and M. T. Record, *Science* **263**, 777 (1994).
[41] J. M. Sturtevant, *Proc. Natl. Acad. Sci. U.S.A.* **74**, 2236 (1977).
[42] E. R. Guinto and E. Di Cera, *Biochemistry* **35**, 8800 (1996).
[43] A. G. Kozlov and T. M. Lohman, *J. Mol. Biol.* **278**, 999 (1998).
[44] D. Xie and E. Freire, *Proteins* **19**, 291 (1994).

given that the binding thermodynamics are nearly the same for the MAbs, the total amounts of accessible surface areas buried when sCD4 binds to the MAbs are also predicted to be similar. The average ΔASA_{total} for the three MAbs is 3400 Å^2 per sCD4 molecule and represents the difference in accessible surface area of unbound sCD4 and MAb versus the sCD4–MAb complex. In the absence of there being any conformational changes coupled to binding (i.e., a hypothetical rigid body association), the calculated value would predict the size of the sCD4–MAb complex interface (one-half the total area buried, or about 1700 Å^2). These calculations tend to predict that only minimal structural change occurs in sCD4 or the MAbs during binding, because the total amount of surface area buried is similar to that found in typical interfaces of protein–protein complexes (625 to 975 Å^2).[45] In one case, the X-ray crystal structure of growth hormone complexed with its receptor,[46] the total interface size of 2030 Å^2 is larger than that predicted here for sCD4–MAbs.

Binding Thermodynamics and Correlation to Conformational Change

Whether conformational changes are coupled to the sCD4–MAb binding reactions can also be assessed qualitatively by analysis of the binding entropy change. Spolar and Record[40] have shown an empirical correlation between binding thermodynamics and coupled conformational changes for cases in which three-dimensional structures are available for free and complexed protein reactants. The extent of conformational change was found to correlate with the entropy of association, ΔS_{assoc}, which was broken down empirically into three terms:

$$\Delta S_{assoc} = \Delta S_{HE} + \Delta S_{rt} + \Delta S_{other} \qquad (7)$$

Here ΔS_{HE} is the entropy due to the hydrophobic effect and ΔS_{rt} is the entropy due to reduction in rotational–translational degrees of freedom. ΔS_{other} is the entropy from all other sources, and is believed to primarily reflect coupled conformational changes. At the temperature T_s, where the entropy of association is zero, Eq. (7) reduces to

$$\Delta S_{other} = -\Delta S_{HE} - \Delta S_{rt} \qquad (8)$$

ΔS_{HE} can be estimated at T_s from $\Delta S_{HE} = 1.35 \, \Delta C_p \ln(T_s/386)$, and ΔS_{rt} was defined empirically as $\Delta S_{rt} = -50$ entropy units (e.u.) for typical "rigid body" macromolecular interactions. The term ΔS_{other} was empirically

[45] J. Janin and C. Chothia, *J. Biol. Chem.* **265**, 16027 (1990).
[46] J. A. Wells and A. M. de Vos, *Annu. Rev. Biophys. Biomol. Struct.* **22**, 329 (1993).

shown[40] to correlate with the number of amino acid residues that "fold," R^{th}, according to $R^{th} = \Delta S_{other}/-5.6$ e.u.

Using Eqs. (7) and (8) with consensus binding thermodynamic parameters (Table II) of $K_d = 0.4$ nM, $\Delta H = -16.0$ kcal/mol, $\Delta C_p = -450$ cal/mol per degree, 21 amino acid residues are predicted to "fold" during the reaction of each sCD4 with the MAbs. Because the statistical uncertainty in these calculations is at least ± 10 residues, we conclude that any conformational changes coupled to sCD4 binding by these MAbs are likely to be limited.

X-ray structural investigations in the literature have shown that conformational changes can occur with MAb–antigen reactions but are usually small and localized to the CDR loops or may involve sliding between the V_H and V_L chains to accommodate antigen.[47–50] Long-range structure changes in Fabs are rare, but one example is reported wherein structure changes were transmitted to the distal portion of the $C_H 1$ domain on complexation with a small ligand antigen.[51] This later case raises the possibility that structure changes could be transmitted to the Fc part of an intact MAb on binding antigen. In contrast to antibodies per se, long-range tertiary and quaternary structure changes have been observed in protein antigens on binding to antibody.[4,52–55]

Interpretation of Binding Kinetics

General Protein–Protein Reaction Kinetics

The kinetics of a protein–protein interaction can also provide insight into the nature of the binding mechanism. Association rate constants for protein–protein interactions are typically in the range of 10^5 to 10^6 M^{-1} sec^{-1}.[56,57] However, unusual association rates can be observed for unique binding mechanisms. For example, interactions dominated by electrostatic

[47] P. N. Colman, W. G. Laver, J. N. Varghese, A. T. Baker, P. A. Tulloch, G. M. Air, and R. G. Webster, *Nature (London)* **326**, 358 (1987).
[48] D. R. Davies and E. A. Padlan, *Curr. Biol.* **2**, 254 (1992).
[49] S. V. Evans, D. R. Rose, R. To, N. M. Young, and D. R. Bundle, *J. Mol. Biol.* **241**, 691 (1994).
[50] I. A. Wilson and R. L. Stanfield, *Curr. Opin. Struct. Biol.* **3**, 113 (1993).
[51] L. W. Guddat, L. Shan, J. M. Anchin, D. S. Linthicum, and A. B. Edmundson, *J. Mol. Biol.* **236**, 247 (1994).
[52] M. J. Allen, R. Jemmerson, and B. T. Nall, *Biochemistry* **33**, 3967 (1994).
[53] D. C. Benjamin, D. C. Williams, S. J. Smith-Gill, and G. S. Rule, *Biochemistry* **31**, 9539 (1992).
[54] A. F. Chaffotte and M. E. Goldberg, *J. Mol. Biol.* **197**, 131 (1987).
[55] B. Friguet, L. Djavadi-Ohaniance, and M. E. Goldberg, *Mol. Immunol.* **21**, 673 (1984).
[56] R. Koren and G. G. Hammes, *Biochemistry* **15**, 1165 (1976).
[57] S. H. Northrup and H. P. Erickson, *Proc. Natl. Acad. Sci. U.S.A.* **89**, 3338 (1992).

steering can have association rate constants upward of $4 \times 10^9 \, M^{-1} \, \text{sec}^{-1}$, while steric and orientation effects reduce association rate constants to $10^3 \, M^{-1} \, \text{sec}^{-1}$.[58-60]

Antibody–Protein Antigen Reaction Kinetics

Association rate constants for protein antigen–MAb interactions typically fall into the range of 10^4 to $10^6 \, M^{-1} \, \text{sec}^{-1}$.[19,33,61-65] Larger association rate constants (greater than $10^8 \, M^{-1} \, \text{sec}^{-1}$) have also been reported for MAb reactions, but only with nonprotein antigens.[66,67] Some antigen–MAb reactions that are coupled to conformational changes may exhibit slow association rates,[30] even though in some cases the equilibrium affinities are the same.[4] In complex binding mechanisms, association kinetics of antigen–antibody reactions can be rate limited by dissociation of dimeric antigen to monomers[54] or folding/unfolding of protein antigens.[52,55]

The similarity of the association and dissociation rate constants in Table III is consistent with the MAbs having the same kinetic sCD4-binding mechanism. The association rate constants of about $10^6 \, M^{-1} \, \text{sec}^{-1}$ are also typical of MAb–protein antigen reactions and provide no evidence of rate-limiting conformational changes.

Implications of Bioenergetics and Kinetics for sCD4-Binding Mechanism of Monoclonal Antibodies

Independent and Identical Two-Site Binding of sCD4 to Monoclonal Antibodies

Both equilibrium and kinetic measurements of sCD4 binding to the individual MAbs were unable to detect either heterogeneity or positive

[58] R. Wallis, G. R. Moore, R. James, and C. Kleanthous, *Biochemistry* **34**, 13743 (1995).

[59] B. W. Pontius, *Trends Biochem. Sci.* **18**, 181 (1993).

[60] C. S. Raman, R. Jemmerson, B. T. Nall, and M. J. Allen, *Biochemistry* **31**, 10370 (1992).

[61] C. A. K. Borrebaeck, A. C. Malmborg, C. Furebring, A. Michaelsson, S. Ward, L. Danielsson, and M. Ohlin, *Bio/Technology* **10**, 697 (1992).

[62] L. Jin and J. A. Wells, *Protein Sci.* **3**, 2351 (1994).

[63] M. Malmqvist, *Curr. Opin. Immunol.* **5**, 282 (1993).

[64] J. C. Mani, V. Marchi, and C. Cucurou, *Mol. Immunol.* **31**, 439 (1994).

[65] R. M. Wohlhueter, K. Parekh, V. Udhayakumar, S. Fang, and A. A. Lal, *J. Immunol.* **153**, 181 (1994).

[66] J. N. Herron, A. H. Terry, S. Johnston, X. He, L. W. Guddat, E. W. Voss, Jr., and A. B. Edmundson, *Biophys. J.* **67**, 2167 (1994).

[67] D. W. Mason and A. F. Williams, *in* "Kinetics of Antibody Reactions and the Analysis of Cell Surface Antigens." Blackwell Scientific Publications, Oxford, 1986.

cooperativity. The SPR kinetic data for sCD4 binding to each of the MAbs were well described by single association and dissociation rate constants, and the ITC data were well described individually by single enthalpy changes and single equilibrium binding constants over a wide range of temperature and pH.[24,25] These findings indicate that sCD4 binding to one Fab domain of the MAb is unaffected by binding at the other, either sterically or by long-range allosteric communication.

Allosteric Communication between Soluble Antigen Binding and Fc Domain

The present study also comments on the question of whether there is intramolecular communication between the Fab and Fc domains of IgG MAbs. This issue is central to MAb engineering and also relates to antigen-induced biological effector functions of IgG MAbs. IgG MAbs mediate various biological effector responses through binding interactions in their Fc domains, including fixation of complement, and phagocytosis and antibody-dependent cytotoxicity of foreign cells. The question arises as to how antigen binding elicits a biological response at the Fc portion of its structure. It is generally believed that much of this response is mediated through multi-valency effects (i.e., aggregation) that originate from the presence of multiple MAbs bound to a multivalent antigen.[68] However, it is conceivable that antigen binding may transmit an intramolecular signal to its Fc domain and regulate its effector functions. In fact, studies of different isotypes of anti-tubulin Fabs indicate that the different constant domains alter the binding bioenergetics and kinetics of antigen binding.[69]

The present study shows equivalent sCD4-binding kinetics and thermodynamics for the IgG_1 MAb CE9.1 and both IgG_4 derivatives. This suggests the absence of allosteric communication between antigen-binding and Fc domains. That is, an existing energetic coupling between antigen-binding and Fc domains in the IgG_1 version of the MAb would likely be disrupted by replacing critical amino acid residue within the allosteric pathway.[70] It is unlikely that replacing all the C_H domains in MAb CE9.1 with IgG_4 C_H domains would preserve precisely all such critical residues. Furthermore, as mentioned above, analysis of the binding thermodynamics and kinetics provided little if any evidence of conformational changes being coupled to sCD4 binding. Thus, bioenergetic analysis of the sCD4-binding interaction with these MAbs is most consistent with only limited local rearrangements

---

[68] H. Metzger, *J. Immunol.* **149,** 1477 (1992).
[69] O. Pritsch, G. Hudry-Clergeon, M. Buckle, Y. Petillot, J. P. Bouvet, J. Gagnon, and G. J. Dighiero, *Clin. Invest.* **98,** 2235 (1996).
[70] G. K. Ackers and F. R. Smith, *Annu. Rev. Biochem.* **54,** 597 (1985).

of CDR loops at the antigen-binding site, and is inconsistent with more extensive allosteric communication between antigen-binding and constant Fc regions of these anti-sCD4 MAbs. However, the different IgG isotypes could impact recognition of cell surface CD4 through differences in steric contraints on the simultaneous binding of both Fab regions, through Fc–Fc self-association interactions, or other factors. In this regard, an Fc effect on recognition of surface antigen is reported for IgG isotypic variants of an antibody directed against the intercellular cell adhesion molecule ICAM-1.[71] Like the CD4 antibodies, the IgG_1 and IgG_4 derivatives of this antibody showed similar binding reactions to a recombinant soluble form of ICAM-1. However, these derivatives and an IgG_2 variant did differ in their avidities for cell surface antigen, an observation attributed to differences in the segmental flexibility of the hinge regions of these molecules, which affected the positioning of the two Fab regions.

Summary

This chapter has described a bioenergetic analysis of the interaction of sCD4 with an IgG_1 and two IgG_4 derivatives of an anti-sCD4 MAb. The MAbs have identical V_H and V_L domains but differ markedly in their C_H and C_L domains, raising the question of whether their antigen-binding chemistries are altered. We find the sCD4-binding kinetics and thermodynamics of the MAbs are indistinguishable, which indicates rigorously that the molecular details of the binding interactions are the same. We also showed the importance of using multiple biophysical methods to define the binding model before the bioenergetics can be appropriately interpreted. Analysis of the binding thermodynamics and kinetics suggests conformational changes that might be coupled to sCD4 binding by these MAbs are small or absent.

Acknowledgments

PRIMATIZED is a registered trademark of IDEC Pharmaceuticals Corporation. We acknowledge SmithKline Beecham and IDEC Pharmaceuticals for coproducing the monoclonal antibodies used in these studies.

---

[71] M. M. Morelock, R. Rothlein, S. M. Bright, M. K. Robinson, E. T. Graham, J. P. Sabo, R. Owens, D. J. King, S. H. Norris, D. S. Scher, J. L. Wright, and J. R. Adair, *J. Biol. Chem.* **269**, 13048 (1994).

[11] Analysis of Interaction of Regulatory Protein TyrR with DNA

By GEOFFREY J. HOWLETT and BARRIE E. DAVIDSON

Introduction

TyrR is a DNA-binding protein that regulates several genes involved in aromatic amino acid biosynthesis in *Escherichia coli*. Regulation occurs via the repression or activation of several unlinked operons that comprise the TyrR regulon.[1,2] Repression varies considerably in magnitude between the different operons and usually involves the coeffectors ATP and tyrosine whereas activation is ATP independent and requires any of the three aromatic amino acids. The DNA sequence elements that bind TyrR (TyrR boxes) have the consensus sequence $TGTAAAN_6 TTTACA$ and may be classified as either strong or weak boxes. Strong boxes bind TyrR strongly *in vitro* in the absence of coeffectors whereas weak boxes have less identity with the consensus sequence and do not bind TyrR unless there is an adjacent strong box and both ATP and tyrosine are present. Most operons have one strong box and one weak box, usually separated (center-to-center) by 23 base pairs (bp). In most operons repressed by tyrosine, overlap of the promoter by the operator is predominantly via the weak box.

The nucleotide sequence of TyrR[3,4] predicts 513 amino acid residues and a subunit molecular weight of 57,640. A multidomain structure is suggested by limited proteolysis experiments[5] and from homology studies that indicate the central domain has a significant degree of sequence similarity with other prokaryotic regulatory proteins such as NtrC, NifA, and XylR.[2] A role for the central domain in ATP binding is indicated by sequences with significant similarity with the ATP-binding site in adenylate kinase. The C-terminal domain contains a putative helix–turn–helix motif

[1] A. J. Pittard and B. E. Davidson, *Mol. Microbiol.* **5**, 1585 (1991).
[2] A. J. Pittard, in *"Escherichia coli* and *Salmonella typhimurium:* Cellular and Molecular Biology (F. C. Neidhardt, J. L. Ingraham, K. Brooks-Low, B. Magasanik, M. Schaechter, and H. E. Umbarger, eds.), p. 368. American Society for Microbiology, Washington, DC, 1987.
[3] E. C. Cornish, V. P. Argyropoulos, A. J. Pittard, and B. E. Davidson, *J. Biol. Chem.* **261**, 403 (1986).
[4] J. Yang, S. Ganesan, J. Sarsero, and A. J. Pittard, *J. Bacteriol.* **175**, 1767 (1993).
[5] J. Cui and R. L. Somerville, *J. Biol. Chem.* **268**, 5040 (1993).

shown to play an essential role in DNA binding[6] while deletion of amino acids in the N-terminal domain generates a protein that can repress but not activate.[4] In the absence of coeffectors, TyrR exists in solution as a dimer but self-associates to hexamer in the presence of ATP (or the nonhydrolyzable analog ATPγS) and either tyrosine or phenylalanine.[7] These observations suggested that tyrosine-induced hexamerization drives TyrR-mediated gene repression and prompted a more detailed analysis of the thermodynamics of effector, TyrR, and DNA interactions. This chapter describes the methodology used to characterize the linkage between the binding of ligands to TyrR and the interaction of TyrR with DNA. The experimental strategies, based primarily on the analysis of sedimentation behavior, provide a quantitative model for the role of tyrosine as a low molecular weight effector of the TyrR regulon.

TyrR–Effector Interactions

Stoichiometry and Binding Constant for ATP and Tyrosine

The binding of low molecular weight ligands to TyrR is conveniently studied using a meniscus depletion sedimentation method originally described for an air-driven ultracentrifuge.[8] This method uses an inert solute such as dextran T40 to stabilize the gradients formed during centrifugation and to prevent stirring during deceleration of the rotor. The method permits rapid separation of free and bound ligand, which, in the case of ATP binding to TyrR, offers the advantage that effects due to the weak ATPase activity of TyrR are minimized. Measurement of the concentration of free ligand as a function of the amount of ligand bound per acceptor yields stoichiometries and binding constants using conventional binding theory.

Use of the meniscus depletion method requires consideration of the potential effects of centrifugation on the equilibrium concentrations of the interacting species. While such an analysis is possible by simulation of the sedimentation velocity behavior of a ligand–acceptor system,[9] a simpler, analytical description is available using sedimentation equilibrium theory. This theory applies to the conditions used in the meniscus depletion method, where the high angular velocities and short column lengths ensure rapid attainment of sedimentation equilibrium.

Consider TyrR (T) with n independent and equivalent sites for a low

[6] J. S. Hwang, J. Yang, and A. J. Pittard, *J. Bacteriol.* **179,** 1051 (1997).
[7] T. J. Wilson, P. Maroudas, G. J. Howlett, and B. E. Davidson, *J. Mol. Biol.* **238,** 309 (1994).
[8] G. J. Howlett, E. Yeh, and H. K. Schachman, *Arch. Biochem. Biophys.* **190,** 809 (1978).
[9] I. Z. Steinberg and H. K. Schachman, *Biochemistry* **5,** 3728 (1966).

molecular weight ligand, A. The various complexes can be described by the following equilibria:

$$T \quad + A \rightleftharpoons TA$$
$$TA \quad + A \rightleftharpoons TA_2$$
$$TA_{i-1} + S \rightleftharpoons TA_i \tag{1}$$
$$TA_{n-1} + A \rightleftharpoons TA_n$$

At sedimentation equilibrium all the species distribute in the centrifuge cell according to

$$C_i(r_1) = C_i(r_2)e^{[\sigma_i(r_1^2 - r_2^2)]} \tag{2}$$

where C_i is the concentration of species i, r_1 and r_2 are radial positions between the meniscus, r_m, and base of the cell, r_b, and σ_i is the reduced molecular weight given by

$$\sigma_i = M_i(1 - \bar{v}_i\rho)\omega^2/2RT \tag{3}$$

M_i and $\bar{v}_i$ are the molecular weight and partial specific volume of species i, ρ is the solution density, ω is the angular velocity, R is the gas constant, and T is the absolute temperature.

For a sector-shaped cell, the total number of moles of species i, Q_i, is related to the average concentration $(C_i)_{av}$ according to

$$Q_i = (C_i)_{av}V = (C_i)_{av}\theta b(r_b^2 - r_m^2)/2 \tag{4}$$

where V is the volume of the cell, θ is the sector angle, and b is the thickness of the cell. The number of moles of species i can also be written as

$$Q_i = \theta b \int_{r_m}^{r_b} C_i(r)r \, dr \tag{5}$$

which on integration and combination with Eqs. (2) and (4) becomes

$$(C_i)_{av} = (C_i)_{r_m} \frac{\{e^{\sigma_i(r_b^2 - r_m^2)} - 1\}}{\sigma_i(r_b^2 - r_m^2)} \tag{6}$$

or

$$(C_i)_{av} = (C_i)_{r_m}X_i \tag{7}$$

where

$$X_i = \frac{\{e^{\sigma_i(r_b^2 - r_m^2)} - 1\}}{\sigma_i(r_b^2 - r_m^2)}$$

Equations (2) and (7), applied to a single solute, allow simulation of sedimentation equilibrium distributions for any given set of experimental parameters. Extension to the case of an interacting ligand acceptor system requires consideration of the equilibrium and conservation of mass conditions. The binding constant for each ligand–acceptor complex is defined by

$$k_i = [TA_i]/[TA_{i-1}][A] \tag{8}$$

where the square brackets denote molar concentrations. Combination with Eqs. (7) and (8) gives

$$\frac{(C_{TAi})_{av}}{(C_{TAi-1})_{av}(C_A)_{av}} = k_i \frac{X_{TAi}}{X_{TAi-1}X_A} \tag{9}$$

Conservation of mass for the acceptor yields

$$(C_{prot})_{av} = (C_T)_{av} + (C_{TA})_{av} + \cdots (C_{TAi-1})_{av} \cdots + (C_{TAn})_{av} \tag{10}$$

where $(C_{prot})_{av}$ is the average molar concentration of acceptor in all of its forms. Because the n binding sites are assumed to be equivalent and independent the various equilibrium constants are related to the intrinsic binding constant, K, which describes the binding to a single site:

$$k_i = (n - i + 1)K/i \tag{11}$$

Combining Eqs. (9)–(11) gives

$$(C_{prot})_{av} = (C_T)_{av}\left[1 + \sum_{i=1}^{n} \frac{(C_A)_{av}^i K^i F_i X_{TAi}}{X_A^i X_T}\right] \tag{12}$$

where F_i is a statistical factor given by

$$F_i = [n(n - 1)(n - 2) \cdots n - (i - 1)]/i! \tag{13}$$

Similar considerations for the average concentration of total ligand, $(C_{lig})_{av}$, leads to

$$(C_{lig})_{av} = (C_A)_{av} + (C_T)_{av} \sum_{i=1}^{n} \frac{i(C_A)_{av}^i K^i F_i X_{TAi}}{X_A^i X_T} \tag{14}$$

Equations (12) and (14) express the conservation of mass and equilibrium condition for a ligand acceptor system at sedimentation equilibrium. The general approach of combining the equilibrium and conservation of mass condition used in the derivation of Eqs. (8)–(14) can be used to simulate sedimentation equilibrium distributions for any system of inter-

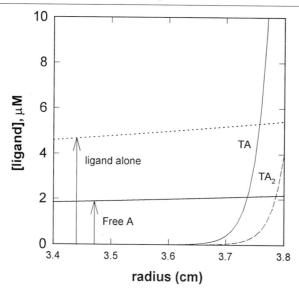

FIG. 1. Meniscus depletion sedimentation equilibrium distributions for the interaction of a ligand A with two sites on an acceptor T. The distributions were calculated using Eqs. (2)–(14) and parameter values, characteristic of the conditions used to study the interaction of ATP with TyrR: $(C_{prot})_{av} = 5\ \mu M$, $(C_{lig})_{av} = 2\ \mu M$, $M_A = 507$, $\bar{v}_A = 0.44$, $M_P = 110{,}000$, $\bar{v}_P = 0.73$, $\omega = 3000$ rad/sec, $r_m = 3.4$, $r_b = 3.8$, $n = 2$, and $K = 0.2\ \mu M^{-1}$. The distributions of acceptor with 1 mol (TA) or 2 mol (TA$_2$) of ligand are shown. The dotted line shows the distribution of ligand calculated for no added acceptor. The proportion of free ligand as a function of total ligands is estimated from measurements of the concentration of ligand at the meniscus in separate samples with and without acceptor as indicated by the arrows.

acting solutes. For an assigned set of experimental parameters the equations may be solved to yield $(C_A)_{av}$ and $(C_T)_{av}$ and hence the distribution of all species defined in Eq. (1). Figure 1 shows simulated distributions for ATP binding to two equivalent and independent sites on TyrR. Under the experimental conditions, the meniscus is depleted of TyrR and the ATP–TyrR complexes leaving unbound ATP in the supernatant.

In practice, the concentration of ligands at the meniscus at the completion of centrifugation is compared with the value obtained for a separate tube containing ligand alone. This corrects for the small amount of sedimentation of the ligand and provides an experimental value for $(C_A)_{av}$. An important consideration is whether centrifugation affects the total amount of free ligand present at sedimentation equilibrium. Combination of Eqs.

(12) and (14) provides an expression for the amount of ligand bound per total acceptor, R, where

$$R = [(C_{\text{lig}})_{\text{av}} - (C_{\text{A}})_{\text{av}}]/(C_{\text{prot}})_{\text{av}} = \frac{\displaystyle\sum_{i=1}^{n} \frac{i(C_{\text{A}})_{\text{av}}^{i} K^{i} F_{i} X_{\text{TA}i}}{X_{\text{A}}^{i} X_{\text{T}}}}{1 + \displaystyle\sum_{i=1}^{n} \frac{(C_{\text{A}})_{\text{av}}^{i} K^{i} F_{i} X_{\text{TA}i}}{X_{\text{A}}^{i} X_{\text{T}}}} \quad (15)$$

When the reduced molecular weight of the ligand is small, X_{T} is approximately equal to $X_{\text{TA}i}$ and X_{A} appoaches 1. Under these conditions Eq. (15) becomes

$$R = nK(C_{\text{A}})_{\text{av}}/[1 + K(C_{\text{A}})_{\text{av}}] \quad (16)$$

Equation (16) is the direct analog of that used for evaluating equilibrium dialysis experiments where "Scatchard" plots in the form $R/(C_{\text{A}})_{\text{av}}$ versus R provide values of K and n from the slope and abscissa intercepts, respectively. The question of interest is how such plots are affected when the reduced molecular weight of the ligand is not negligible, as was assumed in the derivation of Eq. (16). The results in Fig. 2 show simulated Scatchard

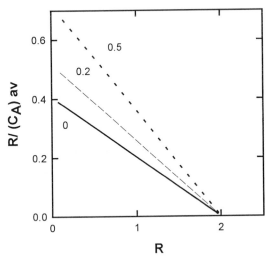

Fig. 2. The binding of a low molecular weight ligand to an acceptor under meniscus depletion sedimentation conditions. The theoretical plots of $R/(C_{\text{A}})_{\text{av}}$ versus R were calculated using Eq. (15) and show the effect on binding data stemming from the molecular weight of the ligand. The following parameters were used: $(C_{\text{prot}})_{\text{av}} = 5\ \mu M$, $(C_{\text{lig}})_{\text{av}} = 2\ \mu M$, $r_{\text{m}} = 3.4$, $r_{\text{b}} = 3.8$, $n = 2$, $K = 0.2\ \mu M^{-1}$, and $\sigma_{\text{P}} = 10$. The values used for σ_{A} are shown. The theoretical curves were virtually independent of σ_{P} for values of σ_{P} greater than 10.

plots calculated using Eq. (15) and for typical experimental parameters with different assumed values of the reduced molecular weight of the ligand, σ_A, ranging from 0 to 0.5. The results in Fig. 2 show that the slope of the plot increases with increasing σ_A while the value of the x intercept is unchanged. Thus a correct value of n is obtained while the value for K is overestimated by approximately 30 and 80% for values of σ_A of 0.2 and 0.5, respectively. The experimental conditions used to generate the sedimentation equilibrium distributions in Fig. 1 correspond to a value of $\sigma_A = 0.06$.

The meniscus depletion sedimentation method has been used to characterize the binding of both ATP and ATPγS to TyrR.[7,10] In a typical experiment, solutions (100 μl) containing TyrR and either [γ-^{32}P]- or [γ-^{35}S]ATP in buffer containing dextran T40 (5 mg/ml) are centrifuged in a Beckman (Fullerton, CA) benchtop TL100 ultracentrifuge and TLA100 rotor at 20° and 140,000g for 4 hr. After centrifugation the concentration of free ligand at the meniscus (upper two 10-μl samples) is determined by scintillation counting. The measured values are corrected for the sedimentation of free ligand measured in a separate tube containing ligand but no acceptor and the results processed to obtain the amount of ligand bound per acceptor, R, as a function of the free ligand concentration, $(C_A)_{av}$. Analysis of data for the binding of [γ-^{32}P]ATP to TyrR using the meniscus depletion sedimentation method indicated a stoichiometry of 1 mol of ATP bound per mole of TyrR subunit and a value for the concentration of ATP for half-maximal saturation of 5 μM. These results are similar to those obtained by steady state dialysis. Analysis of the binding of the nonhydrolyzable analog [γ-^{35}S]ATP to TyrR indicated half-maximal saturation at 4 μM ligand. These ligand affinities predict saturation of TyrR by ATP under normal conditions in the cell, where the concentration of ATP is approximately 3 mM.

The meniscus depletion method has also been used to examine the binding of tyrosine to TyrR.[10] The results indicate cooperative, ATP-dependent binding with a stoichiometry of 1 mol of tyrosine per TyrR subunit with half-maximal binding at approximately 50 μM. The cooperativity of tyrosine binding is linked to the self-association of TyrR.[7,11]

Tyrosine-Induced Hexamerization

Sedimentation Velocity. Sedimentation velocity measurements provide a rapid screening procedure to study the effect of tyrosine and other ligands

[10] V. P. Argaet, T. J. Wilson, and B. E. Davidson, *J. Biol. Chem.* **269**, 5171 (1994).
[11] T. J. Wilson, V. P. Argaet, G. J. Howlett, and B. E. Davidson, *Mol. Microbiol.* **17**, 483 (1995).

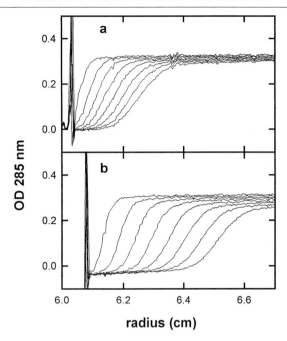

FIG. 3. The effect of tyrosine on the sedimentation velocity of TyrR. The optical density profiles at 285 nm were obtained at 8-min intervals, using an XLA analytical ultracentrifuge at a speed of 40,000 rpm and 20°. The solution conditions were (a) 200 μM ATPγS in 25 mM phosphate buffer, 100 mM KCl, 1 mM EDTA, and 10 mM MgCl$_2$, pH 7.4. (b) Same solution conditions with the addition of 500 μM tyrosine.

on the state of association of TyrR.[7] Optical density scans obtained for TyrR, using an XLA analytical ultracentrifuge and taken at 8-min intervals (Fig. 3), show depletion of protein at the meniscus, generating a solvent plateau and a migrating boundary representing TyrR sedimentation. The rate of sedimentation of TyrR increases significantly in the presence of tyrosine and ATP. Curve fitting of the data using the program SVED-BERG[12] and assuming a single sedimenting species yields the set of sedimentation coefficients shown in Table I. The value for TyrR in the absence of ligands (5.3S) is similar to the values obtained in the presence of saturating concentrations of either tyrosine or ATP and is consistent with a dimeric structure for TyrR. The sedimentation coefficient of TyrR in the presence of both ATP and tyrosine is much higher ($s = 10.5$S) and indicative of self-

[12] J. S. Philo, in "Modern Analytical Ultracentrifugation: Acquisition and Interpretation of Data for Biological and Synthetic Polymer Systems" (T. M. Schuster and T. M. Laue, eds.), p. 156. Birkhauser, Boston, 1994.

TABLE I
EFFECTS OF LIGANDS ON SEDIMENTATION COEFFICIENT
OF TyrR[a]

| Added ligand(s) | Sedimentation coefficient ($s_{20,w}$) | |
	Initial	After 24 hr[b]
None	5.3	5.6
Tyrosine	5.6	—
ATP	5.5	—
ATPγS	5.6	—
ATP + tyrosine	10.4	5.6
ATPγS + tyrosine	10.4	10.6

[a] Sedimentation velocity experiments were performed with 9 μM TyrR solutions, using a Beckman Optima XLA ultracentrifuge at 20° and
an angular velocity of 40,000 rpm.[7] —, Not determined. The concentrations of ligands were
as follows: tyrosine, 500 μM; ATP, 200 μM;
and ATPγS, 200 μM.

[b] After the initial sedimentation velocity run the
samples were kept at 25° for 24 hr and subjected
to a second sedimentation velocity analysis.

association. This increase in sedimentation coefficient in the presence of
ATP and tyrosine is not dependent on the weak ATPase activity of TyrR
because a similar increase in the sedimentation coefficient is observed when
the ATP is substituted with the nonhydrolyzable ATPγS. The reversibility
of the tyrosine plus ATP effect is established by mixing and incubation of
the samples for 24 hr under conditions in which TyrR hydrolyzes ATP.
The sedimentation coefficient of TyrR in a solution initially containing
tyrosine and ATP declines over a period of 24 hr to that observed in a
solution containing TyrR alone, whereas the value in the presence of tyrosine and ATPγS remains unaltered (Table I).

Sedimentation Equilibrium. Analysis of the ATP-dependent, tyrosine-
induced oligomerization of TyrR to determine the molecular weight and
binding constants for the interacting species involved sedimentation equilibrium experiments. This approach offers the advantage that it avoids some
of the uncertainties of sedimentation velocity analysis, where the rate constants for the interactions and the shape of the various complexes affect
the interpretation. Sedimentation equilibrium distributions for TyrR in the
presence of 200 μM ATPγS and in the presence of 200 μM ATPγS plus
500 μM tyrosine give excellent fits to single sedimenting species of molecu-

lar weight 113,000 and 340,000, respectively. Thus, TyrR is a dimer in the presence of saturating ATPγS and a hexamer in the presence of saturating ATPγS and tyrosine.

A feature of sedimentation equilibrium experiments is the capacity to analyze data obtained for multiple interacting systems. For a dimer–hexamer interaction, Eq. (2) written for both species becomes

$$C_D(r) = C_D(r_{ref})e^{\sigma_D(r^2-r_{ref}^2)} \tag{17}$$

$$C_H(r) = C_H(r_{ref})e^{\sigma_H(r^2-r_{ref}^2)} \tag{18}$$

where C_D and C_H are the molar concentrations at radial position r, r_{ref} is an arbitrary reference position, and σ_D and σ_H are the reduced molecular weights of dimer and hexamer, respectively. The total concentration (g/vol) of protein at r, $C_p(r)$, becomes

$$C_p(r) = C_D(r_{ref})e^{\sigma_D(r^2-r_{ref}^2)} + 3K_{DH}C_D(r_{ref})^3e^{3\sigma_D(r^2-r_{ref}^2)} \tag{19}$$

where the dimer–hexamer association constant K_{DH} is defined by

$$K_{DH} = C_H/C_D^3 \tag{20}$$

Implicit in Eq. (19) is the assumption that $3\sigma_D = \sigma_H$ and there is no volume change on association. Equation (19) relates the total concentration at radial position r to K_{DH}, $C_D(r_{ref})$, and σ_d and in principle can be fitted by nonlinear regression to obtain these parameters from the experimental data. The program used to achieve this fitting, SEDEQ1B, imposes an additional constraint on the fitting process to satisfy the condition of signal conservation that eliminates the reference concentration as an independent variable fitting parameter.[13] Application of this procedure to sedimentation equilibrium data obtained using high concentrations of TyrR (30 μM) in the presence of saturating (200 μM) ATPγS yields a value of 2.1×10^8 M^{-2} for K_{DH} (Table II).[11]

Figure 4 presents sedimentation equilibrium data for TyrR in the presence of saturating levels of ATPγS (30 μM) but subsaturating levels of tyrosine (90 μM). Nonlinear least-squares analysis of the data fitted according to a dimer–hexamer model [Eq. (19)] and neglecting the molecular weights of the bound ligand yields the distributions of total dimer and total hexamer shown in Fig. 4. This fit gives a value of 2.4×10^{11} M^{-2} for the apparent equilibrium, $K_{DH}(app)$,[14] where $K_{DH}(app)$ is defined by

[13] A. P. Minton, in "Modern Analytical Ultracentrifugation: Acquisition and Interpretation of Data for Biological and Synthetic Polymer Systems" (T. M. Schuster and T. M. Laue, eds.), p. 81. Birkhauser, Boston, 1994.

[14] M. F. Bailey, B. E. Davidson, A. P. Minton, W. H. Sawyer, and G. J. Howlett, *J. Mol. Biol.* **263**, 671 (1996).

TABLE II
ASSOCIATION CONSTANTS FOR INTERACTIONS OF TyrR[a]

Interaction	Binding constant	Value
TyrR dimer–hexamer	K_{DH}	$2.1 \times 10^8 \ M^{-2}$
TyrR dimer–tyrosine	K_{DY}	$2.8 \times 10^3 \ M^{-1}$
TyrR hexamer–tyrosine	K_{HY}	$4.2 \times 10^4 \ M^{-1}$
TyrR dimer–oligonucleotide	K_{OD}	$9.9 \times 10^6 \ M^{-1}$
TyrR hexamer–oligonucleotide	K_{HO}	$2.0 \times 10^8 \ M^{-1}$

[a] The association constants were determined by the procedures described in text and refer to conditions of 30 μM ATPγS in 25 mM phosphate buffer, 100 mM KCl, 1 mM EDTA, and 10 mM MgCl$_2$, pH 7.4. The oligonucleotide, 42A/42B, corresponds to base pairs 218–259 of the *tyrR* gene. The number of tyrosine-binding sites on TyrR dimer and hexamer are 2 and 6, respectively. TyrR dimer and hexamer have a single oligonucleotide-binding site. [Adapted from Bailey *et al.*[14]]

Eq. (20) in terms of the concentration of all species of liganded and unliganded dimer and hexamer. It should be noted that the data in Fig. 4 also fit to a dimer–tetramer–hexamer model. However, the values of the individual steps in the dimer–tetramer–hexamer model are difficult to resolve uniquely and, furthermore, the various values obtained indicate tetramer is a minor component.[7] Accordingly, the direct dimer–hexamer

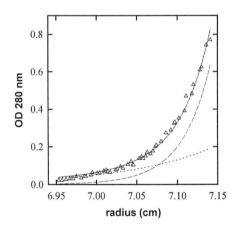

FIG. 4. Sedimentation equilibrium distribution of TyrR in the presence of ATPγS (30 μM) and tyrosine (90 μM). The equilibrium absorbance profile at 280 nm as a function of radial distance (in cm) was obtained at 10,000 rpm and 20°. The solution conditions were similar to those described in Fig. 3. The solid line represents the line of best fit calculated using Eq. (19) and for an apparent dimer–hexamer equilibrium constant K_{DH}(app) of $2.4 \times 10^{11} \ M^{-2}$. The dotted and broken lines refer to the fitted distributions of dimer and hexamer, respectively.

model has been used to describe the effects of tyrosine on the interactions of TyrR.

The induction of hexamer formation by tyrosine in the presence of saturating ATP is attributed to the preferential binding of tyrosine to TyrR hexamers according to a model originally proposed to describe the cooperative binding of ligands to a self-associating acceptor.[15] Values of K_{DH}(app) determined as a function of tyrosine concentration are related to the concentration of free tyrosine, [Y], by Eq. (21)[11]:

$$K_{DH}(\text{app}) = K_{DH}(1 + [Y]K_{HY})^6/(1 + [Y]K_{DY})^6 \qquad (21)$$

where the equilibrium constants refer to conditions of ATPγS saturation, K_{DH} is the hexamerization constant in the absence of tyrosine, and K_{DY} and K_{HY} are the association constants for the binding of tyrosine to TyrR dimers and hexamers, respectively. Analysis of data for K_{DH}(app) as a function of tyrosine concentration[11] yield values of 2800 M^{-1} for K_{DY} and 42,000 M^{-1} for K_{HY} (Table II). The values for K_{DH}, K_{DY}, and K_{HY} shown in Table II adequately describe the cooperative binding of tyrosine to TyrR.[11] Furthermore, the value obtained for K_{HY} is in good agreement with values obtained from direct tyrosine-binding measurements using meniscus depletion sedimentation experiments. Exploring the links between tyrosine-induced hexamerization and TyrR-mediated gene repression involved direct studies of the binding of TyrR dimers and hexamers to DNA.

TyrR–Oligonucleotide Interactions

The following double-stranded, 42-base pair oligonucleotide was designed to investigate the interactions of TyrR dimer and hexamer with DNA.

```
5'-TTTCCGTCTT TGTGTCAATGATTGTTGACAGA AACCTTCCTG   42A
   AAAGGCAGAA ACACAGTTACTAACAACTGTCT TTGGAAGGAC   42B
```

This oligonucleotide, designated 42A/42B, contains a strong TyrR box indicated by the nucleotides shown in boldface and represents base pairs 218–259 from the 5' regulatory region of the *tyrR* gene.[3]

Stoichiometry of Interaction of Dimeric TyrR and DNA

Fluorescence Titrations. Derivatives of 42A/42B with fluorescein attached to the nucleotide at position 7 from the 5' end of the A chain were

[15] L. W. Nichol, W. J. H. Jackson, and D. J. Winzor, *Biochemistry* **6**, 2449 (1967).

used to examine the molar ratios of TyrR–oligonucleotide complexes.[16] The fluorescence anisotropy of fluorescein–7-42A/42B increases from 0.112 to 0.126 on the addition of TyrR. Nonlinear least-squares fitting of fluorescence anisotropy data obtained as a function of added TyrR fits to a molar ratio of one TyrR dimer per oligonucleotide and a dissociation constant of 0.2 μM.[16]

Sedimentation Equilibrium Analysis. While spectroscopic titrations provide information about the molar ratios required for saturation, a more direct measurement of stoichiometry is provided by the determination of the molecular weight of the protein–DNA complexes. Two approaches were used. The first relied on the difference in the spectral properties of 42A/42B and TyrR to resolve the sedimentation equilibrium profiles into the separate contributions of the components present. A mixture of TyrR (3.4 μM) and oligonucleotide (0.5 μM) was centrifuged to sedimentation equilibrium and the optical profile at 260 nm analyzed. On the basis of the results of the fluorescence titrations these conditions correspond to "stoichiometric" concentrations, where the concentrations are significantly above the dissociation constant for the interaction. Under these conditions the ratio of TyrR to oligonucleotide (6.8:1) ensures the oligonucleotide is predominantly bound, such that the main light-absorbing species at 260 nm is the oligonucleotide–TyrR complex.

Analysis of the data, using Eq. (2) and assuming a single solute, yields an estimate of the weight-average reduced molecular mass, σ_{av}, of 44,900. This value is close to that calculated from the addition of the reduced molecular masses of TyrR dimer (29,300) and oligonucleotide (12,200), indicating a 1:1 complex. A second approach used to determine the molecular weight of the TyrR-oligonucleotide complex involved attachment of fluorescein to the 5′ end of the A chain of 42A/42B. This position was chosen to distance the chromophore from the TyrR-binding region of the oligonucleotide. The fluorescein label allows the sedimentation equilibrium profile of the oligonucleotide to be measured at a wavelength of 490 nm, in a spectral region where TyrR does not absorb. Analysis of sedimentation equilibrium data obtained using initial concentrations of 5 μM fluorescein–1-42A/42B and 40 μM TyrR indicated a single sedimenting solute with a weight-average reduced molecular mass of 44,000, consistent with the formation of a 1:1 complex of TyrR dimer and oligonucleotide. The observation of a single binding site for the 42-bp oligonucleotide 42A/42B on TyrR can be accommodated by assuming TyrR has a twofold axis of

[16] M. Bailey, P. Hagmar, D. P. Millar, B. E. Davidson, G. Tong, J. Haralambidis, and W. H. Sawyer, *Biochemistry* **34**, 15802 (1995).

symmetry with each putative helix–turn–helix DNA-binding motif recognizing each arm of the 22-bp recognition sequence within the 42-mer.[4]

Binding Constant for TyrR Dimer–Oligonucleotide Interactions

Determination of the binding constant for the interaction of dimeric TyrR with 42A/42B involved the direct fitting of sedimentation equilibrium data obtained at multiple wavelengths. Preliminary experiments were performed to select conditions so that there were appreciable levels of bound and free species. Sedimentation equilibrium profiles for mixtures of TyrR ($1 \, \mu M$) and 42A/42B ($0.5 \, \mu M$) in the presence and absence of ATPγS were collected at wavelengths of 230, 232, 260, 280, and 285 nm. Oligonucleotide binding to dimeric TyrR (D) was assumed to result in the formation of a single oligonucleotide–dimer complex, OD, governed by the association constant, K_{OD}:

$$K_{OD} = \frac{[OD]}{[O][D]} \tag{22}$$

The total absorbance at a particular radius and wavelength, $A(r, \lambda_j)$, is related to the concentrations of the individual species according to Eq. (2) and their molar extinction coefficients, ε_i:

$$A(r, \lambda_j) = \sum_{i=1}^{i=n} [\varepsilon_{i,j}[i]_{r_m} e^{\sigma_i(r^2 - r_m^2)}] + b_j \tag{23}$$

where n is the total number of species and b_j is the baseline correction at λ_j obtained experimentally by depleting the meniscus of TyrR and oligonucleotide using high angular velocities. Summation of Eq. (23) written for TyrR dimer (D), oligonucleotide (O), and the 1:1 complex (OD) and combination with Eq. (22) yields

$$A(r, \lambda_j) = \varepsilon_{O,j}[O]_{r_m} e^{\sigma_O(r^2 - r_m^2)} + \varepsilon_{D,j}[D]_{r_m} e^{\sigma_D(r^2 - r_m^2)}$$
$$+ \varepsilon_{OD,j} K_{OD} [O]_{r_m} [D]_{r_m} e^{\sigma_{OD}(r^2 - r_m^2)} + b_j \tag{24}$$

The three unknowns in Eq. (24), $[O]r_m$, $[D]r_m$, and K_{OD} can be solved using sets of sedimentation equilibrium data obtained at different wavelengths. Fitting experimental data using Eq. (24) requires the following assignments. The molecular mass of O and D are calculated from their composition and the molar extinction coefficients from absorbance measurements, using solutions of known concentrations of TyrR and oligonucleotide. The molecular mass and molar extinction coefficient of OD were assumed to be the sum of the molecular masses and molar extinction coefficients of O and D, respectively. Experimental support for this assumption was provided in mixing experiments,[14] where the change in absorbance on the binding of TyrR dimer to the oligonucleotide was less than 5% over the relevant wavelength range. The partial specific volume for the oligonucleotide was taken as 0.55 ml/g and the value for TyrR dimer was calculated as 0.731 ml/g from the amino acid composition, using the program SEDNTERP.[17] The weight-average partial specific volume of the complex (0.704 ml/g) was calculated assuming no volume change on association of the oligonucleotide with TyrR.

$$\bar{v}_{OD} = (M_O \bar{v}_O + M_D \bar{v}_D)/M_{OD} \tag{25}$$

The association constant for the formation of the 42A/42B–TyrR dimer complex was obtained from analysis of sedimentation equilibrium profiles obtained at wavelengths of 230, 232, 250, 260, 270, 280, and 285 nm.[14] This range of wavelengths is needed to minimize the difficulties of reproducibly selecting precise wavelengths using the monochromator of the XLA analytical ultracentrifuge. Global analysis of the data using Eq. (24) and the commercial software package SIGMAPLOT (Jandel Scientific, San Rafael, CA) yielded a value of the association constant for the interaction of 42A/42B with TyrR in the absence of ATPγS of 2.8 ($\pm$0.1) $\times$ 10^6 M^{-1}. The presence of ATPγS enhanced the affinity of TyrR for the oligonucleotide by a factor of approximately 3.5 [K_{OD} = 9.9 ($\pm$0.3) $\times$ 10^6 M^{-1}] (Table II).

Stoichiometry of TyrR Hexamer–Oligonucleotide Interactions

Meniscus Depletion Sedimentation Experiments. The number of sites for oligonucleotide binding per hexamer was estimated using as analytical ultracentrifuge and meniscus-depletion experiments. As discussed in relation to Fig. 2, these experiments give reliable estimates of the stoichiometry of ligand–acceptor interactions. High concentrations of TyrR in the presence of saturating ATPγS (200 μM) and tyrosine (550 μM) were used to ensure greater than 95% hexamer. These conditions correspond to "stoichiometric" binding conditions, in which added oligonucleotide, at substoichiometric amounts, is essentially all bound. A mixture of TyrR hexamer (62 μM) and fluorescein–1-42A/42B (5 μM) was centrifuged in the XLA analytical ultracentrifuge at 13,000 rpm in the presence of increasing amounts of nonlabeled competitor (42A/42B). The rotor speed was chosen to deplete TyrR and its DNA complexes from the meniscus, leaving unbound oligonucleotide. Measurements of the optical density at 490 nm near the meniscus indicated there was essentially no free fluorescein-labeled

[17] T. M. Laue, B. D. Shah, T. M. Ridgeway, and S. L. Pelletier, *in* "Analytical Ultracentrifugation in Biochemistry and Polymer Science" (S. E. Harding, A. J. Rowe, and J. C. Horton, eds.), p. 90. Royal Society of Chemistry, Cambridge, 1992.

oligonucleotide in the absence of nonlabeled competitor.[14] The stoichiometry of the TyrR hexamer–oligonucleotide interaction was measured by titrating the amount of nonlabeled oligonucleotide required to displace labeled oligonucleotide from the TyrR–DNA complexes. Likely models were considered to be either one or three oligonucleotide sites on the hexamer. For mixtures of TyrR (62 μM subunit) and fluorescein–1–42A/42B (5 μM), these two models predict either 10 μM total oligonucleotide (one-site model) or 30 μM total oligonucleotide (three-site model) would be required before significant displacement of the labeled oligonucleotide occurs. The one-site model provided the best fit to the data.[14]

Competitive Fluorescence Anisotropy Titration. Comparable studies were also performed in which the amount of bound oligonucleotide (fluorescein–1–42A/42B) was determined by fluorescence anisotropy measurements.[14] Mixtures of TyrR (62 μM) in the presence of saturating ATPγS and tyrosine and 0.9 μM fluorescein–1–42A/42B were titrated with increasing amounts of unlabeled oligonucleotide. Comparison of the experimental data with simulated fluorescence anisotropy data provided additional support for the one-site compared with the three-site model. These studies indicate a stoichiometry of one binding site for 42A/42B oligonucleotide per TyrR hexamer with two sites lost on hexamerization.

Oligonucleotide Binding to TyrR Hexamer

Two-Component Analysis. The effect of tyrosine-induced hexamerization of TyrR on the binding of oligonucleotide was initially analyzed in a model-independent fashion to obtain the separate distributions of the two macromolecular components. This approach is similar to methods previously reported.[18] The absorbance at a specific radius and wavelength, λ_j, is given by

$$A(r, \lambda_j) = \varepsilon_{T,j}[T](r) + \varepsilon_{O,j}[O](r) + b_j \tag{26}$$

where the square brackets denote molar concentrations of total TyrR (T) and total 42A/42B(O). Wavelength scans of TyrR and 42A/42B alone, performed in an XLA analytical ultracentrifuge, were analyzed by the program EXCOEF to provide values for $\varepsilon_{T,j}$ and $\varepsilon_{O,j}$, the molar extinction coefficients of T and O, respectively. Measurement of $A(r, \lambda_j)$ at different wavelengths may be solved for $[T](r)$ and $[O](r)$ using Eq. (26) and the program TWOCOMP to obtain the radial distributions of the TyrR and 42A/42B components.

[18] M. S. Lewis, R. I. Shrager, and S. J. Kim, *in* "Modern Analytical Ultracentrifugation: Acquisition and Interpretation of Data for Biological and Synthetic Polymer Systems" (T. M. Schuster and T. M. Laue, eds.), p. 94. Birkhauser, Boston, 1994.

Sedimentation equilibrium profiles obtained for mixtures of TyrR (1.9 μM) and 42A/42B (0.5, 0.9, and 1.3 μM) in the presence of ATPγS (30 μM) and tyrosine (190 μM) were analyzed using the TWOCOMP procedure, to obtain the total distributions of the two macromolecular components. The ln(concentration) versus r^2 plots for both the oligonucleotide and TyrR components were nonlinear, indicating heterogeneity with respect to molecular mass. The weight-average reduced molecular weight, σ_{av}, for the TyrR component can be estimated from the slope of a line drawn through the entire data (Fig. 5). Values for σ_{av} for total TyrR over the whole cell initially increase from 44,100 in the absence of 42A/42B, to 49,500 in the presence of 0.5 μM 42A/42B, then decrease to approximately 44,500 at 0.9 and 1.3 μM 42A/42B. Thus at low ratios of oligonucleotide to TyrR, binding favors self-association of the protein from dimer to hexamer. However, as this ratio is increased and the pool of free protein is depleted, binding favors dissociation of TyrR from hexamer to dimer. Simulations based on a one-site model, in which the oligonucleotide binds preferentially to the hexamer, provide an explanation for the effect of oligonucleotide

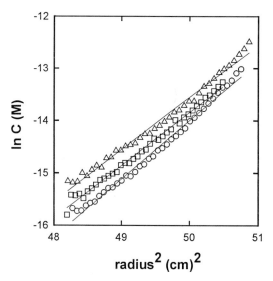

FIG. 5. The effect of oligonucleotide 42A/42B on the sedimentation equilibrium distribution of TyrR. Mixtures of TyrR and 42A/42B were centrifuged at 10,000 rpm in the presence of 30 μM ATPγS and 190 μM tyrosine. The sedimentation equilibrium data at 230, 232, 260, 280, and 285 nm were analyzed using Eq. (26) and the program TWOCOMP to obtain the molar concentration of total TyrR as a function of radial distance. Mixtures were composed of 1.9 μM TyrR alone (△), and with 0.5 μM (O) and 1.3 μM (□) 42A/42B. [Adapted from Bailey et al.[14]]

binding on the weight-average reduced molecular mass of TyrR. These simulations show that at low levels of added oligonucleotide there is an initial increase in the proportion of TyrR hexamers followed by a significant decrease in favor of dimers. This behavior is attributed to the stronger oligonucleotide affinity of the TyrR hexamer dominating at low oligonucleotide levels, while the decreased number of binding sites on hexamerization, and the reduced concentration of free TyrR dimers, favor the formation of dimeric species at high oligonucleotide concentrations. A three-site model, involving equivalent and independent sites on the hexamer, does not predict an initial increase in hexamerization followed by a decrease at higher oligonucleotide concentrations.

Model-Dependent Analysis of Oligonucleotide–TyrR Hexamer Interaction. Sedimentation equilibrium data were also analyzed by direct fitting of multiple wavelength data in a model-dependent fashion, using procedures analogous to those described above to fit oligonucleotide–TyrR dimer interaction [Eqs. (22)–(24)]. The self-association of TyrR from dimer to hexamer (H) in the presence of ATPγS and tyrosine is defined by the apparent association constant $K_{DH}(app)$ [Eq. (20)]. The binding of O to a single site on the hexamer is defined by the association constant, K_{HO}:

$$K_{HO} = \frac{[HO]}{[O][H]} \quad (27)$$

where HO is the molar concentration of the single oligonucleotide–TyrR hexamer complex according to the one-site model. Analysis of oligonucleotide binding to hexameric TyrR was carried out by extending Eq. (24) to account for the presence of species H and HO.

$$
\begin{aligned}
A(r, \lambda_j) = {} & \varepsilon_{O,j}[O]_{r_m} e^{\sigma_O(r^2 - r_m^2)} + \varepsilon_{D,j}[D]_{r_m} e^{\sigma_D(r^2 - r_m^2)} \\
& + \varepsilon_{OD,j} K_{OD}[O]_{r_m}[D]_{r_m} e^{\sigma_{OD}(r^2 - r_m^2)} \\
& + 3\varepsilon_{H,j} K_{DH}(app)[D]_{r_m}^3 e^{\sigma_H(r^2 - r_m^2)} \\
& + 3\varepsilon_{HO,j} K_{HO} K_{DH}(app)[D]_{r_m}^3 [O]_{r_m} e^{\sigma_{HO}(r^2 - r_m^2)} + b_j
\end{aligned}
\quad (28)
$$

Because values and K_{OD} and $K_{DH}(app)$ were available from previous experiments (Table II), Eq. (28) represents a set of equations at a multiple wavelengths that may be solved to obtain values for [O] and [D] at r_m and K_{OH}. The molecular mass and molar extinction coefficient of the oligonucleotide–hexamer complex were assumed to be the sum of the components making up that complex. The weight-average partial specific volume of HO was calculated as 0.723 assuming no volume change on complex formation [Eq. (25)]. The sedimentation equilibrium distributions of the three TyrR–oligonucleotide mixtures described above were collected at 230, 232, 250,

260, 270, 280, and 285 nm, and the data for each mixture were globally fitted to a one-site model in which TyrR dimer and hexamer each possess a single oligonucleotide-binding site. Analysis of the data to determine the binding constant for the 42A/42B–TyrR hexamer interaction involved assigning K_{OD} as the value determined for the TyrR dimer–oligonucleotide interaction in the presence of saturating ATP (Table II) and $K_{DH}(app)$ based the tyrosine-induced self-association of TyrR in the absence of DNA.[7] The approach assumed that at the tyrosine concentration used (190 μM), the binding of tyrosine to dimeric TyrR ($K_d = 330 \ \mu M$)[11] had no effect on the binding of oligonucleotide. The model also assumes the absence of tetrameric TyrR species.[7] Representative plots of the data at wavelengths of 230, 260, and 285 nm are shown for a mixture of 1.9 μM TyrR and 0.9 μM 42A/42B in Fig. 6. The values obtained for K_{HO} from global analysis of the three data sets were all in close agreement, the average value being $2.0 \ (\pm 0.1) \times 10^8 \ M^{-1}$ (Table II). This value indicates that the oligonucleotide is bound approximately 20-fold tighter by TyrR hexamer compared with dimer. Attempts to fit the three data sets to a three-site model, involving three oligonucleotide-binding sites on the hexamer, were unsuccessful.[14]

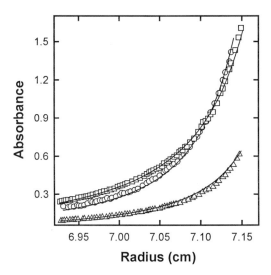

FIG. 6. Sedimentation equilibrium data for a mixture of TyrR (1.9 μM) and oligonucleotide 42A/42B (0.9 μM). Data were collected at 230, 232, 250, 260, 270, 280, and 285 nm and fitted globally, assuming the oligonucleotide binds to one site on TyrR dimer and to one site on TyrR hexamer according to the equilibrium constants summarized in Table II. Representative data collected at (○) 230 nm, (□) 260 nm, and (△) 285 nm together with the fitted lines are shown. [Adapted from Bailey et al.[14]]

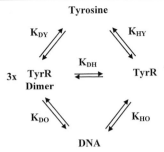

FIG. 7. Thermodynamic model for the interaction of tyrosine and oligonucleotide (42A/42B) with TyrR. Values for the equilibrium constants are summarized in Table II. TyrR dimer has two binding sites for tyrosine and one for oligonucleotide binding while TyrR hexamer has six sites for tyrosine and one oligonucleotide-binding site.

Modeling Tyrosine-Mediated Repression by TyrR

The binding constants and stoichiometries summarized in Table II describe the binding of oligonucleotide 42A/42B to TyrR dimers and hexamers and are the basis of the simple thermodynamic model presented in Fig. 7. The model pertains to conditions of ATP saturation and features preferential binding of tyrosine to TyrR hexamer with an affinity approximately 14 times higher than the binding of tyrosine to TyrR dimer. This preferential binding of tyrosine induces hexamerization of TyrR,[7] underlies the cooperative binding of tyrosine to TyrR,[10] and accounts for the effect of TyrR concentration on tyrosine binding.[11] The model also incorporates preferential binding of oligonucleotide in which the single binding site on the hexamer has an affinity for 42A/42B approximately 20 times higher than the binding site on TyrR dimer. At low oligonucleotide levels the higher affinity for the hexamer favors hexamerization while at a high oligonucleotide level dimerization is favored by the greater number of binding sites per subunit compared with hexamer.[14]

The model shown in Fig. 7 allows examination of the effect of tyrosine on the binding of oligonucleotides to TyrR. Relevant equations have been derived previously[19] to describe the random binding of two ligands to a monomer–tetramer system represented by chorismate mutase/prephenate dehydrogenase from *E. coli*. The interaction of tyrosine (Y) and oligonucleotide (O) with TyrR dimer (D) and hexamer (H) generates the species DY_iO_j and HY_iO_j, where i and j vary from 0 to n and 0 to p for the dimer and from 0 to m and 0 to q for the hexamer. Summation of these species and use of the appropriate statistical factors [Eq. (13)] and the binomial

[19] G. S. Hudson, G. J. Howlett, and B. E. Davidson, *J. Biol. Chem.* **258**, 3114 (1983).

theorem generates expressions for the total amount of TyrR and oligonucleotide. The total molar concentration of TyrR (C_{prot}) in dimer equivalents is given by

$$C_{prot} = (1 + \alpha)^n(1 + \beta)^p[D] + 3(1 + A\alpha)^m(1 + B\beta)^q[H] \qquad (29)$$

Where $\alpha = [Y]K_{DY}$, $\beta = [O]K_{DO}$, $A = K_{HY}/K_{DY}$, and $B = K_{HO}/K_{OD}$ and square brackets represent molar concentrations of the free species. The molar concentration of total oligonucleotide is given by

$$C_{Oligo} = [O] + p\beta(1 + \alpha)^n(1 + \beta)^{p-1}[D] \qquad (30)$$
$$+ qB\beta(1 + A\alpha)^m(1 + B\beta)^{q-1}[H]$$

For given values of the binding constants and stoichiometries (Table II) and the total concentrations of TyrR and oligonucleotide Eqs. (29) and (30) may be solved to determine the concentrations of bound oligonucleotide as a function of the concentration of free tyrosine.

The results presented in Fig. 8 show the fraction of free oligonucleotide

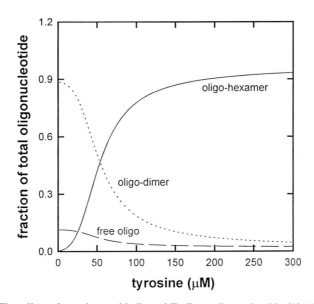

FIG. 8. The effect of tyrosine on binding of TyrR to oligonucleotide (42A/42B DNA). The fraction of free oligonucleotide and oligonucleotide bound to TyrR dimer and hexamer was calculated using Eqs. (29) and (30) according to the model shown in Fig. 7 and for values for the relevant binding constants and stoichiometries summarized in Table II. The calculations are based on a total concentration of TyrR (800 nM dimer equivalents), TyrR-binding sites (1.5 nM), and range of tyrosine concentrations comparable to those existing *in vivo*. The cooperative dependence of DNA–TyrR hexamer formation on tyrosine concentration is characterized by a Hill coefficient of 3.

and oligonucleotide bound to TyrR dimer and hexamer as a function of free tyrosine concentration ranging from 0 to 300 μM under conditions similar to those existing *in vivo*. The concentration of ATP in the cell of about 3 mM ensures TyrR is saturated with ATP. Calculations based on the volume of *E. coli* cells and TyrR content indicate a TyrR subunit concentration in the range 400–800 nM.[7] Similarly, the site concentration of a single TyrR strong box is about 1.5 nM. The results in Fig. 8 show that at low tyrosine concentrations the predominant oligonucleotide species is the dimer–oligonucleotide complex with a small amount of free oligonucleotide. As the concentration of free tyrosine increases there is a decrease in the concentration of both these species and a corresponding sigmoidal increase in the concentration of the oligonucleotide–hexamer complex. This increase in the concentration of oligonucleotide bound to hexamer is highly cooperative and characterized by a Hill coefficient of approximately 3. An alternative expression for cooperativity is the ratio of the concentration of ligand required for 90% compared with 10% of the full change.[20] For a noncooperative change this cooperativity factor is approximately 81. The tyrosine dependence of the oligonucleotide–TyrR hexamer formation shown in Fig. 8 is characterized by a cooperativity factor of 7.9.

Concluding Comments

The finding that hexameric TyrR, composed of three dimers, binds the oligonucleotide 42A/42B with a 1 : 1 stoichiometry is somewhat surprising. The conclusion is based on several observations.

1. The end point of titration data using meniscus depletion experiments
2. Fluorescence anisotropy data using unlabeled competitor oligonucleotide
3. Multicomponent sedimentation equilibrium analysis in which an initial increase in the amount of hexamer is followed by a decrease at higher levels of oligonucleotide (Fig. 5)
4. Global analysis of sedimentation equilibrium data obtained at several different TyrR–DNA ratios (Fig. 6)

There are a number of possible explanations for the loss of two oligonucleotide-binding sites on hexamerization. The binding of the 42A/42B oligonucleotide to a single dimer-binding site on the outside of the hexamer may sterically prevent access to the other two sites. The binding of oligonucleotide to one site on the hexamer may induce such a strong negatively coopera-

---

[20] A. Goldbeter and D. E. Koshland, Jr., *Proc. Natl. Acad. Sci. U.S.A.* **78**, 6840 (1981).

tive effect that the occupation of the other two sites in the hexamer is prevented. A further explanation that cannot formally be excluded at present is that the hexamer has a central cavity containing the oligonucleotide-binding site. Structural studies at the molecular level are required to distinguish these possibilities.

The simple thermodynamic model presented in Fig. 7 provides a quantitative description of the role of low molecular weight effector molecules in regulating the solution interactions of TyrR. An underlying hypothesis is that the binding of TyrR hexamer to specific DNA-binding sites limits the capacity of RNA polymerase to effectively transcribe the message. Extrapolation of the model to the *in vivo* situation requires several important qualifications. The model pertains to the ATP-dependent binding of tyrosine to TyrR that mediates the repression of genes by TyrR.[11] Activation of genes by TyrR involves the binding of aromatic amino acids to an ATP-independent site on TyrR and does not appear to involve hexamerization of TyrR.[11] An additional qualification of the model described in Fig. 7 is based on equilibrium constants derived from studies in dilute solution. The model predicts cooperative binding of TyrR hexamer to DNA over a tyrosine concentration range likely to exist *in vivo*. However, it should be noted that these equilibrium constants can be affected by molecular crowding due to the high concentrations of macromolecules in the cell. In this regard, extrapolation of the data in Fig. 8 to the *in vivo* condition must be considered as only approximate. In addition, the binding of TyrR to oligonucleotide shown in Fig. 8 relates to the binding to a single TyrR box. Operons of the TyrR regulon typically have a strong and weak TyrR box close together. Further studies are required to determine the stoichiometry and binding constants for TyrR binding to more complicated DNA structures containing additional TyrR-binding sites.

Self-association of regulatory proteins may be a more general mechanism for regulating gene expression. Our studies of a constitutive form of the bacterial enhancer-binding protein, NTRC, indicate that ATP induces oligomerization.[21] Laue *et al.*[22] have used analytical ultracentrifugation to study a spectrally enhanced derivative of the λ *c*I repressor containing 5-hydroxytryptophan in place of normal tryptophan. This study showed that λ *c*I repressor self-associates from dimer to octamer, with specific oligonucleotide favoring the octameric species. An important requirement of the procedures described in this chapter for the analysis of DNA interactions is the ability to obtain high-quality data in a concentration range

[21] M. E. Farez-Vidal, T. J. Wilson, B. E. Davidson, G. J. Howlett, S. Austin, and R. A. Dixon, *Mol. Microbiol.* **22,** 779 (1996).
[22] T. M. Laue, D. F. Senear, S. Eaton, and A. J. B. Ross, *Biochemistry* **32,** 2469 (1993).

where there are appreciable levels of the interacting free and bound species. In such cases, the analytical ultracentrifuge provides an important avenue for studying the degree and stoichiometry of the self-association as well as the thermodynamics and stoichiometry of the DNA-protein interactions.

Acknowledgments

The work of our colleagues, in particular Tim Wilson, Michael Bailey, and Bill Sawyer, who helped produce much of the published work referred to in this chapter, is gratefully acknowledged. Computer software referred to in this chapter was generously made available by John Philo (SVEDBERG), Tom Laue (SEDNTERP), and Allen Minton (TWOCOMP, SEDEQ1B, and EXCOEF). These programs are available via anonymous FTP at a site (*bbri.harvard.edu*) kindly maintained by Walter Stafford. The financial support of the Australian Research Council is also gratefully acknowledged.

[12] Proteolytic Footprinting Titrations for Estimating Ligand-Binding Constants and Detecting Pathways of Conformational Switching of Calmodulin

By MADELINE A. SHEA, BRENDA R. SORENSEN, SUSAN PEDIGO, and AMY S. VERHOEVEN

Overview of Experimental Rationale

A major goal of molecular biophysics is to understand quantitatively how proteins interact to regulate complex biological phenomena. Frequently, such regulatory proteins have multiple domains or subunits with many degrees of conformational freedom (see Fig. 1). Although advances in structural biophysics and computational chemistry permit us to visualize proteins at an increasingly fine level of detail, it is not possible to predict function from form or form from sequence.

When we wish to know the degree to which two stable domains interact cooperatively or whether two homologous binding sites are functionally equivalent, it is imperative to make experimental measurements of those properties. However, while thermodynamic and kinetic properties of biological systems comprise the driving forces for cellular events, they are notoriously difficult to measure in a way that allows us to dissect global properties of such systems and understand the contributions from individual domains or subunits of the full assembly.

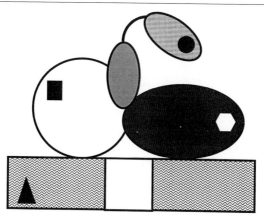

FIG. 1. Schematic diagram of protein assembly, with four chromophores indicated by small polygons.

Residue-Specific Analysis of Cooperative Proteins

One approach to determining molecular mechanisms of cooperativity is to observe whether there are residues that respond in concert to changes in environmental factors such as signaling ligands, metabolites, pH, salt, and other solutes. These concomitant changes elucidate the pathways or circuits of communication within a single protein or oligomeric assembly.

For decades, residues with an intrinsic spectroscopic signal have been used to report on the ligand-linked conformational responses of proteins. However, this approach is limited by the natural frequency of specific residues in the primary sequence (e.g., tyrosine and tryptophan for optical spectroscopy). As indicated in Fig. 1, the distribution of optically active reporter groups may be not be ideal: they may not reflect changes at the joints or hinges between domains or at the protein–protein interfaces in the assembly. Thus, there is no way of knowing in advance whether they will be sensitive to ligand-induced conformational change.

Chemical modification methods make it possible to add reporter groups to reactive side chains (e.g., cysteine or lysine), but that does not expand the repertoire greatly for any given protein. Furthermore, interpretation is complicated by the difficulty in obtaining preparations of uniformly labeled protein. The growing ease of mutagenesis and chemical synthesis allows the substitution of naturally occurring and synthetic amino acids at discrete positions in a protein sequence, providing the option for introducing spectral reporters at will. However, this is still a laborious approach and there are always concerns that the functional properties of a mutated or chemically modified protein will differ from those of the wild-type protein, misleading the investigator.

If one wants to study the allosteric properties of a wild-type protein or protein assembly, it is possible to do this directly by using chemicals or endopeptidases (hereafter, simply referred to as "proteases") to probe the differential susceptibility of peptide bonds affected by ligands or allosteric effectors binding to the protein. This approach builds on the use of partial (or "limited") proteolysis to map domain boundaries within proteins (usually the most susceptible positions are in the "linker" regions with the fewest structural constraints). Expanding on this approach, it is possible to determine not only whether the environment of a peptide bond has changed in the presence or absence of a ligand but also to resolve a titration curve by monitoring fractional changes in proteolytic susceptibility at levels of partial ligand saturation.

In this chapter, we describe advantages and limitations of using enzymatic probes of residue-specific responses of regulatory proteins as they undergo ligand-induced switching transitions over the course of a titration. We provide a detailed description of the proteolytic footprinting titration method[1,2] that we developed to study thermodynamic and structural properties of calmodulin (CaM), a eukaryotic calcium receptor essential for many pathways of signal transduction. The focus of this chapter is the methodological approach rather than a review of the calmodulin literature. However, it is necessary to introduce key elements of the structure and function of calmodulin to explain the rationale and benefits of this technique as it may be applied to other proteins.

Multisite Cooperative Protein: Calmodulin and Its Enigmas

Calmodulin is a small (148 amino acids, 16.7 kDa), highly acidic protein whose secondary structure is primarily α helical. Although similar in size to myoglobin, its organization is more like that of hemoglobin: a dimer of dimers. CaM consists of two homologous domains. Each half-molecule domain contains a pair of calcium-binding sites (see Fig. 2) in a helix–loop–helix or EF-hand motif. These domains form four-helix bundles as shown in Fig. 3. Calcium binding opens a hydrophobic cleft in each domain by changing the angle between interacting pairs of helices (i.e., A and D move as a unit away from B and C).

CaM binds calcium ions with positive cooperativity and generally activates target enzymes by binding to them in its calcium-saturated form and relieving self-inhibition. In most high-resolution structures or three-

[1] S. Pedigo and M. A. Shea, *Biochemistry* **34,** 1179 (1995).
[2] M. A. Shea, A. S. Verhoeven, and S. Pedigo, *Biochemistry* **35,** 2943 (1996).

Calcium Binding Sites

```
  1 2 3 4 5 6 7 8 9 10 11 12 13 14 15 16 17 18 19 20 21 22 23 24 25 26 27 28 29 30 31 32 33 34 35 36 37 38 39
  A D Q L T E E Q I A E F K E A F S L F │D K D G D G T I T T K E│ L G T V M R S L

  40 41 42 43 44 45 46 47 48 49 50 51 52 53 54 55 56 57 58 59 60 61 62 63 64 65 66 67 68 69 70 71 72 73 74 75
  G Q N P T E A E L Q D M I N E V │D A D G N G T I D F P E│ F L T M M A R K

  76 77 78 79 80 81 82 83 84 85 86 87 88 89 90 91 92 93 94 95 96 97 98 99 100 101 102 103 104 105 106 107 108 109 110 111 112
  M K D T D S E E E I R E A F R V F │D K D G N G Y I S A A E│ L R H V M T N L

  113 114 115 116 117 118 119 120 121 122 123 124 125 126 127 128 129 130 131 132 133 134 135 136 137 138 139 140 141 142 143 144 145 146 147 148
  G E K L T D E E V D E M I R E A │D I D G D G Q V N Y E E│ F V Q M M T A K
```

Composition: A/11 C/0 D/17 E/21 F/8 G/11 H/1 I/8 K/7 L/9 M/9 N/7 P/2 Q/6 R/6 S/4 T/12 V/7 W/0 Y/2

Fig. 2. Sequence and amino acid composition of rat calmodulin aligned to show four homologous EF-hand segments; boxes indicate the 12-residue calcium-binding site.

dimensional (3-D) models, both domains of CaM interact with the target protein like hands grasping an α-helical "rope." [There are few known targets such as neuromodulin that selectively interact with apoCaM (i.e., zero calcium ions bound). CaM also can be found as an intrinsic subunit of an oligomer, remaining associated regardless of calcium level.]

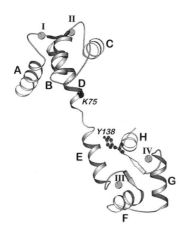

Fig. 3. Ribbon diagram of calcium-saturated CaM (based on Brookhaven PDB file *3cln.pdb* [Y. S. Babu, C. E. Bugg, and W. J. Cook, *J. Mol. Biol.* **204**, 191 (1988)] indicates the similar secondary and tertiary structure of each of the four EF-hand units, two in each half-molecule domain. Eight helices are labeled A through H. The positions of Lys-75 and Tyr-138 are indicated. Ribbon drawing generated by Molscript [P. J. Kraulis, *J. Appl. Crystallogr.* **24**, 946 (1991)]. A narrower ribbon in the "central helix" indicates a region that is nonhelical in solution NMR studies [M. Ikura, S. Spera, G. Barbato, L. E. Kay, M. Krinks, and A. Bax, *Biochemistry* **30**, 9216 (1991)].

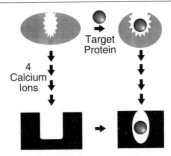

FIG. 4. Thermodynamic linkage scheme of calmodulin binding four calcium ions (from top to bottom) and a target protein or antagonist (from left to right).

The schematic linkage scheme in Fig. 4 indicates that calcium binding to calmodulin and its association with target proteins are processes that are linked energetically. This simplified representation does not depict the complexity of competition among multiple target proteins that occurs in a cell.

The high-resolution crystal structure of calcium-saturated rat calmodulin (*3cln.pdb*) reported by Babu and co-workers[3] (see Fig. 3) was remarkable for many features, including a long (seven-turn) helix, dubbed the "central helix." The protein was described as having a "dumbbell" structure. The two half-molecule domains appeared to be well separated but similar in secondary and tertiary structure. This separation raised as many questions as it answered about signal transduction events regulated by CaM that were believed to require the interaction of both domains with target proteins.

Homologous Domains, Different Affinities

Despite the study of calmodulin for two decades, there are many aspects of the molecular mechanism of cooperative calcium binding that remain poorly understood. For example, because of the high degree of similarity between the primary sequences of the two half-molecule domains of CaM (see Fig. 2), it would have been natural to assume that their affinity for calcium would be similar as well. However, 1-D nuclear magnetic resonance (NMR) studies of stoichiometric calcium titrations[4] and kinetic studies of fragments[5] showed unequivocally that the calcium affinities of the homolo-

---

[3] Y. S. Babu, C. E. Bugg, and W. J. Cook, *J. Mol. Biol.* **204,** 191 (1988).

[4] R. E. Klevit, D. C. Dalgarno, B. A. Levine, and R. J. P. Williams, *Eur. J. Biochem.* **139,** 109 (1984).

[5] S. R. Martin, A. Andersson-Teleman, P. M. Bayley, T. Drakenberg, and S. Forsén, *Eur. J. Biochem.* **151,** 543 (1985).

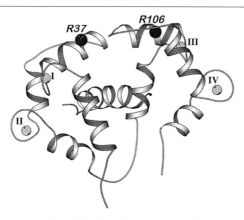

FIG. 5. Calcium-saturated calmodulin bound to a target peptide from rabbit skeletal myosin light-chain kinase resolved by NMR [model 1 from *2bbm.pdb*; M. Ikura, G. M. Clore, A. M. Gronenborn, G. Zhu, C. B. Klee, and A. Bax, *Science* **256**, 632 (1992)]. Graphical depiction using NanoVision (ACS Software).

gous domains were well separated, with sites III and IV (in the C-domain) having an approximately 10-fold higher affinity for calcium. Those two sites were almost completely filled at a calcium concentration barely sufficient to begin saturating sites I and II in the N-domain. That finding suggested that once an atomic resolution model of the calcium-saturated form of CaM was available, there might be some evidence of differences between the structures of the two domains.

However, all the subsequent crystallographic X-ray diffraction (XRD) and NMR determinations of calcium-saturated calmodulin, whether verte-brate or invertebrate, have indicated that the backbone structures of the two domains are nearly identical. There were no obvious "sources" for the energetic differences between domains. (The primary difference between models derived from XRD and NMR was that NMR studies[6] showed a small region of the "central helix" to be disordered.) Similarly, the structures of calcium-saturated CaM interacting with various target peptides (resolved by both NMR and XRD methods; see Ref. 7 for review) showed highly similar domains, each grasping the target peptide with an extensive network of noncovalent interactions as shown in Fig. 5.

Because the structures of Ca^{2+}_4–CaM with and without peptide repre-sent the "ends" of two axes in the linkage scheme (the two species shown on the bottom of Fig. 3), they cannot inform us about the partially saturated

[6] G. Barbato, M. Ikura, L. E. Kay, R. W. Pastor, and A. Bax, *Biochemistry* **31**, 5269 (1992).
[7] K. Török and M. Whitaker, *BioEssays* **16**, 221 (1994).

species of CaM along the pathway. The dichotomy between the binding differences and structural similarities of the domains thus suggested that important aspects of the regulatory mechanism were to be found in properties of the apo state and the process of filling the sites of the protein—not purely in the details of final structures that may be adopted by calcium-saturated CaM.

Information about the distribution of partially saturated states of a multisite protein is always difficult to obtain because these species can rarely be studied as pure preparations. Even if it were possible to isolate a species that had fewer than four calcium ions bound, unless the ligands were "frozen" onto their sites, they would rearrange in number and position. The stoichiometric calcium titrations monitored by 1-D proton NMR had shown that there was a measurable population of a species that had both sites III and IV filled while sites I and II were vacant.[4] The simplest succession of ligation states consistent with this finding is indicated schematically in Fig. 6A.

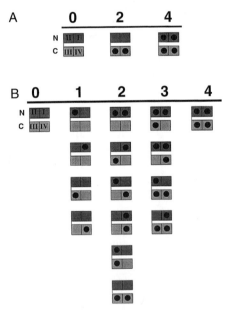

FIG. 6. Schematic diagrams of CaM ligation pathways. The N-domain and C-domain are indicated by horizontal shaded rectangles; squares within them represent individual calcium-binding sites saturated with calcium (●) or vacant. (A) A model for the transition between apoCaM and Ca^{2+}_4–CaM, showing a single intermediate ligation species with sites I and II in the N-domain vacant and sites III and IV in the C-domain occupied (●). (B) Schematic diagram of all 16 possible ligation species for a 4-site molecule.

Determining Number of Combinations of Vacant and Filled Sites

In principle, with four distinct sites for calcium binding, there are 16 possible ligation states of CaM, as shown in Fig. 6B. Thus, there are 14 intermediate species possible between the apo state (zero calcium bound) and the saturated state (four calcium bound). These schematic drawings do not represent the many degrees of conformational freedom available to the protein. For example, there has been no attempt in Fig. 6 to indicate whether the calcium-induced opening of the hydrophobic cleft is driven by saturation of just one or both of the sites in each domain. There is conflicting evidence on this point in the literature.

Although the fractional abundance of intermediates is low, there may be a preferred pathway (i.e., unequal distribution of the probabilities of intermediate states such that some are more populated than others). Defining that pathway and understanding it quantitatively was the challenge that motivated us to develop the proteolytic footprinting titration method. This approach was built on the longstanding observation that ligand binding often changes the susceptibility of a protein to cleavage, usually protecting it. On this basis we expected that each domain of CaM would become progressively more protected from proteolysis as calcium saturation increased, and that the titration curves for each domain would be congruent but offset as indicated in the schematic drawing in Fig. 7A.

The separation in binding affinities of the domains allows for sequential events (i.e., ordered activation of the two domains) to occur as calcium levels rise and fall. In this fashion, enzymes that require interactions with only one domain but not both might be activated at intermediate calcium levels. However, even if there were no cellular role for a partially saturated species of CaM and the apo and calcium-saturated "end states" of CaM were the only forms that interacted with targets, the molecular mechanism of regulating the transition between the end states in response to a cellular wave of calcium is highly significant for the targets of CaM.

By way of analogy, one may consider the mechanism of oxygen binding to hemoglobin. That four-site cooperative protein is not expected to "spend" a significant fraction of its time in the blood at the 50% saturation level. However, the process by which it traverses that point on the binding curve (particularly in the direction of release of ligands from the heme pockets) is of utmost significance to an oxygen-hungry organism.

Questions to Be Addressed

The specific questions about CaM, a multisite, multidomain protein, that drove the development of proteolytic footprinting titrations as a quantitative research technique may be summarized as follows.

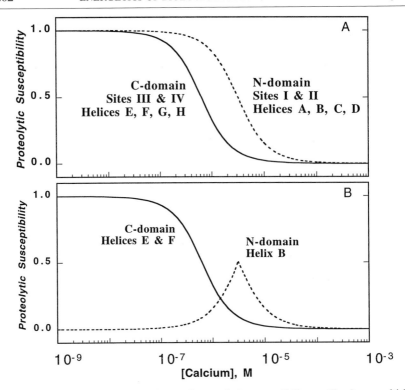

FIG. 7. Schematic diagram of classes of proteolytic susceptibility profiles for a multisite protein. (A) Expected profiles for residues in the two homologous domains of CaM. Ligand binding generally protects proteins from proteolysis; hence, a decrease in signal is expected to be proportional to occupancy. The offset is expected on the basis of the differences in calcium affinity (the free energy of calcium binding to sites I and II is less favorable than to sites III and IV). (B) Observed susceptibility pattern: monotonic for cleavage of residues in helices E and F near site III (C-domain) and biphasic in helix B (N-domain).

Which of the 14 intermediate (partially saturated) states are significantly populated during the transition from apo-to fully saturated CaM?

What are the rules for switching between these states?

Why are two highly homologous domains apparently so different in ligand affinity?

What are the magnitudes and directions of cooperative (allosteric) interactions between sites and domains?

Use of Proteolytic Footprinting Titrations

Questions like those preceding this section are posed for many regulatory proteins. To answer them, we sought to develop a technique that could monitor ligand-induced transitions of individual residues and relate those to the extent of saturation of individual binding sites and the linked conformational change occurring in the intermediate states. Prior to these studies, there were published observations of differences in the proteolytic susceptibility of apo- and saturated CaM when exposed to trypsin or thrombin.[8] Those "plus–minus" studies inspired our effort to develop and validate an approach to resolve protein–ligand titrations analogous to the quantitative DNase footprinting titration method developed by the Ackers laboratory to study site-specific repressor proteins binding to DNA operator regions.[9]

Before describing the experimental details of the proteolytic footprinting titration method, we compare and contrast this approach with others that are commonly applied to studying ligand binding. The arguments presented here may help an investigator determine whether this approach is uniquely suited to answer questions about the molecular mechanism of cooperativity of a particular protein assembly of interest.

Spectroscopic Methods

The sequence of vertebrate CaM poses multiple challenges that could not be addressed by obtaining ever-more-accurate binding data from existing macroscopic binding techniques (e.g., those that rely on optical spectroscopy or flow dialysis to measure average saturation). Because of the amino acid composition and internal repeats within its primary sequence, CaM is not amenable to standard spectroscopic approaches. There is a paucity of probes for optical spectroscopy (no tryptophan residues and the intensity of only one of the two tyrosine residues in vertebrate CaM changes significantly in response to calcium binding). There are no cysteine residues that could be covalently modified to provide spectroscopic reporter groups. The sequences of the four sites are so similar that it is not easy to monitor them individually even when using a sophisticated method such as heteronuclear multidimensional NMR, which is also expensive. Besides the high degree of overlap among the chemical shifts, the C-domain exhibits slow exchange, which requires monitoring peak area or volume rather than simply differences in chemical shift position to correlate with extent of binding.

Mutagenesis

A common approach to probing the function of a regulatory protein is to perturb it through mutagenesis and determine whether properties such

[8] W. Drabikowski and H. Brzeska, *J. Biol. Chem.* **257,** 11584 (1982).
[9] M. Brenowitz, D. F. Senear, M. A. Shea, and G. K. Ackers, *Methods Enzymol.* **130,** 132 (1986).

as assembly of subunits, stability (enthalpy and melting temperature), or ligand affinity (measured by spectroscopy or dialysis) have been affected. Because biological function is determined by global interactions as well as local properties, even single "conservative" substitutions may have unexpected and undesirable effects. This is more of a problem with smaller proteins. CaM is essentially a covalent dimer of two 75-residue domains. Any modification "designed" to have a single purpose may have broad deleterious effects. Thus, it is attractive to have a method that is capable of monitoring residue-specific changes in a wild-type protein (or naturally existing mutant) to determine directly the complex networks of interactions that are difficult to divide into their component elements. Once the basic features of conformational behavior are known for the wild-type protein, mutational perturbation can be a powerful tool to probe the mechanistic foundation of those properties.

Monitoring Residue-Specific Responses

Most of the existing models for molecular mechanisms of cooperativity in biological systems are based exclusively on functional and structural knowledge about the end states of a pathway or cellular process. The properties of the partially liganded intermediate states are generally inferred as averages of the properties of the end states. This approach is used often because it is difficult to monitor the energetics of biochemical transitions in ways that permit discrimination between more complex models.

To do better, it is necessary to observe binding independently to distinct domains, subunits, or sites, even though the degree of saturation or conformational change in one location may depend on changes occurring elsewhere simultaneously in the macromolecular assembly. Simulations of titrations and species population diagrams shown in Figs. 8–10 compare two cases that differ dramatically in the chemical distribution of binding free energy (i.e., intrinsic affinities and cooperativity) among four ligand binding sites.

    Case A (Fig. 8A): All four of the sites (a–d) have equal affinity for the ligand ($k_i = 1 \times 10^6 \ M^{-1}$) and bind ligand independently (i.e., there are no cooperative interactions). Using an individual-site method to monitor the fractional saturation of each site yields four identical titration curves. These also are identical to the macroscopic titration curve (i.e., average degree of saturation) for the whole four-site protein.

    Case B (Fig. 8B): There are two different classes of sites. One pair (sites a and b) consists of two equal and independent sites, with free energies of binding identical to the sites in case A. The other pair

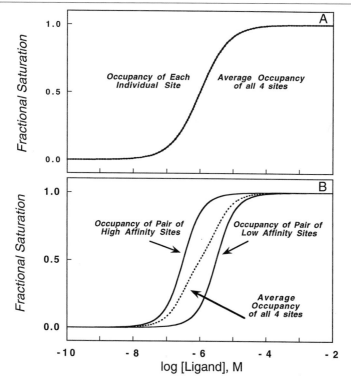

FIG. 8. Simulations of fractional saturation for two molecules having the same total free energy (macroscopic equilibrium constant K_4) for ligand binding at four sites. In case A all sites have equal affinity for ligand and do not interact; in case B two pairs of sites differ in intrinsic association constants by a factor of 10, but both with pairwise cooperativity constants of 10.

of sites (c and d) consists of two sites that are equal to each other but characterized by a lower intrinsic affinity for ligand than that of sites a and b. However, positive cooperative interactions enhance the probability of binding to sites c and d. Monitoring each of the four sites in this protein yields two pairs of isotherms that are well separated in their median ligand activities (~10-fold) and also contrast in shape to the macroscopic titration curve because of differences in cooperativity.

At 1 μM ligand activity, the average degree of saturation is identical (50%) in cases A and B. However, the difference between the fractional saturation of the unequal sites in case B is almost 70% (84.6% for sites a and b versus 15.4% for sites c and d). This is detected easily even with crude methods.

The differences between these two cases are also manifest as large differences between the levels of partially saturated states (those having one, two, or three ligands bound regardless of their distribution on the sites). Figure 9 shows that while the changes in the population of end states (zero or four ligands bound) as a function of ligand activity are similar for both cases, the population of intermediate species varies dramatically. For example, the maximal abundance of doubly saturated species is almost 38% in case A (Fig. 9A) compared with almost 66% in case B (Fig. 9B).

The macroscopic titration curves (e.g., average degree of fractional saturation) for cases A and B are overlaid in Fig. 10A. Their small and systematically varying difference is shown in Fig. 10B. The precision of each of these data sets would need to be better (e.g., the noise would need to be smaller) than the largest difference seen ($<3.5\%$) to be able to detect the difference between these two curves reliably. Thus, cases A and B are essentially impossible to distinguish on the basis of average degree of

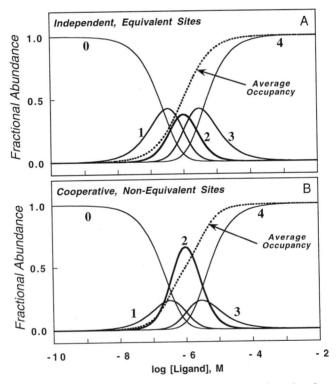

FIG. 9. Comparison of fractional distributions of partially liganded species of a macromolecule that binds four ligands; (A) and (B) correspond to the cases shown in Fig. 8.

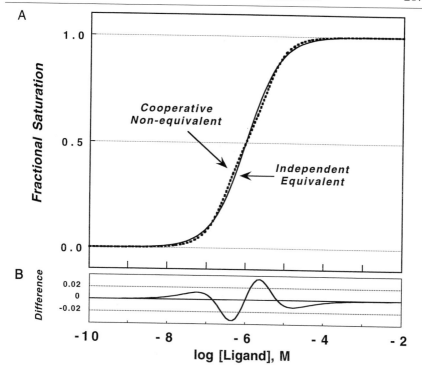

FIG. 10. (A) Comparison of average fractional saturation curves for the two cases illustrated in Figs. 8 and 9. (B) Difference between the curves.

fractional saturation. This leads to the conclusion that improving the precision of a classic biophysical technique or the methods of data analysis applied to it will not suffice. Instead, it is essential to obtain knowledge about the distribution of microstates that can be gleaned directly only by applying experimental techniques that can monitor individual positions of a multisite cooperative complex or that isolate distinct chemical species that differ with respect to the position as well as the number of ligands bound.

When it is possible to distinguish ligand binding at each site of a multisite (multivalent) macromolecule either individually or in subsets of the whole, then we obtain greater insight into the molecular switching mechanism at each level of ligand activity. The following section describes some of the issues to consider when deciding whether conducting proteolytic footprinting titrations will be a good approach for studying the ligand-binding properties of a specific protein.

Selection of Proteases and Conditions

Proteins Amenable to Study

Although the proteolytic footprinting titration method was developed for calmodulin (a soluble small protein that can be obtained in milligram quantities as a homogeneous preparation), it is possible to use this approach on larger proteins and ones that are available only in smaller quantities. The limiting factor is having a strategy that provides a good signal-to-noise (S/N) ratio for generating, detecting, and quantitating peptides resulting from a single cleavage of a full-length protein. This is discussed further below. Although our studies of CaM relied on a sensitive and reliable UV detector to monitor the peptide products of a limited proteolysis reaction separated by high-performance liquid chromatography (HPLC), other approaches could be used depending on how the protein is obtained (e.g., via overexpression or from a tissue) and whether it is amenable to chemical modification.

While some proteins (e.g., those with many disulfide bridges) may be highly resistant to partial proteolysis, most regulatory proteins undergo ligand-induced structural changes that will be evident as ligand-dependent differences in proteolytic susceptibility. Proteins that require detergent for stabilization are also tractable for study by this method [cf. studies on human replication protein A (RPA), the primary eukaryotic single-stranded DNA-binding protein[10]].

Ligand-binding titrations must be conducted under conditions in which the proteases are active; however, because limited proteolysis is the goal, it is not necessary that the buffer conditions be at the pH and temperature optimum of the protease being applied. In some cases, it is an advantage to have a less active protease to minimize or avoid secondary cleavage (cleavage of a peptide rather than the full-length protein). Therefore, the buffer conditions should be selected to match those of other experiments (e.g., electrophoretic mobility shift analysis, spectroscopy, sedimentation, analytical gel-permeation chromatography) with which these titrations will be compared. Typically the conditions controlling exposure to protease (time, molar ratio of protease to substrate, and absolute concentration) can be adjusted to accommodate the preferable buffer conditions.

Essential Conditions of Partial Proteolysis Reactions

In proteolytic footprinting titrations, the overriding goal is to use a protease to obtain a snapshot of the distribution of molecular species pres-

---

[10] X. V. Gomes, L. A. Henricksen, and M. S. Wold, *Biochemistry* **35**, 5586 (1996).

ent in a solution by reporting on ligand-linked conformational change occurring at one or more peptide bonds during a binding titration. For that snapshot to be representative of the whole population of molecules in the solution of protein and ligand, the exposure to protease must satisfy two criteria. Partial proteolysis must not shift the distribution of ligands bound to protein and the peptide products must be primary cleavage products whose abundance will reflect the environment of an individual peptide bond as it was within the native protein (and not that of the same bond within a cleavage product).

Although it may be appealing to use a protease that can "hit" many positions throughout a sequence [e.g., Endoproteinase GluC (EndoGluC) probe of CaM as described in Ref. 1], titrations probed by proteases that cleave at only a few positions are interpreted much more easily because it is much simpler to avoid secondary cleavage products and to obtain clean chromatographic separations of peptides. That, in turn, facilitates unambiguous chemical analysis of the products (i.e., determination of mass and amino acid composition) as well as reproducible quantitation of relative amounts of each peptide at different levels of ligand or allosteric effector. If one is screening binding titrations for a large series of mutant proteins or distinct ligands, this makes it much more likely that preliminary identification of HPLC fragments can be made correctly by visual inspection, with subsequent confirmation by chemical analysis.

Footprinting titrations are conducted in a discontinuous fashion (i.e., individual samples of ligand and protein are equilibrated for analysis to yield each point on the titration curve). Thus, it is straightforward to make a single sample and divide it into aliquots that will be studied by footprinting (using several different proteases) as well as other methods (e.g., spectroscopic techniques) and compared.

Selecting Proteases on Basis of Specificity

There are many annotated compilations of proteases and their enzymatic characteristics; these include data in charts and catalogs from commercial vendors [e.g., Boehringer Mannheim (Indianapolis, IN) and Calbiochem (La Jolla, CA)] as well as accessible articles and technical monographs (see Refs. 11 and 12). These are a useful starting point when selecting enzymes. However, it is important to recognize that the degree to which a protease is regarded as specific or nonspecific in the literature is of little

[11] J. M. Wilkinson, in "Practical Protein Chemistry—A Handbook" (A. Darbre, ed.), p. 121. John Wiley & Sons, New York, 1986.
[12] R. J. Beynon and J. S. Bond, "Proteolytic enzymes: A Practical Approach." IRL Press, Oxford, 1989.

or no importance when selecting ones to use for proteolytic footprinting titrations. A variety of different proteases (5–10) should be screened at several levels of ligand or allosteric effector.

In particular, investigators are cautioned not to depend on trypsin alone, despite its common application in the biochemistry literature on domain mapping. The high likelihood of secondary cleavage products and the common contamination of trypsin with chymotrypsin completely obfuscate the interpretation of titrations. It may be more useful to apply papain (also likely to cut at arginine or lysine), thrombin, or clostripain (which will probe arginine residues) or bromelain (which is likely to cut at lysine residues).

Because proteolytic footprinting titrations are conducted under native (not denaturing) conditions in physiological buffers, the number and position of cleavage products must be determined empirically, rather than relying on sequence-recognition motifs that are reported in the bioanalytical literature. Both thrombin and chymotrypsin provide examples of this for CaM.

Thrombin cuts CaM at two of the six arginine residues in the protein (the bonds R37–S38 and R106–H107); however, neither of these satisfies the canonical definition of a good thrombin cleavage site (i.e., an arginine preceded by glycine and followed by isoleucine or leucine). Furthermore, regardless of the duration of exposure of CaM to thrombin, the secondary cleavage peptide (S38–R106) is never observed.[2] The only products are two sets of primary cleavage products (Fig. 11), emphasizing that the conformational milieu of each bond is important to proteolysis. In contrast, the cloned domains of CaM (which are peptides consisting of residues 1–75 or 76–148) can be footprinted with thrombin.[13] However, this could not be predicted on the basis of primary sequence alone.

Chymotrypsin is expected to cleave after aromatic or large hydrophobic residues. Although there are 10 aromatic residues (8 phenylalanine and 2 tyrosine) and 9 methionine residues in vertebrate CaM, only a small subset of these participate in scissile bonds. This is shown in Tris-tricine sodium dodecyl sulfate–polyacrylamide gel electrophoresis (SDS–PAGE) analysis[14] of chymotrypsin footprinting of a calcium titration of CaM (Fig. 12A). The number of cleavage products visible is much lower than even the 20 that would be predicted on the basis of the phenylalanine and tyrosine residues alone.

It is also notable that there are several classes of response at the scissile bonds probed by chymotrypsin: some positions are susceptible in the ab-

[13] B. R. Sorensen and M. A. Shea, *Biochemistry* **37**, 4244 (1998).
[14] H. Schagger and G. von Jagow, *Anal. Biochem.* **166**, 368 (1987).

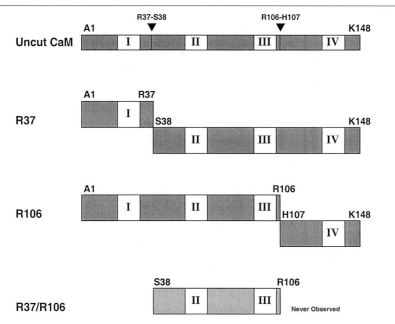

FIG. 11. Schematic diagram of primary cleavage fragments of CaM generated by thrombin, a protease that cuts at two of six arginine residues (R37 in helix B and R106 in helix F). The expected secondary cleavage product (S38–R106) has never been detected by SDS–PAGE or HPLC, even after 24 hr of exposure of apo CaM to thrombin (i.e., a period far exceeding that required to digest all of the whole CaM into primary fragments).

sence of calcium and become progressively protected by calcium binding (analogous to E87 probed by EndoGluC and R106 probed by thrombin) while others are cleaved only at intermediate levels of calcium binding (analogous to E31 probed by EndoGluC, R37 probed by thrombin, and S38 probed by bromelain). This qualitative electrophoretic analysis (Fig. 12A) is shown to illustrate that inspection of the primary sequence alone is insufficient to predict (1) the number and positions of primary cleavage events and (2) classes of response that may be observed at those positions.

Figure 12B illustrates chymotrypsin footprinting of a calcium titration of CaM in the presence of trifluoperazine (TFP), an antipsychotic drug that acts as a CaM antagonist. TFP protects CaM from proteolysis even under apo conditions (no calcium bound), where the C-domain of CaM is normally susceptible. The band that is most abundant corresponds to a fragment that has a mass close to that of a half-molecule (i.e., it represents cleavage between the two domains near the junction of helices D and E); its precise

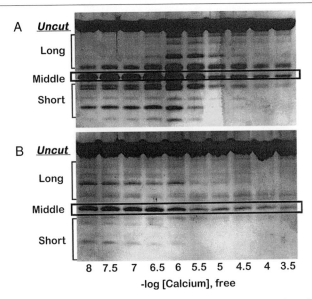

FIG. 12. Chymotrypsin footprinting of equilibrium calcium titrations of rat CaM, showing effect of bound antagonist. (A) Titration in the absence of TFP (trifluoperazine), an antipsychotic drug that binds to the hydrophobic clefts of CaM. (B) Equimolar ratio of TFP to CaM at each point in the titration. Buffer conditions: 50 mM HEPES, 100 mM KCl, 0.5 mM nitrilotriacetic acid (NTA), 0.5 mM EGTA, pH 7.40, 25°. Calcium levels ranged from 10 nM to ~0.3 μM in half-log units. Peptide products were separated by Tris–Tricine SDS–PAGE and detected by staining with silver nitrate.

chemical composition has not yet been determined. It becomes protected as calcium binds, in contrast to the susceptibility of K75 probed by bromelain (see Fig. 19).

Complementary to the studies described above, which were conducted with "specific" proteases, are those conducted with "nonspecific" proteases. Although it may be even more difficult to predict where such proteases will cleave a protein of interest, they will cut reproducibly at a subset of peptide bonds. Thus, proteases such as elastase, subtilisin, papain, and proteinase K may yield a small number of primary cleavage products (≤ 10) that may be monitored to resolve ligand-binding titrations. For any protease, even ones for which it is possible to predict accurately a set of expected scissile bonds, the primary cleavage products must be analyzed chemically to confirm identity (see the next section). Therefore, these "nonspecific" enzymes do not create extra work and provide a large arsenal for probing a broad array of regions in the protein of interest.

It is essential to conduct parallel proteolytic footprinting titrations with multiple proteases to prove that any unusual susceptibility profiles reflect ligand-induced conformational responses in the protein being titrated and are not protease dependent. As noted, because these studies are conducted as discontinuous titrations, it is straightforward to prepare a large sample of protein equilibrated with ligand at a series of activities and then probe individual aliquots with different proteases.

On the basis of the stoichiometric titrations of calcium binding to CaM monitored by NMR,[4] we had expected that monitoring bonds in each domain would yield two sets of congruent proteolytic footprinting titration curves (one set corresponding to the calcium-linked conformational response in each domain), as was shown in Fig. 7A. However, some peptide bonds were susceptible to cleavage only at intermediate levels of calcium saturation,[1,2] as illustrated in Figs. 7B and 12A. This pattern of susceptibility would have been highly suspect had it been observed by probing calcium titrations of CaM with only one protease. However, having observed it by using more than six proteases to footprint calcium-binding titrations to wild-type CaM and knowing that the pattern is different in mutants of CaM, we conclude that the phenomenon is one that informs us about conformational switching in CaM itself. These unexpected patterns demonstrated that helix B responds to binding of calcium at sites III and IV in the C-domain, undergoing a conformational change that is equal and opposite to that which occurs when calcium binds at sites I and II in the N-domain, where helix B is located.

Calibration of Proteases

Several attributes of a protease must be tested to determine whether it will be suitable for proteolytic footprinting titrations.

1. It is important to consider the purity of a protease. Variable contamination of trypsin with chymotrypsin and vice versa are well recognized in commercial sources of these enzymes. Multiple grades are available to control this problem although the purer the enzyme, the more expensive it is. It may be possible to limit the effect of a contaminating protease by including a specific inhibitor for it; however, these do not necessarily bind tightly under all solution conditions, and therefore the inhibitor must be tested under the buffer conditions of the ligand-binding titration.

2. Proteolytic activity varies with lot number and vendor. Adjustments may be needed to obtain sufficient cleavage for detection of primary peptides without getting high levels of secondary cleavage. Small chromophoric substrates (such as the Chromozym series of compounds from Boehringer Mannheim) are helpful for calibrating protease activity independent of the

protein to be titrated. A microtiter plate assay using visible wavelength absorbance is a quick way to measure proteolytic activity under several sets of buffer conditions (see example in Fig. 13).

3. To study variable proteolytic susceptibility induced by a ligand binding to a protein, it is undesirable for the ligand to affect protease activity directly. For example, to study calcium titrations of calmodulin, it was essential to determine whether calcium affected the activity of each protease. This was accomplished by testing the proteolysis of chromophoric substrates as a function of calcium concentration. Figure 13 shows a comparison of three proteases. Trypsin was activated by calcium while thrombin and chymotrypsin were not affected significantly. Although it would be possible to correct numerically for the difference in trypsin activity at each calcium concentration, excessive generation of secondary cleavage products caused us to abandon it in favor of other proteases.

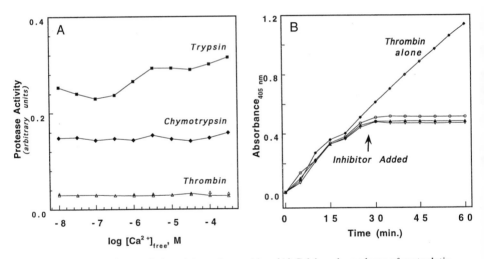

FIG. 13. Tests of proteolytic activity and quenching. (A) Calcium dependence of proteolytic activity tested for trypsin, chymotrypsin, and thrombin. For each enzyme at each calcium concentration, the initial rate of hydrolysis of a chromophoric substrate (purchased from Boehringer Mannheim) was monitored by observing the change in absorbance at 405 nm per minute per microgram of protease per milliliter. Substrate/protease pairs were chromozym-Tyr/trypsin, (succinyl-L-ananyl-L-alanyl-L-prolyl-L-phenylalanyl-4-nitranilide)/chymotrypsin, and chromozym-TH/thrombin; pCa ($-\log[Ca^{2+}]_{free}$) values indicated were rounded to nearest tenth. (B) Time dependence of quenching proteolytic activity of thrombin with a commercial inhibitor, FFRCK, added at 20 min to final concentrations of 0.2 mM ($\triangledown$), 0.1 mM ($\blacktriangle$), and 0.02 mM ($\bigcirc$). The activity of thrombin is low; the rate of hydrolysis of chromozym-TH, a chromophoric substrate, was monitored by observing the change in absorbance at 405 nm per minute per microgram of protease per milliliter.

Exposure of Protein–Ligand Complexes to Protease

It is important to control exposure to protease reproducibly, to limit digestion to a small fraction of the protein molecules and thereby minimize secondary cleavage. This requires precise control of the amount of protease added to each ligand–protein mixture as well as the time interval (initiation and quenching) of proteolysis.

Activity of Protease

Two parameters that are important to consider are the final concentration of protease in the equilibrated protein–ligand solution and the molar ratio of protease to protein. Many lyophilized proteases can be stored indefinitely; however, aliquots in solution may be stable for long intervals only if kept frozen in acidic solutions. Thus, the volume of added protease may have an impact on reaction conditions because the addition of its storage buffer may affect solution pH or salt conditions. Some proteases are expensive, and there is no need to waste enzyme when an equivalent fraction of the protein being titrated may be cleaved by allowing the proteolysis reaction to proceed for a longer time. Therefore, the smaller the addition the better. The simplest approach to finding good conditions is to expose a protein–ligand mixture to several ratios of protease (e.g., 1:100, 1:1000, and 1:10,000) for several time intervals (e.g., 2, 10, and 30 min) and analyze the number and nature of resulting peptides to find conditions such that most of the peptides will be primary cleavage products.

Some proteases are sensitive to the time spent under buffer conditions that are selected to mimic physiological conditions for the ligand titration experiment and may lose activity in a manner that is not entirely uniform or predictable. However, it is possible to include a small chromophoric substrate that serves as a positive control (this was done for EndoGluC[1]). If the cleavage of that small substrate is independent of titrant concentration (in this case, calcium level) and it does not interfere with the protein–ligand interaction under investigation, then its fractional degree of cleavage may be used to normalize for variations in protease activity among many samples.

Interval of Proteolysis

In studies of calcium titrations of CaM, we found that some proteases (such as EndoGluC) required short intervals (5 to 10 min) to limit secondary cleavage whereas others (e.g., thrombin) could be applied for hours with no difference in the ratio of primary cleavage products (see Fig. 14). This must be determined empirically; however, the more common scenario is to require a short interval.

An autosampler unit on an HPLC offers a precise way of initiating

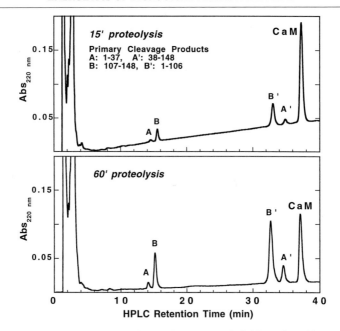

FIG. 14. HPLC separation of products of exposure of CaM to thrombin at pCa 5.99 (1 μM) for a 15- or 60-min period. The injection front represents buffer components and FFRCK, the inhibitor used to quench the reaction. In order of retention time, proteolytic products were as follows: A, 1–37; B, 107–148; B', 1–106; and A', 38–148; whole calmodulin (CaM) eluted last.

cleavage by adding small volumes of a protease solution to small volumes of protein equilibrated with ligand. This approach controls the dosage (duration and concentration) of protease more precisely than manual pipetting and also eliminates a major source of operator fatigue. Another advantage of using an autosampler unit is that each sample can be prepared just prior to its injection, with the same delay between the onset of proteolysis and injection for every sample in the titration curve.

Quenching of Proteolysis

The most simple and satisfactory method of quenching proteolysis is to inject an aliquot of a reaction mixture onto an HPLC column. The acidic conditions and immediate initiation of separation give results equivalent to the addition of a specific inhibitor. This approach eliminates an additional step and expense and it simplifies the resulting chromatogram.

If partial proteolysis for multiple reactions in a discontinuous titration curve must be terminated simultaneously prior to separation, several methods are available. For reactions that will be analyzed by SDS–PAGE, addition of detergent followed by rapid boiling may be sufficient to quench proteolysis. However, this treatment can cause a burst of proteolysis because the protease continues to be active as the substrate (protein under study) unfolds, loses ligand, and becomes more susceptible to proteolysis.

For this and other reasons, it may be preferable to quench proteolysis by addition of a specific inhibitor and then store the samples for subsequent analysis by HPLC. For thrombin, D-phenylalanylphenylalanylarginyl chloromethyl ketone (FFRCK) serves this purpose as shown in Fig. 13. Acid precipitation of peptides and proteins in a vial or microcentrifuge tube can also quench effectively; however, variable solubilization prior to injection can skew the ratios of peaks in the chromatogram and thus introduce noise or a systematic shift in the resolved titration curves.

The extent of proteolysis is kept low to minimize secondary cleavage products. This may be done by limiting the time of exposure or by arriving at conditions under which the enzymatic activity is low. To strike a balance between enhancing signal to noise (having large peptide peaks to measure above baseline noise in the chromatogram) and minimizing secondary cleavage, which will obfuscate the interpretation of all peaks containing a terminus, it is necessary to compare the products of cleavage after intervals of varying lengths.

Figure 14 compares the products that result from exposing CaM to thrombin for 15 and 60 min at a free calcium concentration close to 1 μM, near the peak in susceptibility of the R37–S38 bond. Although the total fraction of CaM cleaved after 1hr is greater, the relative probabilities of cleavage at R37 and R106 are almost identical during the two intervals and no secondary cleavage product is evident. This indicates that partial proteolysis did not change the fractional saturation of the remaining uncut CaM and that all the signals (i.e., the peptides) are coming from a single cut of whole CaM.

Quantitation and Identification of Peptides

Because the intrinsic susceptibility of different peptide bonds varies enormously, it is critical to have sensitive and accurate methods to detect low levels of primary cleavage products. The method that is used for quantitation depends on the purpose of the titration and the level of rigor desired in its interpretation. We have used several approaches, described below.

Separation and Quantitation: SDS–PAGE. SDS–PAGE is a rapid and easy method for separating products of most partial proteolysis reactions. Tris–Tricine gels[14] are used routinely to resolve small peptides obtained in our studies. Examples are given in Fig. 12 (chymotrypsin footprinting of wild-type CaM) and Fig. 15 (thrombin footprinting of E140Q-CaM).

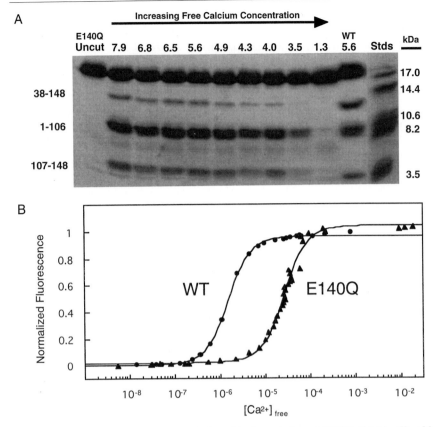

FIG. 15. (A) SDS–PAGE analysis of thrombin footprinting of E140Q-CaM in 50 mM HEPES, 100 mM KCl, 0.5 mM NTA, 0.5 mM EGTA, pH 7.40, 25°; pCa values are noted above the lanes. Lane 1 shows uncut CaM; lane 11 has products of thrombin footprinting of wild-type (WT) CaM at pCa 5.6 for comparison of migration of peptides. Lane 12 has molecular weight standards with sizes annotated. (B) Fluorescence analysis of calcium binding to WT CaM and E140Q-CaM in 50 mM HEPES, 100 mM KCl, 0.5 mM NTA, 0.5 mM EGTA, pH 7.40, 25°. DifluoroBAPTA ($K_d = 1.33 \times 10^{-6}$) was used as an internal indicator of free calcium at each point in the titration. Free energies for binding one (ΔG_1) and two (ΔG_2) calcium ions were resolved from fits of the data to Eq. (6), using NONLIN (WT CaM: $\Delta G_1 = -7.20 \pm 0.26$ kcal/mol, $\Delta G_2 = -15.77 \pm 0.06$ kcal/mol; E140Q-CaM: $\Delta G_1 = -6.13 \pm 0.81$ kcal/mol, $\Delta G_2 = -12.21 \pm 0.09$ kcal/mol). Curves through the data were simulated with the preceding free energies.

1. Staining with silver nitrate is adequate for detecting most peptides. It is sufficient for determining whether there are few or many peptides generated under a particular set of exposure conditions. It also provides a quick assessment of the number and kinds of classes of response to ligand binding (see Figs. 7 and 12). One note of caution is that some of the "quick"

silver-staining methods are not sensitive to small peptides. The protocol of Morrissey[15] is preferable even though it has more steps and takes longer than many other procedures.

It is important to recognize that, although silver nitrate offers sensitivity to a few nanograms of protein, the colorimetric response is not linear. Therefore, unless one wishes to calibrate the staining procedure with a series of samples of known mass, the resolution of binding isotherms is more precise by HPLC separation and peak integration for analysis.

2. Staining the products of partial proteolysis with dyes such as Coomassie blue is not recommended unless large amounts of protein are used. It is less sensitive and although more linear than staining with silver nitrate, still should be calibrated with samples of known mass.

3. Rarely, a peptide product generated during exposure to proteolysis is not visualized by SDS–PAGE. This is the case for a primary cleavage peptide (residues 1–37) generated during thrombing footprinting of CaM. However, it is resolved well with an HPLC instrument. Whether this is because it is difficult to resolve during electrophoresis, diffuses rapidly and is not fixed in the gel, stains poorly, or all of these, we do not know.

Proteolytic footprinting titrations provide detailed information about a wild-type protein sequence and alleviate the need to spend time or money on making or isolating mutants. However, it can be illuminating to apply this method to mutant proteins for comparison. Figure 15 shows the titration of a mutation of rat CaM at E140 (the bidentate glutamate of site IV), using SDS–PAGE (Fig. 15A) and fluorescence (Fig. 15B). These data show that this mutation dramatically reduces the binding affinity of site IV of CaM. Sites I and II in the N-domain now saturate at a lower level of calcium than do sites III and IV in the C-domain. The effect of changing Glu-140 to a glutamine is so severe that complete protection (normally observed at micromolar calcium levels) now requires 100 mM calcium.

Separation and Quantitation: HPLC Column. Using an HPLC (C$_{18}$ column) for separation has been the most reliable method for analyzing peptides containing a few amino acids up to the full length of CaM (148 residues). Because the number and abundance of peptides vary over a titration, it is necessary to determine empirically the volume of a proteolysis reaction mixture to inject so that there would be sufficient protein on the column to ensure a good signal-to-noise ratio at all ligand concentrations.

1. With a dual-wavelength monitor, it is possible simultaneously to monitor peptide bonds and aromatic side chains. The wavelengths 220 nm and 280 nm were selected to minimize contributions from calcium buffer components such as EGTA. Because of the small size of CaM and presence

---

[15] J. H. Morrissey, *Anal. Biochem.* **117,** 307 (1981).

of only two tyrosine residues in the vertebrate sequence (Y99 and Y138), it is possible to distinguish peptides within the sequence 1 to 98 or 139 to 148 on the basis of their low absorbance at 280 nm. Manual integration of peaks has proved more reliable and reproducible than analysis conducted using the automatic peak detection and integration routines that are available as part of the software package that runs the instrument.

2. Postcolumn detection methods are also available, using antibodies or chemical modification reagents to quantitate peptides. These are usually less precise because an additional calibration of modification efficiency is necessary; however, they may increase sensitivity and therefore be necessary for some proteins.

Figure 16 shows an example of susceptibility profiles resolved quantitatively by HPLC. In this case, thrombin footprinting of a calcium titration of H135R-PCaM, a mutant of site IV in *Paramecium* CaM (PCaM), is

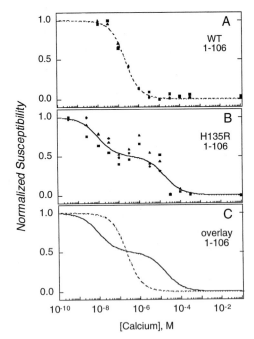

FIG. 16. Thrombin footprinting of (A) WT PCaM and (B) mutant H135R-PCaM. Data shown are of fragment 1–106 generated from cleavage at R106. Reaction conditions were similar to those described previously [B. R. Sorensen and M. A. Shea, *Biochemistry* **37,** 4244 (1998)]. Free energies for binding one (ΔG_1) and two (ΔG_2) calcium ions were resolved from fits of the data to Eq. (6), using NONLIN. Lines through the data were simulated using the above free energies. (C) An overlay of the simulated curves for ease of comparison.

compared to that of wild-type PCaM. The responses of R106 in the C-domain are compared. Remarkably, heterogeneity of sites III and IV is detectable in the mutant.

Identification Methods. Another advantage of a chromatographic system is the option to "collect peaks" (collect small volumes of the effluent) for subsequent chemical analysis. There are many monographs available on modern methods of identification of peptides. The quantity of pure material sufficient for unambiguous identification varies depending on the instruments and expertise available. Common approaches are listed below.

1. Mass spectroscopy [e.g., matrix-assisted laser desorption ionization time-of-flight (MALDI-TOF)] may be sufficient for identifying primary cleavage products if a protease cuts at a few predictable positions.

2. Amino acid composition analysis is a reliable means for obtaining the relative ratios of amino acids within a peptide. For many proteins, this is sufficient to assign the boundaries of fragments. However, even precise results may be ambiguous for a protein that has repeating motifs such that a cleavage product may contain multiple repeats of these. Used in conjunction with mass spectroscopy, amino acid analysis is almost failsafe.

Both of the methods described above are also excellent for determining whether the material in a single chromatographic peak is heterogeneous. This is a particular problem in a protein such as CaM, whose sequence has repeating motifs; similar challenges may arise for proteins with multiple zinc finger domains, for example.

We rely on a combination of mass spectroscopy (MALDI-TOF), amino acid analysis (AAA), and the ratio of absorbance at 280 nm to that at 220 nm. While an approach such as N-terminal sequencing can be reliable chemically, it is often 10 times the cost of AAA and it will not work for blocked peptides. For a protein with an acetylated amino terminus, blocked peptides will comprise half of all primary cleavage products. In such cases, if it is possible to obtain a preparation of the same protein that has been bacterially overexpressed, it may be possible to identify blocked peptides by comparing chromatographic data and then factoring in the direct MS and UV data.

Classes of Observed Susceptibility Profiles

In general, ligand binding makes proteins more rigid and less susceptible to proteases. This is indicated schematically in Fig. 17, where a globular protein undergoes a conformational transition from an apo form that is flexible to a liganded (fully saturated) protein that is rigid.

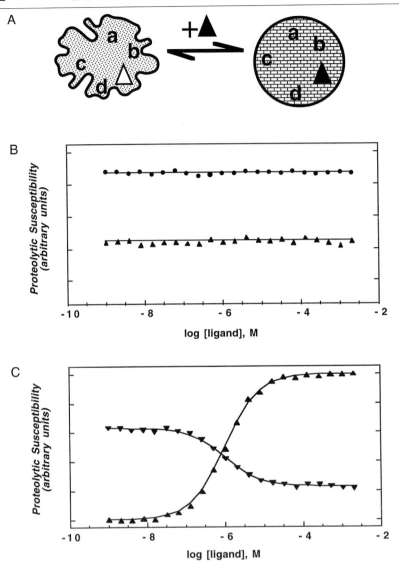

Fig. 17. (A) Different conformations of vacant (apo) and saturated protein with a single site for ligand (▲) binding. *a–d*, Peptide bonds. Classes of proteolytic footprinting titrations expected when binding a single ligand include (B) ligand-independent susceptibility (● and ▲, different degrees of intrinsic susceptibility) and (C) ligand-dependent susceptibility (▼, bonds less susceptible in saturated form; ▲, bonds more susceptible in saturated form).

This phenomenon of ligand-induced stiffening of proteins is often ex-ploited to map the domain structure of a newly identified protein or to "trim" a protein or complex in preparation for crystallization trials. Because there are only two possible ligation states of the protein in Fig. 17, there are a limited set of responses to partial proteolysis if each of these adopts a single conformation or ensemble of similar conformations. (Although both the apo and fully saturated species of the protein may be in conforma-tional equilibrium among many forms, the simplest case is being illustrated here as an example.)

The simplest response to ligand binding is a nonresponse. Ligand-independent positions will maintain a particular level of susceptibility (zero or nonzero) regardless of the degree of saturation of the protein by ligand as indicated in Fig. 17B.

In contrast, a peptide bond that is protected by ligand binding will have a susceptibility profile shown in Fig. 17C (curve ▼–▼). Because at every point in the titration curve there are only two species, apo and saturated, the degree of protection of the peptide bond may be interpreted as the fractional population of the liganded state. Conversely, if a peptide bond is more susceptible in the saturated form (by virtue of being more solvent exposed or less constrained structurally), its susceptibility profile will in-crease monotonically (as shown in Fig. 17C (curve ▲–▲). The absolute degrees of susceptibility (probability of cleavage) may differ for individual peptide bonds or different proteases probing the same peptide bond.

For a protein that binds a ligand at two specific sites (shown in Fig. 18A, an example is presented for an ordered binding process such that there is an intermediate ligation species that has one site occupied and a conforma-tion that is the average of the two end states. This is presented purely for the sake of illustration. It is rare to have binding events so fully separated. In this example, the conformational change linked to ligand binding is shown as being localized.

When a proteolytic footprinting titration is conducted, there may be both ligand-independent and ligand-dependent responses observed de-pending on the conformational switching pathway. As in the example of a protein that binds a single ligand (Fig. 17), some peptide bonds may become protected while others are more susceptible because they are more ex-posed to solvent or because ligand binding disrupts structural constraints on that region of the structure. An example of this kind of variation is seen in calcium titration of CaM, for bromelain footprinting of K75 and E87.

For a multisite protein, additional forms of ligand-dependent responses are possible, related to the number and order of conformational transitions. As described earlier, for CaM, positions in helix B of the N-domain have

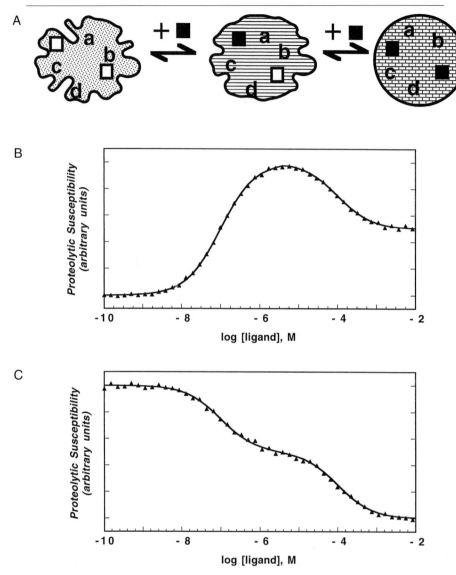

FIG. 18. Schematic diagram of complex patterns of ligand (■)-induced changes in proteolytic susceptibility.

an alternating (biphasic) pattern of susceptibility, such that a peak is reached when CaM is half-saturated. This has been confirmed by multiple proteases (including profiles for E31, R37, and S38). The inescapable conclusion is that calcium binding to the high-affinity sites III and IV in the C-domain affects the conformation of helix B (making it more susceptible to proteolysis) and the subsequent binding of calcium to sites I and II protects helix B.

One of the most perplexing aspects of this finding comes from inspection of the crystallographic structure of Ca^{2+}_4–CaM. The long distance between residues in helix B and the calcium ions in sites III and IV suggests that conformations adopted by the apo form of CaM and some of the intermediate ligation species bring these parts of the molecule into closer proximity or that there is a complex network of conformational effects propagated through the backbone of the protein. As shown in Fig. 20, more than 40 Å

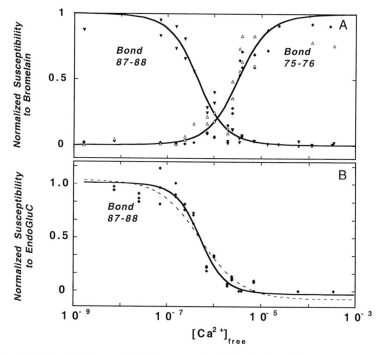

FIG. 19. Quantitative proteolytic footprinting of CaM using (A) bromelain and (B) EndoGluC. The data represent cleavage at positions K75 with bromelain and E87 with EndoGluC and demonstrate two classes of monotonic susceptibility. Reaction conditions were similar to those described previously [S. Pedigo and M. A. Shea, *Biochemistry* **34**, 1179 (1995)].

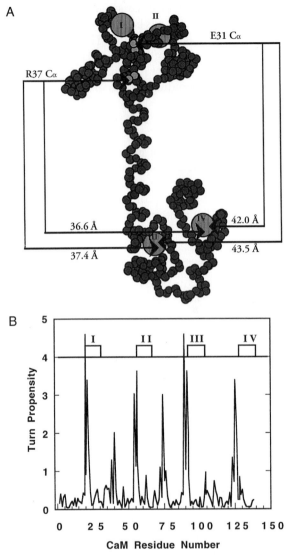

FIG. 20. (A) Ca$^{2+}_4$–CaM (*3cln.pdb*), showing distances between the C$_\alpha$ atoms of E31 and of R37 and the center of calcium ions in sites III and IV (all >35 Å). (B) Chou–Fasman analysis of propensity for turns based on primary sequence.

separates the α-carbons of residues E31 and R37 from the calcium ions in sites III and IV. Chou–Fasman analysis of the propensity for turns in the structure (Fig. 20B) suggested that the middle of the "central helix" would be a flexible region able to undergo conformational rearrangement. Such a "hinge" would accommodate a closer approach of the two domains than that shown in Fig. 3 for calcium-saturated CaM. Subsequent NMR studies indicated that this region indeed is unstructured in solution. However, there are no detectable interdomain nuclear Overhauser effects (NOE) in the studies reported so far.

Protocol for Proteolytic Footprinting Titrations

A summary of the procedure follows, with a primer on the numerical analysis of susceptibility profiles that can be interpreted as titration curves. While some aspects of this method will be specific for the cooperative protein under investigation, the rationale for each procedure has been described above and adaptations for other assemblies are straightforward.

Summary of Experimental Procedure

A discontinuous titration, either stochiometric or equilibrium, is conducted by mixing or dialyzing protein and ligand in separate aliquots. It is important to remember that the only species that can be prepared as homogeneous samples are the vacant and saturated forms of the protein. In the case of CaM, these are the species with zero or four calcium ions bound. This is indicated schematically in Fig. 21. (For the purposes of this chapter we are not incorporating debates about the number and nature of possible "nonspecific" sites of metal binding to CaM at high concentrations of ions.)

For all titrations, it is important to use a protein sample that is of high purity and does not contain contaminating proteins that selectively bind the same ligand (i.e., preexisting degradation products will make accurate interpretation impossible). Concentrations of ligand are selected so that they will provide good definition of the plateaus expected at the end points of the titration curves as well as the transition(s). For the study of a multisite protein having sites of unequal affinities for ligand, the most desirable approach is to screen the ligand-binding properties first at concentrations that are evenly spaced at half-log units over several orders of magnitude {e.g., for CaM the titrations routinely span pCa $[-\log(Ca^{2+})_{free}]$ 9 (nM) to pCa 3 (mM)} and then complete the titration curve by preparing additional samples at more finely spaced intervals if they are needed.

For each ligand concentration that was selected, several independent

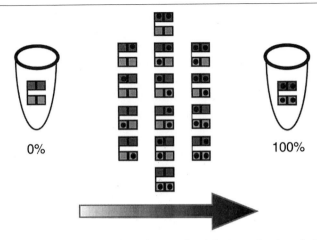

Fɪɢ. 21. Schematic diagram of 16 species populated during a titration of sites I, II, III, and IV of vertebrate CaM. The only species that can be obtained as pure samples are the apo (zero calcium bound) and fully saturated (four calcium bound) forms. At intermediate levels of saturation, a distribution of the remaining 14 species exists.

samples of protein are equilibrated with ligand and exposed to a protease under identical solution conditions. Proteolysis is quenched by an inhibitor or by injection onto the HPLC column. Primary cleavage products are separated by HPLC and their fractional abundance determined by integrating peak areas. Multiple injections of the same sample ($N = 2$ or 3, usually) are conducted to correct for random error in pipetting and loading. Chemical identification of collected samples occurs separately.

Susceptibility is determined by calculating the fractional area of each peak relative to the total mass of CaM (i.e., area of all CaM-derived peaks) injected from an equilibrated mixture. This approach corrects for any variation in the volume of digested sample injected onto the column in each analysis. A spreadsheet program such as Excel (Microsoft, Redmond, WA) is a straightforward way to reduce these data to resolve a proteolytic footprinting titration curve. Additional experimental details for these procedures are described below.

Conducting Stoichiometric Titration

To conduct a stoichiometric titration, aliquots of ligand are added to a sample of protein at high concentration [i.e., above the dissociation constant for the site(s) with the least favorable free energy of binding that ligand]. Accurate knowledge of the protein concentration is important so that the ligand-to-protein ratio is determined accurately. It is also important for determining whether the titration is indeed occurring under stoichiometric or equilibrium conditions. Colorimetric assays, UV absorbance, and amino acid analysis are all standard methods for determining protein concentration. Accuracy and precision vary in the hands of the operator and for reasons related to individual proteins. For all these reasons, it is important to measure a buffer blank to avoid being misled.

Conducting Equilibrium Titration

For an equilibrium binding titration, there are two common approaches to obtaining equilibrated populations of species to probe. One is to add a small amount of ligand to a sample of protein that is at a concentration well below the dissociation constant for ligand. The major difficulty with this approach is that if the protein is at low concentration, the peptides derived from it after partial proteolysis will be even less abundant. However, for precise quantitation, they must be at a level well above the background noise. On a practical level, this means that it is difficult to take this approach for interactions where the dissociation constants are in the micromolar or nanomolar ranges.

Unless it is practical to use an approach such as radioactive end labeling to increase the sensitivity of peptide detection, it is simpler and safer to extensively dialyze a high concentration of protein against buffered solutions of ligand (especially when studying an inexpensive ligand). This has been described for CaM previously.[1,2,13] An advantage is that separate dialysis bags containing different purified proteins (e.g., several mutants and wild type) may be dialyzed simultaneously. The resulting set of dialysates is then characterized by having the same concentration of ligand (i.e., the independent variable is known to be identical). Aliquots of each equilibrated sample are subjected to proteolysis and the resulting data may be compared with great confidence. Although this approach would not be practical for studying titrations where the ligand was a precious protein, it works well for small diffusible ligands such as salts, metals, and drugs.

Measuring Susceptibility, the Dependent Variable

It is more rigorous and chemically more manageable to conduct parallel analyses of a titration with several different proteases, each probing only a few positions, than to attempt to minimize secondary cleavage and interpret the data from footprinting studies conducted with a protease that cleaves at many positions with equal probability. In this section, we describe thrombin footprinting of isolated domains of CaM because this study provides good examples of all the necessary steps in data reduction, from HPLC chromatogram to normalized titration curves. The original report of this study did not include these raw data or level of detail because of space constraints.[13]

Calcium Buffers and Protein Dialysates. For the example to follow, proteolytic reactions are performed in calcium buffers containing 50 mM HEPES, 91.4 ± 0.7 mM KCl, 5 mM nitrilotriacetic acid (NTA), and 0.05 mM EGTA; pH 7.40 ± 0.01 at 22.0 ± 0.2°. The conductivity is measured with a Radiometer (Copenhagen, Denmark) CDM83 conductivity meter and this is used to determine the [KCl] experimentally. The amount of calcium chloride required to vary the free calcium concentration from nanomolar to millimolar in these buffers is calculated on the basis of the association constants of calcium for NTA, EGTA, and calmodulin.[16,17] Corresponding volumes of 1 M CaCl$_2$ are added to the preceding 1× buffer stock to create 15 individual buffers of varied pCa. For buffers with low and intermediate calcium concentrations (high and intermediate pCa values), the resulting free calcium level is determined experimentally with a fluorimetric indicator: difluoro (dif) BAPTA with a K_d of $1.52 \times 10^{-6} M^{-1}$ and BAPTA with a K_d of $1.26 \times 10^{-7} M^{-1}$. For buffers with higher (saturating) levels of calcium, a calcium-selective electrode (F2110Ca) from Radiometer is used to measure pCa.[1] The experimentally determined free calcium levels in these buffers range in pCa from 9.48 (apo) to 3.67 (saturated).

Purified aliquots of protein are exhaustively dialyzed against each of the buffers. Because of the technical simplicity of equilibration, multiple proteins (e.g., wild type and mutant forms) may be dialyzed simultaneously to assure identical ligand activity. In this example, full-length calmodulin is compared with fragments corresponding to the the N-domain (residues 1–75) and C-domain (residues 76–148) by dialyzing aliquots of all three in the same buffer series.

Concentrations of each of the dialyzed protein samples are determined by the bicinchoninic acid (BCA) protein assay (Pierce, Rockford, IL), which has been calibrated by amino acid analysis of protein content in several samples. It is not essential to have the protein concentrations for all samples in a titration be absolutely identical but it is helpful for them to be within a range of ~25% so that signal-to-noise ratio is similar for all samples. In the study illustrated here, the ranges of sample concentrations are as follows: CaM, 0.626–0.773 mg/ml (37.5–46.3 μM); N-domain, 0.612–0.976 mg/ml (73.6–117.4 μM); and C-domain, 0.351–0.384 mg/ml (41.8–45.7 μM).

[16] A. Fabiato and F. Fabiato, *J. Physiol.* **75,** 463 (1979).
[17] L. G. Sillen and A. E. Martell, "Stability Constants of Metal–Ion Complexes: Special Publication No. 25 of the Chemical Society." Alden & Mowbray, Oxford, 1971.

Proteolytic Footprinting Reactions. Dialysates of CaM, N-domain, and C-domain are diluted in the corresponding pCa buffer to 0.2 mg/ml (12 μM CaM, 24 μM N- and C-domains). To initiate proteolysis, 3.5 μl of a 0.2-U/μl stock of thrombin (specific activity, 1000 NIH units/mg of protein) is added to each CaM dialysate in a total volume of 140 μl (0.005 U of thrombin per microliter to 0.2 mg of CaM per milliliter). Although it may be expected that the domains might require less protease for the same degree of cleavage, the reverse is true. To obtain a reasonable signal-to-noise ratio for peak integration of fragments of the domains, the level of thrombin used for proteolytic footprinting of the isolated N-domain is 0.01 U of thrombin per microliter to 0.2 mg of N-domain per milliliter (twice the level used for CaM). For proteolytic footprinting of the isolated C-domain, 0.05 U of thrombin per microliter to 0.2 mg of C-domain per milliliter (10 times the level of thrombin used for CaM) is used. Samples are incubated with thrombin for 60 min at 22.0°; to quench proteolysis, an aliquot of 10 mM FFRCK is added (CaM, 5 μl to make 0.35 mM; N-domain, 9.52 μl to make 0.64 mM; and C-domain, 14 μl to make 0.91 mM).

HPLC Separation of Proteolytic Footprinting Products. Products of limited proteolysis are separated and quantified in a manner similar to that used with EndoGluC titrations of CaM.[1] For each titration, 2 vol of 50 μl (10 μg) from a footprinting reaction at each pCa is injected successively onto a Vydac (Separations Group, Hesperia CA) C_{18}, 220 × 4.6 mm i.d. column by a Gilson (Middleton, WI) HPLC model 231 autosampler. Samples are eluted at 1.0 ml/min and monitored at wavelengths of 220 and 280 nm. The gradient used to separate the fragments of CaM is 35–45% B [acetonitrile–water (80:20, v/v) with 0.06% (v/v) trifluoroacetic acid (TFA)] in 24 min, followed by 45–57% B in 25 min, where Buffer A is 0.06% (v/v) TFA. The gradient used to separate the fragments of the isolated N-domain is 35–40% B in 12 min, followed by 40–53% B in 25 min. The gradient used to separate fragments of the isolated C-domain is 30–35% B in 5 min followed by 35–50% B in 42 min. In all cases, the C_{18} column is washed in 100% B for 2 min and reequilibrated in 35% B prior to loading subsequent samples. Examples of chromatographic profiles for thrombin footprinting of the isolated N-domain and C-domain are shown in Fig. 22.

The proteolytic fragments of the isolated domains are collected from the HPLC effluent and concentrated with a Savant (Hicksville, NY) SpeedVac concentrator; mass spectroscopy is performed to identify fragments. Calculations (based on assuming cleavage positions of R37 and R106[2]) yield values of percent accuracy ([theoretical mass]/[experimental mass]) of

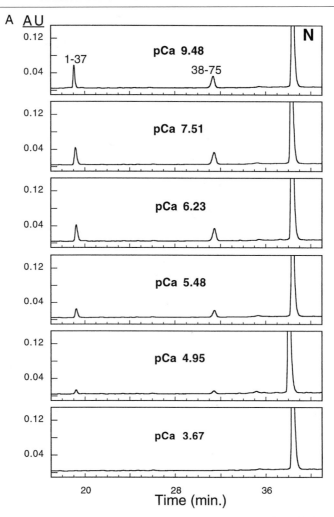

FIG. 22. HPLC chromatograms of fragments generated by thrombin footprinting of calcium titrations of individual domain fragments of CaM. Peak elution monitored by absorbance (AU) at 220 nm. Prior to proteolysis, protein samples were dialyzed in 50 mM HEPES, 91.4 ($\pm$0.7) mM KCl, 5 mM NTA, 0.05 mM EGTA, and increasing calcium concentrations (where pCa = $-$log[Ca^{2+}]$_{free}$); pH 7.40 at 22°. (A) Titration of the N-domain. (B) Titration of the C-domain. Peptides were identified by mass spectrometry and amino acid analysis.

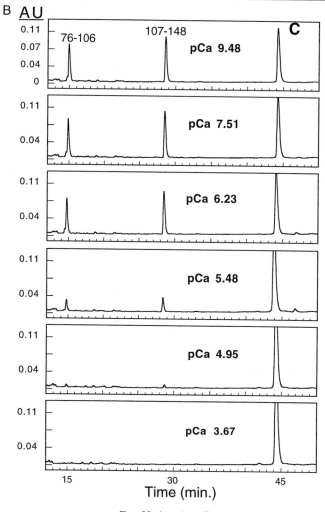

FIG. 22. (*continued*)

fragment identification of 99.7–99.8% for the four fragments (1–37 and 38–75 in the N-domain and 76–106 and 107–148 in the C-domain).

Interpretation of Footprinting Titrations

For many decades, biochemists and biophysicists have reported titration curves of proteins based on the fractional change of a signal from an intrinsic spectral reporter group (e.g., aromatic side chain) that reflects ligand-induced changes in environment (e.g., solvent exposure, quenching, hydro-phobicity of a pocket). Ligand-dependent proteolytic susceptibility similarly provides a sensitive indication of conformational change linked to binding. Both of these approaches are indirect. The degree of ligand-induced change in absorbance, fluorescence, or susceptibility of a peptide bond to proteo-lytic attack is not necessarily proportional to the extent of saturation of a binding site. However, we may interpret such signals as binding isotherms if we can validate them with appropriate controls that give us precise but less detailed information about the macroscopic binding properties. The following section describes the process for reducing the chromatographic data to susceptibility profiles and for analyzing these as binding isotherms when that is warranted.

Resolution of Susceptibility Profiles from HPLC Chromatograms

As described previously,[1,2] susceptibility profiles are generated by man-ual integration of the peak areas of fragments generated in a footprinting reaction. The fractional area of each peptide, A_i, is determined by Eq. (1):

$$A_i = \frac{\text{peak area of peptide}_i}{\text{peak area of uncut protein} + \sum_{j=1}^{n} \text{peak area of peptide}_j} \qquad (1)$$

where the denominator represents the mass of protein exposed to protease and injected on the HPLC column (i.e., the sum of the areas corresponding to the uncut protein and all proteolytic fragments). From the chromato-grams shown for thrombin footprinting of the N-domain of CaM (Fig. 22A), two ratios would be calculated: A_{1-37}, representing the fractional area of the peak for peptide 1–37, and A_{38-75}, representing the fractional area of the peak for peptide 38–75. The ratio of A_{1-37} to A_{38-75} (i.e, the ratio of complementary primary cleavage products) should be the same for every pCa level. For an effluent monitored by absorbance (in this case, 220 nm), the ratio of the two fractional areas will be determined by the difference in the extinction coefficients. For peaks with small areas, even a small contribution from baseline noise can make a large change in the relative value of fractional area. This underscores the need to optimize the signal-to-noise ratio in the development phase of the HPLC method.

Precision of Quantitation and Background Correction

Although most HPLC manufacturers supply software for integrating peaks, it is often not sufficient for deconvoluting overlapping peaks or for making precise corrections for baseline slopes and irregularities. Commercially available software such as PeakFit (Jandel Scientific, San Rafael, CA) and academic software intended for analysis of spectral and chromatographic data are often superior tools. The highest priority is to obtain a consistent assessment of areas for all the peaks that represent fragments and full-length forms of the protein of interest.

Absolute Susceptibility Profiles

The raw or absolute susceptibility values will vary greatly for peptide bonds within a protein, as was indicated in Figs. 17 and 18. These differences will reflect a variety of attributes: the intrinsic flexibility and accessibility of that bond in the protein as well as its conformational response to ligand. Thus, within a region of primary sequence that is identical in two proteins or protein fragments, the absolute susceptibility may differ for the same solution conditions. An example is shown in Fig. 23A, where the susceptibil-

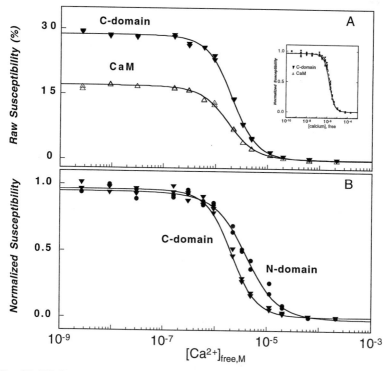

FIG. 23. (A) Comparison of the raw and normalized (*inset*) susceptibility of R106 in wild-type CaM and the isolated C-domain. (B) Comparison of the normalized susceptibility of the isolated N- and C-domains. [Data taken from B. R. Sorensen and M. A. Shea, *Biochemistry* **37**, 4244 (1998).]

ity of the R106–H107 bond in CaM to thrombin proteolysis is compared for whole CaM and a fragment (C-domain) composed of residues 76–148, as a function of calcium ranging over 6 orders of magnitude, from nanomolar to millimolar. The absolute susceptibility (i.e., the fraction of bonds cleaved under fixed conditions of protease exposure) of the two proteins differs. The C-domain is more susceptible, perhaps because its N terminus is now foreshortened by 75 residues. However, the fractional response to calcium is identical, as shown by the comparison of the normalized values in the inset. That is, the same degree of change in conformation is reached at each pCa level, despite the starting structures being different.

Normalized Susceptibility Profiles

To create the susceptibility profiles shown in the inset in Fig. 23A, the calculated fractional areas [Eq. (1)] for each primary cleavage product generated at each point in a titration were normalized from 0 to 1 and plotted as the normalized susceptibility of that primary cleavage site against the free calcium concentration of each buffer. For a given dose of protease (e.g., a specific concentration of protease applied for a fixed interval) at a specific activity of ligand, the fraction of each fragment of the protein of interest should be measurable reproducibly to within 5%.

Resolution of Binding Constants

There are two classes of conclusions that are drawn from the proteolytic footprinting titrations. One is a chemical analysis of the positions that are cleaved and their respective classes of responses to ligand binding. The other is a quantitative interpretation based on treating the susceptibility profiles being representative of occupancy of the neighboring sites.

The fractional population or probability (f_s) of a single species (s) of a macromolecule is given by Eq. (2),

$$f_s = \frac{\exp(-\Delta G_s / RT) \cdot [X]^j}{\sum_{s,j} \exp(-\Delta G_s / RT) \cdot [X]^j} \tag{2}$$

where $[X]$ is the ligand activity, ΔG_s represents the Gibbs free energy $[-RT \ln(K_s)]$ of the species, and j represents the stoichiometry of ligand binding to species s.

For a monomeric ligand that interacts with a single binding site on a macromolecule (see Fig. 17) with an equilibrium constant of k, the probability of the species with a site occupied is equivalent to the fractional satura-

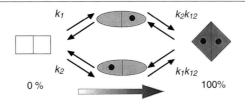

FIG. 24. Linkage scheme for ligand that binds at two sites simultaneously. Microscopic association equilibrium constants: k_1 and k_2 represent intrinsic binding constants for calcium and k_{12} represents cooperativity between two sites in a domain. A dash (–) is a vacant site and a dot (●) is an occupied site.

tion $\overline{Y}$ of the molecule (the ratio of moles of sites occupied to total moles of sites) given by Eq. (3):

$$\overline{Y} = \frac{k[X]^1}{1 + k[X]^1} \tag{3}$$

In the general case shown in Fig. 24 for two sites binding a monomeric ligand, the fractional saturation of site 1 is given by the expressions below. The numerator of Eq. (4a) contains the Boltzmann terms for the two species that have site 1 occupied and the denominator accounts for all the moles of site 1 in the population. Equation (4b) represents the corresponding equation for saturation of site 2.

$$\overline{Y}_1 = \frac{k_1[X]^1 + k_1 k_2 k_{12}[X]^2}{(1 + (k_1 + k_2)[X]^1 + k_1 k_2 k_{12}[X]^2)} \tag{4a}$$

$$\overline{Y}_2 = \frac{k_2[X]^1 + k_1 k_2 k_{12}[X]^2}{(1 + (k_1 + k_2)[X]^1 + k_1 k_2 k_{12}[X]^2)} \tag{4b}$$

For two sites that are independent (i.e., noncooperative), $k_{12} = 1$ and these equations simplify to Eqs. (5a) and (5b). The average fractional saturation for the molecule is given by Eq. (5c).

$$\overline{Y}_1 = \frac{k_1[X]^1}{1 + k_1[X]^1} \tag{5a}$$

$$\overline{Y}_2 = \frac{k_2[X]^1}{1 + k_2[X]^1} \tag{5b}$$

$$\overline{Y}_t = \frac{k_1[X]^1}{2(1 + k_1[X]^1)} + \frac{k_2[X]^1}{2(1 + k_2[X]^1)} \tag{5c}$$

If the two sites shown in Fig. 24 are also equivalent (i.e., have equal intrinsic affinities for ligand), then $k = k_1 = k_2$. In that case, the total average fractional saturation $(\overline{Y}_t)$ is equal to the fractional saturation of each site in a domain (a case that was illustrated in Figs. 8–10) and is equivalent to the simple Langmuir binding isotherm given in Eq. (3). The total free energy (ΔG_2) of ligand binding to both sites in such a system is given by $-2RT \ln k$.

Calcium binding to each domain of calmodulin has been analyzed with a standard linkage scheme for a macromolecule binding two ligands[18] (Fig. 24). Equation (6) allows for the sites to be heterogeneous (unequal intrinsic affinities, $k_1 \neq k_2$) and cooperative ($k_{12} \neq 1$); this general form is referred to as the Adair equation for two sites.

$$\overline{Y}_t = \frac{K_1[X]^1 + 2K_2[X]^2}{2(1 + (K_1)[X]^1 + K_2[X]^2)} \tag{6}$$

The macroscopic equilibrium constant K_1 represents the sum of two intrinsic microscopic equilibrium constants $(k_1 + k_2)$ that are not necessarily equal. K_2 represents the equilibrium constant $(k_1 k_2 k_{12})$ for binding ligand to both sites; it accounts for any positive or negative cooperativity regardless of source or magnitude. This has been deemed most appropriate because all the signals appear to monitor properties of both sites in a single domain rather than reporting on the occupancy of an individual site.

More complex formulations (with and without assumptions about the similarity of sites and possible pairwise or higher order interactions) will apply to more complex systems such as those with additional ligand-binding sites. Furthermore, it may be essential to consider the stoichiometry of the ligand (i.e., monomer–dimer equilibrium).

Fitting for End Points. As noted in a previous section, the raw susceptibility data for each peptide are normalized to the highest and lowest experimentally determined values for A_i. However, with any experimental signal, there will be finite variations in the asymptotes of the resulting normalized data. To account for this in the susceptibility profiles for different peptides, the function $[f(X)]$ used for nonlinear least-squares analysis of the fractional abundance of each peptide is modified by an offset and scale, as shown in Eq. (7):

$$f(X) = Y_{[X]_{low}} + \overline{Y}_t(\text{Span}) \tag{7}$$

where $\overline{Y}_t$ refers to the average fractional saturation as described by the pertinent equation above and $Y_{[X]_{low}}$ corresponds to the value of the suscep-

[18] R. E. Klevit, *Methods Enzymol.* **102**, 82 (1983).

tibility at the lowest ligand (calcium) concentration of the titration being fit.[2] Note that the span is negative for a monotonically decreasing signal and positive for a monotonically increasing signal. The values for all parameters were fit simultaneously (i.e., all were floated) when possible. It is straightforward to resolve parameters for monotonically varying isotherms. However, for biphasic susceptibility profiles, it was generally essential to fix the Span value close to ± 1, as appropriate for the direction of the curve. Changes in that value (± 0.2) were tested for their effect on the resolved parameters. Data for complementary primary cleavage products were fit separately and as combined sets. There should be good agreement between these.

For a single titration, the format of Eq. (7) is equivalent to fitting for a low and high end point without normalizing the data. That convention was not used because in attempting to fit the biphasic curves, it was essential to try functions with the form of Eq. (8), by which each titration may be characterized by a different extent and direction of ligand-induced conformational change (i.e., Span parameter). This formulation requires three parameters that are nonredundant (in contrast to the four parameters that would result for two sets of high and low end points but with two of them identical for a continuous curve).

$$f(X) = Y_{[X]_{\text{low}}} + \overline{Y}_a(\text{Span}_a) + \overline{Y}_b(\text{Span}_b) \qquad (8)$$

Multiple criteria reported by NONLIN[19] were used for evaluating goodness of fit for the parameters that minimized the variance of each fit. These included (1) the value of the square root of variance, (2) the values of asymmetric 65% confidence intervals, (3) the presence of systematic trends in the distribution of residuals, (4) the magnitude of the span of residuals, and (5) the absolute value of elements of the correlation matrix.

Pairwise Cooperativity. It is not possible to determine k_{12}, the intradomain cooperativity constant, analytically from macroscopic binding data alone. However, it may be estimated by assuming that the binding sites have equal intrinsic affinities ($k_1 + k_2 = 2k$). This leads to K_c, the apparent cooperativity constant, as given in Eq. (9); K_c provides a lower limit for the value of k_{12}, the actual microscopic cooperativity constant.

$$K_c = 4K_2/K_1^2 \qquad (9)$$

The corresponding expression for the free energy is $\Delta G_c = \Delta G_2 - 2\Delta G_1 - RT \ln(4)$. There is much confusion about this issue. In the case of a protein like CaM, where the binding sites for a single ligand are similar,

[19] M. L. Johnson and S. G. Frasier, *Methods Enzymol.* **117**, 301 (1985).

it is not misleading to tabulate this parameter. However, it should be recognized that compensatory contributions of cooperative free energy and heterogeneous binding energies are such that for the same cooperative chemical interaction (i.e., k_{12} constant), the value of K_c will increase if there is a decrease in heterogeneity between sites. For example, if the ratio of k_1/k_2 is 10, K_c is $0.33k_{12}$, but for a k_1/k_2 ratio of 2, K_c is $0.89k_{12}$ (see Ref. 20). It is fundamentally important to recognize that the value of K_c cannot indicate positive cooperativity if there is none and is necessarily <1 (i.e., indicates apparent anticooperativity) if the sites are both noncooperative (i.e., $k_{12} = 1$) and also heterogeneous (unequal) in intrinsic affinity.

Note that the precision of the calculation of K_c or ΔG_c depends on the precision of both ΔG_1 and ΔG_2, and the precision of ΔG_1 is generally lower than that of ΔG_2. Therefore, this estimate of cooperativity is used as a diagnostic indicating the nature of intradomain cooperativity, but there is little significance to small numerical differences between resolved values.

Summary

To dissect the chemical basis for interactions controlling regulatory properties of macromolecular assemblies, it is essential to explore experimentally the linkage between ligand binding, conformational change, and subunit assembly. There are many advantages to using techniques that will probe the occupancy of individual binding sites or monitor conformational responses of individual residues, as described here.

Proteolytic footprinting titrations may be used to infer binding free energies for ligands interacting with multiple sites or domains and to detect otherwise unrecognized "silent" interdomain interactions. Microgram quantities of pure protein are required, which is low relative to the hundreds of milligrams needed for comparable discontinuous equilibirum titrations monitored by NMR.[21] By running comparative studies with several proteases, it is easy to determine whether resulting titration curves are consistent, independent of the protease used and therefore representative of the structural response of the protein to ligand binding or other differences in solution conditions (pH, salt, temperature). The results from multiple techniques (e.g., NMR, fluorescence, and footprinting) applied to aliquots from the same discontinuous titration may be compared easily to test for consistency.

Classic methods for determining thermodynamic and kinetic properties of calcium binding to calmodulin include filter binding and equilibrium or

[20] C. G. Caday, P. K. Lambooy, and R. F. Steiner, *Biopolymers* **25**, 1579 (1986).
[21] S. Pedigo and M. A. Shea, *Biochemistry* **34**, 10676 (1995).

flow dialysis (employing the isotope ^{45}Ca), spectroscopic studies of stopped-flow fluorescence, calorimetry, and direct ion titrations. A cautionary note is that many different sets of microscopic data would be consistent with a single set of macroscopic constants determined by classic methods. This was well illustrated in Fig. 9. Thus, while it is important to compare results with those obtained by classic binding methods, they are, by definition, incapable of resolving the microscopic constants of interest. Thus, there is only one "direction" for comparison.

Quantitative proteolytic footprinting titrations applied to studying calmodulin provided the first direct quantitative estimate of negative interactions between domains. Although studies of site-knockout mutants had suggested interactions between domains, this approach gave the first evidence for the pathway of anticooperative interactions between domains by showing that helix B responds structurally to calcium binding to sites III and IV in the C-domain. Despite two decades of study of calmodulin and the application of limited proteolysis studies to the apo and fully saturated forms, this finding emerged only when titration studies were undertaken as described.

This highlights the general observation that while the behavior of the intermediate states in a cooperative switch are the key elements of the transition mechanism, they are the most difficult to observe. The unexpected finding that the isolated domains are nearly equivalent in their calcium-binding properties (Fig. 23 B) leaves us with many of the questions we had at the start: How does the sum of two nearly equivalent domains result in a molecule that switches sequentially rather than simultaneously? But it underscores why it is not yet possible to understand similar proteins by sequence gazing alone.

Acknowledgments

We acknowledge the contributions of many colleagues to the development and application of the quantitative proteolytic footprinting method. Figures that were developed as part of an undergraduate research program include Figs. 12 and 13 (James K. Kranz), Fig. 15 (Maria Hutchins), and Fig. 16 (Laurel Coffeen). Others in our laboratory who contributed were Arthur Chen, Cynthia R. Kephart, Daniel D. Kephart, Wolfgang Schaller, and Mark L. Yates. Asmaa Baker assisted with the preparation of many of the figures. Our efforts have benefited greatly from discussions with others, including Gary K. Ackers, D. Wayne Bolen, Michael D. Brenowitz, Herbert Halvorson, Michael L. Johnson, J. Ching Lee, and Donald F. Senear. The National Science Foundation, the American Heart Association, and the University of Iowa supported this work.

[13] Analysis of Reversibly Interacting Macromolecular Systems by Time Derivative Sedimentation Velocity

By WALTER F. STAFFORD

Sedimentation velocity boundary analysis of interacting systems can provide information about stoichiometries and equilibrium constants that is complementary to information obtained by equilibrium sedimentation analysis. The theoretical basis for the analysis of sedimentation boundaries was first developed in the 1950s by Gilbert and co-workers, who solved the transport equations of the case of no diffusion.[1-3] Extensive numerical simulations of the transport equations for various reversibly interacting systems were carried out by Cann,[4] Cox and Dale,[5] and others in the 1960s, 1970s, and 1980s and have given us a firm basis for the use of transport methods for the analysis of interacting systems.

This chapter discusses the application of time derivative techniques to the analysis of interacting systems by sedimentation velocity. Potential problems and limitations associated with the computation of weight average sedimentation coefficients for both rapidly reversible and kinetically limited systems in general are treated. An example of the application of finite element simulation software to the analysis of an antigen–antibody system at high dilution is given. A potential problem with convective instability of pressure-dependent systems is also discussed and shown not to be a problem if care is taken.

The accessible concentration range for sedimentation velocity has been extended considerably by the time derivative/signal averaging method.[6,7] The combination of taking the time derivative, which eliminates completely the time-independent optical background, and of averaging the time derivative curves results in a considerable increase in precision compared with older methods of analysis. Therefore, because of its sensitivity and selectivity, sedimentation velocity may be the only way the equilibrium constant for an interacting system can be estimated by sedimentation analysis. The

---

[1] G. A. Gilbert, *Proc. R. Soc.* **A250,** 377 (1959).

[2] G. A. Gilbert and R. C. Jenkins, *Proc. R. Soc.* **A253,** 420 (1959).

[3] G. A. Gilbert, *Discuss. Farad. Soc.* **20,** 68 (1955).

[4] J. R. Cann, "Interacting Macromolecules." Academic Press, New York, 1970.

[5] D. J. Cox and R. S. Dale, *in* "Protein–Protein Interactions" (C. Frieden, and L. W. Nichol, eds.), pp. 173–212. John Wiley & Sons, New York, 1981.

[6] W. F. Stafford, *Anal. Biochem.* **203,** 295 (1992).

[7] W. F. Stafford, *Methods Enzymol.* **240,** 478 (1994).

increased sensitivity allows investigation of interacting systems that were previously inaccessible to ultracentrifugal analysis. Reversibly interacting systems whose dissociation constants in terms of mass concentration units are in the range of 5–100 μg/ml can be studied using Rayleigh optics, and those whose dissociation constants, expressed in optical density (OD) units, are less than 0.1 OD unit can be studied using UV optics.

Both kinetically controlled and pressure-dependent interacting systems may be difficult to analyze by sedimentation velocity. It is shown below that kinetically controlled systems can be treated to yield reliable association equilibrium constants from analysis of weight average sedimentation coefficients. Interactions whose reaction times are on the same order as the rates of sedimentation can be identified and treated. Furthermore, it is shown that the rate of reequilibration has little effect on the calculation of equilibrium constants from weight average sedimentation coefficients, as long as the system is at equilibrium at the start of sedimentation. Therefore, systems having both slow forward and slow reverse reaction rates can be studied by sedimentation velocity. This means that sedimentation velocity can be used to analyze some systems that might not be amenable to sedimentation equilibrium methods. For example, an interacting system that has a relaxation time on the order of 12 to 24 hr ($k_r = 5 \times 10^{-4}$ to 1×10^{-5} sec^{-1}, where k_r is the first-order dissociation rate constant) would require more than 10 days of sedimentation to analyze by sedimentation equilibrium but could be analyzed essentially as a quasistatic system by sedimentation velocity. Therefore, successful analysis is possible even when the system is reversible but not fast.

Sedimentation transport of pressure-dependent interacting systems can lead to the generation of negative concentration gradients.[8] In the absence of a stabilizing density gradient, convection can result. One might think that pressure effects would obviate the use of sedimentation velocity to study systems that exhibited any volume change on association because any volume change can in principle produce negative concentration gradients. However, it can be shown that for most interacting systems with even substantial volume changes that the stabilizing density gradient set up by redistribution of buffer components at the usual concentrations is expected to be enough to prevent convection as long as certain precautions are taken.

Two general approaches to the analysis of interacting systems are discussed. The first approach, which involves computing the weight average sedimentation coefficient as a function of concentration, is insensitive to details of boundary shape. The weight average sedimentation coefficient, s_w, is a composition-dependent parameter and, therefore, for any given

[8] G. Kegeles and W. F. Harrington, *Methods Enzymol.* **27**, 306 (1973).

system controlled by mass action, it will be a function of the plateau concentration only, reflecting the composition of the solution at the plateau concentration. Analysis of the functional dependence of s_w on plateau concentration allows the determination of the equilibrium constants describing the system. For some reviews of the analysis of interacting systems see the articles by Stafford,[9] Lee and Rajendran[10], and Rivas et al.[11] The weight average sedimentation coefficient has been used extensively by Timasheff and co-workers to study the self-association of tubulin under various conditions. For review of this work the reader is referred to articles by Na and Timasheff and to references contained therein.[12,13] The weight average sedimentation coefficient computed with the time derivative method[6,9] has been used extensively by Lobert, Correia, and co-workers[14–16] to study the effects of vinca alkaloids and nucleotides on the self-assembly of tubulin. Most recently, the weight average along with the Z and $Z + 1$ averages have been used by Toedt et al. to characterize a tobacco mosaic virus (TMV) coat protein mutant.[17]

The second general approach involves numerical solutions of the Lamm equation to simulate sedimentation boundary profiles for various values of the parameters describing the system. It can be subdivided further into two different methods. The first simulation method involves the generation of a series of sedimentation patterns spanning the suspected range of parameters followed by direct comparison with the observed patterns. This comparison approach was used extensively by Timasheff and co-workers[18–22] in the

[9] W. F. Stafford, in "Modern Analytical Ultracentrifugation: Acquisition and Interpretation of Data for Biological and Synthetic Polymer Systems" (T. M. Schuster and T. M. Laue, eds.), pp. 119–137. Birkhäuser, Boston, 1994.
[10] J. C. Lee and S. Rajendran, in "Modern Analytical Ultracentrifugation: Acquisition and Interpretation of Data for Biological and Synthetic Polymer Systems" (T. M. Schuster and T. M. Laue, eds.), pp. 138–155. Birkhäuser, Boston, 1994.
[11] G. Rivas, W. F. Stafford, and A. P. Minton, "Methods: A Companion to Methods in Enzymology," Vol. 19, pp. 194–212. Academic Press, San Diego, California, 1999.
[12] G. C. Na and S. N. Timasheff, Methods Enzymol. 117, 496 (1985).
[13] G. C. Na and S. N. Timasheff, Methods Enzymol. 117, 459 (1985).
[14] S. Lobert, B. Vulevic, and J. J. Correia, Biochemistry 35, 6806 (1996).
[15] S. Lobert, C. A. Boyd, and J. Correia, Biophys. J. 72, 416 (1997).
[16] S. Lobert, A. Frankfurter, and J. J. Correia, Biochemistry 34, 8050 (1995).
[17] J. M. Toedt, E. H. Braswell, T. M. Schuster, D. A. Yphantis, Z. F. Taraporewala, and J. N. Culver, Protein Sci. 8, 261 (1999).
[18] R. P. Frigon and S. N. Timasheff, Biochemistry 14, 4567 (1975).
[19] R. P. Frigon and S. N. Timasheff, Biochemistry 14, 4559 (1975).
[20] S. N. Timasheff, R. P. Frigon, and J. C. Lee, Fed. Proc. 35, 1886 (1976).
[21] G. C. Na and S. N. Timasheff, Biochemistry 25, 6214 (1986).
[22] G. C. Na and S. N. Timasheff, Biochemistry 25, 6222 (1986).

1970s and 1980s, using schlieren optics in the 1-to 10-mg/ml range to analyze self-association of tubulin.

The second simulation method uses a nonlinear least-squares fitting procedure employing simulation software as a function evaluator. Unless the system can be parameterized in terms of a single dimensionless parameter, the nonlinear curve fitting approach is the better of these last two. Several examples of simulated single-parameter self-associating and hetero-associating systems have been discussed previously.[9] A description of global curve-fitting software for analyzing self-associating and heteroassociating interacting systems using numerical finite element solutions to the Lamm equation to fit to time difference data[23] will be described elsewhere. The first use of finite element simulations for curve fitting to sedimentation velocity data was originally presented by Todd and Haschemeyer.[24] The Todd–Haschemeyer method has been revived and reprogrammed by Demeler and Saber[25] and extended by Schuck,[26] using a moving reference frame to accelerate significantly the computations. More recently Schuck and Demeler[27] have developed a method for fitting for the time-invariant background component when using the Todd–Haschemeyer method for the analysis of both interference and absorbance data. These programs are available from the RASMB (Reversible Associations in Structural and Molecular Biology) FTP site (*ftp://rasmb.bbri.org/*) and can be used to analyze mixture of independently sedimenting species as well as monomer–dimer and monomer–trimer self-associations.

Sedimentation transport experiments are not carried out at thermodynamic equilibrium; therefore, care must be exercised to assure that equilibrium constants derived from weight average sedimentation coefficients actually reflect the equilibrium situation. There are several problems unique to sedimentation velocity analysis that need to be addressed in this regard; they include (1) the effects of radial dilution, which may cause a shift in the equilibrium during sedimentation, (2) hydrodynamic concentration dependence of the sedimentation coefficients for the species making up the system, (3) thermodynamic concentration dependence influencing the diffusion coefficients through the activity coefficient, (4) kinetics of reequilibration, and (5) negative concentration gradients resulting from pressure dependence of the equilibrium constants. The effects of hydrodynamic and thermodynamic nonideality can be minimized by performing experiments at sufficiently low concentration. They may cause significant problems in

[23] W. F. Stafford, *Biophys. J.* **74**, A301 (1998).
[24] G. P. Todd and R. H. Haschemeyer, *Proc. Natl. Acad. Sci. U.S.A.* **78**, 6739 (1981).
[25] B. Demeler and H. Saber, *Biophys. J.* **74**, 444 (1998).
[26] P. Schuck, *Biophys. J.* **75**, 1503 (1998).
[27] P. Schuck and B. Demeler, *Biophys. J.* **76**, 2288 (1999).

the analysis of weakly associating systems, which must be studied at concentrations higher than about 1 mg/ml if their effects are not taken into account. Their effects can be dealt with explicitly in curve-fitting routines and are not discussed further here. However, the other effects, which are addressed below, are relevant to all concentration ranges.

Background Theory for Time Derivative Method

The large increase in sensitivity of sedimentation analysis has been achieved mainly because of the development of on-line, real-time Rayleigh interferometric systems pioneered by Laue, Yphantis, et al.[28-31] These systems have been developed further and optimized for sedimentation velocity analysis.[7,9,31-33] The rapidity of on-line systems has allowed the convenient computation of the time derivative of the concentration distribution resulting in automatic baseline correction of the sedimentation patterns. Moreover, the ability to collect rapidly large amounts of data means the data can be effectively signal averaged to increase the signal-to-noise ratio.

A derivation of the apparent sedimentation coefficient distribution function, $g(s^*)$ versus s^*, from the time derivative has been presented previously.[6] It should be pointed out that the unnormalized apparent distribution function, designated as $g(s^*)$, can be considered as simply the first derivative of the concentration profile with respect to s^*, dc/ds^* versus s^*, and, therefore, contains all the contributions from diffusion. The original derivations of $g(s^*)$ both from the radial derivative[34,35] and from the time derivative[6] were based on the assumption that diffusion was negligible. It was recognized that the apparent distribution approached the "true" distribution either in the limit of high molecular weight (i.e., as D approached zero) or as the patterns were extrapolated to infinite time. In spite of these limitations and nomenclature, the uncorrected, unnormalized,

[28] T. M. Laue, Ph.D. dissertation. University of Connecticut, Storrs, Connecticut, 1981.

[29] M. S. Runge, T. M. Laue, D. A. Yphantis, M. R. Lifsics, A. Saito, M. Altin, K. Reinke, and R. C. Williams, Jr., *Proc. Natl. Acad. Sci. U.S.A.* **78**, 1431 (1981).

[30] D. A. Yphantis, *Biophys. J.* **45**, 324a (1984).

[31] D. A. Yphantis, J. W. Lary, W. F. Stafford, S. Liu, P. H. Olsen, D. B. Hayes, T. P. Moody, T. M. Ridgeway, D. A. Lyons, and T. M. Laue, *in* "Modern Analytical Ultracentrifugation: Acquisition and Interpretation of Data for Biological and Synthetic Polymer Systems" (T. M. Schuster and T. M. Laue, eds.), pp. 209–226. Birkhäuser, Boston, 1994.

[32] W. F. Stafford and S. Liu, *Prog. Biomed. Optics* **2386**, 130 (1995).

[33] W. F. Stafford, S. Liu, and P. E. Prevelige, *in* "Techniques in Protein Chemistry VI" (J. W. Crabb, ed.), pp. 427–432. Academic Press, San Diego, California, 1995.

[34] W. B. Bridgman, *J. Am. Chem. Soc.* **64**, 2349 (1942).

[35] H. Fujita, "Foundations of Ultracentrifugal Analysis." John Wiley & Sons, New York, 1976.

apparent distribution function, $\hat{g}(s^*)$, is a useful analytical tool. Because the $\hat{g}(s^*)$ versus s^* distribution is geometrically similar to the corresponding concentration gradient profile, dc/dr versus r, it contains all the same useful visual and analytical information.[9]

It is shown below that the assumption of no diffusion leads to an approximation in $\hat{g}(s^*)$ that must be taken into account only for low molecular weight materials (<50 kg/mol). Now we consider an exact derivation of $\hat{g}(s^*)$ and show how it is related to the original approximate derivation. The following derivation of dc/ds^* from the time derivative makes no assumptions about diffusion and, therefore, is completely general. The purpose of this derivation is twofold: (1) to present the complete derivation of the time derivative analysis and (2) to show the exact relation between the $\hat{g}(s^*)$ functions derived from the spatial derivative and from the time derivative, respectively.

Considering c as a function of s^* and t, $c = c\,(s^*, t)$, we can write

$$dc = \left(\frac{\partial c}{\partial t}\right)_{s^*} dt + \left(\frac{\partial c}{\partial s^*}\right)_t ds^* \tag{1}$$

where c is the concentration expressed on the c-scale and

$$s^* \equiv \frac{1}{\int_{t=0}^{t} \omega^2(t)\,dt} \ln\left(\frac{r}{r_{\mathrm{m}}}\right)$$

where $\omega^2(t)$ is the angular velocity of the rotor as a function of time.

The factor in the denominator can be replaced by $\omega^2 t_{\mathrm{sed}}$, where, t_{sed} is the effective time of sedimentation and ω is the final angular velocity of the rotor,

$$s^* \equiv \frac{1}{\omega^2 t_{\mathrm{sed}}} \ln\left(\frac{r}{r_{\mathrm{m}}}\right)$$

since

$$t_{\mathrm{sed}} = \frac{\int_{t=0}^{t} \omega^2(t)\,dt}{\omega^2}$$

In all equations, t is meant to be t_{sed}.

Now, dividing by dt and holding r constant we have

$$\left(\frac{\partial c}{\partial t}\right)_r = \left(\frac{\partial c}{\partial t}\right)_{s^*} + \left(\frac{\partial c}{\partial s^*}\right)_t \left(\frac{\partial s^*}{\partial t}\right)_r \tag{2}$$

and after a brisk rearrangement, we have an exact expression for the unnormalized, apparent sedimentation coefficient distribution function:

$$\left(\frac{\partial c}{\partial s^*}\right)_t = \left[\left(\frac{\partial c}{\partial t}\right)_r - \left(\frac{\partial c}{\partial t}\right)_{s^*}\right]\left[\left(\frac{\partial t}{\partial s^*}\right)_r\right] \equiv \hat{g}(s^*) \tag{3}$$

Its computation would require computing both the time derivative of the concentration at constant r as well as the time derivative at constant s^* at each point in the boundary. Computation of the time derivative at constant r results in complete elimination of the time-independent baseline, which is purely a function of r, as we have seen before[6]; however, computation of the time derivative at constant s^* does not remove the baseline contribution. In fact, it can be shown that Eq. (3) is analytically identical to the equation of Bridgman[34] and, therefore, must contain all the baseline components. For comparison, an unnormalized relation of the Bridgman form can be derived simply by considering c as a function of r and t, $c = c(r, t)$:

$$dc = \left(\frac{\partial c}{\partial r}\right)_t dr + \left(\frac{\partial c}{\partial t}\right)_r dt$$

and then, dividing through by ds^* and holding t constant, we have

$$\hat{g}(s^*) \equiv \left(\frac{\partial c}{\partial s^*}\right)_t = \left(\frac{\partial c}{\partial r}\right)_t\left(\frac{\partial r}{\partial s^*}\right)_t$$

For sufficiently high molecular weight species it follows that

$$\left(\frac{\partial c}{\partial t}\right)_{s^*} \cong \left(\frac{\partial c}{\partial t}\right)_p = -2\omega^2 \int_{s^*=0}^{s^*=s^*} s^*\left(\frac{\partial c}{\partial s^*}\right)_t ds^* \tag{4}$$

because

$$\lim_{D\to 0}\left(\frac{\partial c}{\partial t}\right)_{s^*} = \lim_{M\to\infty}\left(\frac{\partial c}{\partial t}\right)_{s^*} = \left(\frac{\partial c}{\partial t}\right)_p \quad \text{for } s > s^* \tag{5}$$

and

$$\lim_{D\to 0}\left(\frac{\partial c}{\partial t}\right)_{s^*} = \lim_{M\to\infty}\left(\frac{\partial c}{\partial t}\right)_{s^*} = 0 \quad \text{for } s < s^* \tag{6}$$

where s is the sedimentation coefficient of the component under consideration; and the subscript p is meant to designate the plateau region, the region centrifugal to the boundary.

This approximation leads to the equation derived previously for the apparent sedimentation coefficient computed from the time derivative for the special case of $D = 0$.[6]

$$\hat{g}(s^*)_t = \left(\frac{\partial c}{\partial s^*}\right)_t = \left[\left(\frac{\partial c}{\partial t}\right)_r + 2\omega^2 \int_{s=0}^{s=s^*} s^* \left(\frac{\partial c}{\partial s^*}\right)_t ds^*\right]\left[\left(\frac{\partial t}{\partial s^*}\right)_r\right] \quad (7)$$

giving validity to the use of this relation, for higher molecular weight molecules, even when diffusion is present.

For low molecular weight components (molar mass less than about 30 kg/mol at 60,000 rpm), diffusion can lead to small but significant systematic errors in the shape of $g(s^*)$ patterns computed using Eq. (7). See below for further discussion.

Simulations

Simulations of sedimentation patterns shown here were carried out on a Digital Equipment Corporation (Beaverton, OR) Alpha Server 1000 4/233 computer using the finite element method of Claverie et al.[36,37] with FORTRAN code executing the basic algorithm kindly supplied by D. J. Cox. The original algorithm was modified as described below to include the mass action calculations at each step of simulation for the rapidly reversible cases and to include the kinetic relaxation calculations at each step for the kinetically controlled systems according to Cox and Dale.[5] Calculations were carried out in a 1600-point grid from meniscus to base with a 1-sec time interval between iterations. After each step of sedimentation and diffusion, the concentrations of all species were recomputed either according to the law of mass action or according to the kinetic equations describing the system.

Two types of rapidly reversible associating systems were treated: a self-associating system and a heterologous interacting system. The self-associating system (a monomer–dimer–tetramer–octamer system meant to simulate myosin minifilament formation) was treated as a single concentration-dependent component whose sedimentation coefficient was taken as the weight average value at each position-dependent concentration value. The diffusion coefficient was taken as the gradient average value at each point in the cell. After each step of sedimentation and diffusion, the concentration of each species at each point along with s_w was interpolated from a look-up table computed once at the beginning of the simulation. The

[36] J.-M. Claverie, H. Dreux, and R. Cohen, *Biopolymers* **14**, 1685 (1975).
[37] J. M. Claverie, *Biopolymers* **15**, 843 (1976).

following relationships were used to treat the reequilibration after each step for the rapidly reversible system as well as for the initial equilibrium condition for the kinetically limited system.

$$2A_1 = A_2 \qquad c_2 = K_2 c_1^2 \qquad (8)$$
$$2A_2 = A_4 \qquad c_4 = K_4 c_2^2 \qquad (9)$$
$$2A_4 = A_8 \qquad c_8 = K_2 c_4^2 \qquad (10)$$

Conservation of mass requires that the total mass concentration, C_{tot}, be given by

$$C_{tot} = c_1 + c_2 + c_4 + c_8 \qquad (11)$$

where c_i is the concentration of species A_i on the c-scale.

The weight average sedimentation coefficient at each point was computed as

$$s_w = \{c_1 s_1 + c_2 s_2 + c_4 s_4 + c_8 s_8\}/C_{tot} \qquad (12)$$

The gradient average value of the diffusion coefficient was computed as follows:

$$D_{ave} = \{(\partial c_1/\partial r)_t D_1 + (\partial c_2/\partial r)_t D_2 + (\partial c_4/\partial r)_t D_4 + (\partial c_8/\partial r)_t D_8\}/(\partial C_{tot}/\partial r)_t \qquad (13)$$

and D_i is the diffusion coefficient of species i ($i = 1, 2, 4, 8$). As explained by Cox and Dale,[5] it is necessary to use the gradient average diffusion coefficient instead of the more common Steiner relation[38] for the average value of D if one is considering position-dependent (i.e., pressure-dependent) equilibrium constants.

For the kinetically limited cases, the rate equations were expanded to first order in the differential of time. The kinetically limited monomer–dimer–tetramer–octamer system was represented by the following three chemical equations and their corresponding linearized rate equations:

$$2A_1 \underset{k_{2,r}}{\overset{k_{2,f}}{\rightleftharpoons}} A_2 \qquad 2A_2 \underset{k_{4,r}}{\overset{k_{4,f}}{\rightleftharpoons}} A_4 \qquad 2A_4 \underset{k_{8,r}}{\overset{k_{8,f}}{\rightleftharpoons}} A_8 \qquad (14)$$

$$\Delta c_1 = [k_{2,r} c_2 - k_{2,f} c_1^2] \Delta t \qquad (15)$$
$$\Delta c_2 = [k_{2,f} c_1^2 + k_{4,r} c_4 - k_{4,f} c_2^2 - k_{2,r} c_2] \Delta t \qquad (16)$$
$$\Delta c_4 = [k_{4,f} c_2^2 + k_{8,r} c_8 - k_{8,f} c_4^2 - k_{4,r} c_4] \Delta t \qquad (17)$$
$$\Delta c_8 = [k_{8,f} c_4^2 - k_{8,r} c_8] \Delta t \qquad (18)$$

[38] R. F. Steiner, Arch. Biochem. Biophys. **49**, 400 (1954).

The value of Δt was adjusted at each radial position so that each relaxation step was carried out with a value of Δc_i that was no more than $0.01c_i$. The concentration changed rapidly at some points, especially near the meniscus and base at the beginning of the run, requiring subdividing each 1-sec transport time step into as many as 1000 kinetic relaxation steps before proceeding to the next step of transport. For most of the more slowly varying cases ($k_r < 10^{-3}$ sec^{-1}), only 10 steps were required.

The rapidly reversible heterologous interacting system (meant to simulate an antigen–antibody reaction) was treated as follows[9]:

$$A + B = C \tag{19}$$
$$C + B = D \tag{20}$$

where A in this example represents a single-chain dimeric antibody molecule having 2 binding sites; B represents the antigen; C represents the singly liganded complex; and D represents the doubly liganded complex.

Sedimentation patterns were simulated for various molar ratios for total A and B and for various values of the equilibrium constants defined by the following equations:

$$K_1 = \frac{[C]}{[A][B]} \tag{21}$$

$$K_2 = \frac{[D]}{[C][B]} \tag{22}$$

where square brackets indicate the molar concentration of each species.

Conservation of mass requires that

$$[A]_0 = [A] + [C] + [D] \tag{23}$$
$$[B]_0 = [B] + [C] + 2[D] \tag{24}$$

Substituting and rearranging, we arrive at

$$[A] = [A]_0/\{1 + K_1[B] + K_1K_2[B]^2\} \tag{25}$$
$$[B] = [B]_0/\{1 + K_1[A] + 2K_1K_2[A][B]\} \tag{26}$$
$$[C] = K_1[A][B] \tag{27}$$
$$[D] = K_1K_2[A][B]^2 \tag{28}$$

The values of [A] and [B] were computed from Eqs. (25) and (26), using a simple iterative routine starting with an initial guess for [B]. The values of [C] and [D] were computed from the values of [A] and [B] with Eqs. (27) and (28).

The total mass concentration, C_t, is given by

$$C_t = c_A + c_B + c_C + c_D \qquad (29)$$

where $c_X = [X]M_X$, X = A, B, C, or D, and M_X is molar mass of species X. The weight average sedimentation coefficient at any point is given by

$$s_w = \{c_A s_a + c_B s_b + c_C s_c + c_D s_d\}/C_t \qquad (30)$$

The value of s_w is a unique function of the total macromolecular concentration for a given ratio of total A to total B and values of K_1 and K_2. In the analysis of the data presented below, the equilibrium constants were assigned values such that the intrinsic binding constant, K_{int}, was the same for all steps, requiring that $K_1 = 4K_2 = 2K_{int}$. This was meant to simulated simple binding with no cooperativity.

Pressure dependence for this system was treated by allowing the intrinsic equilibrium constant to vary with radial position according to

$$K_{int}(r) = K_{int}(r_m) \exp\left(-\frac{\omega^2(r^2 - r_m^2)\rho\Delta V}{2RT}\right) \qquad (31)$$

where $K_{int}(r)$ is the value of K_{int} at radial position r, $K_{int}(r_m)$ is the value of K_{int} at the meniscus ($P \approx 1$ atm), ρ is the solution density, ΔV is the molar volume change in cubic centimeters per mole of complex, R is the gas constant, and T is the absolute temperature.[8,9]

The rate equations for kinetically limited heterologous interactions were treated in terms of the following model, involving two binding sites for B on A,

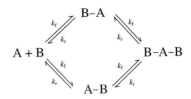

in which k_f/k_r is equal to K_{int}, the intrinsic molar equilibrium constant, and is assumed to be the same for all steps. This system can be represented by the following set of linearized rate equations:

$$\Delta C_A = [-2k_f C_A C_B + k_r C_C]\Delta t \qquad (32)$$
$$\Delta C_B = [-2k_r C_A C_B + k_r C_C - k_f C_B C_C + 2k_r C_D]\Delta t \qquad (33)$$
$$\Delta C_C = [2k_f C_A C_B - k_r C_C - k_f C_B C_C + 2k_r C_D]\Delta t \qquad (34)$$
$$\Delta C_D = [k_f C_B C_C - 2k_r C_D]\Delta t \qquad (35)$$

where C_A, C_B, C_C, and C_D are molar concentrations of species A, B, C, and D, respectively; and $C_C = C_{AB} + C_{BA}$.

The relationships between the intrinsic rate constants, k_i, and the macroscopic equilibrium constants, K_i, are given by

$$K_1 = \frac{2k_f}{k_r} = 2K_{int} \quad \text{and} \quad K_2 = \frac{k_f}{2k_r} = 0.5 K_{int} \qquad (36)$$

In simulations, the rate constants were chosen so that $K_1 = 2K_{int}$ and $K_2 = 0.5K_{int}$ with $K_{int} = 1 \times 10^7 \, M^{-1}$, $[A]_0 = 1 \times 10^{-7} \, M$, and $[B]_0 = 2[A]_0$, where K_1 and K_2 are the macroscopic equilibrium constants. The values of $[B]_0 = 2[A]_0$ correspond to the experimental situation to be presented below (Fig. 3). These kinetic equations were tested by showing that for values of $k_r > 10^{-2} \, sec^{-1}$, the sedimentation profiles for all species were essentially the same as for those computed from the mass action equations above. It was also demonstrated that for values of $k_r < 10^{-9} \, sec^{-1}$, the profiles corresponded to the summation of the same independently sedimenting species.

Kinetically Limited Interacting Systems and s_w

The weight average sedimentation coefficient is a composition-dependent parameter, and therefore, for an interacting system, its dependence on total protein concentration can be analyzed to determine association constants and stoichiometries. To answer the question of what effect a kinetically limited reequilibration would have on the computation of s_w, a series of sedimentation runs were simulated for various values of forward and reverse rate constants corresponding to a given value of the equilibrium constant. We consider two systems, the antigen–antibody system mentioned above and a monomer–dimer–tetramer–octamer self-associating system for which the stepwise molar equilibrium constants are equal. The equilibrium constants and initial condition were chosen so that there were equal amounts of all species present at the start of the run. The rate constants for the self-associating system were varied effectively from zero to infinity with all the forward rate constants made equal to each other and all the reverse rate constants made equal to each other for all steps.

The heterologous interacting (i.e., the antigen–antibody) system was also simulated for various values of the rate constants corresponding to a given value of the intrinsic equilibrium constant. The system was simulated as described above for $K_{int} = 1 \times 10^7 \, M^{-1}$ so that $K_1 = 2K_{int}$ and $K_2 =$

$0.5K_{int}$ and $[B]_0 = 2[A]_0$ with $[A]_0 = 1 \times 10^{-7}\ M$. The value of s_w for each simulated curve was computed using the following relationship:

$$s_w = \frac{\int_{s*=0}^{s*=s_p^*} s*g(s*)ds*}{\int_{s*=0}^{s*=s_p^*} g(s*)ds*}$$

where s_p^* is a value of $s*$ in the plateau region at which $g(s*) = 0^9$.

For computational purposes this relation can be recast in the following discrete form:

$$s_w = \frac{\sum_{i=i}^{i=n} s_i^* g(s_i^*)}{\sum_{i=1}^{i=n} g(s_i^*)}$$

where n corresponds to a point in the plateau region where $g(s*) = 0$.

Effect of Diffusion on Distribution Patterns

As mentioned above, for low molecular weight components (molar mass less than about 30 kg/mol at 60,000 rpm), diffusion can lead to small but significant systematic errors in the shape of $g(s*)$ patterns computed using Eq. (7). This error, which results from the approximation used in going from Eq. (3) to Eq. (7), can cause a shift in the peak position of $g(s*)$ from the correct value of s toward smaller values. To determine the magnitude of the factors contributing to this error, a series of simulations were carried out over a range of M and s values. The simulations showed that for a single species, the peak position of the $g(s*)$ versus $s*$ curve slightly underestimates the true value of s by an amount that is dependent on the molecular weight and square of the speed but independent of the value of the sedimentation coefficient. This error is greatest at the beginning of the run and approaches zero in the limit of infinite time. For a boundary whose midpoint is located near the middle of the cell, the correction factors are given in Table I. These correction factors should be applied if accurate shape information is to be obtained from the $s*$ values obtained from the peak position. However, in most situations the correction is mainly an esthetic refinement to the observed patterns, the primary interest being in the general characteristics of the boundary shape and how they change under various conditions of self-association, complex formation, or ligand binding.

TABLE I
CORRECTION FACTORS FOR PEAK VALUE OF s^* IN $g(s^*)$ PLOTS
AS FUNCTION OF MOLECULAR WEIGHT AND SPEED[a]

s^*_{peak}/s_{true}	M (kg/mol)[b]	Speed (rpm)[c]	$(M \cdot rpm^2) \times 10^{-12}$
0.99	60.0	60,000	216
0.98	35.0	46,000	129
0.97	25.0	39,000	90
0.96	18.0	33,000	65

[a] These factors become important and should be taken into account if accurate shape information is to be obtained from s^*_{peak}. The factors s^*_{peak} and s_{true} are dependent on the product $M\omega^2$.
[b] Assuming $r_m = 5.9$ and $r_{mid} = 6.5$ cm. At 60,000 rpm.
[c] Assuming $r_m = 5.9$ and $r_{mid} = 6.5$ cm. M = 60.0 kg/mol, where r_m is the meniscus, r_{mid} is the radius of the midpoint of the boundary, and M is molar mass.

Antigen–Antibody System

The interaction between a single-chain antibody construct, 741F8 (sFv')$_2$, and the extracellular domain (ECD) of the c-erbB-2 oncogene product was studied in the 0.1 to 10 μM range to determine the equilibrium constant for the interaction. This system had been partially characterized at higher concentrations.[39,40] Data from those earlier studies are plotted in Figs. 1[31,32,39,41] and 2. The 741F8 (sFv')$_2$ sedimented at $s_{20,w} = 3.4S$ and the ECD, at $s_{20,w} = 4.3S$, while the 1:1 mixture behaved as a reaction boundary with a peak position at about 7.0S. When excess ECD was added to drive the reaction toward completion, the value of s^* corresponding to the peak in the reaction boundary moved to about 7.9S (Fig. 2). From this observation, a value of 8.0S was taken as the value for the doubly liganded complex for the purposes of simulation and curve fitting. Under these conditions (2–4 μM) the 2:1 complex is predominantly populated.

To observe the dissociation process and to obtain data to be used for the estimation of the equilibrium constants, the 1:2 mixture was run in the range of 0.03 to 0.3 μM, where significant disssociation could be observed.

[39] G. P. Adams, J. E. McCartney, M. S. Tai, H. Oppermann, J. S. Huston, W. F. Stafford, M. A. Bookman, I. Fand, L. L. Houston, and L. M. Weiner, Cancer Res. 53, 4026 (1993).
[40] J. E. McCartney, M. S. Tai, R. M. Hudziak, G. P. Adams, L. M. Weiner, D. Jin, W. F. Stafford, S. Liu, M. A. Bookman, A. A. Laminet, I. Fand, L. L. Houston, H. Oppermann, and J. S. Huston, Protein Eng. 8, 301 (1995).
[41] S. Liu and W. F. Stafford, Anal. Biochem. 224, 199 (1995).

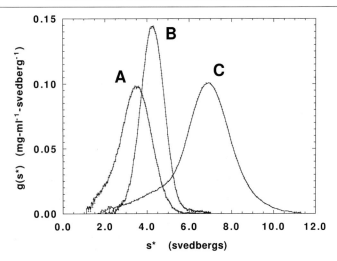

Fig. 1. Demonstration of a heterologous interaction between a single-chain antibody and its antigen. (A) 741F8 (sFv′)$_2$ (4.5 μM)(t_{sed} = 6547 sec); (B) ECD (3.4 μM); (C) mixture of 741F8 (sFv′)$_2$ (1.89 μM)(t_{sed} = 6552 sec) and ECD (3.4 μM)(t_{sed} = 4132 sec). Recombinant extracellular domain (ECD) of the c-ErbB-2 oncoprotein and recombinant dimeric single-chain antibody 741F8 (sFv′)$_2$ were prepared as described previously.[39] The experiments were carried out in an AN-F Ti rotor on a Beckman Instruments model E equipped with on-line Rayleigh optics as described previously,[31] using either three or four 12-mm cells and run at 56,000 rpm at 20°. Experiments were also carried out on a Beckman Optima XL-A equipped with on-line Rayleigh optics as described previously,[32] using four cells in an AN-60 titanium rotor at 20°. Temperature calibration was carried out with ethanolic cobalt chloride solutions as an optical thermometer.[41]

The samples were run at nominal loading concentrations of [(sFv′)$_2$] = $3.0 \times 10^{-7}, 1.0 \times 10^{-7}, 0.3 \times 10^{-7}$ M corresponding to total protein concentrations of 70, 23, and 8 μg/ml, respectively. Figure 3 shows the experimental curves obtained from this dilution series. Along with the experimental curves are plotted the simulated curves generated for a value of 1.0×10^7 M^{-1} for K_{int} and the actual loading concentrations. The experimental loading concentrations were estimated by integrating the $g(s^*)$ curves to get the plateau concentration and then corrected to the initial concentration taking radial dilution into account. These values of the initial concentrations were then used as input parameters for the simulated curves. There is some deviation at higher s^* values at the highest loading concentration, suggesting possible further aggregation, possibly self-association of the complex. Molecular weights higher than that expected for the 1:2 complex were also seen in sedimentation equilibrium runs at higher concentrations (W. F. Stafford, unpublished observations, 1994). The value of $K_{\text{int}} = 1 \times 10^7$ M^{-1} agrees well with the value of $K_{\text{eq}} \sim 2 \times 10^7$ M^{-1} obtained by surface plasmon resonance.[39]

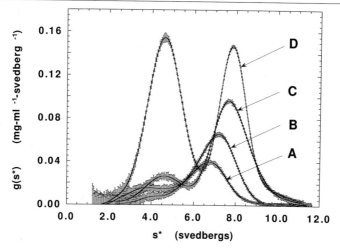

FIG. 2. Demonstration of equilibrium binding of ECD by 741F8 (sFv')₂. The two components were mixed in various ratios to determine the stoichiometry of binding (A) 0.5 mol of ECD/mol of (sFv')₂, (B) 1.0 mol of ECD/mol of (sFv')₂ (C) 2.0 mol of ECD/mol of (sFv')₂ and (D) 4.0 mol of ECD/mol of (sFv')₂. The 1 : 1 complex is expected to have a sedimentation coefficient of about 5.9–6.0S based on the interaction of an anti-c-ErbB-2 Fab (3.5S) with the ECD (data not shown here).

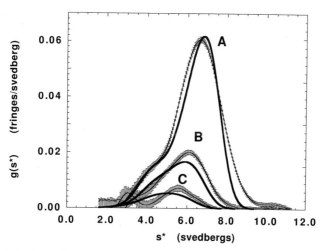

FIG. 3. Dilution series of ECD by 741F8 (sFv')₂ at stoichiometric ratio of [ECD]/[(sFv')₂] = 2.0. The 741F8 (sFv')₂ (molecular mass of 52 kDa) was combined with 2 mol of the ECD (90 kDa) so that $[ECD]_0 = 2[(sFv')_2]_0$. The solid curves are simulated data using $K_{int} = 1 \times 10^7 M^{-1}$, using values of $[(sFv')_2]_0$ corresponding to the values loaded into the cell, which were determined by integrating the $g(s^*)$ curves, corrected for radial dilution, to obtain $c_0 = \int g(s^*)ds^*$. Several guesses for K_{int} spanning 0.5×10^{-7} to $2.0 \times 10^{-7} M^{-1}$ were tried. Visual comparison with the data led to the choice shown.

Effect of Kinetically Limited Reequilibration on Computation of s_w

To address the question of the effect that kinetically limited reequilibration would have on the computation of equilibrium constants from the analysis of weight average sedimentation coefficients, extensive simulations were carried out for a range of forward and reverse rate constants. It was found that kinetically limited reequilibration during sedimentation would not significantly affect the computation of reliable values of s_w for an interacting system as long as the system was at equilibrium at the start of the run. The simulations shown in Fig. 4 for the monomer–dimer–tetramer–octamer system gave the same value of s_w to within the usual experimental error over the entire range of rate constants from infinitely fast to 10^{-9} sec^{-1} for the first-order dissociation rate constants for the same value of $K_{int} = 1 \times 10^7\ M^{-1}$. The systems were simulated for values of the equilibrium constant selected such that the systems were poised at the most concentration-sensitive conditions with nearly equal amounts of all species present at the start of the run. Parameters estimated from these systems show little sensitivity to radial dilution.

The systems were compared at the same sedimentation time for different values of the rate constants as well as at different times for the same value of the rate constants. The weight average sedimentation coefficient computed for either the self- or heteroassociating system was independent

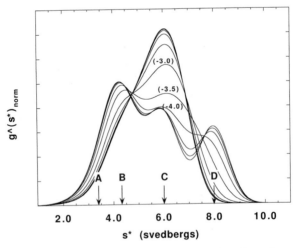

FIG. 4. Effect of different rates of reequilibration on the time derivative patterns and on the computation of s_w for a simulated antigen–antibody system corresponding to the sFv–ECD system described in Figs. 1–3. A, B, C, and D indicate the sedimentation coefficients of the species in equilibrium: 3.4, 4.3, 6.0, and 8.0S, respectively. See text for description of the kinetic model and other details of the reactions.

of either time or the rate constants, with a variation of only about 3% over the entire range of time and rate constant values. One might expect a decrease in weight average sedimentation coefficient as a function of time for two reasons: First, for a noninteracting system, s_w decreases with time because the faster moving components will have experienced greater radial dilution at any given time than the more slowly moving ones. Therefore, the plateau composition will change with time because the more rapidly sedimenting species will be removed more rapidly from the system than the more slowly sedimenting species. Second, for a rapidly reversible, interacting system, the weight average sedimentation coefficient will decrease because of the shift of the equilibrium by mass action as the plateau concentration decreases.

However, it can be shown that neither one of these effects results in serious error in the estimation of s_w. Sedimentation patterns were generated for various values of the rate constants for the heterologous system, both as a function of time and for various values of the reverse rate constant, k_r, for the values of the parameters shown in Fig. 4. Similar analysis was carried out for the monomer–dimer–tetramer–octamer system (Fig. 5). Figure 6 shows time series for a wide range of values of k_f and k_r (10^{-9} sec^{-1} < k_r < ∞) corresponding to the given equilibrium constants.

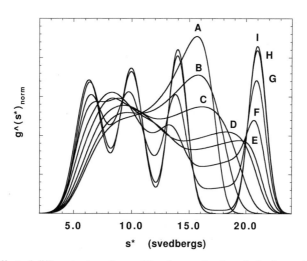

FIG. 5. Effect of different rates of reequilibration on the time derivative patterns and on the computation of s_w for a simulated monomer–dimer–tetramer–octamer system. Curves A through H correspond to values of k_r of infinite, 10^{-2}, 3×10^{-3}, 10^{-3}, 7×10^{-4}, 3×10^{-4}, 10^{-4}, 10^{-5}, and 10^{-9} sec^{-1}, respectively. (s_1 = 6.4S, D_1 = 1.26F, s_2 = 10.0S, s_3 = 14.0S, and s_4 = 21.0S); C_{tot} = 4 g/liter, k_f/k_r = K_{12} = K_{24} = K_{48} = 1.0 liter/g.) See text for further details and Table II for values of s_w computed from these curves. The values chosen here for K and C_{tot} forces the weight fraction of each species to be 0.25 at the start of the run.

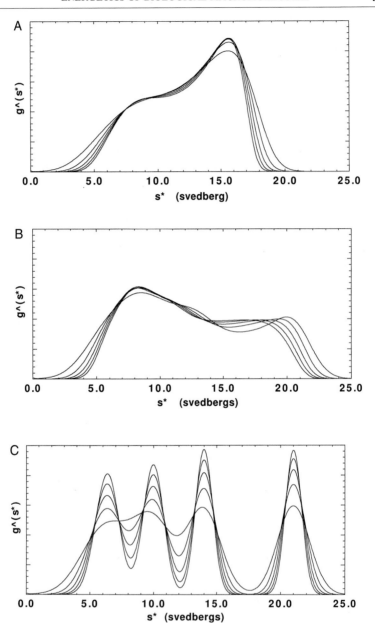

FIG. 6. Effects of radial dilution on computation of s_w for the monomer–dimer–tetramer–octamer system at various rates of reequilibration; the system is the same as that shown in Fig. 5: Time series for (A) $k_f = \infty$, $k_r = \infty$; (B) $k_f = 10^3/\text{liters-g}^{-1} \text{ sec}^{-1}$, $k_r = 10^{-3} \text{ sec}^{-1}$; and (C) $k_f = 10^{-9} \text{ liter g}^{-1} \text{ sec}^{-1}$, $k_r = 10^{-9} \text{ sec}^{-1}$. Times of sedimentation were 1440, 2400, 3360, 4320, and 5280 sec, respectively. (*Note:* Narrower curves are from later times.)

TABLE II

s_w COMPUTED AFTER VARIOUS TIMES OF SEDIMENTATION FOR
MONOMER–DIMER–TETRAMER–OCTAMER SYSTEM[a]

t_{sed} (sec)	$s_{w,true}$[b] ($k_f = k_r = \infty$)	s_w, computed				$s_{w,true}$[d] ($k_f = k_r = 0$)
		$k_f = \infty$ $k_r = \infty$	$k_f = 10^{-3c}$ $k_r = 10^{-3}$	$k_f = 10^{-4}$ $k_r = 10^{-4}$	$k_f = 10^{-9}$ $k_r = 10^{-9}$	
1440	12.67S	12.23	12.45	12.56	12.59	12.74
2400	12.55	12.41	12.50	12.54	12.59	12.67
3360	12.43	12.39	12.31	12.48	12.54	12.60
4320	12.32	12.35	12.28	12.41	12.48	12.53
5280	12.20	12.30	12.21	12.33	12.42	12.46

[a] As described in text.

[b] The value of $s_{w,true}$ was computed from the law of mass action at the plateau concentration and represents the instantaneous "true" value of s_w as opposed to the "cumulative average," which is the value computed from the $g(s^*)$ versus s^* curves.

[c] Units of k_f are liters g^{-1} sec^{-1} and the units of k_r are sec^{-1}. The values of k_f and k_r are the same for all steps so that $K_{eq} = 1.0$ liter/g and $C_0 = 4.0$ g/liter. This forces the weight fraction of all four species to be 0.25 at the start of sedimentation.

[d] Computed from radial dilution relation, $c_p/c_0 = \exp(-2\omega^2 st)$, for each component in the mixture. It corresponds to the instantaneous "true" value of s_w as described above. The value of s_w corresponding to the initial loading concentration is 12.85S and is the same for all cases. The values of s and D for the individual species are $s_1 = 6.4S$, $D_1 = 1.26F$, $s_2 = 10.0S$, $s_4 = 14.0S$, $s_8 = 21.0S$ (speed, 34,000 rpm).

Note: The maximum variation of s_w over the entire range shown is 4%.

The most interesting result of these simulations is that the value of s_w is essentially independent of the values of the rate constants for particular values of the equilibrium constants as long as the system is at equilibrium at the start of the run. Moreover, they show that, because the values of s_w are essentially independent of time, radial dilution has only an insignificant effect on the estimation of s_w. Table II summarizes the results of four simulations for various rates of reequilibration of the monomer–dimer–tetramer–octamer system. In each case the system was initially at equilibrium. The heterologous system represents a less extreme case than the monomer–dimer–tetramer–octamer system and was found to be even less sensitive to radial dilution.

Effects of Radial Dilution on Computation of s_w

Radial dilution arises because of the sectorial shape of the centrifuge cell. As material sediments, it is transported into regions of larger cross-sectional area and, therefore, larger volume. The dilution it experiences is proportional to the square of the distance traveled from the center of rotation. The plateau concentration decreases about 21% as the boundary position moves from 5.9 to 6.5 cm. This dilution will cause a shift in the equilibrium concentrations of sedimenting species, leading to a decrease in s_w with time. It can be shown, in spite of radial dilution, that for many systems the effects of radial dilution on the computation of s_w do not introduce serious errors into the determination of equilibrium constants computed from s_w. For the case of the simulated monomer–dimer–tetramer–octamer system (Figs. 5 and 6), the effects of radial dilution on s_w were seen to be insignificant (Table II) over a wide range in sedimentation time after the boundary had cleared the meniscus.

Pressure-Dependent, Reversible Interactions

Negative concentration gradients produced by pressure-dependent, reversibly interacting systems can be sustained only in the presence of a stabilizing gradient of an additional component. Density gradients established by redistribution of buffer components during sedimentation are often sufficient to stabilize moderate negative concentration gradients. To demonstrate this source of stabilization, the density gradient set up by the redistribution of 0.1 M NaCl at 56,000 rpm was simulated using $\bar{v}_3 = 0.33$ cm^3/g, $s = 0.24$S, and $D = 150 \times 10^{-7}$ cm$^2 \cdot$ sec^{-1}. The density distribution set up by the macromolecular system was added to the density distribution of NaCl to compute the total density gradient from the following relationship: $\rho(r) = \rho_0 + (1 - \bar{v}_2\rho_0)c_2 + (1 - \bar{v}_3\rho_0)c_3$, where c_i $(i = 2, 3)$ is the concentration in units of g/cm^3 of protein and NaCl, respectively, as a function of position, and $\rho_0 = 0.9982$ at 20°.

The macromolecular system used for the simulation of pressure effects was the antigen–antibody system described above but with a volume change of +1000 cm^3/mol on association for each step (Fig. 7).[9] If the simulation

FIG. 7. Stabilization of pressure effects. Salt and buffer redistribution can stabilize negative macromolecular concentration gradients. (A) The apparent sedimentation coefficient distribution function for the system A + 2B = D for various values of ΔV at $\theta = [A]_0 K_{int} = 1.0$. (a) $\Delta V = 10,000$ (b) $\Delta V = 1000$, (c) $\Delta V = 0$ cm^3/mol. (B) Simulation of salt redistribution to stabilize negative concentration gradients generated by pressure-dependent complex formation with a volume change of association of +1000 cm^3/mol. (a) Density distribution of 0.1 M NaCl after 1800 sec of sedimentation with instantaneous acceleration, (b) total density distribution, (c) protein concentration distribution. (C) Comparison of plots of total density distributions reached after an equivalent sedimentation time of 600 sec at 56,000 rpm (i.e., same value of $\int \omega^2 dt = 2.06 \times 10^{10}$ sec^{-1}). Final speed, 56,000 rpm, with (a) instantaneous acceleration to final speed, (b) acceleration at 5000 rpm/min, and (c) acceleration at 2000 rpm/min.

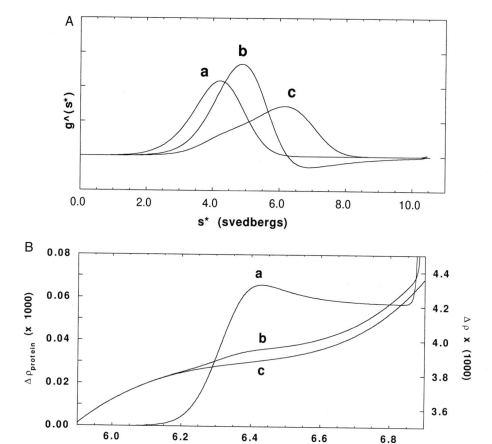

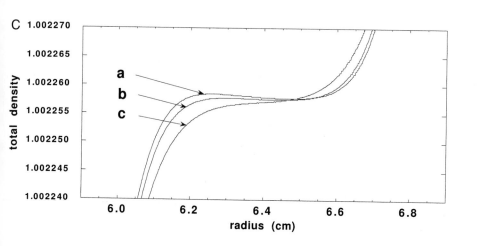

was allowed to start instantaneously at full speed (56,000 rpm), it was found that a negative gradient was set up initially and dissipated after about 15 min of sedimentation time as the NaCl redistributed. If, however, acceleration was allowed to take place at the rate of 5000 rpm/min, no overall negative gradient was established, and at the rate of 2000 rpm/min, no negative gradients were established before sufficient salt redistribution had taken place to stabilize the system.

The volume change of $+1000 \ cm^3/mol$ used here is a rather extreme case for this type of system (simple complex formation); therefore, one may conclude that convection should not be a problem for most dilute interacting systems exhibiting a volume change on association as long as normal amounts of salt are present and acceleration rates are kept below 5000 rpm/min. On the other hand, for large macromolecular assemblies like myosin filaments,[42] for which the molar volume change is much larger, density gradients established during sedimentation by salt redistribution usually will not be sufficient to stabilize the system against convection.

Discussion

Methods for the analysis of interacting systems by analytical ultracentrifugation using the time derivative of the concentration profile have been discussed. The time derivative of the concentration profile can be converted into an apparent sedimentation coefficient distribution function, $g(s^*)$ versus s^*, that is geometrically similar to the gradient plots of dn/dr versus r that are produced by schlieren optics. Therefore, these patterns, which are of much higher precision than the corresponding schlieren patterns, can be used to study interacting systems at much lower concentrations than could be done previously by sedimentation velocity using refractometric optics. Typically, schlieren optics in 12-mm cells limited the range of concentration to 1 mg/ml and above. With interference optics and time derivative methods, the concentration range has been expanded to the 1- to 10-μg/ml range. An antigen–antibody system having an intrinsic equilibrium constant of $1 \times 10^7 \ M^{-1}$ was studied at protein concentrations ranging from 8 to 70 μg/ml, corresponding to 1.3×10^{-8} to $3 \times 10^{-7} \ M$ complex. The equilibrium constant was obtained for this system by comparison of simulated sedimentation patterns with the observed patterns.

Weight average sedimentation coefficients can be computed from the time derivative patterns. Because weight average sedimentation coefficients are composition dependent, depending only on the plateau concentrations of the species comprising a particular system, they can be used to obtain

[42] R. Josephs and W. F. Harrington, *Biochemistry* **7**, 2834 (1968).

thermodynamic information such as stoichiometries and association equilibrium constants. Several questions concerning the effects of kinetics of reequilibration and the effect of radial dilution on the computation of s_w were considered. It was concluded that neither kinetically limited reequilibration nor radial dilution would be expected to introduce serious errors into the computation of s_w as long as the system is at equilibrium at the start of the run.

Pressure-dependent systems were also considered. It had been previously shown, in the absence of stabilizing density gradients, that the potential for convection exists for any interacting system that has a finite volume change on association. By use of simulations of buffer redistribution at typical salt concentrations, it is shown here that these negative concentration gradients can be stabilized by redistribution of buffer components under appropriate conditions, allowing the analysis of systems with moderate volume changes by sedimentation velocity.

Work is currently underway to incorporate these simulation routines into a nonlinear fitting algorithm to provide greater flexibility in treating interacting systems by time derivative methods.[23] A program called ABCDFitter is available from the author for fitting to time difference data to analyze one- or two-step complex formation. A detailed description of this algorithm and software will appear elsewhere.

Acknowledgments

I thank A. Garcia, T. Laue, and M. Jacobsen for detailed critical reading of this manuscript.

[14] Kinetic, Equilibrium, and Thermodynamic Analysis of Macromolecular Interactions with BIACORE

By DAVID G. MYSZKA

Introduction

The popularity of using biosensors to characterize biomolecular interactions continues to grow. These instruments are capable of providing detailed information about the energetics of macromolecular interactions without

METHODS IN ENZYMOLOGY, VOL. 323

labeling. While commercial instruments such as BIACORE (Biacore AB, Uppsala, Sweden) are relatively easy to use, extracting accurate rate constants for a reaction requires careful experimental technique and robust methods of data analysis. Previous chapters in this series have described methods for improving the experimental design for biosensor experiments[1] and qualitative methods for analyzing polymerization reactions.[2] This chapter illustrates how biosensors can be used as a biophysical tool to determine equilibrium constants and kinetic rate constants for binding events. In addition, by measuring reactions at different temperatures, it is possible to extract thermodynamic information about the system. The model system used in this study involves the interaction of interleukin 2 (IL-2) ligand with the α subunit of its receptor.[3-6]

Experimental Methods

Materials

All biosensor data are collected on a BIACORE 2000 optical biosensor. The instrument, CM5 research-grade chips, coupling reagents, and P20 are from Biacore AB (*www.biacore.com*). Ethylenediamine is from Sigma (St. Louis, MO), and *m*-maleimidobenzoyl-*N*-hydroxysulfosuccinimide ester (sulfo-MBS) is from Pierce (Chicago, IL). Recombinant IL-2 ligand is expressed in *Escherichia coli*, refolded, and purified as described.[7,8] The extracellular domain of the α subunit of the IL-2 receptor (αIL-2R) is expressed in insect cells via recombinant baculovirus infection and immunoaffinity purified as described.[3] All proteins were judged to be greater than 95% pure on the basis of reverse-phase high-performance liquid chromatography (HPLC).

[1] T. A. Morton and D. G. Myszka, *Methods Enzymol.* **295,** 268 (1998).

[2] D. G. Myszka, S. J. Wood, and A. L. Biere, *Methods Enzymol.* **309,** 386 (1999).

[3] Z. Wu, S. F. Eaton, T. M. Laue, K. W. Johnson, T. R. Sana, and T. L. Ciardelli, *Protein Eng.* **7,** 1137 (1994).

[4] Z. Wu, K. W. Johnson, Y. Choi, and T. L. Ciardelli, *J. Biol. Chem.* **270,** 16045 (1995).

[5] Z. Wu, K. W. Johnson, B. Goldstein, Y. Choi, S. F. Eaton, T. M. Laue, and T. L. Ciardelli, *J. Biol. Chem.* **270,** 16039 (1995).

[6] D. G. Myszka, P. G. Arulanantham, T. W. Sana, Z. Wu, T. A. Morton, and T. L. Ciardelli, *Protein Sci.* **5,** 2468 (1996).

[7] B. E. Landgraf, D. P. Williams, J. R. Murphy, K. A. Smith, and T. L. Ciardelli, *Proteins Struct. Funct. Genet.* **9,** 207 (1991).

[8] B. E. Landgraf, B. Goldstein, D. P. Williams, J. R. Murphy, T. R. Sana, K. Smith, and T. L. Ciardelli, *J. Biol. Chem.* **267,** 18511 (1992).

Assay Design

The wild-type ectodomain of the αIL-2 receptor contains a free thiol at position Cys-192, which may be modified without influencing ligand-binding activity.[9] While a portion of the purified receptor oxidizes to form a covalent dimer,[10] the remaining free form of the protein can be used to immobilize the receptor on the biosensor surface, employing maleimide-mediated coupling.[6,11] This coupling procedure produces a homogeneous and stable surface, which is essential for detailed kinetic analysis.

Immobilization

The first two flow cells on a standard CM5 sensor chip are prepared for thiol coupling. Receptor is immobilized to flow cell 1 and flow cell 2 is planned as a reference surface. The reference surface is treated in the same way as the reaction surface, i.e., under the same coupling conditions, to normalize the chemistries between the two flow cells. The carboxymethyl-dextran matrix in each flow cell is first activated with a mixture of 0.4 M N-ethyl-N'-(3-dimethylaminopropyl) carbodiimide and 0.1 M N-hydroxy-succinimide in water for 7 min. A 1 M (pH 8.0) aqueous solution of ethylene-diamine is injected for 7 min to create an amino-functionalized surface. A solution of sulfo-MBS (50 mM in NaHCO$_3$, pH 8.5) is injected for 7 min to form a reactive maleimide group. Purified receptor is diluted to a concentration less than 25 nM in sodium acetate buffer (10 mM, pH 5.0) and injected in 5-sec pulses over flow cell 1 until the desired surface density is achieved. The receptor density is kept low [30 resonance units (RU) total] to avoid mass transport effects, crowding, and aggregation.[1] The remaining activated groups on both flow cells are blocked by injecting a solution of cysteine (100 mM, pH 4.0) for 7 min.

Binding Conditions

All binding experiments are performed in buffer containing 10 mM sodium phosphate (pH 7.4), 150 mM sodium chloride, and 0.005% (v/v) P20. Bovine serum albumin (BSA, 0.1 mg/ml) is added to the buffer to minimize loss of protein at dilute concentrations. IL-2 is injected through flow cells 1 and 2 with the instrument operating in serial mode. A flow rate of 100 μl/min is used to minimize mass transport effects and deliver a

[9] C. M. Rusk, M. P. Neeper, L. M. Kuo, R. M. Kutny, and R. L. Robb, *J. Immunol.* **140,** 2249 (1988).
[10] K. Kato and K. A. Smith, *Biochemistry* **26,** 5359 (1987).
[11] D. J. O'Shannessy, M. Brigham-Burke, and K. Peck, *Biochemistry* **205,** 132 (1992).

consistent sample plug.[1] IL-2 is injected at concentrations of 233, 78, 26, 8.6, 2.9, and 0 nM. Each injection is repeated four times in random order. The dissociation phase is monitored for 2 min. Because the responses return to baseline within 20 sec, no regeneration step is required. Each set of binding experiments is repeated with the instrument equilibrated at 5, 15, 25, and 35°. It is faster to reequilibrate the instrument at the new temperature by going from low to high temperatures.

Data Processing

The quality of the biosensor data can be dramatically improved by processing the data to remove system-dependent artifacts. Figure 1A shows the raw data obtained for an injection of IL-2 at 233 nM over the reaction and reference surface. At these low surface densities it is somewhat difficult to discern the response that is attributed to the IL-2 ligand–receptor interaction, as system artifacts such as bulk refractive index changes and injection noise are at the same magnitude as the binding response itself. Most of these systematic artifacts are easily removed by data processing, thereby generating high-quality data.

Zeroing. The first step in data processing is to zero the response on the y axis just prior to the start of the association phase. Normally it is best to subtract an average of the response 10 sec before the injection. These results are shown in the plot in Fig. 1B. It is also important to set the start of the injection on the time axis (x axis) to zero. When collecting data with the instrument operating in serial mode, the injection time for each flow cell must be set separately to account for the short delay between flow cells. With the expanded scale in Fig. 1B, it is easier to see the differences between the reaction and reference surface data, as well as their similarities. Many of the injection artifacts, such as abrupt jumps and drift in the signal, are identical between these surfaces.

Reference Subtraction. Subtracting the reference surface data from the reaction surface data eliminates the refractive index change and injection noise as shown in Fig. 1C. It is easier to see the binding response attributed to the IL-2–receptor reaction in this plot. The differences in the position of the reaction and reference flow cells can cause sharp spikes at either the beginning or end of the association phase. During data analysis, these outlier data points need to be eliminated in order to obtain an accurate representation of the quality of the fit.

Replication Overlays. In Fig. 1D, the corrected responses have been overlaid for four replicate injections of the same IL-2 concentration as well as a set of blank injections. Looking closely at the blank injections, a small systematic drift is noticeable in the data about 10 sec into the association

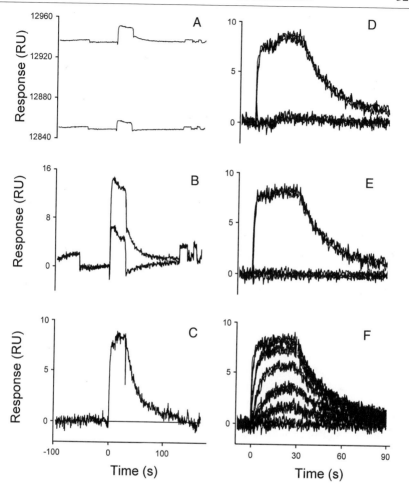

Fɪɢ. 1. Processing biosensor data. (A) Raw biosensor data for IL-2 injected at a concentration of 233 nM. The top and bottom sensorgrams are from the receptor and reference surfaces, respectively. (B) Data sets were zeroed on the y and x axis just prior to the start of the injection. (C) Data from the reference surface were subtracted from the data from the receptor surface. (D) Overlay of four replicate injections for IL-2 at 233 nM, as well as a buffer blank. (E) Corrected sensorgrams after subtracting the average of the buffer blank injections (double referencing) from IL-2 injection and buffer blank data. (F) Overlay of corrected sensorgrams for a series of IL-2 concentrations (233, 78, 26, 8.6, 2.9, and 0 nM).

phase. The signal drifts up about 1 RU and then it stabilizes. Note that the exact same drift is observed in all the IL-2 responses. While the magnitude of the drift is small, at these low surface densities it makes a significant contribution to the binding response. These minor systematic deviations are often observed when working on low-capacity surfaces and they are specific for each flow cell.

Double Referencing. The systematic deviations observed on low-capacity surfaces are extremely reproducible and occur equally in all the analyte and blank injections. Therefore, the response from the blank injections can be used to remove these artifacts from all the data. By subtracting the average of the response of the blank data from Fig. 1D, the corrected plots shown in Fig. 1E are generated. Notice the baseline data are completely flat and the responses from the sample data are stable once they reach equilibrium. This double-referencing procedure dramatically improves the quality of data, making it possible to collect reliable data on BIACORE 2000 below 2 RU. It is particularly useful for collecting data on low molecular weight analytes, because the binding responses are small to begin with.

Overlay All Analyte Responses. The final step in preparing the sensor data for analysis is to overlay the response from the different analyte concentrations as shown in Fig. 1F. These IL-2 responses represent an example of high-quality biosensor data. Sensorgrams were collected over a wide IL-2 concentration range (2.9 to 233 nM). The binding capacity was kept low to avoid mass transport effects. Replicate injections demonstrate that the responses are reproducible. And the responses come to equilibrium and dissociate back to baseline.

Figure 2 shows similar data sets collected for the IL-2 reaction with the instrument equilibrated at 5, 15, 25, and 35°. It is clear from comparing the responses between data sets that the dissociation rate increases with increasing temperature. By globally fitting the response data at each temperature, it is possible to accurately interpret the binding rate constants as described below.

Kinetic Data Analysis

The goal of data analysis is to determine the values for the rate constants that best describe the binding interaction. The first step is to choose the appropriate reaction model. The IL-2–receptor interaction data were fit to a simple reversible reaction mechanism.

$$A + B \underset{k_d}{\overset{k_a}{\rightleftharpoons}} AB \qquad (1)$$

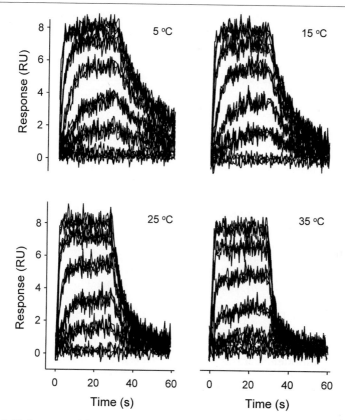

Fig. 2. IL-2–receptor biosensor data collected at different temperatures. IL-2 concentrations were 233, 78, 26, 8.6, 2.9, and 0 nM. Each injection was replicated four times. Running buffer contained 10 mM sodium phosphate (pH 7.4), 150 mM sodium chloride, 0.005% (v/v) P20, and bovine serum albumin (0.1 mg/ml). Binding data were collected with the BIACORE instrument equilibrated at 5, 15, 25, and 35° as shown.

A represents IL-2 in solution and B represents the immobilized receptor. k_a and k_d represent the association and dissociation rate constants, respectively.

Global Fitting. The association and dissociation phase data for each IL-2 concentration were fit simultaneously using a global data analysis program.[1,12,13] The differential rate equations that describe the binding

---

[12] D. G. Myszka and T. A. Morton, *Trends Biochem. Sci.* **23,** 149 (1998).

[13] CLAMP: Biosensor data analysis program is available at *www.hci.utah.edu/groups/ interaction.*

TABLE I

KINETIC PARAMETERS FOR INTERLEUKIN 2–RECEPTOR INTERACTION[a]

| Temperature (°C) | Standard deviation (RU) | | k_a (M^{-1} sec^{-1}) | k_d (sec^{-1}) | K_D (nM) |
	Replication	Residual			
5	0.352	0.376	4.66 (4) $\times$ 10^6	0.0420 (2)	9.0 (1)
15	0.355	0.395	6.53 (6) $\times$ 10^6	0.0682 (4)	10.4 (1)
25	0.370	0.410	8.77 (9) $\times$ 10^6	0.114 (8)	13.0 (9)
35	0.365	0.409	1.01 (1) $\times$ 10^7	0.213 (2)	21.1 (3)

[a] The value in parentheses represents the standard error in the last significant digit.

reaction, Eq. (2), were numerically integrated using semi-implicit extrapolation.[14]

$$d[A]/dt = 0$$
$$d[B]/dt = -k_a[A][B] + k_d[AB] \qquad (2)$$
$$d[AB] = k_a[A][B] - k_d[AB]$$

The concentration of IL-2 in the bulk flow ([A]) was assumed to be constant during the association phase. The amount of free receptor sites on the surface ([B]) and the response attributed to the IL-2–receptor complex ([AB]) were modeled in response units (RU). The modeled data were fit to the experimental data using nonlinear least-squares fitting procedures.[1] Parameter values were adjusted automatically by Levenberg–Marquardt routines to achieve the best fit.[14] Global analysis provides a stringent test for the reaction model and better parameter estimates.[15,16]

The global fits to the IL-2 biosensor data are shown in Fig. 3 (see color insert). The modeled data overlay with the experimental data for each concentration of IL-2, indicating a simple 1:1 interaction model describes the data well. It should be noted that the binding reactions reach equilibrium and completely dissociate back down to baseline, demonstrating that it is possible to fit complete biosensor binding profiles with simple reaction models.

Assessing Residuals. The residual standard deviations for the four data sets are close to their respective replication standard deviations as reported in Table I. The replication standard deviation is a model-independent as-

[14] W. H. Press, S. A. Teukolshy, W. T. Vetterling, and B. P. Flannery, "Numerical Recipes in C." Cambridge University Press, Cambridge, 1992.

[15] T. A. Morton, D. G. Myszka, and I. M. Chaiken, *Anal. Biochem.* **227,** 176 (1995).

[16] D. G. Myszka, H. Xiaoyi, M. Dembo, T. A. Morton, and B. Goldstein, *Biophys. J.* **75,** 583 (1998).

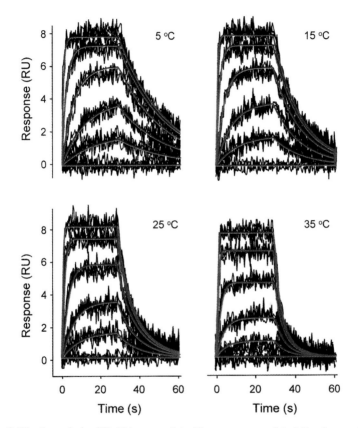

FIG. 3. Kinetic analysis of IL-2 biosensor data. The sensorgrams (black lines) were globally fit to a simple reversible interaction model [Eq. (1)]. The red lines show the best fit obtained from nonlinear regression analysis.

sessment of the total experimental noise. The average of the residual standard deviation is only 0.037 RU greater than the average of the replication standard deviation, which indicates that there is little information left in the residuals that is not accounted for by the reaction model.

Rate Constants. The rate constants that were returned from the kinetic analysis are also reported in Table I. The standard errors in the association and dissociation parameters are low as a result of globally fitting the response data.[15] These rate constants correspond well to the values determined previously for the IL-2–receptor interaction using BIACORE.[6] The association rate increases 2-fold in going from 5 to 35°, while the dissociation rate increases 4-fold. Accurate estimates for these fast association rate constants were possible because data were collected on low-density receptor surfaces, minimizing the effects of mass transport.[16,17]

Equilibrium Analysis. Equilibrium constants may be determined from biosensor data if the reactions reach a steady response during the association phase. The binding responses for the IL-2 data collected at 35° (Fig. 2) fit this criterion. The response value at equilibrium was determined by averaging the signal from 20 to 25 after the start of each IL-2 injection. These values represent the amount of IL-2–receptor complex ($[AB]_{eq}$) formed at each IL-2 concentration ($[A]$). The equilibrium dissociation constant (K_D) can be determined from nonlinear least-squares curve fitting of the data to

$$[AB]_{eq} = AB_{max}(1/(1 + K_D/[A])) \tag{3}$$

AB_{max} represents the maximum capacity of the surface in response units (RU). It should be noted that the equilibrium data on BIACORE are collected for binding reactions without the need to separate bound and free analyte.

The binding isotherm for the IL-2–receptor equilibrium data is shown in Fig. 4. The data fit well to a single-site model yielding a K_D of 21 ± 1 nM. This matches the K_D of 21.1 ± 3 nM calculated from the ratio of the kinetic rate constants,

$$K_D = k_d/k_a \tag{4}$$

Obtaining similar equilibrium dissociation constants, determined by equilibrium analysis and kinetic analysis, increases confidence in a simple, single-site model used to determine kinetic rate constants. Because the IL-2 reactions below 35° did not reach equilibrium in the allotted time, for these temperatures, equilibrium constants were calculated from the kinetic rate

---

[17] D. G. Myszka, T. A. Morton, M. L. Doyle, and I. M. Chaiken, *Biophys. Chem.* **64,** 127 (1997).

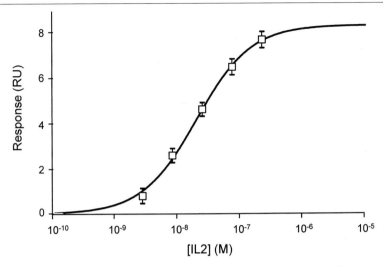

Fig. 4. Equilibrium analysis of IL-2–receptor biosensor data. Data were fit to a simple binding isotherm [Eq. (3)] using nonlinear regression analysis.

constants (see Table I). The affinity of the IL-2–receptor interaction varies approximately 2-fold, from 9 to 21 nM, in going from 5 to 35°.

Thermodynamics

Free Energy. The Gibbs free energy change (ΔG) for a reaction is related to the equilibrium dissociation constant by

$$\Delta G° = RT \ln K_D \tag{5}$$

R is the gas constant (1.987 cal mol K^{-1}) and T is the absolute temperature in degrees Kelvin. Using Eq. (5) the $\Delta G°$ at 25° was calculated to be -10.73 ± 0.03 kcal/mol. The Gibbs free energy change for the reaction describes the energetic difference between reactants and products relative to the reference standard state of 1 M, but does not on its own provide information about the kinetic pathway of the reaction.

Transition State Free Energy. The temperature dependence of the rate constants was analyzed according to transition state theory,[18,19] which relates the rate constant of a reaction to an equilibrium constant between the reactants and the transition state, a transient high-energy species that

---

[18] H. Eyring, *Chem. Rev.* **17,** 65 (1935).
[19] A. Ferscht, "Enzyme Structure and Mechanism." W. H. Freeman & Company, New York, 1984.

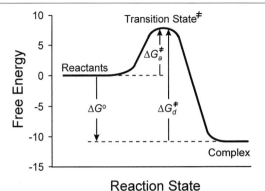

Reaction State

FIG. 5. Free energy profile for the IL-2–receptor interaction. $\Delta G_a^\ddagger$ and $\Delta G_a^\ddagger$ are the change in Gibbs free energy required for the formation of the transition state starting from reactants and complex, respectively. $\Delta G°$ is the change in Gibbs free energy between reactants and complex.

decays to form product.[20] The free energy of activation ($\Delta G^\ddagger$) for the formation of the transition state is given by

$$\Delta G^\ddagger = -RT \ln(kh/k_B T) \qquad (6)$$

where k is the rate constant for a reaction, h is the Planck constant $(1.584 \times 10^{-34}$ cal sec), k_B is the Boltzmann constant $(3.3 \times 10^{-24}$ cal $K^{-1})$, R is the gas constant $(1.987$ cal mol $K^{-1})$, and T is the absolute temperature. The transmission coefficient[18] was assumed to be unity[19] and could be ignored.[20] Using the rate constants obtained at 25°, the Gibbs free energies for the formation of the transition state for the association $(\Delta G_a^\ddagger)$ and dissociation reactions $(\Delta G_d^\ddagger)$ were determined to be 7.98 ± 0.01 and 18.74 ± 0.01 kcal/mol, respectively.

Figure 5 shows the relationship between the transition state and equilibrium free energies for the IL-2–receptor interaction. During the association process activation energy is required for the formation of the transition state. The larger the barrier the slower the association rate. Dissociation of the complex to form the transition state requires even greater energy. The difference between the energy required to achieve the transition state for the association and dissociation processes corresponds to the equilibrium free energy.

van't Hoff Model. van't Hoff analysis is a noncalorimetric approach to determining the enthalpy change for a binding reaction. A key assumption in using the van't Hoff relationship is that the binding reaction involves a

[20] E. M. De La Cruz and T. D. Pollard, *Biochemistry* 35, 14054 (1996).

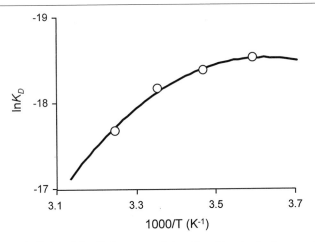

FIG. 6. van't Hoff plot of the IL-2–receptor interaction. The standard errors were smaller than the data symbols and are therefore not shown. The curved line was determined from nonlinear fitting of the integrated van't Hoff equation [Eq. (8)] directly to the equilibrium constants, which were weighted by the inverse of the square root of the variance.

single equilibrium throughout the temperature range studied. For protein reactants, it may be necessary to establish that they do not aggregate or change conformation over the temperature range studied.

For a simple equilibrium reaction, the van't Hoff enthalpy change is calculated from the temperature dependence of a given equilibrium dissociation constant.

$$\ln K_D = \Delta H^\circ / RT + \Delta S^\circ / R \tag{7}$$

The equilibrium constants for the IL-2–receptor interaction were plotted in Fig. 6 as a van't Hoff plot (ln K_D versus $1/T$). This plot would be linear if ΔH° and ΔS° were independent of temperature. However, for protein–protein interactions the ΔH° and ΔS° values are expected to deviate from linearity according to the heat capacity change, ΔC_p. The curvature in the data on the van't Hoff plot suggests the reaction is indeed accompanied by a change in heat capacity, ΔC_p.

To more accurately interpret the thermodynamic parameters, the equilibrium data were thus fit directly to an integrated form of the van't Hoff equation, which for a finite, constant heat capacity change ΔC yields

$$K = K^\circ e^{-\Delta H^\circ / R(1/T - 1/T^\circ)} e^{\Delta CT^\circ / R(1/T - 1/T^\circ)} \left(\frac{T}{T^\circ}\right)^{\Delta C/R} \tag{8}$$

ΔH° is the enthalpy change and K° represents the equilibrium dissociation constant for the reaction at a reference temperature T°.

TABLE II
EQUILIBRIUM THERMODYNAMICS FOR INTERLEUKIN 2–RECEPTOR INTERACTION[a]

Parameter	ΔG° (kcal/mol)	ΔH° (kcal/mol)	ΔS° (cal/mol K)	ΔC_p° (cal/mol K)
Equilibrium	−10.73 (3)	−6.2 (8)	15 (3)	−290 (90)

[a] At 25°. The values in parentheses represent the standard error in the last significant digit.

The curvature in the fit to the van't Hoff plot (see Fig. 6) reflects the change in heat capacity of -280 ± 90 cal mol^{-1} K^{-1}. At 25°, complex formation is driven by both a favorable enthalpy change ($\Delta H^\circ = -6.2 \pm 0.8$ kcal/mol) and entropy change ($\Delta S^\circ = 15$ cal/mol K), as calculated from $\Delta G^\circ = \Delta H^\circ - T\Delta S^\circ$. Table II summarizes the thermodynamic parameters determined from van't Hoff analysis of the IL-2 ligand–receptor interaction at 25°. The magnitudes of these ΔH, ΔS, and ΔC_p terms are somewhat small relative to other protein–protein interactions, and suggest the interaction may proceed in a rigid body-like manner.

Figure 7 shows the temperature dependence of the thermodynamic parameters determined from the van't Hoff analysis. The values of ΔH and ΔS change an order of magnitude across the experimental temperature range. At low temperatures the reaction is driven by favorable entropy

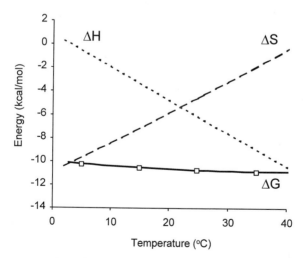

FIG. 7. Temperature dependence of the thermodynamic parameters. ΔG, ΔH, and $-T\Delta S$ are represented by the solid, dotted, and dashed lines, respectively. Open squares on the ΔG plot show the temperatures at which the binding reactions were studied.

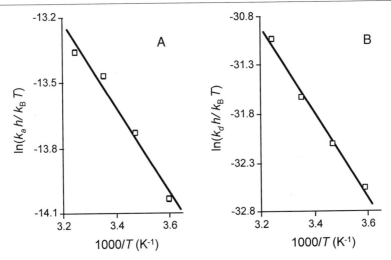

FIG. 8. Temperature dependence of IL-2–receptor kinetics. Eyring plots of the association (A) and dissociation (B) rate constants. Solid lines are the weighted best fit to Eq. (9).

whereas at high temperatures enthalpy makes the most significant contribution. The opposite temperature dependence of the enthalpy and entropy largely cancel each other out, so that there is only a minor change in the overall $\Delta G°$ from -10.24 to -10.82 kcal/mol in going from 5 to 35°.

Eyring Analysis. Using the temperature dependence of the reaction rate constants, the free energy of activation can be broken down into its enthalpy and entropy components.[21] Expanding the equation for the free energy of activation [Eq. (3)] to include the enthalpy and entropy yields

$$\ln\left(\frac{kh}{k_{\mathrm{B}}T}\right) = \left(\frac{-\Delta H^{\ddagger}}{R}\right)\left(\frac{1}{T}\right) + \frac{\Delta S^{\ddagger}}{R} \tag{9}$$

The enthalpy of activation ($\Delta H^{\ddagger}$) is obtained from a plot of the left-hand side of Eq. (9) versus $1/T$ (Eyring plot) as shown for the IL-2 data in Fig. 8. While the intercept of the plot is $\Delta S/R$ it is often more accurate to calculate the entropy of activation ($\Delta S^{\ddagger}$) from $\Delta G^{\ddagger} = \Delta H^{\ddagger} - T\Delta S^{\ddagger}$. The transition state thermodynamics for the IL-2–receptor interaction are given in Table III. This information can be used to break down the free energy barriers into their enthalpy and entropy components.[21]

---

[21] H. Roos, R. Karlsson, H. Nilshans, and A. Persson, *J. Mol. Recognition* **11**, 204 (1999).

TABLE III
TRANSITION STATE THERMODYNAMICS FOR INTERLEUKIN 2–RECEPTOR INTERACTION[a]

Interaction	$\Delta G^{\ddagger}$ (kcal/mol)	$\Delta H^{\ddagger}$ (kcal/mol)	$\Delta S^{\ddagger}$ (cal/mol K)
Association	7.98 (1)	3.9 (4)	−14 (1)
Dissociation	18.74 (1)	8.6 (6)	−33 (2)

[a] The values in parentheses represent the standard error in the last significant digit.

BIACORE and Calorimetric Thermodynamics. As described above, extracting thermodynamic information from biosensor analysis requires van't Hoff analysis to determine the enthalpy change. While biosensors can quantitate the amount of complex formed in the presence of reactants, it should be stressed that biosensors measure only complex formation in time, compared with calorimetry, which directly measures the enthalpy of the reaction. The van't Hoff enthalpy change will equal the calorimetric enthalpy change only if the binding reaction follows a two-state transition between free and complexed molecules.[22] Systematic differences in the van't Hoff and calorimetric enthalpies may be an indication that the reaction involves multiple states or major conformational changes, water reorganization, or protonation. A comparison between calorimetric and biosensor thermodynamics could provide complementary information about the reaction energetics.[21,23] When the amounts of reagents are not available for microtitration calorimetry, biosensors may be useful in providing some insight into the reaction thermodynamics.

Conclusion

BIACORE is a powerful tool used to measure the reaction kinetics and equilibrium constants for biological molecules. Improved methods for data collection and processing make it possible to dramatically improve the quality of biosensor data. The ability to describe binding data with simple interaction models validates the technology as a tool for kinetic analysis. Recording the temperature dependence of reaction parameters makes it possible to extract some thermodynamic information about the binding reaction. Together, kinetic and thermodynamic data will increase our understanding of biomolecular recognition events.

[22] I. Jelesarov and H. R. Bosshard, *J. Mol. Recognition* **12,** 3 (1999).
[23] G. Zeder-Lutz, E. Zuber, J. Witz, and M. H. V. van Regenmortel, *Anal. Biochem.* **246,** 123 (1997).

Acknowledgments

I thank Thomas Ciardelli (Dartmouth Medical School) for providing the IL-2 ligand and receptor used in this study, as well as Michael Doyle (SmithKline Beecham) for helpful comments on the manuscript. This work was supported by the Huntsman Cancer Institute.

[15] Structure–Function Relationships in Two-Component Phospholipid Bilayers: Monte Carlo Simulation Approach Using a Two-State Model

By István P. Sugár and Rodney L. Biltonen

Introduction

Important functions of cell membranes such as permeability of small water-soluble molecules and signaling through in-plane chemical reactions[1] are strongly related to and affected by the lateral organization of the multi-component lipid matrix (Ref. 2 and references therein). The functional significance of membrane lateral heterogeneity is emphasized by discoveries in membrane cell biology. For example, the ability of glycosphingolipids to organize compositional clusters in biological membranes is postulated to be a key feature, not only in their own intracellular sorting and trafficking, but also in the sorting and trafficking of proteins with covalent glycosylphos-phatidylinositol anchors. Glycosphingolipid clusters, also called DIGs (detergent-resistant, glycosphingolipid-enriched complexes) or "glycolipid rafts," may serve as target sites for the fusion/budding of envelope viruses with eukaryotic cells. Most of what is currently postulated about glycosphingolipid organization in cell membranes originates from studies of model membrane systems.[3–5]

[1] T. E. Thompson, M. B. Sankaram, R. L. Biltonen, D. Marsh, and W. L. C. Vaz, *Mol. Membr. Biol.* **12,** 157 (1995).
[2] O. G. Mouritsen and R. L. Biltonen, *in* "Protein–Lipid Interactions, New Comprehensive Biochemistry" (A. Watts, ed.), p. 1. Elsevier, Amsterdam, 1993.
[3] T. Thompson, Y. Barenholz, R. Brown, M. Correa-Freire, W. Young, and T. Tillack, *in* "Enzymes of Lipid Metabolism II" (L. Freysz, ed.), p. 387. Plenum Press, New York, 1986.
[4] T. E. Thompson and R. E. Brown, *in* "New Trends in Ganglioside Research: Neurochemical and Neuroregenerative Aspects" (R. Ledeen, E. Hogan, G. Tettamati, A. Yates, and R. Yu, eds.), Fidia Res. Series, Vol. 14, p. 65. Liviana Press, Padova, Italy, 1988.
[5] T. E. Thompson and T. W. Tillack, *Annu. Rev. Biophys. Chem.* **14,** 361 (1985).

Computer simulations of lipid bilayers makes it possible to study the relationships between the microscopic level membrane configurations and the phenomenological level membrane functions. For example, it was proposed and demonstrated by computer simulations for one-component lipid membranes that there is a linear relationship between the bilayer permeability for small molecules and the length of the gel–fluid interface.[6–8]

There exists a broad range of models of phospholipid bilayers from the most detailed all-atom models[9–11] to Pink's 10-state model.[12] The 10-state model has been used frequently to simulate phenomena related to phospholipid mono- and bilayers[13] such as permeability, protein–lipid interaction, effects of sterols, and effects of drug binding.

However, we have seen from a growing number of examples that simple two-state models are able to simulate heat capacity curves and fluorescence recovery after photobleaching (FRAP) threshold temperatures of phospholipid bilayers in quantitative agreement with the observed differential scanning calorimetry (DSC) and FRAP data.[14–18] The simplicity of these models is exemplified by the gel–fluid transition of one-component lipid bilayers, where it is assumed that each hydrocarbon chain exists in either a gel or fluid state, and only nearest-neighbor interactions between the chains need to be considered. These models are so-called minimal models, making assumptions that are physically plausible and absolutely necessary for the correct simulation of the observed excess heat capacity curves. As a consequence the number of model parameters is minimal and the parameters have explicit physical meaning. The unique feature of this approach is that experimental data are used to estimate the values of the parameters.

In the first part of this chapter we describe a two-state, two-component "minimal" model of dimyristoylphosphatidylcholine/distearoylphosphatidylcholine (DMPC/DSPC) mixtures developed by Sugar et al.[18] After the

[6] D. Marsh, A. Watts, and P. F. Knowles, *Biochim. Biophys. Acta* **858,** 161 (1976).

[7] M. I. Kanehisa and T. Y. Tsong, *J. Am. Chem. Soc.* **100,** 424 (1978).

[8] L. Cruzeiro-Hansson and O. G. Mouritsen, *Biochim. Biophys. Acta* **944,** 63 (1988).

[9] H. L. Scott, *Biochim. Biophys. Acta* **469,** 264 (1977).

[10] H. L. Scott, *Biochemistry* **25,** 6122 (1986).

[11] H. L. Scott, E. Jacobson, and S. Subramaniam, *Comput. Phys.* **12,** 328 (1998).

[12] A. Caille, D. A. Pink, F. de Verteuil, and M. J. Zuckermann, *Can. J. Phys.* **58,** 581 (1980).

[13] O. G. Mouritsen, *Chem. Phys. Lipids* **57,** 178 (1991).

[14] T. Heimburg and R. L. Biltonen, *Biophys. J.* **70,** 84 (1996).

[15] T. Heimburg and D. Marsh, Thermodynamics of the interaction of proteins with lipid membranes. *In* "Biological Membranes: A Molecular Perspective from Computation and Experiment" (K. M. Merz and B. Roux, eds.), p. 405. Birkhauser, Boston, 1996.

[16] R. Jerala, P. F. F. Almeida, and R. L. Biltonen, *Biophys. J.* **71,** 609 (1996).

[17] I. P. Sugar, R. L. Biltonen, and N. Mitchard, *Methods Enzymol.* **240,** 569 (1994).

[18] I. P. Sugar, T. E. Thompson, and R. L. Biltonen, *Biophys. J.* **76,** 2099 (1999).

estimation of the model parameters we demonstrate that the model describes correctly the observed calorimetric data. In the second part the properties of the bilayer configurations, such as cluster size distribution, cluster number, and percolation frequency of gel and fluid clusters at different mole fractions and temperatures, are calculated. We make a comparison between the calculated average size of gel clusters and the average linear size of the gel clusters observed by atomic force microscopy. We then point out the correlation between the calculated percolation temperature of gel clusters and the threshold temperature observed by FRAP at different DMPC/DSPC mole fractions. Finally, based on our simulation results we make a comparison between the predictions of the two existing theories of excess membrane permeability for small molecules.

Lattice Model of Dimyristoylphosphatidylcholine / distearoylphosphatidylcholine Bilayers

Two-component lipid bilayers are the simplest model systems to study membrane lateral heterogeneity and its effect on membrane function. Among these model systems DMPC/DSPC mixtures have been studied most extensively experimentally[19-32] and theoretically.[18,33-41]

[19] S. Mabrey and J. M. Sturtevant, *Proc. Natl. Acad. Sci. U.S.A.* **73,** 3862 (1976).
[20] P. W. M. Van Dijck, A. J. Kaper, H. A. J. Oonk, and J. De Gier, *Biochim. Biophys. Acta* **470,** 58 (1977).
[21] D. A. Wilkinson and J. F. Nagle, *Biochemistry* **18,** 4244 (1979).
[22] W. Knoll, K. Ibel, and E. Sackmann, *Biochemistry* **20,** 6379 (1981).
[23] D. Lu, I. Vavasour, and M. R. Morrow, *Biophys. J.* **68,** 574 (1995).
[24] M. B. Sankaram and T. E. Thompson, *Biochemistry* **31,** 8258 (1992).
[25] M. B. Sankaram, D. Marsh, and T. E. Thompson, *Biophys. J.* **63,** 340 (1992).
[26] R. Mendelsohn and J. Maisano, *Biochim. Biophys. Acta* **506,** 192 (1978).
[27] T. Brumm, K. Jørgensen, O. G. Mouritsen, and T. M. Bayerl, *Biophys. J.* **70,** 1373 (1996).
[28] W. L. C. Vaz, E. C. C. Melo, and T. E. Thompson, *Biophys. J.* **56,** 869 (1989).
[29] V. Schram, H. -N. Lin, and T. E. Thompson, *Biophys. J.* **71,** 1811 (1996).
[30] B. Piknova, D. Marsh, and T. E. Thompson, *Biophys. J.* **71,** 892 (1996).
[31] S. Pedersen, K. Jørgensen, T. R. Bœkmark, and O. Mouritsen, *Biophys. J.* **71,** 554 (1996).
[32] C. Gliss, H. Clausen-Schaumann, R. Gunther, S. Odenbach, O. Rand, and T. M. Bayerl, *Biophys. J.* **74,** 2443 (1998).
[33] P. H. Von Dreele, *Biochemistry* **17,** 3939 (1978).
[34] J. H. Ipsen and O. G. Mouritsen, *Biochim. Biophys. Acta* **944,** 121 (1988).
[35] E. E. Brumbaugh, M. L. Johnson, and C. Huang, *Chem. Phys. Lipids* **52,** 69 (1990).
[36] E. E. Brumbaugh and C. Huang, *Methods Enzymol.* **210,** 521 (1992).
[37] R. Priest, *Mol. Cryst. Liq. Cryst.* **60,** 167 (1980).
[38] I. P. Sugar and G. Monticelli, *Biophys. J.* **48,** 283 (1985).
[39] N. Jan, T. Lookman, and D. A. Pink, *Biochemistry* **23,** 3227 (1984).
[40] K. Jørgensen, M. M. Sperotto, O. G. Mouritsen, J. H. Ipsen, and M. J. Zuckermann, *Biochim. Biophys. Acta* **1152,** 135 (1993).
[41] J. Risbo, M. Sperotto, and O. G. Mouritsen, *J. Chem. Phys.* **103,** 3643 (1995).

The "minimal" model of symmetric DMPC/DSPC bilayers described in this section is a straightforward generalization to two-component bilayers of our two-state model for one-component systems.[14,16,17] Monte Carlo methods are used to drive the model systems toward thermal equilibrium with the surrounding. After attaining equilibrium the actual bilayer configurations, produced by thermal fluctuations and lateral diffusion of the molecules, follow a Boltzmann distribution.

Lattice Geometry, States, and Configuration

A monolayer of the DMPC/DSPC bilayer is modeled as a triangular lattice of N lattice points (coordination number $z = 6$). Each lattice point is occupied by one acyl chain. The acyl chains of DMPC and DSPC molecules represent component 1 and component 2, respectively. In the lattice model nearest-neighbor pairs of similar acyl chains are interconnected, forming either DMPC or DSPC molecules. $N_1/2$ and $N_2/2$ are the number of DMPC and DSPC molecules, respectively. Every lattice point can exist in two states corresponding to the gel (g) and liquid crystalline or fluid state (l).

The actual lattice configuration can be characterized by a square matrix S and by a connection vector c, each containing N elements. Each matrix element S_{ij} represents a lattice point. In accordance with the triangular lattice geometry of the monolayer the following six matrix elements are the nearest neighbors of the ijth matrix element S_{ij}: $S_{i-1,j-1}$, $S_{i-1,j}$, $S_{i,j-1}$, $S_{i,j+1}$, $S_{i+1,j}$, and $S_{i+1,j+1}$. The possible values of a matrix element 1, 2, 3, and 4 refer to component 1 in the g state, component 2 in the g state, component 1 in the l state, and component 2 in the l state, respectively. Vector c lists the lattice positions of the chemically connected pairs of acyl chains, i.e., c_k is the location of the acyl chain connected to the acyl chain in the kth lattice point. The index ij of the S matrix elements and the index k of the c vector elements are in the following relationship: $k = (j - 1)N^{1/2} + i$.

In modeling the transition of one-component phospholipid bilayers, it is not necessary to take into consideration the connections between the acyl chains of the molecules.[16] However, in the case of two-component phospholipid bilayers the connections must be considered in order to calculate correctly the mixing entropy and percolation threshold concentration of the system. Note, that although by using the independent chain model we could calculate the excess heat capacity curve of a DMPC/DSPC mixture in agreement with the experimental curve, the values of the model parameters were significantly different from those given in Table I.

TABLE I
MODEL PARAMETERS OF TWO-STATE TWO-
COMPONENT (DMPC/DSPC) BILAYER MODEL

Parameter	cal/mol · chain
ΔE_1	3028
ΔE_2	5250
w_{11}^{gl}	323.45
w_{22}^{gl}	352.32
w_{12}^{gg}	135
w_{12}^{ll}	80
w_{12}^{gl}	370
w_{12}^{lg}	410
	cal/mol · chain/deg
ΔS_1	10.19378
ΔS_2	16.01689

By analysis of the configuration matrix **S**, one can obtain the number of ith component in mth state (N_i^m) and the number of certain pairs of nearest-neighbor acyl chains (N_{ij}^{mn}), where one of the chains is of component i in state m and the other chain is of component j in state n.

There are simple relationships between the quantities defined above:

$$N = N_1 + N_2$$
$$N = N^g + N^l \tag{1}$$
$$N_1 = N_1^l + N_1^g$$
$$N_2 = N_2^l + N_2^g$$

Periodic boundary conditions are utilized in order to eliminate the effects of the lattice edges.[42] These boundary conditions result in five additional relationships:

$$\frac{z}{2}N = N_{11}^{gg} + N_{11}^{gl} + N_{11}^{ll} + N_{12}^{gg} + N_{12}^{gl} + N_{12}^{lg} + N_{12}^{ll} + N_{22}^{gg} + N_{22}^{gl} + N_{22}^{ll} \tag{2}$$

$$zN_1^g = 2N_{11}^{gg} + N_{11}^{gl} + N_{12}^{gg} + N_{12}^{gl}$$
$$zN_1^l = 2N_{11}^{ll} + N_{11}^{gl} + N_{12}^{ll} + N_{21}^{gl}$$
$$zN_2^g = 2N_{22}^{gg} + N_{12}^{gg} + N_{22}^{gl} + N_{21}^{gl} \tag{3}$$
$$zN_2^l = 2N_{22}^{ll} + N_{12}^{gl} + N_{12}^{ll} + N_{22}^{gl}$$

[42] K. Huang, in "Statistical Mechanics," p. 336. John Wiley & Sons, New York, 1963.

Configurational Energy and Degeneracy

Let E_i^m be the intrachain energy of an acyl chain of component i in state m. In this model E_i^m is assumed to be constant and independent of the location and orientation of the rotational isomers in the acyl chain. It follows then that the energy levels E_1^l and E_2^l are highly degenerate. The degeneracy of the energy level of component i in state m is f_i^m.

E_{ij}^{mn} is the interaction energy between component i in state m and component j in state n. Only nearest-neighbor interactions between the lattice points are considered, because van der Waals interactions between the acyl chains are short range.* Since the interaction energies are assumed to be unaffected by the location and orientation of the rotational isomers in the interacting chains, the interaction energies are also degenerate; the degeneracy of interaction energy E_{ij}^{mn} is f_{ij}^{mn}.

The energy of one layer of the bilayer in configuration $\mathbf{S}$ is

$$
\begin{aligned}
E(\mathbf{S}) = {} & E_1^g N_1^g + E_1^l N_1^l + E_2^g N_2^g + E_2^l N_2^l + E_{11}^{gg} N_{11}^{gg} + E_{11}^{gl} N_{11}^{gl} + E_{11}^{ll} N_{11}^{ll} \\
& + E_{12}^{gg} N_{12}^{gg} + E_{12}^{gl} N_{12}^{gl} + E_{21}^{gl} N_{21}^{gl} + E_{12}^{ll} N_{12}^{ll} + E_{22}^{gg} N_{22}^{gg} \\
& + E_{22}^{gl} N_{22}^{gl} + E_{22}^{ll} N_{22}^{ll}
\end{aligned}
\tag{4}
$$

The degeneracy of configuration $\mathbf{S}$ is calculated from the degeneracies of intrachain energies and the interchain energies

$$
\begin{aligned}
f(\mathbf{S}) = {} & (f_1^g)^{N_1^g}(f_2^g)^{N_2^g}(f_1^l)^{N_1^l}(f_2^l)^{N_2^l}(f_{11}^{gg})^{N_{11}^{gg}}(f_{11}^{gl})^{N_{11}^{gl}}(f_{11}^{ll})^{N_{11}^{ll}} \\
& (f_{12}^{gg})^{N_{12}^{gg}}(f_{12}^{gl})^{N_{12}^{gl}}(f_{21}^{gl})^{N_{21}^{gl}}(f_{12}^{ll})^{N_{12}^{ll}}(f_{22}^{gg})^{N_{22}^{gg}}(f_{22}^{gl})^{N_{22}^{gl}}(f_{22}^{ll})^{N_{22}^{ll}}
\end{aligned}
\tag{5}
$$

Configurational Probability

The probability of configuration $\mathbf{S}$ in the thermodynamic equilibrium is

$$
p(\mathbf{S}) = \frac{f(\mathbf{S})e^{-E(\mathbf{S})/kT}}{Q(N_1/2, N_2/2, T, V)} = \frac{e^{-\chi(\mathbf{S})/kT}}{Q(N_1/2, N_2/2, T, V)}
\tag{6}
$$

where $Q(N_1/2, N_2/2, T, V)$ is the partition function of the canonical ensemble of the lattice model, T is the absolute temperature, k is the Boltzmann constant, and V is the volume of the monolayer.† The function $\chi(\mathbf{S})$ is defined by

$$
\chi(\mathbf{S}) = E(\mathbf{S}) - kT \ln f(\mathbf{S})
\tag{7}
$$

---

* Long-range dipole–dipole interactions between the molecules were originally incorporated into our model. However, in the case of DMPC/DSPC mixtures the effect of the dipole–dipole interaction on the calculated excess heat capacity curves was negligibly small.

† In our model the volume change and the respective change in the volume energy associated with the gel-to-liquid crystalline transition are neglected.[43]

[43] J. F. Nagle and H. L. Scott, Jr., *Biochim. Biophys. Acta* **513**, 236 (1978).

The configuration-dependent part of this function plays a central role in the Monte Carlo simulation. To reduce the number of model parameters we substitute Eqs. (1)–(5) into Eq. (7) and obtain $\bar{\chi}$, the configuration-dependent part of $\chi(\mathbf{S})$:

$$\bar{\chi}(\mathbf{S}) = N_1^l(\Delta E_1 - T\Delta S_1) + N_2^l(\Delta E_2 - T\Delta S_2)$$
$$+ w_{11}^{gl}N_{11}^{gl} + w_{22}^{gl}N_{22}^{gl} + w_{12}^{gg}N_{12}^{gg} + w_{12}^{ll}N_{12}^{ll} + w_{12}^{gl}N_{12}^{gl} + w_{21}^{gl}N_{21}^{gl} \qquad (8)$$

where

$$\Delta E_i = [E_i^l + (z/2)E_{ii}^{ll}] - [E_i^g + (z/2)E_{ii}^{gg}] \qquad (9)$$

$$\Delta S_i = k \ln f_i^l - k \ln f_i^g \qquad (10)$$

$$w_{ij}^{mn} = [E_{ij}^{mn} - (E_{ii}^{mm} + E_{jj}^{nn})/2] - kT \ln \left[\frac{f_{ij}^{mn}}{(f_{ii}^{mm} f_{jj}^{nn})^{1/2}} \right] \qquad (11)$$

Steps in Monte Carlo Simulations

Each simulation can start from either the all-gel or all-fluid state or any state in between. Initially component 1 is assigned to the first N_1 lattice points, while component 2 is assigned to the remaining N_2 lattice points. Initially the molecules are similarly oriented, i.e., the acyl chains on the first and second lattice points represent a phospholipid molecule, and in general the acyl chains on the $(2k - 1)$th and $2k$th lattice points are connected, i.e., $c_{2k-1} = 2k$. This "standard configuration," introduced first by Kasteleyn,[44] can be easily generated. Note that Jerala *et al.*[16] started simulations from random orientations of the molecules. However, the generation of these initial random orientations was complicated because it was necessary to eliminate vacancies appearing in these configurations.

Generation of Trial Configurations

During the Monte Carlo simulation, trial configurations of the two-component phospholipid bilayers are generated by means of three different elementary steps.

Local State Alteration. In this step the trial configuration is generated by changing the state of a randomly selected acyl chain from gel to fluid or from fluid to gel. This trial configuration generation, the Glauber method,[45] is essential for the simulation of gel-to-fluid transitions of lipid bilayers.

[44] P. W. Kasteleyn, *J. Math. Phys.* **4**, 287 (1963).
[45] R. J. Glauber, *J. Math. Phys.* **4**, 294 (1963).

Exchanging Different Molecules. In this step two randomly selected molecules of different lipid components are exchanged. Although this elementary, nonphysical step is different from the Kawasaki method,[46] in which nearest-neighbor molecules are exchanged, the rate of attainment of equilibrium of the lateral distribution of the bilayer components is improved.[47]

Reorientation of a Pair of Nearest-Neighbor Molecules. In this step a pair of nearest-neighbor molecules is randomly selected. If the positions of the selected acyl chains define the nodes of a rhombus, then one of the chains and the chain on the opposite node are exchanged.[16] This exchange involves a rotation of the respective molecules by ±60 degrees. A series of these elementary reorientations leads to the equilibrium distribution of the orientation of the molecules. Note that like the exchange of different molecules, the reorientation step results in the lateral movement of the molecules. Thus a series of reorientation steps is able to drive the system to the equilibrium of the lateral distribution of the molecules. However, as mentioned above, the nonphysical steps of exchanging different molecules are also used to substantially accelerate convergence to the equilibrium distribution.[47]

Decision Making

A trial configuration, S_{trial}, once generated, is acceptable when the following inequality holds:

$$RAN \leq \exp[-\{\bar{\chi}(S_{trial}) - \bar{\chi}(S_{orig})\}/kT] \tag{12}$$

If it is not, the original configuration S_{orig} is retained. In Eq. (12) RAN is a pseudorandom number, distributed homogeneously in the interval $(0, 1)$. This method of decision making drives the system toward thermodynamic equilibrium, the Boltzmann distribution over all the possible configurations, independently of the choice of the initial configuration and the choice of the actual path toward equilibrium.[47]

Defining Monte Carlo Cycle

In a Monte Carlo simulation a certain chain of elementary steps, generating trial configurations, are repeated. During this chain of elementary steps, the Monte Carlo cycle, the system has the opportunity of realizing all of its configurations one time. N elementary steps of local state altera-

[46] K. Kawasaki, *in* "Phase Transitions and Critical Phenomena" (C. Domb and M. S. Green, eds.), Vol. 2, p. 443. Academic Press, London, 1972.
[47] H. Sun and I. P. Sugar, *J. Phys. Chem.* **101**, 3221 (1997).

tion give the opportunity of realizing any of the 2^N acyl chain states of the lattice.

By exchanging different molecules $(N/2)!/[(N_1/2)!(N_2/2)!]$ different arrangements of the molecules can be created. Any one of these arrangements can be realized by repeating the elementary steps of exchange of different molecules $N_1/2$ times (or $N_2/2$ times if $N_2 > N_1$).

An acyl chain at the ith lattice point is connected with one of the six nearest-neighbor acyl chains. Assuming independent orientations of the molecules, $3^{N/2}$ is the total number of different orientations in the lattice. In reality the orientations of the molecules are coupled and thus the number of possible orientations in the bilayer is much smaller. The probability that any selected lipid molecule has a neighbor that can participate in a reorientation step is 0.75 on a lattice with randomly oriented molecules.[16] After $N/2/0.75$ reorientation steps, about 50% of the molecules has an opportunity to change orientation by $+60°$, while the other half of the molecules can change orientation by $-60°$. Each orientation is accessible for each molecule at least once after performing $2(N/2/0.75) = 4N/3$ reorientation steps.

Global State Alteration

In principle the equilibrium distribution of the system is attainable after many Monte Carlo cycles. In practice, however, the system could be trapped during the time of the simulation in one of the local free energy minima dependent on the initial configuration.

Accelerated convergence to the equilibrium distribution can be obtained by incorporating nonphysical, shuffling operations into the algorithm.[47] We incorporate a shuffling operation at the end of each Monte Carlo cycle. In this step the trial configuration is generated by altering the state of every acyl chain from gel to fluid and from fluid to gel.

The number of Monte Carlo cycles needed to attain the equilibrium depends on the lattice size, the actual values of the model parameters, temperature, mole fraction, and the types of steps generating trial configurations. To monitor convergence the membrane energy is calculated at the end of each Monte Carlo cycle. Starting the simulation from an all-gel state the membrane energy drifts toward larger values as the simulation progresses and after a certain number of Monte Carlo cycles it begins to fluctuate around a stationary value. This point signifies the attainment of equilibrium. In our Monte Carlo simulations, at every temperature and mole fraction, the equilibrium is attained at fewer than 6000 Monte Carlo cycles. After attaining the equilibrium another 6000 cycles are performed, and the snapshots are analyzed after each cycle to extract quantities characterizing the membrane configuration such as lattice energy, cluster number, and cluster size. From the distribution of these quantities thermodynamic averages are calculated. To calculate the excess heat capacity, a thermodynamic average that is directly measurable by scanning calorimetry, two different methods are utilized. After determining the average lattice energy at 50 different temperatures, the excess heat capacity curve is obtained from the numerical derivative of the calculated energy curve. In an alternative method the variance of the energy at different temperatures is calculated in order to obtain directly the excess heat capacity curve.[17] This method is more time consuming, however, because convergence to the equilibrium value of the energy variance is an order of magnitude slower than convergence to the energy average.[48] Note that in the case of our simulation protocol the above two methods resulted in practically the same excess heat capacity values.

Determination of Model Parameters

The $\bar{\chi}(S)$ function, Eq. (8), contains 10 unknown model parameters: ΔE_1, ΔE_2, ΔS_1, ΔS_2, w_{11}^{gl}, w_{22}^{gl}, w_{12}^{gg}, w_{12}^{ll}, w_{12}^{gl}, w_{21}^{gl}. The model parameters have been determined by the following strategy. ΔE_1 and ΔE_2 are estimated by means of the integral of the measured excess heat capacity curves of single-component DMPC and DSPC multilamellar vesicles (MLVs), respectively.

To obtain the maxima of the calculated excess heat capacity curves of the one-component systems at the respective measured temperatures of the heat capacity maxima, T_{m1} and T_{m2}, the model parameters should satisfy the following two constraints:

$$\Delta S_1 \approx \Delta E_1 / T_{m1} \tag{13}$$
$$\Delta S_2 \approx \Delta E_2 / T_{m2} \tag{14}$$

These approximate relationships can be derived when the parameters w_{ii}^{gl}, where $i = 1$ or 2, are equal to zero and $2kT_{mi} \ll \Delta E_i$.[18] In the case of a DSPC bilayer the above approximation results in an error of about 3% in the value of ΔS_2.

The remaining two parameters of the one-component systems, w_{11}^{gl} and w_{22}^{gl}, were estimated by comparing the experimental excess heat capacity at T_{mi} of each one-component system with a series of excess heat capacities calculated at different values of the respective w_{ii}^{gl} parameter. The parameters resulting in a good fit are listed in Table I.

In Table I the parameters $w_{11}^{gl} = 323.45$ cal/mol · chain and $w_{22}^{gl} = 352.32$

[48] M. Mezei and D. L. Beveridge, *Ann. N.Y. Acad. Sci.* **482**, 1 (1986).

cal/mol · chain are of the same magnitude as the value of the parameter obtained from the similar analysis of the excess heat capacity curve of DPPC small unilamellar vesicles (SUVs).[14,16,18] The somewhat larger value obtained here is the result of higher cooperativity associated with the gel–fluid transition of MLVs as described by the narrower excess heat capacity function.

The positions of the high- and low-temperature peaks in the excess heat capacity curves for the binary mixtures are strongly related to the values of the parameters w_{12}^{gg} and w_{12}^{ll}. In the case of a 60/40 mixing ratio, the calculated low- and high-temperature peaks in the excess heat capacity curves were obtained in agreement with the observed peak positions by assuming $w_{12}^{gg} = 135$ cal/mol · chain and $w_{12}^{ll} = 80$ cal/mol · chain. This result shows that the pure fluid phase is closer to an ideal mixture than the pure gel phase, as expected.

The values of the remaining two parameters, w_{12}^{lg} and w_{12}^{gl}, were estimated simultaneously, by comparing the calculated excess heat capacity curve, obtained at different pairs of the parameter values, with the experimental excess heat capacity curve at a mixing ratio of 60/40. The parameters obtained from the parameter estimation are listed in Table I. These sets of parameters were used in all the subsequent simulations. In estimating these parameters, it was assumed that the w_{ij}^{kl} parameters are independent of temperature. We note that the method suggested by Ferrenberg and Swendsen[49] of accelerating multiparameter fitting is not practical in our Monte Carlo simulations because the tabulation of the distribution function of N_1^l, N_2^l, N_{11}^{gl}, N_{22}^{gl}, N_{12}^{gg}, N_{12}^{ll}, N_{12}^{lg}, N_{12}^{gl}, in an eight-dimensional matrix requires a prohibitively large memory.

Results and Discussion

System Size and Type of Transition

All calorimetric scans were performed on a home-made high-sensitivity calorimeter[50] at scan rates from 0.1 to 5°/hr and lipid concentrations of 20–22 mM. The obtained excess heat capacity curves were scan rate independent at these slow scanning rates. The excess heat capacity curves of one-component DMPC and DSPC bilayers showed a sharp, symmetric peak with a heat capacity maximum of $C_p(T_{m1} = 297.044$ K$) = 40,000$ cal/mol · chain/deg and $C_p(T_{m2} = 327.779$ K$) = 24,000$ cal/mol · chain/

[49] A. M. Ferrenberg and R. H. Swendsen, *Phys. Rev. Lett.* **61,** 2635 (1988).
[50] J. Suurkuusk, B. R. Lentz, Y. Barenholz, R. L. Biltonen, and T. E. Thompson, *Biochemistry* **15,** 1393 (1976).

deg, respectively. By using the parameters, listed in Table I, the simulated excess heat capacity maxima agree with the respective experimental data if the lattice size is large enough. Figure 1A shows that the calculated excess heat capacity maxima become independent from the system size at a threshold linear system size of 100 and 250 for DSPC and DMPC, respectively. In Fig. 1B the threshold linear lattice sizes are shown at different DMPC/DSPC mole fractions. To eliminate finite size effects one must perform simulations for lattice sizes that are larger than these threshold values. In the present simulations the following lattice sizes are utilized: 350×350 for DMPC; 300×300 for DSPC; 100×100 for DMPC/DSPC mixtures of mixing ratios 10/90, 20/80, 90/10, and 80/20; and 40×40 for DMPC/DSPC mixtures of mixing ratios 30/70, 40/60, 50/50, 60/40, and 70/30.

By varying the system size not only quantitative but also qualitative changes take place in the transition properties of the simulated bilayer, i.e., change in the type of the transition. In general, the type of transition can be characterized by the distribution of the fluctuating extensive parameters of the system taken at the midpoint of the transition. If the distribution is unimodal the transition is continuous (or second-order transition), otherwise it is a phase transition (or first-order transition). In our model, membrane energy is the only fluctuating extensive parameter of the system (canonical ensemble). At the midpoint of the gel–fluid transition the calculated energy distribution is bimodal or unimodal if the system size is below or above the threshold size. In Fig. 1C and D the energy distributions, calculated above and below the threshold size, are shown at 70/30 and 0/100 DMPC/DSPC mole fractions, respectively. Thus with increasing lattice size the gel–fluid transition changes at a threshold size from first-order to second-order transition, i.e., at the thermodynamic limit the transition is continuous (second-order transition) at every mole fraction.

Excess Heat Capacity Curves and Melting Curves

In Fig. 2 the experimental and calculated excess heat capacity curves are shown at different DMPC/DSPC mixing ratios. The excess heat capacity was calculated from the energy fluctuation according to Eq. (20). Each simulation was performed using the model parameters listed in Table I. There is an excellent agreement between the calculated and experimental excess heat capacity curves for DMPC/DSPC mole fractions of 60/40, 50/50, 40/60, 30/70, 20/80, and 10/90 while for the other mole fractions, although the location of the calculated peaks is correct, the heights of the low-temperature peaks are significantly smaller than those of the experi-

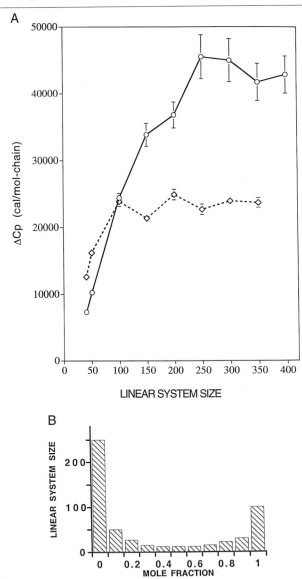

Fig. 1. Finite size effects. (A) Excess heat capacity maxima, calculated by using Eq. (20) at different linear system sizes. Solid line: DMPC, $T = 297.044$ K; dashed line: DSPC, $T = 327.779$ K. The error bars were calculated from the result of eight computer experiments started with different seed numbers for random number generation. In each computer experiment the number of Monte Carlo cycles was 10^5, and the system was equilibrated during the first 6000 cycles. (B) Threshold linear system sizes at different mole fractions of DMPC/DSPC mixtures. (C) and (D) Energy distributions calculated at 70/30 and 0/100 DMPC/DSPC mole fractions, respectively. Each distribution is labeled by the respective linear system size.

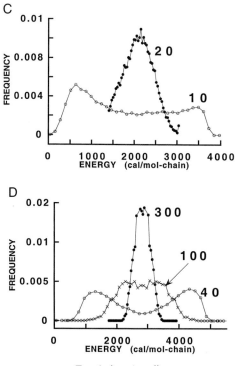

FIG. 1. (*continued*)

mental peaks. These deviations may be related to the fact that the experimental DMPC gel–fluid transition enthalpy, 3028 cal/mol · chain, underestimates the true transition enthalpy. In our experiment, to ensure equilibrium-transition of DMPC, a particularly slow scanning rate, 0.1°/hr, was utilized within a temperature range of only 0.6° and the transition enthalpy was determined by integrating the excess heat capacity curve over this short temperature range. The existence of large "wings" on the high- and low-temperature side of the heat capacity curves of one-component phospholipid bilayers, noted first by Mouritsen,[13] however, would require integration over a longer temperature range for the better estimation of the transition enthalpy. In the case of DSPC a three times larger temperature range was scanned and thus the integration over this temperature range gives a better estimate of the true transition enthalpy.

The experimental excess heat capacity curves are commonly used to construct so-called phase diagrams of the two-component lipid bilayers. The solidus and liquidus curves of the diagram are created by plotting the

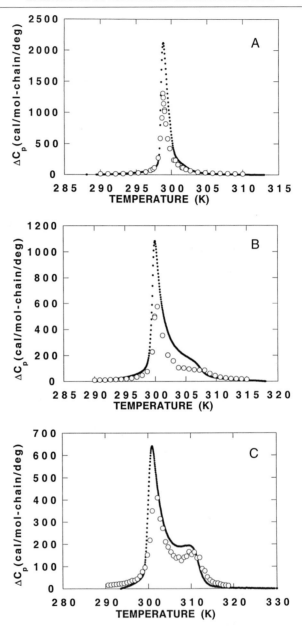

FIG. 2. Excess heat capacity curves. Solid line: Experimental excess heat capacity curves; open circles: excess heat capacity curves calculated by means of Eq. (20). DSPC mole fractions are (A) 0.1, (B) 0.2, (C) 0.3, (D) 0.4, (E) 0.5, (F) 0.6, (G) 0.7, (H) 0.8, (I) 0.9.

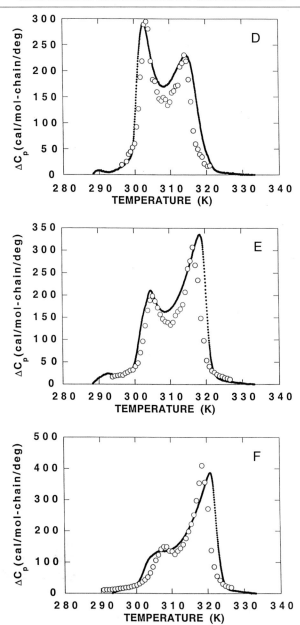

FIG. 2. (*continued*)

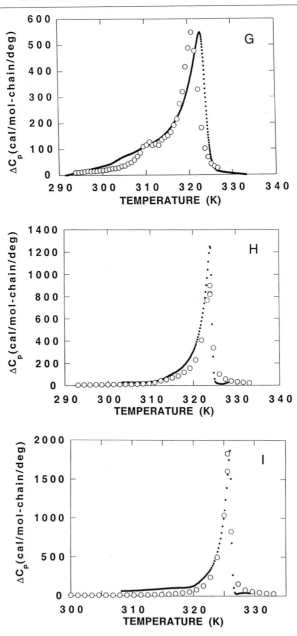

Fig. 2. (*continued*)

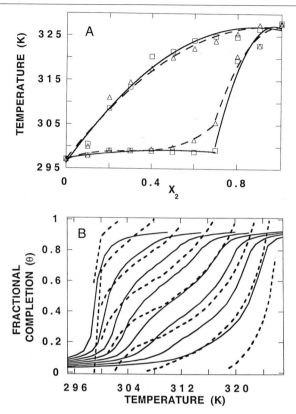

FIG. 3. Phase diagrams and melting curves. (A) Phase diagrams constructed from the excess heat capacity curves in Fig. 2. The open triangles and open squares, representing, respectively, the calculated and experimental onset and completion temperatures of the gel-to-fluid transition at different DSPC mole fractions, were used to construct the solidus and liquidus lines of the phase diagrams. (B) Comparison of the melting curves calculated directly from the simulations [see Eq. (16)] (solid lines) and by using the phase diagram in (A) and the lever rule [see Eq. (15)] (dashed lines). The curves from left to right belong to the following DSPC mole fractions: 0.1, 0.2, 0.3, 0.4, 0.5, 0.6, 0.7, 0.8.

onset and completion temperatures, respectively, against the mole fraction.* Figure 3A shows two phase diagrams constructed from the experimental and calculated excess heat capacity curves.

It is important to emphasize, however, that these are not phase diagrams in a strict thermodynamic sense because the gel–fluid transition of DMPC/

---

* A straight line is fitted to the inflection point of the excess heat capacity curve close to the completion of the transition and its intercept with the baseline defines the completion temperature. A similar procedure close to the onset of the transition defines the onset temperature.

DSPC mixtures is not a first-order phase transition. At a given temperature, T, the solidus and liquidus curve of a real phase diagram define the compositions in the coexisting solid (X^g) and liquid (X^l) phase regions, respectively, and these compositions remain constant when the total mole fraction (X) is changed. These properties of the first-order phase transitions and real phase diagrams result in the lever rule:

$$\theta(T) = [X - X^g(T)]/[X^l(T) - X^g(T)] \qquad (15)$$

where $\theta(T)$ is the fractional completion of the transition at temperature T. Because the gel–fluid transition in DMPC/DSPC mixtures is not a first-order phase transition the diagrams in Fig. 3A are not real phase diagrams and the lever rule is not applicable to obtain the fractional completion of the transition. It is shown in Fig. 3B that a mechanical application of the lever rule to the phase diagrams in Fig. 3A results in serious errors in the estimation of the fractional completion of the gel–fluid transition (see dashed lines). Solid lines show the correct fractional completion curves calculated from the simulated data as follows:

$$\theta(T) = [\langle N_1^l(T)\rangle + \langle N_2^l(T)\rangle]/N \qquad (16)$$

Domain Structure of Membranes

The good agreement between the observed and calculated excess heat capacity curves increases our confidence in the simple two-state bilayer model and thus we perform computer experiments to study the thermodynamic averages that are characteristic of the configuration of two-component bilayers. A membrane domain or cluster is a group of lipids in lateral proximity sharing a certain property. For example, a compositional cluster is a cluster of similar lipid molecules existing in either the gel or fluid state. On the other hand, a gel cluster is formed by gel-state hydrocarbon chains of any lipid components.

The reason for cluster formation lies in the lateral heterogeneity of the membranes.[51] The number, size, and shape distribution of the clusters are related to the physical conditions such as temperature, pressure, and mole fraction, and also to the interactions between the components of the membrane. In the case of one-component bilayers the effects of the three different interchain interactions (g–g, g–l, and l–l) on the gel or fluid cluster formation can be characterized by a single parameter, w_{11}^{gl}, which is a combination of the interchain interaction energies and degeneracies [see Eq. (11)].

When $w_{11}^{gl}/kT \approx 0$, gel- and fluid-state molecules are randomly distrib-

---

[51] O. G. Mouritsen and K. Jørgensen, *BioEssays* **14**, 129 (1992).

uted; the average number of gel-state hydrocarbon chains in the proximity of a gel-state hydrocarbon chain is determined solely by chance (by the concentration of gel-state hydrocarbon chains). The cluster shapes are irregular, ensuring high entropy, which minimizes the free energy of the system. With increasing values of w_{11}^{gl}/kT the average size of the gel and fluid clusters increases, while their average number decreases. Also, with increasing value of w_{11}^{gl}/kT the cluster shapes become less irregular and are getting closer to circular.

In the case of DMPC/DSPC mixtures the effect of 10 different interchain interactions ($1g$–$1g$, $1g$–$2g$, $2g$–$2g$, $1l$–$1l$, $1l$–$2l$, $2l$–$2l$, $1g$–$1l$, $1g$–$2l$, $2g$–$1l$, and $2g$–$2l$) on the cluster formation can be characterized by six w_{ij}^{kl} parameters [see Eq. (11)]. At room temperature $0 < w_{ij}^{kl}/kT < 1$ (see Table I). Because of the relatively small values of the w_{ij}^{kl}/kT parameters one can expect small, irregular shape clusters of the minor phase. Small, irregular shape gel and fluid clusters are shown in Fig. 4A and C, respectively. These snapshots of equimolar DMPC/DSPC mixtures were simulated by using our simple two-state model. The average characteristics of these small clusters can be determined by means of cluster statistics.

Cluster Statistics

The snapshots were analyzed after every Monte Carlo cycle, using a cluster-counting algorithm[17,52] to obtain the cluster size distributions, cluster numbers, and percolation frequencies. The cluster-counting algorithm labels each cluster in a snapshot with a different number and then the labeled clusters are analyzed and sorted according to certain properties such as size, number, and shape.

Figure 5A–C shows the size distributions of gel clusters at three different temperatures for an equimolar mixture of DMPC/DSPC. The size distribution of the gel clusters is unimodal above a certain threshold temperature and the bilayer contains only "small" gel clusters. These so-called percolation threshold temperatures of the gel clusters, T_{perc}^g, are listed in Table II at different mole fractions. "Large" gel clusters appear in the bilayer below the percolation threshold temperature and the cluster size distribution is bimodal. In this case the position of the minimum between the two peaks of the bimodal distribution separates the small clusters from the large ones.

The situation is just the opposite for the size distribution of fluid-phase clusters (see Fig. 5D–F). Below the percolation threshold temperature of fluid clusters, T_{perc}^l (Table II), the size distribution is unimodal with a peak at small cluster size. Above the percolation temperature, however, an addi-

[52] D. Stauffer, *in* "Introduction to Percolation Theory," p. 110. Taylor and Francis, London, 1985.

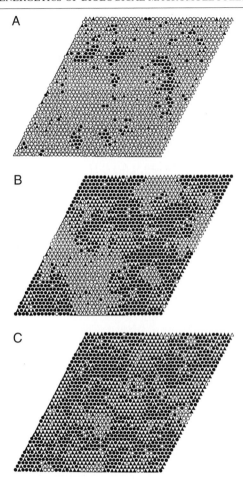

Fɪɢ. 4. Snapshots from an equimolar mixture of DMPC/DSPC. (A) $T = 321$ K, (B) $T = 307$ K, (C) $T = 302$ K. Closed circles, DSPC chain in gel state; open circles, DSPC chain in fluid state; closed triangles, DMPC chain in gel state; open triangles, DMPC chain in fluid state.

tional peak appears at large cluster sizes, i.e., the cluster size distribution becomes bimodal.

Cluster Number

The integral of the size distribution provides the average number of clusters in the lattice. In Fig. 6 the average number of gel and fluid clusters is plotted against the temperature. With decreasing temperature the number of gel-phase clusters increases up to a maximum, at 322 K. Below this temperature the number of gel clusters starts to decrease because the clusters coalesce, forming eventually a "large" gel cluster. With increasing temperature the number of fluid-phase clusters increases up to a maximum, at 302 K. Above this temperature the coalescence of fluid-phase clusters dominates over the cluster formation and thus the number of clusters starts to decrease. It should be noted that at a temperature at which it is generally assumed that the system exists in a single structural phase (e.g., 280 K for the gel and 350 K for the fluid phase) the average number of clusters of the minor phase is still significant. This is indicative of lateral density heterogeneities existing far from the transition range.[2,51]

Cluster Percolation

When the cluster size distribution is bimodal, on average there is only one "large" cluster in the lattice. A large cluster is percolated if it spans the lattice either from the top to the bottom or from the left to the right edge. The frequency of the appearance of a percolated cluster at the end of each Monte Carlo cycle is the percolation frequency. In Fig. 7A and B the percolation frequencies of fluid and gel clusters are plotted against the temperature at different mixing ratios. The percolation threshold temperatures of the fluid and gel clusters, T_{perc}^l and T_{perc}^g, listed in Table II, are in good agreement with the peak positions of the excess heat capacity curve T_{peak1} and T_{peak2}, respectively.

Direct observation of cluster percolation is not available. Fluorescence recovery after photobleaching (FRAP) provides indirect information on the connectedness of clusters. Recovery takes place when fluorescent molecules diffuse from the unbleached area of the membrane to the photobleached area. There is practically no recovery in pure gel phase because the lateral diffusion of the fluorescent labeled lipid molecules is 1000 times slower in gel phase than in fluid phase. In a gel–fluid mixed phase the recovery suddenly increases from a threshold temperature, T_{FRAP}. FRAP threshold temperatures were measured by Vaz et al.[28] at different mole fractions of DMPC/DSPC mixtures. It is assumed that the FRAP threshold is related to the percolation threshold temperature of either the gel or the fluid clusters. The long-range lateral diffusion of the fluorescent probe becomes blocked when the percolation of gel clusters takes place or the long-range lateral diffusion of the probe becomes possible at the percolation of fluid clusters.* The correct interpretation of the FRAP data can be made by

---

* It is important to note that percolation of gel and fluid clusters is not mutually exclusive. It is possible, for example, that a gel cluster spans horizontally the upper part of the lattice while the lower part of the lattice is spanned horizontally by a fluid cluster. Thus there is a temperature range where both gel and fluid clusters can be percolated. For example, according to the calculated cluster size distributions at 307 K there are both large gel and large fluid clusters (see Fig. 5B and E). The percolation frequency of these large clusters is 1.0 and 0.46 for gel and fluid clusters, respectively.

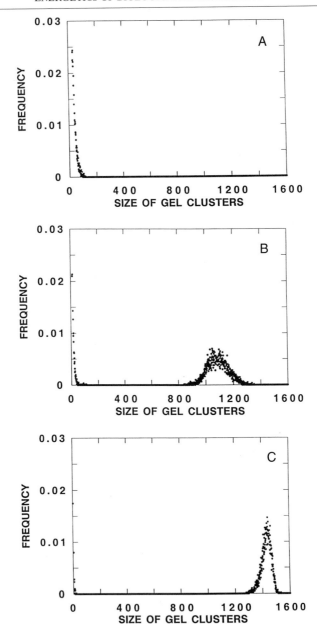

FIG. 5. Cluster size distributions in an equimolar mixture of DMPC/DSPC. (A–C) Gel clusters; (D–F) fluid clusters. Temperatures: (A, D) 321 K; (B, E) 307 K; (C, F) 302 K.

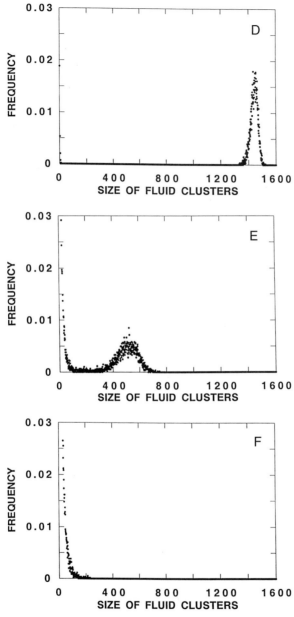

FIG. 5. (*continued*)

TABLE II

CALCULATED AND EXPERIMENTAL PERCOLATION THRESHOLD
TEMPERATURES OF GEL-TO-FLUID TRANSITIONS[a]

	X_2				
	0.3	0.4	0.5	0.6	0.7
T_{peak1}	302.3	303.8	304.9	307.8	310.4*sh*
T_{peak2}	310.4	314.4	316.9	318.9	321.3
T^l_{perc}	300.8	302.2	303.9	308.5	313.1
T^g_{perc}	309.1	313.8	317.5	320.6	323.0
$T^g_{perc0.36}$	307.1	311.9	315.9	318.9	321.5
T_{FRAP}	306.6	312	316	319	321.5

[a] In DMPC/DSPC lipid bilayers at different DSPC mole fractions, X_2. *sh* marks the temperatures at shoulders of the excess heat capacity curves; T_{peak1}, calculated temperature at the low-temperature peak of the excess heat capacity curve; T_{peak2}, calculated temperature at the high-temperature peak of the excess heat capacity curve; T^l_{perc}, calculated percolation threshold temperature of fluid clusters; T^g_{perc}, calculated percolation threshold temperature of gel clusters; $T^g_{perc0.36}$, temperature at which the percolation frequency of gel clusters is 0.36; T_{FRAP}, threshold temperature from the FRAP experiment.[28] The determination of the percolation threshold temperature is similar to that of the completion temperature of the transition. A straight line is fitted to the inflection point of the percolation frequency curve (Fig. 7A and B) and its intercept with the zero frequency line defines the percolation threshold temperature.

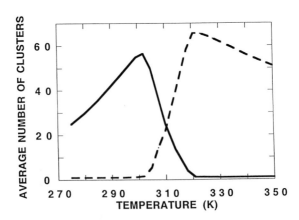

FIG. 6. Average cluster numbers in an equimolar mixture of DMPC/DSPC. Solid line, fluid clusters; dashed line, gel clusters.

using the results of our simulations. In Fig. 7C the calculated percolation threshold temperatures of gel and fluid clusters are plotted against the measured FRAP threshold temperatures. The correlation is weak for fluid clusters but there is a strong positive correlation for gel clusters with a constant difference of 1.8° between the calculated and measured threshold temperatures. However, if we plot the temperatures, $T^g_{\text{perc0.36}}$, where the percolation frequency of the gel clusters is 0.36 against the FRAP threshold temperatures the two sets of temperatures are completely identical. In conclusion, at the FRAP threshold temperature the percolation probability of the gel clusters is 0.36, and gel clusters cease to block efficiently the long-range diffusion of the fluorescence probe molecules.

Small Clusters

Because the small clusters are so small their direct detection is difficult.[25,31] In 1998 Gliss et al.[32] obtained estimates of the average linear size of gel clusters by using neutron scattering and atomic force microscopy. The neutron diffraction measurements of an equimolar DMPC/DSPC mixture at 38 and 41° resulted in an average center-to-center distance between adjacent gel domains of 5–10 nm. Atomic force microscopy studies supported the above estimate for the average size of gel domains and showed rather irregular cluster shapes.

What is the average size of the small gel clusters in our simulations? By using the cluster size distribution one can calculate the weighted average of the size of the small clusters as follows:

$$\langle s \rangle = \left[\sum_{i=1}^{i\text{th}} i^2 P(i) \right] \Big/ \left[\sum_{i=1}^{i\text{th}} i P(i) \right] \qquad (17)$$

where $P(i)$ is the probability of finding a cluster of size i (i.e., the number of i-size clusters to the total number of clusters); ith is the threshold cluster size separating the small clusters from the large ones. The threshold cluster size is defined by the local minimum between the two maxima of the bimodal cluster size distribution. In the case of unimodal cluster size distribution ith $= N$, where N is the number of hydrocarbon chains in a layer of the bilayer.

What is the meaning of the above-defined weighted average? Let us pick hydrocarbon chains randomly from the lipid layer. Every time the hydrocarbon chain is an element of a small cluster the respective cluster is selected. The average size of the selected clusters is the weighted average of the small clusters. This definition implies that larger clusters are selected more frequently than smaller ones, i.e., the average is weighted by the

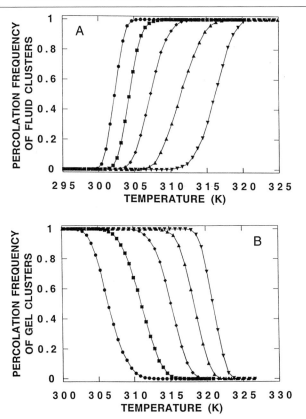

FIG. 7. Percolation frequency of gel and fluid clusters in DMPC/DSPC mixtures. (A) Calculated percolation frequency of fluid clusters versus temperature. (B) Calculated percolation frequency of gel clusters versus temperature. DSPC mole fractions are as follows: 0.3 (●), 0.4 (■), 0.5 (◆), 0.6 (▲) 0.7 (▼). (C) Four characteristic temperatures of the percolation curves are plotted against the FRAP threshold temperatures measured at different DMPC/DSPC mole fractions. (●) Percolation threshold temperatures of gel clusters; (○) percolation threshold temperatures of fluid clusters; (□) temperatures at 0.36 percolation frequency of fluid clusters; (■) temperatures at 0.36 percolation frequency of gel clusters (the slope of the solid, regression line is 1.00 ± 0.02, while the linear correlation coefficient is $r = 0.9994$).

cluster size. We introduced this weighted average because the observed average linear size of the clusters is a similarly weighted average.[32]

In Fig. 8 the weighted average size of the clusters is plotted against the temperature at different mole fractions of DMPC/DSPC. Each curve has a sharp maximum superimposing to a broad hump. In Fig. 8A the average size of the small gel clusters approaches one at low temperature. At this temperature the membrane is close to an all-gel state, and only small fluid

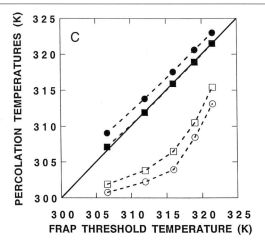

FIG. 7. (*continued*)

clusters can be present. Within a small fluid cluster of size ≥6 chains a small gel cluster of size $i \approx 1$ chain can form, while the probability of finding gel clusters of size $i = 2$ or 3 is much smaller. Thus at low temperature the size distribution function of small gel clusters is $P(1) \approx 1$ and $P(i) \approx 0$ for $i > 1$, which results in $\langle s \rangle \approx 1$ for the weighted average size of the small gel clusters [see Eq. (17)].

The broad hump disappears on approaching equimolar mixing because the two peaks overlap each other. The height and the half-width of the sharp peaks is 110 ± 20 chains and 2°, respectively. The height and the half-width of the broad hump (25 ± 5 chains and 15°, respectively) can be estimated from those curves where the broad hump is separated from the sharp peak.

The average linear size of the irregular shape, small clusters $\langle l \rangle$ is estimated by using the average cluster size $\langle s \rangle$ as follows:

$$\sqrt{4\langle s \rangle A_0/\pi} < \langle l \rangle < \langle s \rangle \sqrt{4A_0/\pi} \tag{18}$$

where the cross-sectional area of a chain is $A_0 = 20$ Å^2 in gel phase and $A_0 = 31$ Å^2 in fluid phase.[53] In Eq. (18) the lower and upper limits assume circular and linear shape clusters, respectively. By using Eq. (18) the linear size of the clusters at the sharp peaks in Fig. 8 is 5 nm $< \langle l \rangle <$ 50 nm for small gel clusters and 6 nm $< \langle l \rangle <$ 62 nm for small fluid clusters. The

[53] J. F. Nagle, R. Zhang, S. Tristram-Nagle, W. Sun, H. I. Petrache, and R. M. Suter, *Biophys. J.* **70**, 1419 (1996).

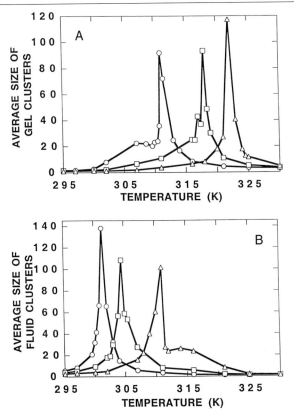

FIG. 8. Average cluster size of small clusters. Temperature dependence of the weighted average cluster size calculated by using Eq. (17). (A) Gel clusters; (B) fluid clusters. DSPC mole fractions: 0.3 (open circles), 0.5 (open squares), 0.7 (open triangles).

linear cluster size at the broad hump in Fig. 8 is 2.5 nm $< \langle l \rangle <$ 13 nm for gel clusters and 3 nm $< \langle l \rangle <$ 15 nm for fluid clusters.

The average linear size of the gel clusters observed directly for an equimolar mixture of DMPC/DSPC at three different temperatures, 38, 41, and 55°, are 5–10, 5–10, and 0 nm.[32] These linear cluster sizes are in agreement with the calculated cluster size values. From Fig. 8A the average size of gel clusters for equimolar mixture at 38, 41, and 55° are 11, 18, and 0 chains, and by using Eq. (18) the respective average linear sizes of the gel clusters are 1–6, 2–9, and 0 nm.

By using Fig. 8B it is interesting to follow the formation of fluid clusters in the DMPC/DSPC (30/70) mixture. At 300 K small, DMPC-rich fluid clusters are present. With increasing temperature the number and size of the fluid clusters increase. At 311 K, as a result of the coalescence of many small fluid clusters, a few larger clusters form, each containing about 25 hydrocarbon chains. Some of these larger clusters are close to each other, and they can easily connect with each other to form an even larger fluid cluster containing about 100 hydrocarbon chains. A slight further increase in the temperature leads to the permanent presence of one large cluster, which rarely spans the whole length of the membrane layer. At 311 K the percolation of the fluid clusters is almost zero (Fig. 7A). The permanent appearance of this large cluster transforms the unimodal cluster size distribution into bimodal distribution. Once a cluster becomes permanently large it does not count in calculating the average size of the small clusters. Thus the average size of the small clusters drops to about 25 chains. This is the size of the larger clusters among the small clusters. These clusters are so much separated from each other that they cannot immediately coalesce with each other. When the temperature increases smaller clusters attach to the larger ones, increasing slightly the average size of the small clusters. From about 314 K, however, the increasing larger clusters get near to the similarly increasing large cluster. When the larger ones among the small clusters start to coalesce with the large cluster the average size of the small clusters starts to decrease.

Membrane Permeability

The permeability of one- and two-component phospholipid bilayers to small water-soluble molecules may be larger at temperatures at which gel and fluid phases coexist than when the bilayers are in either all-gel or all-fluid phase.[6,54–60] Similar to the construction of the excess heat capacity curve from the heat capacity curve one can create the excess permeability curve from the permeability curve. In the cases of one-component phospholipid bilayers and close to ideal mixtures of dimyristoylphosphatidylcholine/ dipalmitoylphosphatidylcholine (DMPC/DPPC)[60] the excess permeability curve possesses one maximum at the midpoint temperature of the gel–fluid

[54] D. Papahadjopoulos, K. Jacobson, S. Nir, and T. Isac, *Biochim. Biophys. Acta* **311**, 330 (1973).
[55] M. C. Blok, E. C. Van der Neut-Kok, L. L. Van Deenen, and J. De Gier, *Biochim. Biophys. Acta* **406**, 187 (1975).
[56] P. Van Hoogevest, J. De Gier, and B. De Kruijff, *FEBS Lett.* **171**, 160 (1984).
[57] E. M. El-Mashak and T. Y. Tsong, *Biochemistry* **24**, 2884 (1985).
[58] A. Georgallas, J. D. McArthur, X. P. Ma, C. V. Nguyen, G. R. Palmer, M. A. Singer, and M. Y. Tse, *J. Chem. Phys.* **86**, 7218 (1987).
[59] J. Bramhall, J. Hofmann, R. DeGuzman, S. Montestruque, and R. Schell, *Biochemistry* **26**, 6330 (1987).
[60] S. G. Clerc and T. E. Thompson, *Biophys. J.* **68**, 2333 (1995).

transition. There are no data available for the excess permeability of nonideal binary mixtures, such as DMPC/DSPC.

Two-dimensional membrane models attributed the excess permeability to the increased fluctuations of the membrane area close to the gel–fluid transition.[43,61] Three-dimensional membrane models emphasize the role of free volume in membrane permeability and thus one may explain the excess permeability by the increased fluctuations of the membrane volume close to the gel–fluid transition.

According to another idea the acyl chain packing mismatch at the interfacial region between gel and fluid domains is the main reason for the excess permeability.[6-8] In this chapter we refer to these models of membrane permeability as the "area/volume fluctuation" model and "phase mismatch" model, respectively. Excess permeability curves observed for one-component membranes and for the ideal mixtures of DMPC/DPPC are in agreement with the predictions of both of these models.[6-8,43,61]

In this section, by using our simulated data we point out that in the case of the nonideal mixture of DMPC/DSPC there is a discrepancy between the excess permeability curves predicted by these two models.

According to the phase mismatch model the excess permeability is proportional to the average length of the gel–fluid interface $\langle L \rangle$. A normalized average length of the gel–fluid interface can be calculated from our simulated thermodynamic averages as follows:

$$\langle L \rangle / L_0 = [\langle N_{11}^{gl} \rangle + \langle N_{12}^{gl} \rangle + \langle N_{21}^{gl} \rangle + \langle N_{22}^{gl} \rangle]/3N \qquad (19)$$

where L_0 is a normalization constant and $3N$ is the number of nearest-neighbor chain–chain interactions.

The normalized average length of the gel–fluid interface is calculated from the Monte Carlo simulations and plotted in Fig. 9 against the temperature at different DMPC/DSPC mixing ratios. According to the phase mismatch model these curves should be similar to the excess permeability curves measured at the respective mixing ratios.

What kind of excess permeability curves are predicted by the area fluctuation model? It was found experimentally that excess heat and excess volume are proportional functions in the temperature range of the chain-melting transition of one- and two-component lipid membranes.[62,63] As a consequence of this proportionality Heimburg[63] pointed out that the fluctuation of the excess heat and the fluctuation of the excess volume are proportional functions, too, and postulated a similar proportionality between the fluctuation of the excess heat and the fluctuation of the excess

[61] S. Doniach, *J. Chem. Phys.* **68**, 4912 (1978).

[62] F. H. Anthony, R. L. Biltonen, and E. Freire, *Anal. Biochem.* **116**, 161 (1981).

[63] T. Heimburg, *Biochim. Biophys. Acta* **1415**, 147 (1998).

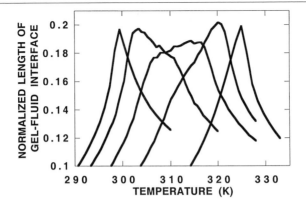

FIG. 9. Average length of gel–fluid interface. The average normalized length of the gel–fluid interface [calculated by Eq. (19)] is plotted against the temperature. The curves from left to right belong to the following DSPC mole fractions: 0.1, 0.3, 0.5, 0.7, 0.9.

membrane surface area. Thus according to the area fluctuation model the excess permeability, ΔP, should be a proportional function of $\Delta C_P RT^2$:

$$\Delta P \sim \langle (\Delta A - \langle \Delta A \rangle)^2 \rangle \sim \langle (\Delta V - \langle \Delta V \rangle)^2 \rangle \sim \langle (\Delta H - \langle \Delta H \rangle)^2 \rangle = \Delta C_P RT^2 \quad (20)$$

The phase mismatch model and the area fluctuation model of excess membrane permeability predict qualitatively different excess permeability curves for certain DMPC/DSPC mixtures. For example, in the case of an equimolar mixture of DMPC/DSPC the predicted excess permeability curve possesses one and two local maxima according to the phase mismatch and the area fluctuation model, respectively. Which model is incorrect? An experimental determination of the excess permeability curve of equimolar DMPC/DSPC membrane would answer this question.

Conclusions

Lateral membrane heterogeneity is related to and affects important membrane functions. In the present chapter a simple two-state model of two-component lipid bilayers (DMPC/DSPC) was utilized to study and demonstrate these structure–function relationships. By using Monte Carlo methods the model simulated the excess heat capacity curves at different DMPC/DSPC mixing ratios in quantitative agreement with the DSC data. The analysis of the calculated membrane energy distribution functions reveals that the gel–fluid transitions of DMPC/DSPC MLVs are continuous transitions and thus in a strict thermodynamic sense one cannot generate a phase diagram of a DMPC/DSPC mixture. The phase diagram commonly constructed from the analyses of the excess heat capacity curves cannot be used to estimate the fractional completion of the transition, θ. We demonstrated that the mechanical application of the lever rule for this phase diagram leads to serious errors in the estimation of θ.

The onset of chemical signaling in membranes can be measured by fluorescence recovery after photobleaching (FRAP). By analyzing the relationship between the FRAP threshold temperatures and the average membrane configurations at different DMPC/DSPC mixing ratios, we found a strong positive correlation between the percolation threshold temperatures of gel clusters and the FRAP threshold temperatures, while the respective correlation was weak for the percolation threshold temperatures of fluid clusters. At the FRAP threshold temperature the probability of percolation for gel clusters was found to be 0.36 at every mixing ratio.

Formation of lipid microdomains, clusters, or "rafts" in cell membranes has been linked to important cell biological processes such as the trafficking and lateral segregation of proteins involved in cellular signal transduction. By simulating the configurations of DMPC/DSPC bilayers the average sizes of the gel and fluid microdomains have been determined at different temperatures and mole fractions. Depending on the temperature and mole fraction the average size of the microdomains varies from 1 to 60 lipid molecules. The calculated average cluster sizes are in agreement with neutron diffraction and atomic force microscopy data.

The "minimal" model of two-component bilayers was used to test the predictions of the existing ideas for excess membrane permeability. The excess membrane permeability for small water-soluble molecules in the gel–fluid coexistence region is explained either by the enhanced membrane area fluctuations or by the mismatch at the interface between the gel and fluid regions. By using our simulations we calculated the excess membrane permeability curves predicted by the above-mentioned two theories. At certain mole fractions the two theories predicted qualitatively different excess permeability curves.

Acknowledgments

We thank Dr. Barbora Piknova, Dr. Thomas Heimburg, Dr. Anne Hinderliter, and Mr. Kim Thompson for providing excess heat capacity data, and Mr. Kevin Kelliher for computational support. Dr. Sugar acknowledges the generous support of Pfizer, Inc., and also thanks Mrs. Lawrence Garner. This work was also supported by grants from the NIH and NSF to Dr. Biltonen.

[16] Parsing Free Energies of Drug–DNA Interactions

By Ihtshamul Haq, Terence C. Jenkins, Babur Z. Chowdhry, Jinsong Ren, and Jonathan B. Chaires

Introduction

A number of clinically important small molecules appear to act by binding directly to DNA, and subsequently inhibiting gene expression or replication by interfering with the enzymes that catalyze these functions.[1-5] The structures of a large number of DNA–drug complexes have been determined at high resolution by X-ray crystallography and nuclear magnetic resonance (NMR), with over 100 structures on deposit in the Nucleic Acid Database.[6] One goal that is being actively pursued by a number of laboratories is the rational design of small molecules that can selectively bind to a chosen DNA sequence with high affinity, thereby providing the means of targeting and selectively inhibiting the expression of any chosen gene. While knowledge of the structures of DNA–drug complexes is of undisputed importance in such a rational design strategy, it is also essential to understand in detail the underlying thermodynamics of the binding processes. Thermodynamics can provide quantitative information that can help elucidate the principal driving forces for the interaction, which can provide insight to guide possible chemical modifications that might enhance both DNA–drug binding affinity and base sequence specificity.

A general problem of considerable current interest is to understand the molecular contributions to ligand–macromolecule binding free energies.[7-12]

[1] E. F. Gale, E. Cundliffe, P. E. Reynolds, M. H. Richmond, and M. J. Waring, "The Molecular Basis of Antibiotic Action," 2nd Ed. John Wiley & Sons, New York, 1981.

[2] L. H. Hurley (ed.), "Advances in DNA Sequence Specific Agents," Vol. 1. JAI Press, Greenwich, Connecticut, 1992.

[3] L. H. Hurley and J. B. Chaires (eds.), "Advances in DNA Sequence Specific Agents," Vol. 2. JAI Press, Greenwich, Connecticut, 1996.

[4] C. L. Propst and T. J. Perun (eds.), "Nucleic Acid Targeted Drug Design." Marcel Dekker, New York, 1992.

[5] T. R. Krugh, *Curr. Opin. Struct. Biol.* **4,** 351 (1994).

[6] H. M. Berman, A. Gelbin, and J. Westbrook, *Prog. Biophys. Mol. Biol.* **66,** 255 (1996).

[7] M. K. Gilson J. A. Given, B. L. Bush, and J. A. McCammon, *Biophys. J.* **72,** 1047 (1997).

[8] T. J. Marrone, J. M. Briggs, and J. A. McCammon, *Annu. Rev. Pharmacol. Toxicol.* **37,** 71 (1997).

[9] J. H. Böhm, *Prog. Biophys. Mol. Biol.* **66,** 197 (1996).

[10] K. P. Murphy and E. Freire, *Pharmaceutical Biotechnology* **7,** 219 (1995).

One aim of this chapter is to offer a perspective on how the binding free energies of DNA–drug interactions might be parsed into the component molecular contributions, using experimental thermodynamic data and the most recent theoretical and empirical estimates for the free energy contributions of specific parameters. As will be seen, this exercise requires accurate experimental determination of heat capacity changes (ΔC_p) for DNA–drug binding reactions by calorimetry and the computation of changes in solvent-accessible surface areas. A practical description of the protocols used in our laboratories to obtain such information follows a brief description of the framework used to parse free energies. More detailed descriptions of specific aspects of the free energy contributions to DNA–drug binding have been presented elsewhere.[13,14]

Models for Drug Binding to DNA

There are two principal binding modes for the noncovalent or reversible interaction of small molecules with duplex-form DNA: intercalation and minor groove binding.[15] Intercalation is a process in which the planar, aromatic portion of a ligand is inserted between contiguous DNA base pairs, with a concomitant lengthening and unwinding of the host DNA helix. In contrast, in the groove-binding mode the structure of the DNA double helix is largely unperturbed, and the ligand simply fits into a binding site within the minor groove conduit. Thus, by analogy with enzyme–substrate interactions, DNA intercalation can be viewed as an "induced fit" mechanism whereas groove binding can be thought of as a "lock-and-key" mechanism.

A conceptual model for the intercalation process is shown in Fig. 1. Formation of an intercalation complex requires at least three distinct steps. In the first step, the DNA must undergo a conformational change to form the intercalation site (Fig. 1A), requiring the axial separation of adjacent base-pair planes to form a cavity into which the intercalating chromophore fits. As the base pairs separate, the DNA is unwound, and the phosphate group spacing in the backbone is increased at the intercalation site. This latter change reduces the local charge density, and results in the release of

---

[11] W. F. Van Gunsteren, P. M. King, and A. E. Mark, *Q. Rev. Biophys.* **27,** 435 (1994).

[12] E. Freire, *Arch. Biochem. Biophys.* **303,** 181 (1993).

[13] J. B. Chaires, *Biopolymers* **44,** 201 (1998).

[14] J. B. Chaires, *Anti-Cancer Drug Des.* **11,** 569 (1996).

[15] W. D. Wilson, *in* "Nucleic Acids in Chemistry and Biology" (G. M. Blackburn and M. J. Gait, eds.), pp. 297–336. IRL Press, Oxford, 1990.

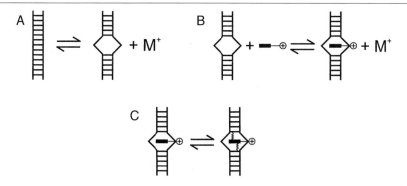

Fɪɢ. 1. Conceptual model for intercalation reactions. (A) An obligatory DNA conformational change to form the intercalation site, with separation of contiguous base pairs, and the concomitant unwinding of the duplex and release of condensed counterions. (B) The hydrophobic transfer of drug from solution into the intercalation site. Further condensed counterions are released if the drug is positively charged. (C) Anchoring of the drug to DNA by the formation of specific molecular interactions (e.g., hydrogen bonds).

condensed counterions.[16,17] In the second step, there is a transfer of the intercalator from solution into the formed intercalation site (Fig. 1B). This step may be thought of as a hydrophobic transfer process, because the nonpolar, planar aromatic ring system typical of an intercalator is removed from solution and becomes buried within the DNA helix. This step is accompanied by additional counterion release if the intercalator is positively charged, a process well accounted for by the polyelectrolyte theories of Manning[18] and Record and co-workers.[19] In the final step, the intercalator is anchored to the intercalation site by the formation of a variety of non-covalent molecular interactions. Specific hydrogen bonds between the bound molecule and DNA may result, together with van der Waals interactions, specific electrostatic bonds, and a variety of stacking interactions. Each of these three steps will contribute to the overall binding free energy for the DNA–drug binding interaction. It should be emphasized that the scheme presented in Fig. 1 is hypothetical, and should not be taken to represent either a detailed or proved reaction mechanism.

Minor groove-binding interactions differ from intercalation processes in at least one significant way. Existing structural evidence suggests that little, if any, change in the DNA conformation is induced by a groove-binding agent and, hence, that the binding sites are essentially preformed to accom-

[16] W. D. Wilson and I. G. Lopp, *Biopolymers* **18**, 3025 (1979).
[17] R. A. Friedman and G. S. Manning, *Biopolymers* **23**, 2671 (1984).
[18] G. S. Manning, *Q. Rev. Biophys.* **11**, 179 (1978).
[19] M. T. Record, Jr., C. F. Anderson, and T. M. Lohman, *Q. Rev. Biophys.* **11**, 103 (1978).

modate an isohelical ligand. A conceptual model for groove binding would consist of only two steps: (1) the hydrophobic transfer of the ligand molecule from solution into the host DNA groove, and (2) the formation of non-covalent molecular contacts between the ligand and DNA substituents accessible in the occupied groove. Hydrogen bonding and van der Waals interactions are probably the major stabilizing factors to be considered.

Free Energy Contributions to Intercalation and Groove-Binding Reactions

For the interactions of an intercalator or groove binder with its DNA-binding site, the free energy $\Delta G°$ is given by the standard Gibbs relationship:

$$\Delta G° = -RT \ln K_b$$

where R is the gas constant, T is the temperature in degrees Kelvin, and K_b is the intrinsic binding constant. How to extract the binding constant from experimental data has been described in detail.[20] The free energy may be partitioned into its enthalpic and entropic contributions by the standard equation:

$$\Delta G° = \Delta H° - T\Delta S°$$

if the enthalpy can be determined by van't Hoff analysis from the temperature dependence of the binding constant, or (preferably) directly by calorimetry. Fundamental thermodynamic relationships are summarized in Table I.

Various schemes exist for parsing the binding free energy into contributions from the processes that lead to complex formation (reviewed in Ref. 13). These schemes all assume the additivity of free energy contributions. For DNA–drug interactions, in view of the conceptual model outlined above, it is reasonable to assume that the binding free energy results from the contributions of (at least) six component terms:

$$\Delta G_{obs} = \Delta G_{conf} + \Delta G_{t+r} + \Delta G_{rot} + \Delta G_{hyd} + \Delta G_{pe} + \Delta G_{mol}$$

The terms have the following meanings. ΔG_{obs} is the experimentally observed binding free energy, estimated from the association constant (K_b) by the standard Gibbs relationship. ΔG_{conf} is the free energy contribution from conformational changes in the DNA and drug induced on complexation, the step shown in Fig. 1A. ΔG_{t+r} is the free energy cost resulting from losses in translational and rotational degrees of freedom on bimolecular complex formation. ΔG_{rot} is the free energy cost due to loss of rotation

[20] J. J. Correia and J. B. Chaires, *Methods Enzymol.* **240**, 593 (1994).

TABLE I
SOME BASIC THERMODYNAMIC RELATIONSHIPS[a]

$$\left(\frac{d\ln K_b}{dT}\right)_P = \frac{\Delta H_{vH}}{RT^2} \text{ (van't Hoff equation)}$$

$$\Delta G° = -RT\ln K_b$$

$$\Delta G° = \Delta H° - T\Delta S°$$

$$\Delta C_p° = d\Delta H°/dT$$

$$\Delta H° = \Delta H_r° + \Delta C_p(T - T_r)$$

$$\Delta S° = \Delta S_r° + \Delta C_p \ln(T/T_r)$$

$$\Delta G° = \Delta H_r° - T\Delta S_r° + \Delta C_p[(T - T_r) - T\ln(T/T_r)]$$

[a] R, the gas constant = 1.9872 cal/(K · mol); K_b, equilibrium constant; T, temperature, in degrees Kelvin (r, reference value); $\Delta G°$, $\Delta H°$, $\Delta S°$, and $\Delta C_p°$, standard changes in free (Gibbs) energy, enthalpy, entropy, and heat capacity, respectively.

around specific bonds in either the drug or the DNA on complex formation. ΔG_{hyd} is the free energy for the hydrophobic transfer of drug from solution into its DNA-binding site [either into the intercalation site (Fig. 1B) or into the minor groove for a groove-binding agent]. ΔG_{pe} is the polyelectrolyte contribution to the binding free energy, arising from the release of condensed counterions from the DNA on formation of the ligand-bound complex.[14] The meaning and experimental determination of ΔG_{pe} are described in detail in Ref. 14. ΔG_{mol} is the free energy contribution resulting from noncovalent intermolecular contacts between the bound drug and the DNA duplex (Fig. 1C). These molecular interactions include hydrogen bond formation, van der Waals interactions, specific electrostatic bond formation, dipole–dipole interactions, and all other possible weak, noncovalent interactions between the drug and the DNA. Table II summarizes these contributions and indicates, where possible, their approximate magnitudes.

These six contributions appear to be a minimal set to begin to parse the free energies of drug binding. Reasonable semiempirical or theoretical estimates for the magnitude of each contribution are available in the published literature, although there is often considerable disagreement over these values in particular cases. A perspective on the magnitude of each contribution for intercalation or groove-binding reactions has been proposed,[13] and should be consulted for a more detailed discussion of each contribution. Formation of a DNA–drug complex results from the balance of large opposing forces. The contributions from conformational changes, the losses of translational and rotational degrees of freedom, and the loss

TABLE II

CONTRIBUTIONS TO FREE ENERGY OF DRUG BINDING TO DNA

Parameter	Description	Approximate value	Refs.
ΔG_{conf}	Free energy cost of DNA and drug conformational changes	Groove binders ≈ 0 kcal/mol Intercalators $\approx +4$ kcal/mol	a–e
ΔG_{r+t}	Entropic cost of bimolecular complex formation due to loss of rotational and translational degrees of freedom	$T\Delta S_{r+t} \approx +15.0$ kcal/mol	f–l
ΔG_{rot}	Entropic cost for restriction of rotational freedom around bonds	$\approx 0.6N$ kcal/mol (N = number of bonds)	m
ΔG_{hyd}	Free energy for the hydrophobic transfer of drug from solution into the DNA-binding site	$(80 \pm 10)\Delta C_p$ or $25.6\Delta A_{np} - 11.2\Delta A_p$[p]	n, o
ΔG_{pe}	Polyelectrolyte contribution to binding free energy	$\Delta G_{pe} = \mathbf{SK}RT \ln[M^+]$ $\{\mathbf{SK} = (\delta \ln K_b / \delta \ln[M^+])\}$[s]	q, r
ΔG_{int}	Contributions from the formation of specific molecular interactions between the drug and DNA (e.g., van der Waals interactions, hydrogen bonds, ionic bonds, etc.)	Problematic; evaluate by systematic changes in drug and DNA substituents thought to be involved in specific molecular interactions	t–x

[a] M. E. Nuss et al., J. Am. Chem. Soc. 101, 825 (1979).

[b] R. L. Ornstein and R. Rein, Biopolymers 18, 1277 (1979).

[c] D. Malhotra and A. J. Hopfinger, Nucleic Acids Res. 8, 5289 (1980).

[d] M. Prabhakaran and S. C. Harvey, Biopolymers 27, 1239 (1988).

[e] A. S. Benight et al., Adv. Biophys. Chem. 5, 1 (1995).

[f] M. Page and W. P. Jencks, Proc. Natl. Acad. Sci. U.S.A. 68, 1678 (1971).

[g] M. I. Page, Angew. Chem. Int. Ed. Engl. 16, 449 (1977).

[h] A. V. Finkelstein and J. Janin, Protein Eng. 3, 1 (1989).

[i] J. Janin, Proteins Struct. Funct. Genet. 24, i (1996).

[j] K. P. Murphy et al., Proteins Struct. Funct. Genet. 18, 63 (1994).

[k] A. Holtzer, Biopolymers 35, 595 (1995).

[l] M. K. Gilson et al., Biophys. J. 72, 1047 (1997).

[m] J. Novotny et al., J. Mol. Biol. 268, 401 (1997).

[n] R. S. Spolar and M. T. Record, Science 263, 777 (1994).

[o] M. T. Record et al., Methods Enzymol. 208, 291 (1991).

[p] ΔC_p is the heat capacity change for the binding reaction, $\Delta C_p = d(\Delta H)/dT$. ΔA_{np} and ΔA_p are the changes in solvent-accessible surface area on binding for nonpolar (np) and polar (p) atoms, respectively.

[q] J. B. Chaires, Anti-Cancer Drug Design 11, 569 (1996).

[r] M. T. Record et al., Q. Rev. Biophys. 11, 103 (1978).

[s] K_b is the binding constant. M^+, monovalent cation; $\mathbf{SK}$, the slope of linear fits to data plotted as $\ln K$ versus $\ln[M^+]$.

[t] J. B. Chaires, Biopolymers 44, 201 (1998).

[u] P. R. Connelly, Curr. Opin. Biotechnol. 5, 381 (1994).

[v] P. D. Ross and S. Subramanian, Biochemistry 20, 3096 (1981).

[w] M. Eftink and R. L. Biltonen, in "Biological Microcalorimetry" (A. E. Beezer, ed.), pp. 343–412. Academic Press, London, 1980.

[x] V. A. Bloomfield, in "Thermodynamics" (E. Freire, ed.), "On-Line Biophysics Textbook" (http://biosci.umn.edu/biophys/OLTB/thermo.html). Biophysical Society, Bethesda, MD, 1998.

of bond rotational freedom are energetically unfavorable. These terms must be balanced and overwhelmed by favorable contributions from the remaining energy terms in order for a stable complex to form. Such an energetic balance sheet defines the key forces that drive the association reaction.

It is becoming increasingly clear that the predominant favorable free energy contribution to DNA–drug binding free energies arises from the hydrophobic contribution, ΔG_{hyd}. The primary focus of the remainder of this chapter is to detail the methodologies involved in obtaining estimates for the ΔG_{hyd} value. The approach used is based on a semiempirical treatment of data for the transfer of small hydrocarbons between aqueous and nonaqueous phases and for a variety of binding and unfolding reactions involving proteins.[21–28] Both heat capacity changes ($\Delta C_p = d\Delta H/dT$) and changes in solvent-accessible surface areas ($\Delta SASA$) may be used to estimate the free energy contribution for the hydrophobic transfer process.[21–28] Ha and co-workers[25] derived the empirical relationship: $\Delta G_{hyd} = (80 \pm 10)\Delta C_p$. If an experimental measure of the heat capacity change is available for the DNA binding of an intercalator or groove binder, this relationship allows the hydrophobic contribution to the binding free energy to be estimated in a simple and direct way. Unfortunately, experimental estimates of ΔC_p are sparse, in part because such values are difficult to obtain. If ΔC_p is small in magnitude, van't Hoff plots are unreliable for accurate determinations of heat capacity changes.[29] Fortunately, isothermal titration calorimetry techniques allow for a more direct determination of ΔC_p, as is described below.

If a value for the heat capacity change is unavailable but the structure of the complex is known, ΔG_{hyd} may be estimated from the change in solvent-accessible surface area on complex formation.[21–28] Janin[21] and Record and co-workers[21–25] have shown that $\Delta G_{hyd} = -(22 \pm 5)\Delta SASA$. While there is some debate over the exact value of the coefficient,[26–28] this equation nevertheless directly relates a key thermodynamic contribution to a structural property. However, the calculation of the binding-induced

[21] J. Janin, *Structure* **5**, 473 (1997)

[22] R. S. Spolar, J. H. Ha, and M. T. Record, Jr., *Proc. Natl. Acad. Sci. USA* **86**, 8382 (1989).

[23] J. R. Livingstone, R. S. Spolar, and M. T. Record, Jr., *Biochemistry* **30**, 4237 (1991).

[24] R. S. Spolar, J. R. Livingstone, and M. T. Record, Jr., *Biochemistry* **31**, 3947 (1992).

[25] J. H. Ha, R. S. Spolar, and M. T. Record, Jr., *J. Mol. Biol.* **209**, 801 (1989).

[26] K. A. Sharp, A. Nicholls, R. Friedman, and B. Honig, *Biochemistry* **30**, 9686 (1991).

[27] K. A. Sharp, A. Nicholls, R. F. Fine, and B. Honig, *Science* **252**, 106 (1991).

[28] K. A. Sharp, *Curr. Opin. Struct. Biol.* **1**, 171 (1991).

[29] J. B. Chaires, *Biophys. Chem.* **64**, 15 (1997).

TABLE III
OUTLINE PROTOCOL FOR PARSING ΔG, AND COMPARISON OF ΔC_p WITH BINDING-INDUCED
CHANGES IN SOLVENT-ACCESSIBLE SURFACE AREA

Biophysical characteristics of DNA and ligand
(structure, conformation, stability and solubility, DNA melting profile, aggregation, etc.)
↓
If possible, establish mode of DNA–ligand interaction (e.g., intercalation or groove binding)
↓
Determine binding stoichiometry for the interaction
(e.g., using continuous variation or Scatchard-type fitting methods)
↓
Conduct equilibrium binding (UV, CD, or fluorescence) studies at different (but fixed) ligand
concentrations by varying the DNA concentration
↓
Examine the salt dependence of K_b in order to determine ΔG_{pe} and ΔG_t
↓
ITC experiments to obtain ΔH and K_b as function of temperature (and, hence, ΔG and ΔS).
Use $d(\Delta H)/dT$ plot to obtain ΔC_p
↓
Calculate the ΔH and $T\Delta S$ contributions to ΔG at a given temperature for molecular insight
↓
Use all thermodynamic information to parse the experimental ΔG into the component terms
↓
Examine any possible link between thermodynamic properties and structure.
Correlate ΔC_p with binding-induced changes in solvent-accessible surface area (SASA):

Structures of DNA, drug and complex known	One or more structures unknown
	↓
	Computer-aided modeling
↓	↓

Compute SASA terms (in Å^2) for each reactant and the DNA–drug complex:
nonpolar (hydrophobic) ΔA_{np} and polar (hydrophilic) ΔA_p components
[calculated $\Delta C_p = (0.32 \pm 0.04)\Delta A_{np} - (0.14 \pm 0.04)\Delta A_p$ cal/(mol·K)]
↓
Draw conclusions from the overall energetics, and the separate ΔC_p versus SASA analysis
↓
Repeat, using structural analogs of the ligand.
Repeat, using different DNA base sequences, or conformations/structures

ΔSASA is not straightforward and requires a careful consideration of several factors, as is described below.

One algorithm for the analysis and interpretation of thermodynamic data obtained for DNA–drug binding interactions is presented in Table III.

Isothermal Titration Calorimetry

Isothermal titration calorimetry (ITC) is a powerful and versatile technique that can be used to measure directly the interaction enthalpy (ΔH), in a model-independent manner, for almost any bimolecular process.[30] Within certain limits, and depending on experimental design, this technique can also be used to obtain the equilibrium constant (K_b) and the stoichiometry (n) for an intermolecular binding interaction. In principle, it is therefore possible to obtain a complete thermodynamic profile for a DNA–drug binding reaction as a function of temperature. Modern computer-controlled ITC instrumentation often makes data acquisition a relatively simple process. However, to obtain accurate and reliable results a number of important factors must be considered when designing the experiment. In this section we discuss these points, how they interrelate, and how they can be addressed so that users of ITC can construct an effective protocol for their experiments.

Instrumentation

The term "isothermal microcalorimeter" is commonly used for calorimeters designed to work in the nanowatt (nW) to microwatt (uW) range under essentially isothermal conditions. Most isothermal titration microcalorimeters are either heat conduction [Calorimetry Sciences Corporation (CSC) (Spanish Fork, UT; www.calscorp.com)] or power compensation instruments [CSC and MicroCal (Northampton, MA; www.microcalorimetry. com)]. They are designed as twin-vessel (reaction and reference vessels) differential instruments with either "insertion vessels" (CSC) or permanently mounted vessels (CSC and MicroCal) in the heat-sensitive zone of the calorimeter.

A number of manufacturers now produce calorimeters with sufficient sensitivity to determine enthalpies associated with the interactions of biological molecules. However, most studies reported in the biochemical/biophysical literature have used instruments sourced from either CSC or MicroCal. The titration calorimeters from each company are of comparable sensitivity. The use and applications of the MicroCal instrument have been adequately described elsewhere.[30] The first commercially available high-sensitivity titration calorimeter from MicroCal was the Omega unit introduced in 1989.[30] This instrument was subsequently modified and automated, and the ITC unit was incorporated into a combined microcalorimetry system (MCS), which also includes a high-sensitivity differential scanning calorimeter. Some of the advantages and disadvantages of this system are summarized as follows:

Advantages

1. Computer control and automation make the MCS user friendly, and instrument operation is straightforward.

---

[30] T. Wiseman, S. Williston, J. F. Brandts, and L.-N. Lin, *Anal. Biochem.* **179,** 131 (1989).

2. Experiments can be carried out relatively quickly, typically in 60–90 min, because of rapid equilibration.
3. Baselines are stable, typically 0.02 μcal/sec constancy over a 15-min period, with stirring
4. In principle the MCS ITC has a useful temperature range of 2–80°. However, this may depend on the ambient room temperature.
5. The Origin software used for data acquisition and analysis is powerful and versatile.

Disadvantages

1. Cell volume ($\approx$1.3 ml) and comparatively high sample concentrations require relatively large amounts of macromolecule.
2. The instrument can be sensitive to vibrations and electrical interference from other devices. This can lead to noisy baselines.
3. The desiccant is sometimes ineffective. Thus, working at low temperatures can be problematic because of condensation within the instrument, especially if ambient humidity is high.
4. The plunger in the injection syringe loosens over time. This can make the motorized injection assembly unable to locate the syringe plunger at the start of a titration.

Sample Preparation and Analysis

To successfully use ITC to determine thermodynamic parameters, including heat capacity changes, it is essential to ensure that both the DNA of interest and the candidate ligand are adequately characterized and prepared. Both components to be used in the ITC experiment are described below.

DNA. As both the preparation and purification of synthetic oligonucleotides (both DNA and RNA) have now become relatively straightforward procedures, these aspects are not discussed in the present chapter. High-performance liquid chromatography (HPLC) purification should be used to ensure homogeneity, and the samples should be analyzed to ensure greater than 99% purity. In this respect, techniques such as capillary electrophoresis (CE), particularly where denaturing conditions are used to remove secondary structure effects, are particularly recommended. Clearly, the presence of any contaminating molecules could lead to difficulties in carrying out ITC experiments. After the DNA sample has been prepared it can be dissolved in an appropriate aqueous buffer. The choice of buffer conditions will necessarily depend on the experimental goals but, in general, the buffer should be of sufficient ionic strength and buffering capacity to ensure that the DNA is conformationally stable at the pH and temperature range to be used in the ITC experiment. When the DNA is made up into solution it should be dialyzed for at least 24 hr against a large volume of the dialysis buffer, to ensure that the ionic conditions are precisely known. After dialysis it is generally necessary to anneal the DNA; this is conveniently achieved by heating the solution to ~95° and then slowly cooling it back to room temperature over an 18- to 24-hr period. This can be accomplished by using either a standard laboratory water bath or a programmable heating block. It is important to anneal the DNA in order to optimize formation of the particular DNA species under study (e. g., duplex, triplex, or tetraplex DNA). For example, if the structure to be examined were an intramolecular hairpin duplex then formation of such a species could be aided by heating to 95° and rapidly quenching ("snap cooling") the DNA solution in dry ice.

It is necessary to accurately quantify the DNA and ligand solutions so that the ITC experiment can be designed to produce a complete binding isotherm. DNA concentrations are usually determined spectroscopically using calculated molar extinction coefficients. These values can be determined by a number of methods. Most commonly the extinction coefficients for the 4 mono- and 16 dinucleotide possibilities are used to calculate the overall extinction coefficient for the oligonucleotide.[31] However, calculated extinction coefficient values have often been shown to differ from experimentally determined values by as much as 20%.[32] It is therefore desirable to accurately determine an extinction coefficient for an oligonucleotide by examining the hyperchromicity resulting from exhaustive hydrolysis with enzymes such as P1 nuclease or snake venom phosphodiesterase/DNase I.[32] Only small quantities of DNA are required for this assay and the experiments are straightforward.

If enzyme-based methods cannot be used, then, as a minimum, the calculated extinction coefficient should be applied to absorbance values measured at temperatures well above the melting transition for the DNA, because these values refer only to the single-stranded species. The reduction in optical absorbance that occurs as the DNA reanneals can be used to calculate the hyperchromic change and an appropriate correction can then be applied to the computed extinction coefficient to enable absorbance measurement for the duplex at room temperatures. For example, consider the AT-tract 12-mer sequence 5'-CGCAAATTTGCG. Using data from Refs. 31 and 33, the extinction coefficient calculated at 260 nm is 1.12 ×

[31] P. N. Borer, in "Handbook of Biochemistry and Molecular Biology, Nucleic Acids" (G. D. Fasman, ed.), 3rd Ed., pp. 589–590. CRC Press, Boca Raton, Florida, 1975.
[32] G. Kallansrud and B. Ward, *Anal. Biochem.* **236,** 134 (1996).
[33] C. R. Cantor, M. M. Warshaw, and H. Shapiro, *Biopolymers* **9,** 1059 (1970).

10^5 M(strands)$^{-1}$ cm^{-1}. If this value were to be used to determine single-strand concentrations at room temperature from absorbancies of duplex DNA, the result would be incorrect as it does not take into account the reduction in absorbance on self-association (i.e., duplex formation). In fact, the melting of this particular self-complementary duplex results in a hyperchromism of 14%; hence, the corrected value for the extinction coefficient at room temperature is instead 9.83 $\times$ 10^4 M(strands)$^{-1}$ cm^{-1}.

Depending on the exact base sequence of the DNA oligomer being used it may also be possible for the DNA to exist in a number of alternative salt- and concentration-dependent conformations. If the sequence studied is capable of forming two or more conformations then quantitation of the binding enthalpy would be difficult because the concentration of the conformer examined will be unknown. For example, the widely studied self-complementary dodecanucleotide d(CGCGAATTCGCG) can exist in solution as an equilibrium mixture of monomolecular hairpin and bimolecular duplex species.[34] Clearly, it is desirable to characterize such equilibria prior to their use in ITC studies. If conditions can be established that will minimize the formation of hairpin structure then a more accurate enthalpy for duplex binding (under those conditions only) can be measured as possible nonspecific binding to the hairpin will be eliminated. In this example, high salt and DNA strand concentrations favor duplex formation. Even sequences of the type 5'-CGCAAAAGCG can form equilibrium mixtures of single-stranded hairpin species and bulged duplexes. Furthermore, in studies in which the enthalpy is to be measured over a large temperature range it is necessary to determine an upper temperature limit for these experiments. This limit is indicated by the start of the DNA thermal melting transition. Both the examination of competing structures and the estimation of high-temperature limits can be readily achieved by conducting UV melts of the DNA prior to any use in ITC. Biphasic melting of the sequence under study is indicative of the presence of competing structures and/or contaminants. Further, carrying out UV melts as a function of both strand and salt concentration will allow optimal conditions for ITC to be established.

Polynucleotides or natural DNA sequences are often used in binding experiments, particularly as part of preliminary or screening studies. These less-defined or "pseudo-random" systems may lead to complex stoichiometries and/or multiple binding affinities. However, synthetic polydeoxynucleotides and natural DNA sequences are normally less expensive than

[34] L. A. Marky, K. S. Blumenfeld, S. Kozlowski, and K. J. Breslauer, *Biopolymers* **22**, 1247 (1983).

synthetic oligonucleotides and as such their use in DNA–drug binding studies is justified. The choice of DNA will ultimately depend on the information sought from the ITC experiment.

Ligand. In comparison with spectrophotometric techniques the ligand concentrations required for ITC are normally relatively high. The exact concentration used depends on factors such as the enthalpy change being evaluated, the magnitude of K_b, the concentration of macromolecule in the calorimeter cell, and the binding stoichiometry. The high concentration of ligand often required can be problematic for two main reasons. First, many ligands of biological interest are of only limited aqueous solubility. It is essential that both the ligand and DNA can separately be dissolved in the same buffer, normally the dialysis buffer that was used to anneal the DNA prior to the ITC experiment. The salt concentration used in the buffer may have profound effects on ligand solubility; as a consequence, many candidate drugs are excluded from binding studies using ITC techniques. In most cases it is impractical to conduct titrations in heterogeneous solvent mixtures, not least because the physical mixing of even slightly different concentrations of solvent usually leads to a large heat signal that masks the smaller heats of binding.

Second, many drugs that are soluble at the concentrations required for ITC exhibit either aggregation or some measure of self-association. This is frequently the case for drugs containing planar aromatic chromophores, for example, "classic" DNA intercalators. Self-association can become evident from titration of a solution of the drug into buffer solution (i.e., dilution) using ITC. If the heat signal is dependent on the ligand concentration it is probable that self-association is occurring. The self-association process can also be assessed qualitatively and quantitatively by examining the effect of drug concentration on absorbance or fluorescence. It may be possible to use such experiments to define an upper limit for drug concentration such that the compound remains monomeric throughout the titration experiment. Of course, ITC itself can be used to examine aggregation phenomena and to quantify the enthalpies of drug dissociation. If aggregation is a problem then, depending on the exact nature of the interaction, it may be possible to use the compound at lower concentrations in the cell and instead to inject the DNA from the syringe (i.e., a "reverse-titration" experiment).

As with DNA it is essential to accurately determine the ligand concentration before use. Wherever possible this should be done spectroscopically, for example, by using either a published extinction coefficient or by determining the extinction coefficient experimentally. Additional problems may also arise owing to nonspecific adsorption of the ligand to glass or plasticware surfaces. This may be avoided either by using only surfaces to which the drug will not adsorb or by precoating vials and UV cells with siliconizing solutions (e. g., Sigmacote; Sigma, St. Louis, MO).

Finally, having prepared both the DNA and the ligand it is important to remove any dissolved gas from the reagent solutions, either by using a vacuum chamber or by applying a vacuum and stirring. After degassing (deaeration), the sample solutions should be reweighed to ensure that the concentrations of materials have not altered. Degassing the samples will avoid problems due to the formation of gas bubbles during the course of a titration and is particularly important if the solutions have been refrigerated prior to their use.

Planning Isothermal Titration Calorimetry Experiment

Consideration of Determined Binding Constant

Several methods can be used to obtain K_b values, differing in the experimental quantity that is measured, the requirements of the DNA and drug samples, and the range of association constants that can be addressed. If possible, it is advisable to use more than one method to obtain the K_b value. Using ITC, the binding constant can be accurately determined only if the drug (titrant) is added to a fixed and constant concentration $[D_0]$ of DNA (titrate) such that $[D_0] \ll 1/K_b$. When $[D_0] \gg 1/K_b$ an accurate determination of the K_b value is not possible because of the sensitivity limit of ITC instruments.[30]

DNA Concentration

Considering the simplest case, where the DNA has a single set of identical binding sites, a calorimetric titration should result in a sigmoidal binding isotherm from which ΔH, K_b, and n can be obtained. The precise shape of the binding curve is dictated by a unitless parameter known as the c value,[30] defined as the compound product of K_b, the total DNA concentration in the cell and the stoichiometry (n) for the interaction. For titrations with $c < 1$ the binding curve will be essentially linear and featureless, whereas the binding isotherm becomes rectangular if $c > 500$. When designing the ITC experiment and choosing appropriate DNA and drug concentrations it is desirable to avoid both these extremes. A low c value is indicative of weak binding and under these circumstances it is necessary to increase the DNA concentration in the cell in order to obtain a higher c value. The converse is true if the c parameter is high (i.e., tight binding); in this case, the DNA concentration used in the experiment should be decreased to produce a less sharp binding isotherm. However, the DNA concentration can be lowered only to a critical point defined by the sensitivity of the calorimeter, and typically 5–10 μcal of heat per injection.

An appropriate DNA concentration should be used such that the c value is in the 10 to 500 range. Clearly, the exact value of c will be difficult to define in initial experiments unless the value for K_b is known. In general, it must be remembered that ITC can reliably measure binding constants only in the 10^4–10^7 M^{-1} range. In practice it is often necessary to perform trial experiments using different concentration regimens until a situation is reached in which the c value is within the optimal working range. If either the value of K_b or its approximate magnitude is known then the task of choosing the working DNA concentration is simpler. For example, if the interaction to be studied has an approximate K_b of 10^6 M^{-1} and the stoichiometry is 1:1, then the range of macromolecule concentration that could normally be used is 10 μM ($c = 10$) to 0.5 mM ($c = 500$). Of course, the exact concentration used will necessarily depend on factors such as availability of materials, the magnitude of the enthalpy, and the solubilities of the two reactants. The ligand solubility must always be considered when selecting the initial DNA concentration; for example, if the DNA concentration is fixed at 0.5 mM the ligand would then have to be 5 mM in the syringe and would be problematic for many established DNA-binding ligands. In practice, the DNA concentration is kept to a minimum for reasons of ligand solubility and (importantly) economy with the nucleic acid.

Ligand Concentration, Number of Injections, and Injection Volume

Once the appropriate DNA concentration has been selected, the concentration of the ligand titrant required is dependent on the molar amount to be injected, and is dictated by the concentration, total volume, and number of injections. Most commercial ITC instruments are supplied with a selection of syringes ranging in volume from 50 to 250 μl. To obtain a complete binding isotherm, the final total concentration of ligand at the end of the experiment should generally be twice the initial concentration of binding sites used. Hence, the concentration of ligand can be determined using the following relationship:

$$[L](I_v/V) = 2n[S]$$

where [L] is the concentration of ligand in the syringe, I_v is the total volume of ligand injected into the cell, V is the (fixed) cell volume, n is the binding stoichiometry, and [S] is the total concentration of binding sites. The total volume of ligand to be injected into the cell will depend on both the size of the injection syringe used and the ligand solubility. For example, if [S] is set at 30 μM and the ligand concentration is kept to a minimum it would be appropriate to add 240 μl of ligand solution to the cell during the course of the titration. Assuming a cell volume of 1.4 ml and a 1:1 stoichiometry, [L] would then have to be 0.35 mM. In this case the titration would be set up using a 250-μl syringe that is programmed to make 16 serial injections of 15-μl aliquots. Alternatively it is also possible to inject a total of 108 μl of ligand solution; here, [L] would be 0.78 mM and the titration could be carried out with 18 serial injections of 6 μl.

Other Considerations

Cleaning, Degassing. There are a number of minor (but important) practical steps that should be taken when performing ITC experiments. Regular cleaning of the calorimeter cells should be undertaken using either mild detergents or by following the manufacturer instructions. Prior to loading the sample cell with the DNA solution, the cell should be thoroughly dried using either a nitrogen gas stream or a vacuum line. This will avoid alterations in the sample concentration during the loading step. Care should be exercised when loading the cells not to trap air bubbles as their presence may lead to a noisy baseline and/or occasional data spikes. If the thermostat temperature of the calorimeter is altered, for example as required for ΔC_p heat capacity determinations, the instrument should be allowed to equilibrate for 24 hr prior to use to prevent untoward experimental baseline drift. Carrying out low-temperature experiments can be problematic if the ambient room temperature and humidity are high. This problem can be alleviated to some extent by using a desiccant to ensure that the gas flow within the calorimeter is dry, and by precooling samples to the required experimental temperature.

Instrument Calibration. There is currently no agreed international standard for the calibration of titration calorimeters. However, common practice is to carry out both electrical and chemical calibration of the instrument on a regular basis. Electrical calibration is carried out by measuring the area under a heat pulse of known applied power and comparing it with the calculated area necessary for that pulse; the two values should agree to within 0.1%. In addition, the ITC instrument can be used to measure standard reactions for which the enthalpies have previously been reported. The ionization of Tris (i.e., tris[hydroxymethyl]aminomethane) is a standard reaction for which the enthalpy values have been reported over a wide range of temperature.[35] For example, the protonation of a 0.02 M solution of ultrapure Tris at 25° using standard 0.0997 M hydrochloric acid should

---

[35] I. Grenthe, H. Ots, and O. Ginstrup, *Acta Chem. Scand.* **24**, 1067 (1970).

give an enthalpy change of -11.34 kcal/mol. If the measured enthalpies should disagree with reported values, then the calibration constants for the instrument must be adjusted until satisfactory agreement is achieved. Other test reactions for calibration have also been suggested. The neutralization of NaOH by HCl is an especially useful calibration reaction. Both reagents are available as highly pure standarized solutions from Aldrich (St. Louis, MO). An enthalpy value of 13.63 kcal/mol at 20° has been accurately determined.[35] For the calibration test, a 0.0995 N volumetric standard solution of NaOH in water is used as the titrate in the cell. To this is added 3-μl injections of a volumetric standard solution of 0.1036 N HCl in water. Data for up to 20 injections in a single run are collected and averaged for comparison with the known enthalpy value.

Possible Errors in Isothermal Titration Calorimetry Method

Calorimeters are uniquely sensitive to systematic errors as practically all processes are accompanied by heat effects. In fact, errors resulting from effects such as evaporation, condensation, adsorption, and incomplete mixing can be particularly problematic. When carrying out ITC studies, to determine standard thermodynamic parameters, it is good practice to repeat the experiments several times in order to obtain consistent results for ΔH, K_b, and n. This allows the standard deviation of the data about the mean to be calculated and used to obtain the error in determined values for binding parameters. The errors for each individual experiment should also be expressed. Error values are obtained directly from the fitting procedure and are effectively a measure of the goodness of fit between the experimental and model-fitted data.

Alternative Protocols

Reverse Titrations. If the interaction is a "simple" 1:1 binding process, it makes little difference whether the ligand is in the syringe or the cell. However, the concentrations of each will obviously have to be adjusted accordingly if a reverse titration is used. For a 1:1 interaction it may be appropriate to carry out a reverse titration if either the ligand is not very water soluble or if self-association is a problem, because lower concentrations can then be used in the calorimeter cell.

Titration in Presence of Excess Binding Sites. In many cases it is useful, even preferable, to titrate the ligand into a solution containing a large excess of binding sites. In such a protocol, ITC is used to obtain the binding enthalpy as directly as possible, and without recourse to model-dependent curve fitting. In such a procedure, a high concentration of DNA (typically 1 mM in base pairs) is placed in the cell to ensure that the $c \geq 1000$. Under such conditions, binding is stoichiometric and all added ligand is bound. Integration of each peak in a series of injections should yield a constant value for the heat per injection. Normalization of the heat per injection with respect to the total number of moles of ligand added per injection yields independent estimates for the binding enthalpy (ΔH_b) without any assumptions regarding the binding model and without recourse to fitting procedures. Values of ΔH_b accumulated from one or more titrations provide a substantial sample size for averaging, yielding statistically reliable error estimates. Such error estimates are crucial when attempting to determine heat capacity changes from the temperature dependence of ΔH_b.

Example of Calorimetric Data Acquisition and Analysis

Figure 2A shows the results of an ITC experiment carried out at 14.8° for the binding of the minor groove dye Hoechst 33258 to a 12 base-pair

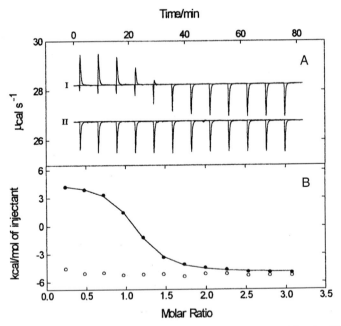

FIG. 2. (A)I: Raw calorimetric data for the titration of 0.29 mM Hoechst 33258 into 28 μM(duplex) A3T3 solution at 14.8°. II: Analogous titration of the identical solution ligand into a buffer containing 8 mM Na$_2$HPO$_4$, 1 mM EDTA, 185 mM NaCl, pH 7.0. (B) Integration of the data shown in Fig. 2A(II) produces the dilution enthalpies for the Hoechst 33258 dye (open circles). Similarly, integration of the peaks in Fig. 2A(I) gives the corrected binding isotherm (closed circles). The solid line represents the fit of the experimental data obtained using nonlinear least-squares fitting to a single set of identical binding sites model (see text).

duplex DNA, d(CGCAAATTTGCG)$_2$, termed A3T3. This experiment used a MicroCal MCS calorimeter. The results of two separate ITC runs are shown: part I of Fig. 2A corresponds to the titration of the Hoechst 33258 into a DNA solution, and part II of Fig. 2A shows the titration for the addition of an identical ligand solution into the phosphate buffer alone. In this example, the duplex solution had been carefully dialyzed against this same phosphate buffer. The peaks correspond to the power required to maintain the sample and reference cells at identical temperatures on each serial injection. The experiment shown in Fig. 2A was set up using a solution of A3T3 in the sample cell at 28 μM(duplex), and 0.29 mM dye solution in the syringe. On the basis of these concentrations, the titration was set up to make 12 serial injections of 20 μl at 300-sec intervals. These concentrations were selected assuming an estimated lower limit of $K_b \sim 10^6 M^{-1}$ for the binding constant, and a 1 : 1 stoichiometry for complexation (A3T3 contains only one binding site for this ligand)[36]; hence, the calculated c value is ~28 and in the optimal 10 to 500 range for ITC.

Having obtained the raw data, the next step is to proceed through each titration peak in turn to establish a baseline from which the peak will be integrated. The limits and shape of the baseline should be manually adjusted until the peak integration is an accurate reflection of the area under the curve. Anomalous spikes, noise, or deviations can be excluded from the area calculations. The baseline added to the raw data in the present example is also shown in Fig. 2A.

The integral with respect to time for each peak corresponds to the enthalpy of interaction, measured in units of microcalories per injection. This value is then divided by the concentration of titrant in the syringe, and by the volume of each injection; the result of this calculation is the interaction enthalpy measured in calories per mole of injectant. As the DNA concentration in the cell is known, the enthalpy can now be plotted against molar ratio to give the binding isotherm on a molarity basis. This procedure has been carried out on the data shown in Fig. 2A, and the resulting binding curve and the ligand dilution curve are shown in Fig. 2B. The enthalpies associated with ligand dilution (Fig. 2B, open circles) are averaged and this value is subtracted from the experimental binding isotherm. The heats associated with dilution of the host DNA should also be determined in a separate experiment and similarly subtracted from the binding curve. However, these enthalpy values are normally small and require only a small correction term.

Once the corrected binding isotherm has been obtained it can be fitted

---

[36] I. Haq, J. E. Ladbury, B. Z. Chowdhry, T. C. Jenkins, and J. B. Chaires, *J. Mol. Biol.* **271**, 244 (1997).

using nonlinear least-squares analysis for a model that adequately describes the interaction. In this example, there is no indication of cooperativity and the sigmoidal binding isotherm can be fitted using a model that assumes a single set of identical binding sites. Because the concentration of DNA used is expressed in terms of duplex the binding stoichiometry should be equimolar. Qualitative examination of the binding curve provides confirmation, as the midpoint of the curve is at an approximate 1 : 1 ratio. Quantitative fitting of these data is achieved by making initial estimates of $\Delta H°$, K_b, and n, and then computing the change in heat content of the cell (ΔQ_i) after each injection (allowing for displaced volume following an injection) and comparing these values with the measured heat for the corresponding injection. The fits are improved by using iterative Marquardt-type algorithms which change the values of the variables on the basis of the results of the previous step. Multiple iterations are continued until there is no further improvement to the fit. For our A3T3–dye example, the result of this fitting procedure is shown in Fig. 2B together with the experimental data points. Models have been developed to fit binding data for more complex interactions that may involve multiple or interacting binding sites. These fitting procedures are beyond the scope of this review, but have been detailed elsewhere.[30,37,38]

Applying this procedure to the data shown in Fig. 2B, using Origin 3.5 software (MicroCal), the following thermodynamic profile is obtained: $\Delta H° = +9.6 \pm 0.1$ kcal/mol, $K_b = (1.1 \pm 0.1) \times 10^6 \ M(\text{duplex})^{-1}$, and $n = 1.04 \pm 0.03$.[36] Using these parameters, values can be determined for $\Delta G°$ and $\Delta S°$ by applying the standard thermodynamic relationships. To measure the binding-induced heat capacity change this experiment must be repeated at a number of temperatures involving a spanned range of at least 15 to 20°. The highest temperature that can be used for the ITC experiments will depend on the inherent thermal stability of the host DNA, usually dictated by the onset of the melting transition. The slope of a plot of ΔH versus T for the experimental enthalpies determined at each temperature gives the ΔC_p value directly.

Figure 3A shows an example of an ITC titration using the "excess site" protocol. In this experiment, the CSC calorimeter was used. Ethidium bromide was titrated into a 1 mM(bp) solution of calf thymus DNA, a concentration high enough to ensure a large excess of potential binding sites. Figure 3B shows the distribution of ΔH values obtained from three separate titration experiments, each with 20 injections. The total of 60

[37] J. Wyman and S. J. Gill, "Binding and Linkage: Functional Chemistry of Biological Macromolecules." University Science Books, Sausalito, California, 1990.
[38] E. Freire, O. L. Mayorga, and M. Straume, *Anal. Chem.* **62**, 950 (1990).

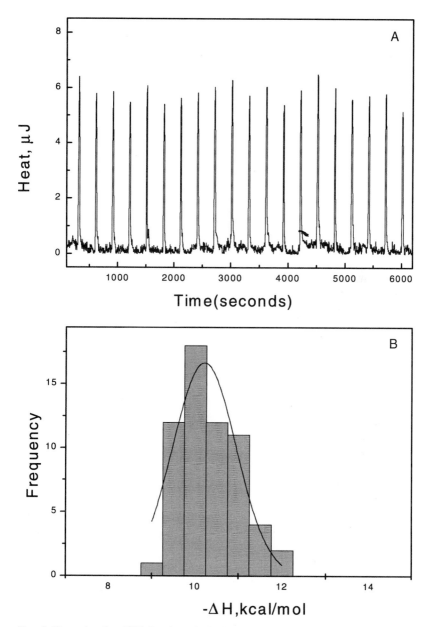

Fig. 3. Example of an ITC titration of ethidium into a solution containing an excess of DNA-binding sites. (A) Primary data from an experiment at 35° in which a series of 3-μl injections of a 1 mM ethidium solution were made into 1 ml of 1 mM(bp) calf thymus DNA in a buffer containing 8 mM Na_2HPO_4, 1 mM EDTA, 185 mM NaCl, pH 7.0. Heats of dilution for ethidium under these conditions were found to be negligible. (B) Distribution of enthalpy values obtained after integration of peaks and normalization for the total moles of ethidium added per injection. Data accumulated from three titration experiments are shown.

independent ΔH values was used to determine the mean and standard deviation, and inspection of the histogram in Fig. 3B confirmed that the values followed the expected Gaussian distribution. A value of $\Delta H° = -10.3 \pm 0.3$ kcal/mol was obtained for the interaction of ethidium with calf thymus DNA at 35°.

Figure 4 shows the determination of ΔC_p for the DNA binding of Hoechst 33258 and ethidium. The linear least-squares slopes of the lines yield ΔC_p values of -330 (±50) and -139 (±10) cal/mol per degree for the DNA binding of Hoechst 33258 and ethidium, respectively. From the relation $\Delta G_{hyd} = (80 \pm 10)\Delta C_p$,[25] estimates for the free energy of hydro-

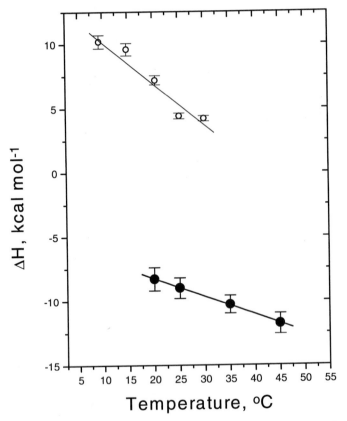

FIG. 4. Determination of ΔC_p from the temperature dependence of $\Delta H°$. Data for Hoechst 33258 (open circles) and ethidium (closed circles) are shown. The lines are the least-squares linear fits to the data, yielding slopes of -330 (±50) and -139 (±10) cal/(mol-deg) for Hoechst 33258 and ethidium, respectively.

phobic transfer of −26.4 (±7.0) and −11.1 (±2.1) kcal/mol are obtained for the Hoechst 33258 dye and ethidium, respectively.

Determination of Solvent-Accessible Surface Areas

The calculation of binding-induced changes in solvent-accessible surface area (ΔSASA) requires that reliable structural information be available for each of the DNA and ligand reactants, and for the formed DNA–drug complex. Such data are usually obtained from X-ray crystallographic or solution NMR studies, where the necessary information is deposited in on-line databases for published structures. Molecular modeling techniques can also be used to obtain plausible structures, particularly where structural data are unreported for one or more system components. In the absence of coordinates for an unbound ligand, for example, the free molecule may often be approximated by separating the ligand from a reported complex. As detailed below, several modern computer programs include a facility to compute SASA values from input coordinate files for any molecule using published algorithms. However, it is also necessary to partition the solvent (water)-accessible molecular surface into the component hydrophilic and hydrophobic surfaces. One analytical procedure used successfully in our laboratories for a number of DNA–drug interactions is described.

Obtaining Structures from Databases

High-quality structural coordinates can be obtained from on-line data-bases for a large number of nucleic acids, DNA/RNA–ligand complexes, and candidate ligand molecules, where the structures have been deposited after either X-ray crystal or NMR solution refinement. Such databases can be searched for DNA type and length, base sequence, or the name/structure of the compound of interest. If available, an atomic coordinate file can then be downloaded to a local computer together with any relevant addi-tional information. Reliable sources include the Nucleic Acid Database[39] (NDB; *ndbserver.rutgers.edu*), the European Bioinformatics Institute (EBI; *www.ebi.ac.uk*), and the Chemical Database Service (CSSR; *cds1.dl.ac.uk*). The Nucleic Acid Database is particularly recommended as the deposited information has been rigorously checked both to verify the structures and to standardize the atomic naming system used for the DNA/RNA compo-nents.[39]

[39] H. M. Berman, W. K. Olson, D. L. Beveridge, J. Westbrook, A. Gelbin, T. Demeny, S. H. Hsieh, A. R. Srinivasan, and B. Schneider, *Biophys. J.* **63**, 751 (1992).

Coordinates are normally deposited only for the nonhydrogen (heavy) atoms present in the structure, unless these were established by refinement at the resolution used or were derived explicitly from an ^{1}H NMR study. In such cases, either molecular modeling or computer-based methods can be used to restore the missing hydrogen atoms, using construction techniques based on standard bond length and angle geometries. Most modern molecular graphics programs, particularly those capable of handling large assemblies (e.g., >500 atoms), include facilities to correctly generate missing hydrogen atoms for database structures. Recommended commercial software packages, operating on a wide variety of computer platforms, include MacroModel (Columbia University, New York, NY; *www.columbia.edu/ cu/chemistry/mmod*), HyperChem (Hypercube, Gainesville, FL; *www. hyper.com*), Insight/Discover and QUANTA (Molecular Simulations, San Diego, CA; *www.msi.com*), SYBYL (Tripos, St. Louis, MO; *www.tripos. com*), and X-PLOR (*http://xplor.csb.yale.edu*).[40] Such programs interpret the base sequence information present in the coordinate file to restore any missing atoms by comparison with "quality-standard" base and sugar residues.[41] Care should be taken to ensure that the DNA or RNA strands are appropriately terminated with, for example, 5′-OH and 3′-OH (and 2′-OH for RNA) sugar hydroxyl groups.

Hydrogen atom generation techniques must similarly be used for the free and bound ligand molecules in any complex, using a molecular graphics package. One recommended alternative is to use the versatile Babel program, a nongraphics package available for all platforms (P. Walters and M. Stahl, University of Arizona, Tucson, AZ; *http://mercury.aichem. arizona.edu/babel.html*), to generate these atoms by protocols that evaluate both the molecular connectivity and atomic valencies.

Finally, crystallographic structures are often deposited in only partial form where, for example, the two parts of the structure are simply related by a symmetry operator. In such situations, the whole coordinate set is obtained by an import procedure within the graphics package that uses the matrix and/or space group information embedded in the downloaded file.

Generating Model Structures for the DNA, Ligand,
and DNA–Ligand Complex

It is often not possible to obtain coordinates from a structural database when the DNA–ligand complex is either unknown or (more usually) one or

[40] A. T. Brünger, "X-PLOR, version 3.1: A System for X-Ray Crystallography and NMR." Yale University Press, New Haven, Connecticut, 1992.
[41] G. Parkinson, J. Vojtechovsky, L. Clowney, A. T. Brünger, and H. M. Berman, *Acta Crystallogr. Sect. D. Biol. Crytallogr.* **52,** 57 (1996).

more component has not been reported. Fortunately, most macromolecular modeling programs include a facility to generate structures for model nucleic acid duplexes of any given base sequence, assuming either a regular A- or B-type backbone geometry. It is usual to select A-RNA and B-DNA conformations as aqueous solution conditions are used for ITC experiments, although models can also be generated for the nonstandard structures where appropriate or for comparison. Dedicated computer programs are also available to generate DNA or RNA structures that incorporate additional features (e.g., chimeric structures, as in GENHELIX-2: *t.c.jenkins@ bradford.ac.uk*).

Model coordinate files for an unbound ligand can be generated by regular molecular construction methods that use standard geometries, or by adaptation of deposited structures for closely similar molecules. It is recommended that such models be subjected to an energy minimization procedure before use to regularize the overall geometry and remove untoward steric clashes. All-atom molecular mechanics refinement to a final root mean square (rms) energy gradient of ≤ 0.1 kcal/(mol · Å) should suffice.

It is clearly a straightforward exercise to obtain reliable structures for both the native DNA (or RNA) duplex and the free ligand molecule. However, if X-ray crystal or NMR coordinates are unavailable for the DNA–ligand complex it will be necessary to generate appropriate model structures using graphics-based molecular modeling techniques. Docking procedures suitable for DNA intercalation[42] or minor groove[43,44] binding models have been detailed elsewhere. The suitability of any proposed model will necessarily be dependent on any biophysical data to support the binding mechanism, stoichiometry, and location of the bound ligand. In the absence of supportive information, such model complexes must necessarily be viewed as only qualitative. Nevertheless, these structural models can be evaluated in order to rank or predict binding-induced changes in SASA for comparison with experimental thermodynamic data. In certain cases, this process may be used to eliminate unlikely modes of ligand complexation.

Computer Methods for Calculation of Solvent-Accessible Surface Areas

Several computer packages are available to compute the solvent-accessible surface areas (SASAs) for molecules, including GRASP (Ref. 45;

[42] S. Neidle and T. C. Jenkins, *Methods Enzymol.* **203**, 433 (1991).
[43] T. C. Jenkins and A. N. Lane, *Biochim. Biophys. Acta* **1350**, 189 (1997).
[44] A. N. Lane, T. C. Jenkins, and T. A. Frenkiel, *Biochim. Biophys. Acta* **1350**, 205 (1997).
[45] A. Nicholls, K. Sharp, and B. Honig, *Proteins Struct. Funct. Genet.* **11**, 281 (1991).

http://trantor.bioc.columbia.edu/grasp), HyperChem, SURFNET (R. Laskowski, University College London; *www.biochem.ucl.ac.uk*), and NACCESS (S. J. Hubbard, UMIST Manchester; *sjh@sjh.bi.umist.ac.uk*). The versatile Unix-based GRASP program is particularly recommended to calculate surface areas for whole molecules and subsets of selected atoms, and includes flexible menu-driven tools for on-screen graphics manipulation and display.[45]

The solvent-accessible surface area represents the molecular portion that can be accessed by solvent molecules, and is determined by rolling a small spherical probe simulating the solvent molecule over the van der Waals surface. Water access is simulated by a probe radius of 1.4 Å. Most programs use the original method of Lee and Richards,[46] and calculate the accessibility of the whole molecule submitted in a PDB file by taking thin z-axis slices through the molecule and calculating the exposed arc lengths for each atom in each slice, and then summing the arc lengths to the final area over all z values. The z width parameter controls accuracy and calculation speed, with a typical value being 0.05 Å. The surface calculation is acutely sensitive to the van der Waals atomic radii that are employed, normally taken from the standard Lennard–Jones values.[47]

The calculation of ΔC_p for an interaction requires the determination of binding-induced changes in both the nonpolar (hydrophobic) and polar (hydrophilic) contributions to the SASA relative to the system reactants, denoted by ΔA_{np} and ΔA_p, respectively. These parameters are given by the following relationships:

$$A = A_{np} + A_p \tag{1}$$
$$\Delta A = \Delta A_{np} + \Delta A_p = A^{complex} - (A^{DNA} + A^{ligand})^{free} \tag{2}$$
$$\Delta A = [A_{np}^{complex} - (A_{np}^{DNA} + A_{np}^{ligand})^{free}] + [A_p^{complex} - (A_p^{DNA} + A_p^{ligand})^{free}] \tag{3}$$

Thus, the solvent-accessible surfaces for the complex and for both the unbound DNA and ligand must be partitioned in terms of the hydrophobic/hydrophilic characters of the underlying atoms that map to form the overall surface. As the determination of total accessible surface area is straightforward, it is necessary to establish only the partial surface contribution from either of the two atomic subsets. The remaining contribution is then obtained by difference.

At this stage, consideration must be given to defining the character of every atom present in the molecular assembly. Surface of carbon, carbon-bound hydrogen and phosphorus atoms are conveniently defined

---

[46] B. Lee and F. M. Richards, *J. Mol. Biol.* **55,** 379 (1971).
[47] C. Chothia, *J. Mol. Biol.* **105,** 1 (1976).

as nonpolar (hydrophobic), whereas those of other atoms may be classified as polar (hydrophilic). Superior parameterization could be implemented at this stage, for example, using fractional hydrophobicity/hydrophilicity data obtained by rigorous self-consistent reaction field methods.[48] However, such detailed techniques are beyond the scope of the present chapter, and it is often sufficient to rely on the simplest atomic descriptors (i.e., polar or nonpolar) for each molecule in the DNA–ligand ensemble. The coordinate file can then be edited to include this information, or a template file can be constructed for use within the application to provide a detailed map of the atomic complexion used to generate the surface. Two SASA calculations are then performed for each of three structures: (1) the DNA–ligand complex, (2) the unbound host DNA duplex, and (3) the free ligand molecule. In addition, the coordinate set for the complex should be separated into two files representing the bound forms of the DNA and the ligand. These structures can be illuminating as to induced distortion or perturbation effects relative to the native species.

As described earlier, predictive algorithms relating ΔC_p and surface area burial have been developed empirically for the solvent transfer of small molecules, and for protein systems. Heat capacity change values can be calculated from the SASA information using the relationship[49]

$$\Delta C_p = (0.32 \pm 0.04)\Delta A_{np} - (0.14 \pm 0.04)\Delta A_p \text{ cal/(mol} \cdot \text{K)} \qquad (4)$$

where ΔA_{np} and ΔA_p are the binding-induced alterations in nonpolar and polar surface area, in units of Å^2, respectively.

Example: Calculation of SASA Changes and Estimation of ΔC_p
    for DNA–Dye Complex

The 1 : 1 binding of Hoechst 33258 to the d(CGCAAATTTGCG)$_2$ (here termed A3T3) duplex is used to illustrate the procedure involved in determining both the ΔA_{np} and ΔA_p contributions to the burial of water-accessible surface area, and hence an estimate for the heat capacity change obtained solely from structural information.

Two X-ray crystal structures are available for the A3T3–dye complex, deposited as Nucleic Acid Database entries gdl028 and gdl026 (Brookhaven PDB entries 296D and 264D, respectively).[50,51] In addition, a crystal struc-

[48] F. J. Luque, X. Barril, and M. Orozco, *J. Comput.-Aided Mol. Des.* **13**, 139 (1998).
[49] R. S. Spolar and M. T. Record, Jr., *Science* **263**, 777 (1994).
[50] N. Spink, D. G. Brown, J. V. Skelly, and S. Neidle, *Nucleic Acids Res.* **22**, 1607 (1994).
[51] M. C. Vega, I. García-Sáez, J. Aymamí, R. Eritja, G. A. van der Marel, J. H. van Boom, A. Rich, and M. Coll, *Eur. J. Biochem.* **222**, 721 (1994).

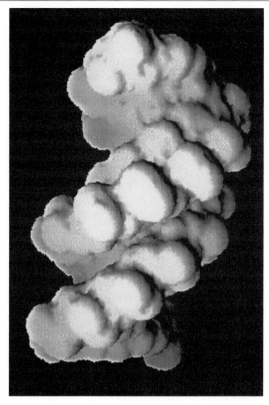

Fig. 5. Solvent (water)-accessible surface for the native d(CGCAAATTTGCG)$_2$ duplex generated using the GRASP program[45] and viewed looking into the core AT-tract minor groove. Coordinates were taken from the published X-ray crystal structure.[52]

ture for the native, drug-free A3T3 duplex is available as NDB entry bdl038 (PDB entry 1D65).[52] Crystal coordinates are not available for the free ligand, although the geometry of the bound molecule differs only negligibly from that found in other reported DNA–Hoechst 33258 complexes and closely resembles that for a minimized model ligand constructed using standard bond and angle geometries. Thus, it can be assumed that binding has little effect on the dye and that the coordinates for the DNA-bound molecule can instead be used.

After downloading the required files and editing to remove the surplus counterion and water coordinates, all the necessary hydrogen atoms were

[52] K. J. Edwards, D. G. Brown, N. Spink, J. V. Skelly, and S. Neidle, *J. Mol. Biol.* **226,** 1161 (1992).

generated using routines within the X-PLOR[40] and Babel computer programs. Each file was next edited to include atomic tags denoting the appropriate polar or nonpolar character, as described earlier. At this stage, the coordinate files were thoroughly checked to remove introduced errors. Solvent-accessible surface areas were then computed for each structure using the GRASP program[45] for (1) all atoms, and (2) the subset of hydrophobic atoms that can be mapped to the total surface. These values were calculated at the highest precision level available using this package. Examples of solvent-accessible surfaces are shown in Figs. 5–7. In Fig. 5, the surface of uncomplexed d(CGCAAATTTGCG)$_2$ is shown, from a view-

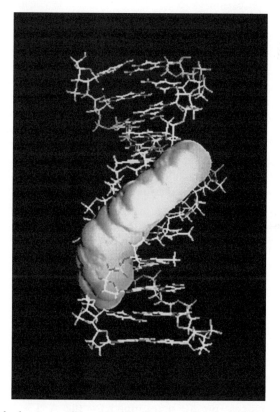

Fig. 6. Partial solvent-accessible surface for the bound Hoechst 33258 dye molecule in the AT-tract minor groove of the host A3T3 DNA duplex. Structural coordinates were taken from the reported crystal structure of the 1 : 1 complex.[51] The surface was generated using a 1.4-Å probe with the GRASP program[45] and illustrates the physical size of the introduced ligand.

FIG. 7. Solvent (water)-accessible surface for the entire d(CGCAAATTTGCG)$_2$–Hoechst 33258 complex, showing the effective surface burial for both the minor groove-bound dye (indicated) and the DNA host. Coordinates were taken from the reported crystal structure for the 1:1 complex.[51] The surface was again generated with the GRASP program[45] with a 1.4-Å probe radius.

point looking into the minor groove. In Fig. 6, the surface of Hoechst 33258 bound in the minor groove is shown. Finally, the surface of the 1:1 DNA–drug complex is shown in Fig. 7. In this latter view, the Hoechst 33258 ligand molecule is seen to be nearly completely buried, with only a small fraction of its surface edge remaining accessible to solvent.

Table IV shows the SASA values and binding-induced ΔA_{np} and ΔA_p surface area changes (burial) determined for the A3T3–dye system, together with the areas measured for canonical A- and B-type DNA duplexes with this 12-mer base sequence. The surface areas determined for the complex show that binding results in a ~20% loss of SASA relative to the initial system components, mostly (>83%) due to hydrophobic

TABLE IV
SASA BURIAL FOR d(CGCAAATTTGCG)$_2$–HOECHST 33258 DYE COMPLEX

Structure	Solvent-accessible surface area (Å^2)^a		
	Nonpolar SASA (A_{np})	Polar SASA (A_p)	Total SASA (A)
Native A3T3^b	2588	2230	4818
A-DNA A3T3^c	2464	2121	4585
B-DNA A3T3^c	2643	2156	4799
A3T3–dye complex^d	2323, 2342	2171, 2196	4494, 4538
Bound DNA	2512, 2492	2219, 2226	4731, 4718
Bound ligand	679, 643	128, 150	807, 793
Induced changes^e	$\Delta A_{np} = -944, -889$	$\Delta A_p = -187, -184$	$\Delta A = -1131, -1073$
Calculated $\Delta C_p{}^f$	$-276 \pm 38, -259 \pm 36$ cal/(mol · K)		

^{a} Solvent-accessible surface areas determined for all-atom coordinate sets using the GRASP program,[45] and with a 1.4-Å radius spherical probe.
^{b} From reported crystal structure for the drug-free, native A3T3 duplex.[52]
^{c} Canonical, computer-generated A- and B-type DNA 12-mer duplexes of this sequence.
^{d} Using coordinates for two crystal structures deposited in the Nucleic Acid Database for the 1:1 A3T3–Hoechst 33258 complex.[50,51]
^{e} Binding-induced alterations in the SASA terms calculated using Eqs. (1)–(3).
^{f} Heat capacity changes calculated using the empirical relationship in Eq. (4).[49]

surface burial. Application of the empirical relationship in Eq. (4)[49] gives a calculated heat capacity change ΔC_p of -276 ± 38 or -259 ± 36 cal/ (mol · K) for the two reported DNA–dye crystal structures. These values compare favorably with our experimental ΔC_p value of -330 ± 50 cal/ (mol · K) determined by ITC (Fig. 4). It is notable that the surface areas calculated for the native[52] and complexed DNA[50,51] differ only negligibly, with a small ~2% reduction in exposed DNA surface. This supports the suggestion of a "lock-and-key" fitting model for the groove-binding process where the conformations of both the DNA host and dye molecule are unaltered, and hence that the groove site is essentially preformed to accommodate the ligand. Further, the SASA values calculated for the computer-generated A- and B-DNA 12-mer sequences (Table IV) illustrate that the duplex is well approximated by the canonical B-DNA conformation and reinforce crystallographic conclusions.[50,51] For the interested reader, a more detailed analysis of both the individual surface components and the binding-induced effects on the system reactants is available elsewhere.[36]

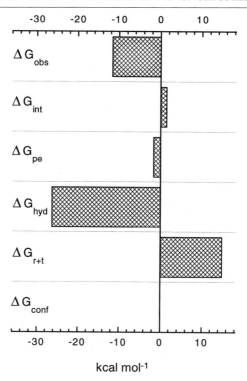

$$kcal\ mol^{-1}$$

FIG. 8. Parsing the free energy of Hoechst 33258 binding to d(CGCAAATTTGCG)$_2$. The estimated contributions to the observed free energy change (ΔG_{obs}) from the five sources discussed in text are shown. Note that the contribution from ΔG_{conf} is assumed to be 0 for this groove-binding agent.

Parsing Free Energy of DNA–Drug Binding. Example: Hoechst 33258 Binding to DNA

The most ambitious attempt to parse the free energy of a DNA–drug interaction to date involves the minor groove-binding Hoechst 33258 dye.[36] A complete, experimentally determined thermodynamic profile was obtained for the interaction of Hoechst 33258 with a self-complementary oligonucleotide containing a single binding site, 5'CGCAAATTTGCG-3'. At 25°, an association constant of $3.2 \times 10^8\ M^{-1}$ was determined, with $(\delta \ln K_b/\delta \ln[M^+]) = -1.0$. The observed binding free energy was found to be -11.7 kcal/mol, with $\Delta H° = +4.3$ kcal/mol and $\Delta S° = +40$ cal/(mol-deg). Best estimates for the energetic contributions to the overall binding free energy are shown in Fig. 8. For this groove binder, the free energy contributions from conformational changes (ΔG_{conf}) and for loss of bond rotations (ΔG_{rot}) were assumed to be 0, an assumption justified after careful examination of the structures of the reactants and the final complex. Figure 8 shows that the unfavorable entropic cost of bimolecular complex formation (ΔG_{t+r}) is overcome predominantly by a large (-26.4 kcal/mol) contribution from the hydrophobic transfer process (ΔG_{hyd}), a value calculated from the experimentally determined heat capacity change. A small favorable free energy contribution was found to arise from the polyelectrolyte effect (ΔG_{pe}). An unexpected finding was that molecular interactions appear to contribute little to the observed binding free energy, and that the overwhelming driving force for complex formation is the hydrophobic effect. While at first glance this conclusion would appear to contradict structural studies that have identified three to five hydrogen bonds that stabilize the complex,[50–52] it must be recalled that Hoechst 33258 displaces a spine of hydration in the minor groove. If this is the case, there is a kind of exchange reaction in which the hydrogen bonds of the bound water are broken, and replaced by hydrogen bonds involving the dye ligand. The thermodynamic analysis indicates that the exchange process is nearly isoenergetic (actually slightly unfavorable), and that the hydrophobic and polyelectrolyte effects are the primary driving forces for association. Such a conclusion would not be likely to emerge from a consideration of reported structural data alone.

Summary and Prospects

The power of thermodynamics is that it provides quantitative information that is independent of the details for the underlying molecular processes. Advances in instrumentation have enabled the acquisition of reliable thermodynamic data for reactions of biological interest. In parallel, significant advances have been made in the interpretation of thermodynamic data, making it possible to begin to parse free energy values into the contributions from a variety of forces and interactions. Considerable insight into the forces that drive association reactions can be gleaned from such interpretations. A framework and specific details for the application of these methods to DNA–drug binding reactions have been described here. While only relatively few compounds have thus far been examined by these approaches, it is clear from these initial successes that valuable new molecular insights will emerge as more DNA-binding drugs are studied in detail.

Acknowledgments

Supported by grant CA35635 from the National Cancer Institute (J.B.C.). We are grateful to the University of Greenwich (B.Z.C.), the University of Sheffield (I.H.), and Yorkshire Cancer Research (program grant to T.C.J.) for support. Thanks also to Dr. Jose Portugal for critically reading the manuscript.

[17] Direct Measurement of Sodium Ion Release from DNA on Binding of Cationic Ligands

By CHARLES H. SPINK and THOMAS P. WHITE

Introduction

The thermodynamic linkage between the binding of charged ligands to DNA and the release of territorially bound counterions has been recognized for a wide variety of cases.[1–10] Study of the nature of this linkage is important not only to understanding the factors that affect DNA stability and conformation, but also to the elucidation of the determinants of binding equilibria involving charged ligands. The fact that ion release from the highly charged DNA duplex accompanies many processes, ranging from chain melting to drug binding, testifies to the need to understand the fundamental relationships between cation association with DNA and the binding of charged ligands to duplex or triplex structures.

Some of the first direct studies of ion-exchange processes on DNA were carried out by Record and co-workers using NMR relaxation methods.[1–6] It was shown that a simple two-state model of sodium ion exchange can suitably describe counterion association with the negative charges of DNA.[5,6] Studies of the exchange of various alkali metal and alkaline earth cations with sodium ion as the initial counterion indicate that not only the charge on the cation, but also cation size affects the magnitude of exchange parameters.[5] In addition, it was found the DNA conformation, B versus

[1] S. Padmanabhan, W. Zhang, M. W. Capp, C. F. Anderson, and M. T. Record, *Biochemistry* **36,** 5193 (1997).

[2] S. Padmanabhan, V. M. Brushaber, C. F. Anderson, and M. T. Record, *Biochemistry* **30,** 7550 (1991).

[3] H. Eggert, J. Dinesen, and J. P. Jacobson, *Biochemistry* **28,** 3332 (1989).

[4] L. Nordenskiold, D. K. Chang, C. F. Anderson, and M. T. Record, *Biochemistry* **23,** 4309 (1984).

[5] C. F. Anderson, M. T. Record, and P. A. Hart, *Biophys. Chem.* **7,** 301 (1978).

[6] M. L. Bleam, C. F. Anderson, and M. T. Record, *Proc. Natl. Acad. Sci. U.S.A.* **77,** 3085 (1980).

[7] M. Casu, G. Saba, A. Lai, M. Luhmer, A. Kirsch-De Mesmaeker, C. Moucheron, and J. Reisse, *Biophys. Chem.* **59,** 133 (1996).

[8] E. Rowatt and R. J. William, *J. Inorg. Biochem.* **46,** 87 (1992).

[9] T. M. Lohman, L. B. Overman, M. E. Ferrari, and A. G. Kozlov, *Biochemistry* **35,** 5272 (1996).

[10] J. H. Ha, M. W. Capp, M. D. Hohenwalter, M. Baskerville, and M. T. Record, *J. Mol. Biol.* **228,** 252 (1992).

 0076-6879/00 $30.00

Z, does affect correlation times of sodium ion in the vicinity of the duplex, but that base composition seems not to affect interaction parameters.[4] Studies of the binding of multicharged ligands, such as the polyamines[2,8–13] or positively charged oligopeptides,[1,11] show similar exchange patterns; that is, a simple two-state exchange model indicates that exchange of sodium occurs as a consequence of binding of these ligands, and that exchange parameters are somewhat sensitive to the size and structure of the ligand. The interaction of intercalating, positively charged acridine derivatives with DNA also shows that duplex conformation can play a role in the characteristics of exchange of ions.[3] To account for changes in relaxation rates on binding of these drugs, it was necessary to account for changes in charge density induced by the intercalation of the drug in the base region.

Although difficult to analyze theoretically, protein binding to DNA frequently is sensitive to the salt concentration in solution, sometimes showing a 500-fold change in the equilibrium constant for a severalfold change in salt concentration.[9,10] Protein binding is complicated by the observation that anions, as well as cations, can affect the values of binding constants, and the osmotic condition of the solution complicates the equilibrium binding to DNA sites.[9] Usually, the association constant of protein with DNA is measured as a function of the sodium ion concentration, and from the dependence of the equilibrium constant on sodium ion, inferences can be drawn about the exchange of counterions.

There has been interest in the binding of cationic lipids to DNA because condensed aggregates of the DNA–lipid complex are useful for transfection experiments.[13–15] Several studies of the binding equilibria suggest that association between DNA and the positively charged ligands occurs via a combination of electrostatic interactions as well as hydrophobic self-association of the nonpolar regions of the lipid.[16–19] The equilibria involve exchange of counterions with the cationic lipid in the electrostatic domain of binding, but it is not clear how cooperative interactions among the hydrophobic

[11] W. Zhang, J. P. Bond, C. F. Anderson, T. M. Lohman, and M. T. Record, *Proc. Natl. Acad. Sci. U.S.A.* **93,** 2511 (1996).

[12] A. Delville, P. Laszlo, and R. Schyns, *Biophys. Chem.* **24,** 121 (1986).

[13] P. L. Felgner, T. R. Gadek, M. Holm, R. Roman, H. W. Chan, M. Wenz, J. P. Northrop, G. M. Ringold, and M. Danielson, *Proc. Natl. Acad. Sci. U.S.A.* **84,** 7413 (1987).

[14] R. W. Malone, P. L. Felgner, and I. M. Verma, *Proc. Natl. Acad. Sci. U.S.A.* **86,** 6077 (1989).

[15] P. Pinnaduwage, L. Schmitt, and L. Huang, *Biochim. Biophys. Acta* **985,** 33 (1989).

[16] K. Hayakawa, J. P. Santerre, and J. C. T. Kwak, *Biophys. Chem.* **17,** 175 (1983).

[17] K. Shirahama, K. Takashima, and N. Takisawa, *Bull. Chem. Soc. Jpn.* **60,** 43 (1987).

[18] C. H. Spink and J. B. Chaires, *J. Am. Chem. Soc.* **119,** 10920 (1997).

[19] S. Z. Bathaie, A. A. Moosavi-Movahedi, and A. A. Saboury, *Nucleic Acids Res.* **27,** 1001 (1999).

groups affect counterion exchange, particularly near the region where condensation occurs.[18] In this kind of system it is useful to have a method that directly measures the free sodium ion activity, so that the stoichiometry of exchange can be determined without assumptions regarding the state of sodium ion territorially bound to the DNA.[12] This chapter illustrates the use of potentiometric determination of sodium ion, using the Ross specific ion electrode for the measurements. The Ross electrode is a combination electrode that uses an internal reference electrode that improves the stability and reliability of measurements with glass electrodes. Applied to the study of the binding of cationic lipids to DNA, it is shown that this type of electrode provides a direct route to the measurement of the release of sodium ions that are territorially bound to the DNA when the cationic ligands bind. We demonstrate the procedures for calibration and use of the electrode for ion exchange of a series of alkyltrimethylammonium bromides (ATAB) with sodium ion that is bound to calf thymus DNA samples. For the smaller solutes it is shown that the results can be consistently analyzed by the two-state ion-exchange model.

Potentiometry with Ion-Selective Electrodes

Ion-selective electrodes have been widely used in studies of ionic equilibria.[20–24] pH measurements are the classic application of glass electode membrane technology for potentiometric determination of ions in solution. Glass membranes have been developed for sensing ions other than H^+, including sodium, lithium, and silver ions.[22] In general, these electrodes behave similarly to pH electrodes, requiring standardization and careful attention to solution conditions, particularly the ionic strength of the medium. Using the typical two-electrode configuration, glass membrane and a reference electrode, the potentiometric determination of alkali metal ions has been somewhat difficult because of problems with the characteristics of these electrode systems, a tendency not to have long-term stability, and the measurements being sensitive to solution conditions. At least part of the instability of the measurements has to do with how the reference electrode solution is physically isolated from the analyte solution. Glass or ceramic frits, as well as frosted glass junctions, have been used in reference electrodes to provide solution contact with the external solution. These junc-

[20] M.-A. Rix-Montel, H. Grassi, and D. Vasilescu, *Biophys. Chem.* **2,** 278 (1974).
[21] R. L. Solsky, *Anal. Chem.* **62,** 21R (1990).
[22] M. S. Frant, *Analyst* **119,** 2293 (1994).
[23] A. De Robertis, P. Di Giacomo, and C. Foti, *Anal. Chim. Acta* **300,** 45 (1995).
[24] J. W. Chen and J. Georges, *Anal. Chim. Acta* **177,** 231 (1985).

tions have often been a problem in long-term stability because they become obstructed either by components of the external solution or by the insoluble salts that are used in reference electrode chemistry. In addition, reference electrodes that use insoluble salts tend to respond slowly to temperature changes because of the slow reequilibration of solubility with temperature. The solution junctions between reference and analyte solutions are also a source of systematic error because of the development of liquid junction potentials.[24] These potentials arise because of gradients in electrolyte concentrations at the junction of the two solutions. If the ionic mobility of cation and anion are different at these junctions, significant (many millivolts) potentials can result as the ions diffuse across the solution boundary. Concentrated solutions of inactive electrolytes, such as potassium or ammonium chlorides, minimize junction potentials because the ionic mobilities of the cation and anion are similar. However, caution is necessary when there are significant differences in the ionic strengths of solutions used for standardization and those of the analyte solution. Often ionic strength adjusters are added to the solutions to compensate for differences.

Some of the problems associated with glass membrane measurements have been minimized in the Ross ion selective electrodes. [The electrode we use is the model 86-II from Orion Research (Beverly, MA) which is a combination electrode.] First, the internal reference electrode eliminates the insoluble salts associated with the calomel or AgCl reference electrodes. The internal reference is a platinum electrode in contact with the I_2/I^- half-cell. This reference is virtually independent of temperature over a wide range, and thus eliminates drift from temperature variations. Second, for the sodium ion electrode the glass used in the membrane is a special high-temperature glass that responds rapidly to changes in sodium ion concentration, so that equilibration time at the membrane surface is fast. Also, with the combination electrode a unique design makes it possible to isolate the internal reference electrode through a bridge of 2 M NH$_4$Cl, which makes contact with the external solution through a spring-loaded sleeve of glass. This assures a constant internal reference potential, and also makes it simple to clean the junction between the internal and analyte solutions.

It is important to recognize that a potentiometric cell actually measures activity of the analyte ion, not its concentration. The potential of a glass membrane combination electrode is related to ionic activity by the following relationship:

$$E = k + S \log A_i \qquad (1)$$

where A_i is the activity equal to $C_i\gamma_i$ (C_i and γ are concentration and activity coefficient of the ion), k is a constant containing the reference electrode potential and other internal membrane potentials relating to the glass electrode design, and S is the Nernstian slope equal to RT/nF, which for a monovalent ion at 25° has a value of 59.18 mV per decade of activity change. In practice the value of S is somewhat smaller than the theoretical number, often in the range of 57 mV/decade. Because it is activity that is being measured, to determine concentration the ionic strength of the medium must be adjusted to be the same in both the standards and the analyte solution, or values of the activity coefficients must be known for the ions at the particular ionic strengths of the solutions being measured.

Model for Sodium Ion Release from DNA on Binding Cationic Ligands

The general ion exchange of sodium ions on binding cationic ligands to DNA can be formulated by the following equation[5,11,13]:

$$L^+ + nDNA[P_i^-]Na^+ \rightarrow DNA[P_i^-]_nL^+ + nNa^+ \qquad (2)$$

The sodium ions that are originally territorially bound to the negatively charged duplex are released as free ions in the solution when L^+ associates with the DNA. Thus, n represents the number of sodium ions per ligand released on binding to the duplex. In this formalism we define the association as per phosphate site on the DNA to normalize to the concentration of DNA sites on the duplex. This does not mean that the sodium ion or ligand are specifically bound to individual sites, but rather are bound in the region of the negatively charged polyelectrolyte ion.

An ion-exchange constant that relates concentrations of ions in solution to those bound by DNA can be written[5]

$$D = \frac{[L^+]_b[Na^+]^n}{[L^+][Na^+]_b^n} \qquad (3)$$

Here $[Na^+]_b$ and $[L^+]_b$ are the moles of sodium ion and ligand ion bound per mole of phosphate sites, while $[Na^+]$ and $[L^+]$ are the moles per phosphate site of free sodium ion and ligand in the solution. From the following material balance relationships it is possible to derive an expression relating the total added ligand per phosphate site to the released sodium ion on binding of cationic ligand:

$$[Na^+]_T = [Na^+] + [Na^+]_b \qquad (4)$$
$$[L^+]_T = [L^+] + [L^+]_b \qquad (5)$$
$$\Delta[Na^+] = n[L^+]_b \qquad (6)$$
$$[Na^+] = [Na^+]_o + \Delta[Na^+] \qquad (7)$$

$[Na^+]_T$ and $[L^+]_T$ are the total sodium ion and ligand ion concentrations per phosphate site, $\Delta[Na^+]$ is the moles of sodium ion released per mole of phosphate in the DNA, and $[Na^+]_o$ is the original free sodium ion in solution per phosphate prior to addition of ligand. Combining these relationships, and defining $[Na^+]_b^0$ as the number of moles of bound sodium ion per phosphate prior to addition of ligand, leads to

$$[L^+]_T = \frac{\Delta[Na^+]([Na^+]_o + \Delta[Na^+])^n}{nD([Na^+]_b^0 - \Delta[Na^+])^n} + \frac{\Delta[Na^+]}{n} \tag{8}$$

Thus, if $\Delta[Na^+]$ can be measured, this equation can be used to determine n and D by nonlinear least-squares curve fitting.[25] The value of $[Na^+]_o$ is measured before addition of ligand, and $[Na^+]_b^0$ can be specified from values defined by counterion condensation theory (the value is thought to be about 0.76 per phosphate),[26,27] or this number can be made a parameter in the fitting procedure. Thus, the measure of released sodium ion becomes the essential variable in determining the ion-exchange properties of the cationic ligand. Next, we show that using potentiometric determination of sodium ion release provides an avenue to determining the ion-exchange parameters defined above.

Measurement of DNA Sodium Ion Release on Binding Alkyltrimethylammonium Ions

In principle, sodium ion release from DNA can be determined in any medium for which there is a measurable change in the sodium ion concentration. We have measured release in a medium that originally contained 0.001 M Na$^+$ and 0.002–0.006 M DNA phosphate sites, as well as in solutions of higher sodium ion content. The protocols outlined below for carrying out a set of measurements were done on samples of calf thymus DNA that had been sheared by sonication, so that the average size of fragments was about 1000 bp, as determined by gel electrophoresis. These DNA samples were dissolved in deionized water, and in some cases dialyzed against large volumes of deionized water. A solution containing about 6 mM phosphate sites was used as a stock solution, and the ATAB solutions were 5 mM in deionized water. It is necessary to calibrate the electrode with standard sodium ion solutions prior to titration of the DNA with ATAB. This is done by titrating into 3.5 ml of deionized water a standard sodium ion

[25] W. H. Press, B. P. Flannery, S. A. Teukolsky, and W. T. Vetterling, "Numerical Recipes," Chapter 14. Cambridge University Press, New York, 1986.
[26] M. T. Record, C. F. Anderson, and T. M. Lohman, *Q. Rev. Biophys.* **11**, 103 (1978).
[27] G. S. Manning, *Q. Rev. Biophys.* **11**, 179 (1978).

solution (0.0434 M) so that the range of sodium ion content is similar to that in the DNA samples. The electrode junction is first formed by pressing on the top of the spring-loaded sleeve to allow a small amount of the internal solution to flow out of the electrode. The electrode is then rinsed with deionized water, carefully removing the last drop of water with a tissue. (Orion suggests rinsing the electrode with their rinsing solution, but we found that water worked fine, and we wanted to avoid contaminating the samples with the ammonium chloride rinse solution.) The cell used for the titration is an 80 × 18 mm test tube set up with a small magnetic stirrer. The electrode is then inserted into the deionized water sample with a magnetic stirrer so that the level of the water covers the sleeve junction on the electrode. A Teflon delivery tube is then inserted into the water, and titration begun by injection of 0.05 ml of the standard sodium ion solution from a 1-ml Hamilton syringe. The first injection generally takes a longer time to reach an equilibrium value than subsequent measurements. (This is likely due to the rinsing of the membrane with deionized water, which temporarily depletes some sodium ion from the membrane.) Thus, we wait about 30 min before subsequent additions. Titration in 0.1-ml increments is continued until 1 ml of standard has been added with a 1-min wait after injection to allow mixing, and then 10 min to reach an equilibrium value. Generally, equilibrium (constant to within 0.1–0.2 mV) is reached within 5 min after the injection. After collection of the calibration data the electrode is carefully removed from the calibration solution, gently rinsed with deionized water, and then inserted into the DNA solution, which is 3.5 ml of a solution of the DNA in deionized water. Prior to the first injection the system is again allowed to equilibrate for 30 min. Injections of ATAB solutions are made from a 2.5-ml Hamilton syringe, generally in 0.1- to 0.2-ml increments. After 1 min of mixing time, the measurements are monitored and the value after 10 min is generally again constant to within 0.1–0.2 mV. For the larger ATAB solutes the solution is observed for evidence of condensation of the DNA, which is characterized by the formation of large, dense particles.

To be certain that the voltage measurements are not affected by the presence of cationic lipids in solution, a 10 mM solution of sodium ion was titrated with 5.65 mM CTAB at 0.1-ml increments. Titration up to a concentration of over 1 mM CTAB yielded a value of the number of micromoles of Na^+ of 38.6 ± 0.5, indicating no substantial effect of CTAB on the measurement of the sodium ion concentration. The effect of DNA concentration on the measurements was also investigated. An approximate threefold dilution of a DNA solution from 6.6 mM showed that the corresponding measured sodium ion was diluted by the same factor within 2% accuracy.

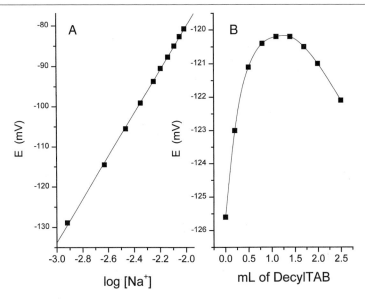

FIG. 1. (A) Calibration curve for sodium ion electrode. Solid line is the regression line through the data. (B) Raw data for the titration of 3.5 ml of 2.56 mM DNA phosphate sites with 5.27 mM decylTAB.

As an example of the raw data from a typical calibration and titration run with cationic ligand, Fig. 1 shows the calibration data and the titration data for 5.27 mM decylTAB with 3.5 ml of 2.56 mM DNA phosphate sites from an undialyzed calf thymus DNA sample. The solid line through the calibration data is the linear regression line, showing good correlation of the measured voltage with log[Na$^+$]. Figure 1B shows the raw data for the titration of DNA with decylTAB. It should be pointed out that the E value of the cell is responding to molar concentration (or activity), and that in addition to sodium ion release to the solution (causing E to become less negative), there is also dilution of the solution. Dilution would cause the voltage to become more negative. It is thus more informative to calculate sodium release on the basis of the number of moles per phosphate, as are all the terms used in the data analysis. The following section summarizes how the calculations are handled in order to use Eq. (8) to determine the ion-exchange parameters.

Calculation of Sodium Ion Release and Determination of n and D

The concentration terms in Eqs. (2) and (8) are expressed as the number of moles per mole of phosphate sites on the DNA, which can be determined spectrophotometrically. (The molar absorptivity per nucleotide is known for most types of DNA; e.g., 6421 liters mol^{-1} cm^{-1} for calf thymus.) Thus, from the potentiometric data the first task is to obtain the equation relating E to sodium ion concentration, using linear regression with Eq. (1) to obtain k and S values (see Fig. 1). The calibration curve can be expressed in terms of concentration rather than activity provided the ionic strength in the DNA sample is similar to that of the standards. If, for example, the DNA samples were in buffers containing higher sodium ion concentrations, then the standard curve should be determined in the same buffer so that the activity coefficients are similar in the two media, or alternatively, corrections can be calculated from known activity coefficient data for sodium chloride solutions.

From the known regression equation the next step is to calculate the sodium ion concentrations in the DNA solutions from the measured potentials. Once these are known the moles of free sodium ion are calculated from the individual volumes of the solutions and the known concentrations. To determine $\Delta[Na^+]$, the original number of moles of Na$^+$ per phosphate site in the starting solution is subtracted from the moles determined during the course of the titration with cationic lipid. This latter number is the moles of released Na$^+$ per phosphate site, and is plotted in Fig. 2 versus

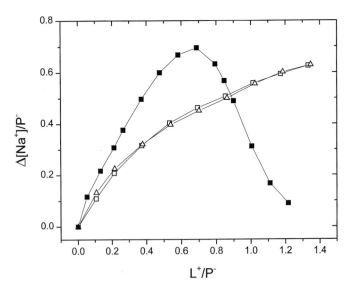

FIG. 2. Sodium ion release curves for methylTAB ($\triangle$), decylTAB ($\square$), and hexadecylTAB ($\blacksquare$) titration of calf thymus DNA. The ordinate represents moles of sodium ion released per DNA phosphate site, and the abcissa represents moles of added ligand per phosphate site.

the total moles of cationic compound added per phosphate site, $[L^+]_T$. Figure 2 shows data for methylTAB, decylTAB, and hexadecylTAB, all of which show a steady initial increase in $\Delta[Na^+]$ with increasing addition of the cationic solute. According to the counterion condensation theory,[26,27] approximately 0.76 Na$^+$ ion is territorially bound per phosphate site along the DNA duplex chain. Thus, the exchange of cationic lipid for Na$^+$ should lead to values of $\Delta[Na^+]$ that approach that theoretical limit when complete exchange has occurred. The rate of increase with added cationic lipid depends on the exchange constant, D, as defined in Eq. (2). Furthermore, Eq. (8) allows for the evaluation of n and D by nonlinear least-squares curve fitting procedures. This process is described next, but it is also interesting to look at sodium release data for a cationic compound that leads to condensation of the DNA. Figures 2 and 3 show data for hexadecylTAB, whose release profile is quite different from the smaller alkyl chain derivatives. Here the curve increases rather steeply, but when the total lipid-to-phosphate ratio is about 0.6, there is a sudden change in the sodium release behavior. The change occurs at the point at which condensation is evident by the formation of large dense particles. The condensation point for hexadecylTAB has been observed by other techniques, and found again to be near the value of $[L^+]_T = 0.6$. It is interesting to note that the condensation of the DNA leads to uptake of sodium ion. That is, the decreasing $\Delta[Na^+]$

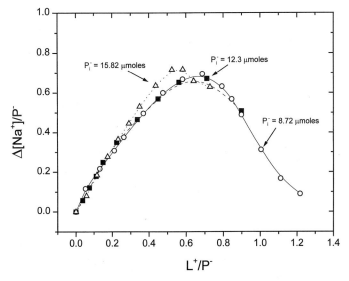

FIG. 3. Sodium ion release curves for hexadecylTAB titrations of calf thymus DNA at different numbers of moles of DNA phosphate sites: (○) 8.72 μmol of sites, (■) 12.3 μmol of phosphate sites, (△) 15.82 μmol of phosphate sites.

after condensation suggests that the aggregate structure that forms reabsorbs sodium ions, either because of a conformational change that has a high negative charge density or because lipid is redistributed such as to expose negative sites to sodium ion. More work is needed in order to characterize this effect. Figure 3 shows that the reproducibility of the sodium ion release curves is quite good for the condensing lipid. Shown are three different release curves at different DNA concentrations. The features of the curves are almost identical, all showing the decrease in $\Delta[Na^+]$ at the condensation point.

With the data for $\Delta[Na^+]$ now available it is possible to use Eq. (8) for determination of the values of n and D in the ion-exchange process. To do this, a nonlinear least-squares procedure was adopted, using the Levenberg–Marquardt algorithm to determine the unknown parameters in Eq. (8).[25] The Origin software program (MicroCal Software, Northampton, MA) has a version of this algorithm. In Eq. (8) values of $[L^+]_T$, $\Delta[Na^+]$, and $[Na^+]_o$ are known; $[Na^+]_o$ is the original sodium ion concentration per phosphate site prior to addition of alkyltrimethylammonium ion, $\Delta[Na^+]$ is the independent variable, the measured sodium ion release per phosphate site, and $[L^+]_T$ is the total cationic lipid added per phosphate site. The other parameter is the original moles of bound sodium ion per phosphate site, $[Na^+]_b^o$. This number can be fixed to correspond to the value specified by counterion condensation theory (0.76) or can be treated as a parameter that can be fitted in the least-squares procedure. If this latter procedure were done, the method would require a three-parameter fit, which for the number of data points in these data sets is difficult to rationalize. Instead, we have chosen to try the fitting by two different approaches. First, $[Na^+]_b^o$ is fixed at 0.76, and the n and D parameters determined by the fitting procedures. As is shown in Table I, the values of n for the methyl-

TABLE I

PARAMETERS DETERMINED FROM CURVE FITTING TO EQ. (8)

Ligand	$[Na^+]_b^o = 0.76$[a]		$n = 0 \leftrightarrow 1$[b]		
	n	D	n	D	$[Na^+]_b^o$
Methyl	0.94 ± 0.05	7.5 ± 0.5	1 ± 0.2	5.3 ± 3	0.82 ± 0.18
Decyl	0.98 ± 0.04	7.6 ± 0.3	1 ± 0.12	6.5 ± 2	0.79 ± 0.09
Hexadecyl	1.43 ± 0.03	243 ± 17	1.5 ± 0.07^c	92 ± 58	0.83 ± 0.07

[a] The original bound sodium was assigned a value of 0.76 for this fitting procedure.
[b] The value of n was constrained to fall between 0 and 1, and $[Na^+]_b^o$ allowed to vary.
[c] This value was obtained by allowing n to vary, since convergence was not possible without doing so.

and decylTAB salts are all close to one when fitting with these two parameters. With this in mind, a second approach was to allow D and $[Na^+]_B^0$ to vary, and constrain n in the fitting procedure to fall within 0 and 1. This was done in order to facilitate convergence. Table I shows that n converges on 1, and the values of D are close to those obtained when the original bound sodium ion was fixed at 0.76 per phosphate. Interestingly, when the value of the $[Na^+]_B^0$ is allowed to vary in the fitting procedure, its average value for the three smaller alkylTAB salts is 0.80 ± 0.07, which is within the experimental error of the theoretical value predicted by counterion condensation theory.[26,27] Figure 4A shows the experimental and calculated release curves for the decylTAB using the constrained three-parameter fit. The agreement is quite good, and it is expected that with a larger number of data points, fitting could be improved.

The hexadecylTAB case is clearly different from the other salts because of the occurrence of condensation. However, it is informative to examine

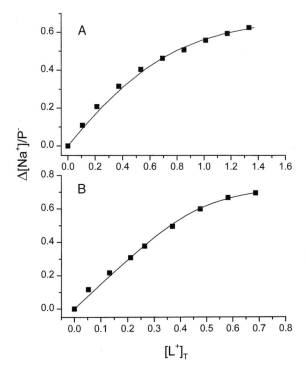

FIG. 4. Results of curve fitting of data with Eq. (8), using the three-parameter fit, but constraining n to vary between 0 and 1. (A) DecylTAB; (B) hexadecylTAB (see also Table I).

the data early in the titration (below total ligand concentrations of 0.65) with the ion-exchange model. Fitting according to the first approach (n and D varying, $[Na^+]_b^0 = 0.76$) leads to values of $n = 1.4$ and $D = 240$, an indication of much stronger binding to the duplex. Allowing the original bound sodium ion parameter to vary, and not constraining n, leads to values of $n = 1.5$, $D = 92$, and $[Na^+]_b^0 = 0.83$, as shown in Table I. The agreement between calculated and experimental curves for this case is shown in Fig. 4B. There is evidence from other measurements of hexadecylTAB that binding to DNA is a cooperative process, a result of a contribution from hydrophobic association among the alkyl chains of the salt.[16-18] While it is unlikely that the above simple ion-exchange process appropriately describes binding that has cooperativity, the values obtained for n and D being larger than for the other salts seems to indicate a shift in the character of the binding process. Cooperativity could explain the steeper change in $\Delta[Na^+]$ during the initial stages of binding, as compared with the smaller cations. To properly describe this kind of binding, however, would require revision of the simple ion-exchange model that includes consideration of an additional binding term that accounts for cooperative association. We are working on this model presently.

Conclusions

Because DNA is a polyelectrolyte with a high charge density along the duplex chain, there are important thermodynamic issues associated with long-range electrostatic interactions between associated counterions or cationic ligands and the DNA. The overall conformation and binding characteristics are thus linked to counterion association, and thus binding equilibria, particularly with cationic ligands, will be importantly dependent on counterion-exchange processes. Information about the exchange of sodium ion as interaction with other cationic ligands occurs is thus a vital link to understanding some of the thermodynamic consequences of the long-range interactions between the polyelectrolyte and ligand. The results presented in this chapter show that the potentiometric determination of sodium ion can provide useful information about the nature of cationic ligand binding to DNA. The counterion-exchange process can be quantified and the simple ion-exchange equilibrium model is confirmed for cases for which there is little or no hydrophobic interaction. More work is needed in order to clarify the role of hydrophobic association among ligand chains in the stability of the complexes between DNA and cationic lipid, as well as what changes occur in polyelectrolyte ion association as the condensation limit is approached.

This approach to studying ion release from DNA (or ion uptake) has _r_ ....g ...ormation about the linkage between sodium ion binding and the binding of other cationic ligands, including proteins, cationic peptides, or positively charged drugs. The method could also be of value in examining the role of ion condensation in the stability of various conformational forms of DNA, including triple helical complexes.

Acknowledgments

The authors acknowledge the American Chemical Society for providing a grant to support the initial stages of this work (Grant PRF-32413-B4). We also thank Brad Chaires for insightful discussions and comments about this project.

[18] Enthalpy–Entropy Compensations in Nucleic Acids: Contribution of Electrostriction and Structural Hydration

By Luis A. Marky and Donald W. Kupke

Introduction

For the complete understanding of how nucleic acids carry out their biological role, it is imperative to have detailed physical information on their interaction with ligands. While intensive investigations have been carried out on the interaction of nucleic acids with a wide variety of ligands, the most pervasive and least understood of these is water itself. We know that water plays a fundamental role in the stability of nucleic acids.[1] Experimental evidence indicates that helical nucleic acid structures are heavily hydrated.[1-3] Sequence-independent DNA hydration sites exist near the edges of the bases facing the grooves, in the vicinity of the sugars, and around the phosphate groups. The fact that the hydration of DNA is sequence dependent has been inferred from X-ray analysis of the crystal structures of oligonucleotides. Examples include the well known "spine of hydration" in the B-DNA dodecamer d(CGCGAATTCGCG),[4] the water pentagons in the A-type octamer d(GGTATACC),[5] and in the minor

[1] W. Saenger, *in* "Principles of Nucleic Acid Structure," p. 368. Springer-Verlag, New York, 1983.
[2] J. E. Hearst, *Biopolymers* **3,** 57 (1965).
[3] S. Hanlon and B. Wolf, *Biochemistry* **14,** 1661 (1975).
[4] H. Drew and R. E. Dickerson, *J. Mol. Biol.* **151,** 535 (1981).
[5] O. Kennard, W. B. T. Cruse, J. Nachman, T. Prange, Z. Shakked, and D. Rabinovich, *J. Biomol. Struct. Dyn.* **3,** 205 (1986).

groove spine of hydration in the Z-type d(CGCGC   rs.
In general, the crystal structure and conformation   lar
duplex of known sequence is sensitive to the hum   ron-
ment.[7] In solution, the hydration of a nucleic acid   nposi-
tion,[8,9] and hydration of sequence-independent si   tionale
for the conformational plasticity of DNA.[1] The pa   ation of
a duplex (A, B, or Z) depends directly on the degree of hydration[10]; hence,
changes of the water activity can shift the equilibrium to favor one duplex
conformation over the others.[1,11] The main difficulties in approaching the
interaction of water with nucleic acids are first, the actual measurement of
the physical properties of bound water, and second, the interpretation
of results. The first problem is being approached through the continued
developments of high-sensitivity densimetric and acoustical techniques,[12,13]
making it possible to determine the change in volume, apparent molar
volume, and adiabatic compressibilities on samples containing 1 mg or less
of solute. The hydration of a nucleic acid duplex as a function of sequence
can then be investigated using molecules of defined sequence. The results
obtained so far suggest that B-DNA is more hydrated than A-DNA or
RNA,[10] homopurine sequences are more hydrated than alternating purine/
pyrimidine sequences,[14,15] and that $dA \cdot dT$ base pairs are more hydrated
than $dG \cdot dC$ base pairs.[9] The negative volume change accompanying the
formation of a DNA duplex from two complementary strands is interpreted
as reflecting net changes in the electrostriction of water dipoles and/or
changes in "structural" or "hydrophobically" ordered water. Both of these
interactions contribute significantly to the ΔV effect.[16]

We are currently investigating sequence effects on the hydration of
nucleic acids. The approach we use is to measure the reaction volume or
volume change, ΔV, that accompanies association reactions. These ΔV

[6] A. H.-J. Wang, G. J. Quigley, F. H. Kolpak, J. L. Crawford, J. H. van Boom, G. van der Marel, and A. Rich, *Nature* (*London*) **282,** 680 (1979).

[7] Z. Shakked, G. Guerstein-Guzikevich, M. Eisenstein, F. Frolow, and D. Rabinovich, *Nature* (*London*) **342,** 456 (1989).

[8] M. J. B. Tunis and J. E. Hearst, *Biopolymers* **6,** 1325 (1968).

[9] V. A. Buckin, B. I. Kankiya, N. V. Bulichov, A. V. Lebedev, I. Ya Gukovsky, V. P. Chuprina, A. P Sarvazyan, and A. R. Williams, *Nature* (*London*) **340,** 321 (1989).

[10] D. Rentzeperis, D. W. Kupke, and L. A. Marky, *Biopolymers* **33,** 117 (1993).

[11] V. I. Ivanov, L. E. Minchekova, A. K. Schyolkina, and A. I. Poletayev, *Biopolymers* **12,** 89 (1973).

[12] G. T. Gillies and D. W. Kupke, *Rev. Sci. Instrum.* **59,** 307 (1988).

[13] V. Buckin, H. Tran, V. Morozov, and Luis A. Marky, *J. Am. Chem. Soc.* **118,** 7033 (1996).

[14] L. A. Marky and D. W. Kupke, *Biochemistry* **28,** 9982 (1989).

[15] D. Rentzeperis, D. W. Kupke, and L. A. Marky, *Biopolymers* **32,** 1065 (1992).

[16] K. Zieba, T. Chu, D. W. Kupke, and L. A. Marky, *Biochemistry* **30,** 8018 (1991).

results, together with standard thermodynamic parameters (ΔG, ΔH, and ΔS) obtained with the same sample solutions, may then be correlated with detailed structural information drawn from other methods, such as nuclear magnetic resonance (NMR) and X-ray crystallography. Therefore, this complete thermodynamic and structural description allow us to inspect, with suitable thermodynamic cycles, the hydration changes that take place in these systems.

In this chapter we first describe our investigation on the formation of four duplexes from mixing its complementary single strands: two fully paired duplexes and two decamer duplexes containing a bulge (see Fig. 1).[16] We obtained complete thermodynamic profiles (ΔG, ΔH, ΔS and ΔV) for the formation of each duplex at 20°. By setting up appropriate thermodynamic cycles, for pairs of reactions, we obtained differential thermodynamic profiles that correspond to specific changes such as the inclusion of an extra base pair, or a nonpaired base or bulge, in a fully paired decamer duplex. The comparison of the sign of the resulting differential enthalpy–entropy compensation with the sign of the differential volume change allows us to distinguish the type of hydration contribution, electrostricted versus structural, that is responsible for the particular change. Additional examples are presented; these include the hydration effects that take place in the interaction of minor groove ligands to dA · dT synthetic polymers,[14] inclusion of mismatches in linear duplexes,[17] the relative hydration of parallel versus antiparallel DNA,[18] and straight versus bent oligomer sequences.[19] One important correlation resulting from these studies is that we have estimated the molar volume of electrostricted water to be 15.5 ml/mol for the water of hydration in the vicinity of five dA · dT base pairs in the B conformation. This result allows us to estimate the overall electrostricted hydration in nucleic acid duplexes.

Experimental Procedures for Measuring Thermodynamic Profiles

Materials

All oligonucleotide samples are synthesized on an ABI PCR-Mate model 391 automatic DNA synthesizer (Applied Biosystems, Foster City, CA) using standard phosphoramidite chemistry, purified by high-performance liquid chromatography (HPLC), and desalted on a Sephadex G-10 exclusion chromatography column. The concentration of the oligo-

[17] M. Zhong, L. A. Marky, N. R. Kallenbach, and D. W. Kupke, *Biochemistry* **36**, 2485 (1987).
[18] D. Rentzeperis, D. W. Kupke, and L. A. Marky, *Biochemistry* **33**, 9588 (1994).
[19] K Alessi, Ph.D. dissertation. New York University, New York, 1995.

Decamer:

```
5'-C-G-C-C-T-A-A-T-C-G-3'
   . . . . . . . . . .
   G-C-G-G-A-T-T-A-G-C
```

Undecamer:

```
5'-C-G-C-C-T-A-T-A-T-C-G-3'
   . . . . . . . . . . .
   G-C-G-G-A-T-A-T-A-G-C
```

T-Bulge:

```
5'-C-G-C-C-T-A-T-A-T-C-G-3'
   . . . . . .   . . .
   G-C-G-G-A-T---T-A-G-C
```

A-Bulge:

```
5'-C-G-C-C-T-A---A-T-C-G-3'
   . . . . . .   . . .
   G-C-G-G-A-T-A-T-A-G-C
```

16mer Duplex:

```
5'-G-A-G-G-A-C-C-A-A-C-A-G-C-C-T-G-3'
   . . . . . . . . . . . . . . . .
   C-T-C-C-T-G-G-T-T-G-T-C-G-G-A-C
```

T-T Mismatch:

```
5'-G-A-G-G-A-C-C-T-A-C-A-G-C-C-T-G-3'
   . . . . . . .   . . . . . . . .
   C-T-C-C-T-G-G-T-T-G-T-C-G-G-A-C
```

aps-D1•D3:

```
5'-A-A-A-A-A-A-A-A-A-A-T-A-A-T-T-T-T-A-A-A-T-A-T-T-3'
   . . . . . . . . . . . . . . . . . . . . . . . .
3'-T-T-T-T-T-T-T-T-T-T-A-T-T-A-A-A-A-T-T-T-A-T-A-A-A-5'
```

ps-D1•D2:

```
5'-A-A-A-A-A-A-A-A-A-A-T-A-A-T-T-T-T-A-A-A-T-A-T-T-3'
   . . . . . . . . . . . . . . . . . . . . . . . .
5'-T-T-T-T-T-T-T-T-T-T-A-T-T-A-A-A-A-T-T-T-A-T-A-A-A-3'
```

[d(GA4T4C)]2:

```
5'-G-A-A-A-A-T-T-T-T-C-3'
   . . . . . . . . . .
   C-T-T-T-T-A-A-A-A-G
```

[d(GT4A4C)]2:

```
5'-G-T-T-T-T-A-A-A-A-C
   . . . . . . . . . .
   C-A-A-A-A-T-T-T-T-G
```

Fig. 1. Sequences of deoxyoligonucleotide duplexes and their designations.

mers is determined in water using appropriate extinction coefficients at 260 nm and high temperature (80–90°). These extinction coefficients in single strands are calculated at 25° using the tabulated values of the dimers and monomer bases,[20] and estimated at high temperatures by extrapolation to 25° of the upper portions of the melting curves, which correspond to the UV–temperature dependence of the single strands.[16] Stock oligomer solutions are prepared by dissolving the dry and desalted oligomers in the appropriate buffer and then dialyzed against the same buffer using membranes with 500–1000 Da molecular mass cutoff. The polymer samples, poly[d(AT)] · poly[d(AT)] and poly (dA) · poly(dT), are purchased from commercial sources. Both polymers are obtained in dried form and are used without further purification. The samples are dissolved in buffer, sonicated, and dialyzed exhaustively against the same buffer. The concentration of each compound in solution is determined spectrophotometrically using their known molar extinction coefficients. The minor groove ligands netropsin and distamycin are obtained from Serva (Heidelberg, Germany) and used without further purification. All salts and other chemicals are reagent grade. Unless otherwise specified, most measurements are performed in buffer solutions consisting of 10 mM sodium phosphate, adjusted to pH 7 and to an overall NaCl concentration of 0 or 100 mM.

Measuring Complete Thermodynamic Profiles

Complete thermodynamic profiles include the determination of the volume change, ΔV, thus complementing those of standard thermodynamic profiles (ΔG, ΔH, and ΔS). These complete parameters characterize in a more comprehensive way the two types of processes described here: duplex formation from mixing its complementary single strands and ligand binding to nucleic acid helices, allowing us to inspect the role of water in these systems.[14,16] What follows is a brief description of the techniques used to obtain each thermodynamic parameter, precautions taken, and additional corrections that are necessary to obtain meaningful parameters.

Magnetic Suspension Densimetry

The volume change measurements are performed in a magnetic suspension densimeter; this is a home-made instrument that has been described previously.[12] This instrument is simple to duplicate and relatively inexpensive, but extreme care must be exercised during the measurement of densities. In particular, one needs to avoid even tiny evaporation of the samples and to make sure that true mechanical and chemical equilibrium has been

[20] C. R. Cantor, M. M. Warshaw, and H. Shapiro, *Biopolymers* **9,** 1059 (1970).

obtained. The calculation of the volume differences, Δv, is done by measuring the mass, with a microbalance, and the equilibrium density of solutions of the reactants and product, according to the equation

$$\Delta v = m_{\text{mix}}/\rho_{\text{mix}} - (m_{\text{A}}/\rho_{\text{A}} + m_{\text{B}}/\rho_{\text{B}}) \tag{1}$$

where m is the mass in grams and ρ is the density of the solution in grams per milliliter. The two terms within parentheses refer to the initial volume of reactants before mixing. The value of Δv is then reduced to that per mole of the limiting reagent to give ΔV. This instrument requires less than 100 μl per measurement, equivalent to ~0.5 mg of total DNA sample. With repetitive samples the density is measured with a precision of better than 5×10^{-6} g/ml, allowing volume difference measurements with a precision of less than 0.5 nl.[12,14-16] It is also important to keep the set temperature, normally 20°, within 1 millidegree.

Isothermal Titration Calorimetry

The measurement of the heat of forming a DNA duplex from mixing its complementary strands, or the ligand interacting with a DNA polymer, isothermically are carried out with an Omega titration calorimeter from MicroCal (Northampton, MA). A detailed description of this instrument has been presented elsewhere.[21] We use a 100-μl syringe for the titrant and complete mixing is effected by stirring this syringe at 400 rpm. We use a titrant concentration 20–40 times higher than the concentration of the DNA oligomer (or polymer) in the reaction cell. Calorimetric titrations are designed only to obtain the overall interacting enthalpies. Typically 5–10 injections of 5 μl each are performed in a single titration. The reference cell of the calorimeter is filled with water or buffer and it only acts as a thermal reference to the sample cell. The calorimeter is calibrated by means of a known standard electrical pulse. The precision of the resulting heat from each injection is less than 0.5 μcal.

Melting Curves

The helix–coil transition of each DNA molecule is characterized by two types of complementing techniques: differential scanning calorimetry (DSC) and UV temperature-dependent profiles. The heat of the helix–coil transition of each duplex is measured with an MC-2 differential scanning calorimeter from MicroCal. Typically a DNA solution is scanned against a buffer solution, normalized for the heating rate, and a buffer versus buffer scan subtracted. Integration of the resulting curve, $\int \Delta C_{\text{p}} dT$, and normalizing for the number of moles yields ΔH_{dsc}, which is independent

---

[21] T. Wiseman, S. Williston, J. F. Brandts, and L.-N. Lin, *Anal. Biochem.* **179**, 131 (1989).

of the nature of the transition; ΔS_{dsc} is obtained by a similar integration, $\int (\Delta C_p / T) dT$. The free energy at a temperature T [the temperature used in isothermal titration calorimetry (ITC) and density experiments], $\Delta G^{\circ}_{dsc}(T)$, is obtained using the relation: $\Delta G^{\circ}_{dsc}(T) = \Delta H_{dsc} - T \Delta S_{dsc}$, which assumes similar heat capacities for the duplex and random coil states. However, these $\Delta G^{\circ}_{dsc}(T)$ values are corrected by the enthalpy factor, $\Delta H_{ITC} / \Delta H_{dsc}$, according to $\Delta G(T) = \Delta^{\circ} G_{dsc}(T)[\Delta H_{ITC} / \Delta H_{dsc}]$, to take into consideration the contribution of single-strand stacking interactions[17,18]; ΔH_{ITC} is the heat obtained in titration calorimetry experiments. The $T \Delta S$ terms at T can be calculated from the Gibbs relationship: $T \Delta S = \Delta H_{ITC} - \Delta G(T)$.

Absorbance-versus-temperature profiles for the oligomer duplexes (at several strand concentrations), unligated polymer duplexes, and saturated ligand–DNA complexes in buffer solutions are measured at 260 nm with a thermoelectrically controlled Perkin-Elmer (Norwalk, CT) 552 spectrophotometer, which is interfaced to a PC-XT computer for acquisition and analysis of experimental data. The temperature is scanned at a heating rate of 1°/min. These melting curves allow us to measure the temperature of the midpoint of the order–disorder transition of these molecules, T_M, and ligand-binding affinities, K, using standard procedures reported elsewhere.[14,22,23] In addition, we obtain melting curves as a function of salt concentration, at a fixed strand concentration, ranging from 0 to 100 mM Na^+. For a particular duplex, these experiments yield the slope of the plots of T_M versus $\log[Na^+]$ that in combination with the enthalpy and T_M values of DSC experiments allows us to determine the counterion release, Δn_{Na^+}, for the helix–coil transition of each duplex, according to the relation[24] $dT_M / d \ln[Na^+] = 0.9(RT_M^2 / \Delta H_{dsc}) \Delta n_{Na^+}$. The Δn_{Na^+} terms were also corrected by the enthalpy factor $\Delta H_{ITC} / \Delta H_{cal}$ to include the contribution of the association of additional counterions to single strands at lower temperatures.[17,18] The salt-dependent melting curves of the saturated ligand–DNA complexes yield the slopes of $\ln K$ versus $\ln Na^+$, which are proportional to the number of moles of counterions displaced on binding of 1 mol of ligand.

Complete Thermodynamic Profiles for Formation of Short DNA Duplexes

We have used a combination of density, calorimetric, and spectroscopic techniques to obtain complete thermodynamic profiles for the formation

[22] L. A. Marky and K. J. Breslauer, *Biopolymers* **26,** 1601 (1987).
[23] D. M. Crothers, *Biopolymers* **10,** 2147 (1971).
[24] M. T. Record, C. F. Anderson, and T. M. Lohman, *Q. Rev. Biophys.* **11,** 103 (1978).

of oligodeoxynucleotide duplexes, top four oligomer duplexes shown in Fig. 1, from the mixing of its two complementary strands, according to the reaction

$$SS_A(aNa^+, mH_2O) + SS_B(bNa^+, nH_2O) + xNa^+(zH_2O)$$
$$+ yH_2O \rightarrow duplex(cNa^+, oH_2O) \quad (2)$$

Each of the thermodynamic parameters reported in Table I refers to the preceding reaction at 20°. Each single strand and duplex state will have associated sodium ions and bound water molecules. Formation of a duplex will result in the binding of counterions with a change in the overall hydration state of each of the DNA species involved. Each of the relevant parameters for this reaction has been measured or calculated from experimental data.

To compare our thermodynamic results directly, we present in Table I tabulated values of the observed independent variables, ΔV and ΔH, along with the calculated dependent variables, corrected $\Delta G°$ and Δn_{Na^+}, and $T\Delta S$. All values refer to the temperature used in ITC and volumetric experiments.

TABLE I

THERMODYNAMIC PROFILES OF DUPLEX FORMATION

Duplex	$\Delta G°$ (kcal/mol)	ΔH (kcal/mol)	$T\Delta S$ (kcal/mol)	ΔV (ml/mol)	Δn_{Na^+} (mol Na$^+$/mol P$_i$)
Fully Paired and Bulged Duplexes[a]					
Decamer	−6.0	−55.5	−49.5	−167	0.140
Undecamer	−6.5	−61.1	−54.6	−208	0.148
T-bulge	−3.9	−53.2	−49.3	−285	0.168
A-bulge	−2.2	−36.0	−33.8	−312	0.152
Inclusion of Mismatched Base Pair[b]					
16-Mer duplex	−10.4	−70	−59.6	−99	NA
T·T mismatch	−8.0	−61	−53.0	−118	NA
Antiparallel versus Parallel DNA[c]					
aps-D1·D3	−15.0	−116.6	−101.6	−340	0.087
ps-D1·D2	−8.2	−74.5	−68.5	−83	0.084

[a] Sodium phosphate buffer (10 mM, pH 7), 0.1 mM Na$_2$EDTA, and 0.1 M NaCl.

[b] Sodium cacodylate buffer (20 mM, pH 7), 0.1 M NaCl, and 1 mM MgCl$_2$.

[c] Sodium cacodylate buffer (10 mM, pH 7), 0.1 mM Na$_2$EDTA, and 0.1 M NaCl.

In the experiments for determining the equilibrium volume change, ΔV, the molar ratio of one strand relative to the complementary strand was 1.1, 10% excess of one of the strands. Each ΔV entry represents an average of two to four determinations. The results of Table I indicate that the formation of all oligomer duplexes is accompanied by a negative ΔV, and indicate a volume contraction or an uptake of water molecules. The ΔH values are obtained directly from titration calorimetry and may include contributions from single-strand base stacking. We observed exothermic enthalpies for the formation of each duplex and their magnitude depends on the particular duplex as described in the next section.

We obtained favorable free energies for the formation of each duplex, resulting from a partial compensation of favorable enthalpies with unfavorable entropies. These ΔG terms in general will include two unfavorable contributions, a nucleation due to the entropy loss of a bimolecular association and a combinatorial factor due to the complementarity of the oligomers; a favorable free energy of chain propagation, which corresponds to the sum of the nearest-neighbors base-stacking interactions derived from its primary sequence. The first two terms are identical for each set of duplexes so that the differences in the free energies will depend primarily on the oligomer sequence. For instance, the free energy difference between the undecamer duplex and the decamer duplex corresponds to the disruption of one AA/TT base-pair stack and the formation of two base-pair stacks: one AT/TA and one TA/AT.

The observed enthalpies for the formation of each duplex also show differences. These measured enthalpies will comprise endothermic contributions from the disruption of the single-stranded stacking interactions and exothermic contributions that includes base-stacking interactions, specific H bonding, base-pair stacking and probably also an exothermic contribution from an increase in hydration. For instance, the formation of an extra $dA \cdot dT$ base-pair in going from the decamer to the undecamer duplex increases the exothermicity of the enthalpy by 5.6 kcal/mol at 20°.

In a similar fashion the derived entropy change is equal to the sum of the following contributions: $\Delta S_{\text{molecularity}}$ (the loss of entropy due to a bimolecular association reaction), $\Delta S_{\text{symmetry}}$ (the loss in entropy due to the complementary nature of the oligomers), $\Delta S_{\text{conformation}}$ (changes in the oligomer configuration in going from single strand to duplex), ΔS_{ions} (uptake or release of counterions), and $\Delta S_{\text{hydration}}$ (uptake of water molecules). Each of the first four entropy contributions is presumed to be identical for the formation of each duplex in a given set. The fifth term is substantially different for these sets according to our ΔV results. In the last column of Table I, we have included the uptake of counterions, Δn_{Na^+} (per phosphate), on formation of each duplex. We obtained Δn_{Na^+} values characteristic of short duplexes.

The significant observation in each of these profiles is that the more favorable $\Delta G°$ of forming a given duplex reflects a favorable heat of formation that is not quite compensated by the observed loss in entropy. This loss in entropy is consistent with an apparent increase in hydration as indicated by the larger volume contraction and with an uptake of sodium ions as determined in the salt-dependence studies. In the formation of a DNA duplex from mixing its component strands, two dominant but opposing hydration effects take place: (1) the higher immobilization of structural water by the single strands, which is due to the greater exposure of the nucleotide chain to solvent; and (2) the higher immobilization of electrostricted water by the double helix, which is due to its higher charge density. Therefore, the observed differences in hydration may well correspond to an uptake of electrostricted water or structural water or to a substitution of electrostricted for structural water molecules. To distinguish between these three possibilities, the measured ΔV values are correlated with standard thermodynamic profiles for similar reactions of duplex formation. The reason is that heat is released in the immobilization of electrostricted water[25,26] while this energetic contribution is nearly zero or slightly positive for the immobilization of structural water. The main hypothesis is that structuring water to improve packing around hydrophobic groups eliminates void spaces, rather than electrostriction that compresses the water dipoles.[27] At the same time, this reorganization of water molecules increases the order of the system, yielding an overall unfavorable entropy.

Differential Profiles for Inclusion of Single Base Pairs and Bulges

When the observed thermodynamic profiles for the fully paired duplexes are considered in terms of a differential effect, that is, if each thermodynamic parameter of the decamer duplex is subtracted from the corresponding parameter of the undecamer duplex, we form an extra $dA \cdot dT$ base pair in the middle of an existing TAAT/ATTA stretch, according to the reaction

$$\text{Decamer} \rightarrow \text{undecamer} \qquad (3)$$

This assumes that the coil states are similar, the single strands are different by just one nucleotide, which may contribute an additional uncertainty of 3% or less. We obtained a small $\Delta\Delta G$ term of -0.5 kcal/mol, which is the result of the nearly total compensation of a favorable $\Delta\Delta H$ of -5.6 kcal/

[25] J. E. Hearst, *Biopolymers* **6**, 1325 (1968).
[26] A. I. Gasan, V. Ya Maleev, and M. A. Semenov, *Stud. Biophys.* **136**, 171 (1990).
[27] B. E. Conway, "Ionic Hydration in Chemistry and Biophysics." Elsevier, New York, 1981.

mol and an unfavorable $\Delta(T\Delta S)$ of -5.1 kcal/mol; this last term correlates in sign with the $\Delta\Delta V$ of -41 ml/mol of duplex (see Table III). Because the overall counterion binding is similar in the formation of these two duplexes, this result strongly suggests that the addition of an extra $dA \cdot dT$ base pair to an existing AT stretch is accompanied by an uptake of water molecules.

To inspect the differential effect of placing a bulge (A or T) in the middle of the TAAT/ATTA stretch of the decamer duplex, each thermodynamic parameter of the decamer plus undecamer duplexes is substracted from the corresponding parameter of the T-bulge plus A-bulge duplexes; this corresponds to the reaction

$$\text{Decamer} + \text{undecamer} \rightarrow \text{T-bulge} + \text{A-bulge} \qquad (4)$$

In this thermodynamic cycle the single strands cancel out exactly. We obtained an unfavorable $\Delta\Delta G$ term of 6.4 kcal/mol, which is the result of the partial compensation of an unfavorable $\Delta\Delta H$ of 27.4 kcal/mol and a favorable $\Delta(T\Delta S)$ of 21 kcal/mol; the positive sign of this differential enthalpy–entropy compensation is opposite to the sign of the differential volume change, $\Delta\Delta V$ of -222 ml/mol of duplex (see Table III). The overall counterion binding is similar in the formation of these four duplexes; this result strongly suggests that the inclusion of the A and T bulges to an existing AT stretch is also accompanied by an uptake of water molecules.

The significant observation from these thermodynamic cycles is that the similar signs of the $\Delta\Delta H - \Delta(T\Delta S)$ compensation (which is actually similar in sign to the $\Delta\Delta G°$ term) with the $\Delta\Delta V$ term for the inclusion of a $dA \cdot dT$ base pair reflect electrostriction effects, while the opposite signs of these quantities for the inclusion of bulges indicate hydrophobic effects. Before we can generalize these results we present additional examples in the following sections.

Interaction of Minor Groove Ligands with $dA \cdot dT$ Synthetic Polydeoxynucleotides

We have obtained complete thermodynamic profiles for the association of minor groove ligands to polynucleotide duplexes[14,15] according to the following reaction:

$$\text{DNA} + \text{ligand} \rightarrow \text{complex} + \Delta n_{\text{Na}^+} + \Delta m\text{H}_2\text{O} \qquad (5)$$

Volume Changes

We used approximately equal volumes of each polynucleotide and of ligand at low concentrations ($1\text{--}2$ mM) and molar ratios of polymer (as phosphate) per ligand of 10–12, or one ligand per binding site (~5 base pairs). Signals of sufficient magnitude were observed in the magnetic suspension densimeter so that reliable results could be obtained on the ~100-μl samples containing 0.1–0.2 μmol of the limiting reactant species. The results are shown in the ΔV column of Table II. The binding of netropsin to poly[d(AT)] · poly[d(AT)] generated a volume contraction and to poly(dA) · poly(dT) a volume expansion, while binding of distamycin generated a larger volume contraction for the poly[d(AT) · poly[d(AT)] duplex. Thus, at this low ionic strength, we may infer that a net hydration accompanies the binding to these polymer duplexes as reflected by the apparent compression of water molecules. The values for these volume changes indicate that the net hydration is strongly dependent on the sequence of the polynucleotide duplex. Specifically, the volume contractions were much larger with the poly[d(AT)] · poly[d(AT)] duplex than for the compositionally identical poly(dA) · poly(dT) duplex. The relatively large volume changes in aqueous systems, as observed here, reflect primarily changes in

TABLE II

THERMODYNAMIC PROFILES FOR ASSOCIATION OF MINOR GROOVE LIGANDS TO A·T DNA POLYMERS AND TO SHORT DUPLEXES WITH A·T RUNS[a]

Duplex	$\Delta G°$ (kcal/mol)	ΔH (kcal/mol)	$T\Delta S$ (kcal/mol)	ΔV (ml/mol)	$d \ln K/d \ln[Na^+]$ (mol Na$^+$/mol)
			Netropsin[b]		
d(AT)·d(AT)	−13.1	−12.2	0.9	−16	−1.64
d(A)·d(T)	−12.3	−0.4	11.9	97	−1.81
			Distamycin[b]		
d(AT)·d(AT)	−12.8	−19.1	−6.3	−128	−1.02
d(A)·d(T)	−11.3	−4.0	7.3	−29	−0.93
		Netropsin Binding to Short Duplexes with dA·dT Runs[c]			
[d(GA$_4$T$_4$C)]$_2$					
First site	−9.9	1.2	11.1	−100	NA
Second site	−9.9	−4.0	5.9	37	NA
[d(GT$_4$A$_4$C)]$_2$					
First site	−10.3	−11.1	−0.8	−83	NA
Second site	−10.3	−9.7	0.6	15	NA

[a] At 20°.
[b] Sodium phosphate buffer (10 mM), 0.1 mM Na$_2$EDTA at pH 7.
[c] Sodium cacodylate buffer (10 mM, pH 7) and 0.1 M NaCl.

solvent structure because we have no reason to suppose that the DNA solutes contain void spaces that can contribute significant volume contractions on drug binding to the heteropolymers relative to those for the corresponding homopolymers shown in Table I.

Calorimetry

To maintain an equivalence of states, we carried out titration calorimetric experiments on the same solutions and under the same conditions as used for the densimetric measurements. In these experiments and before the DNA polymers reach saturation, the heats obtained for each injection were independent of the total concentration of added ligand. After a small correction for ligand dilution heats, molar binding enthalpies were calculated by averaging the heat evolved of the initial injections that correspond to the complete binding of ligand. These values are listed for each of the polynucleotides under the ΔH column in Table II. Inspection of this column shows that all the polymer reactions with each ligand resulted in exothermic enthalpies. The enthalpy changes were also found to be independent of the amount of bound ligand. The following trend was apparent for these enthalpy changes: The binding of netropsin and distamycin to alternating poly[d(AT) · poly[d(AT)] was accompanied by larger exothermic enthalpies relative to the homopolymer, poly(dA) · poly(dT). The measured enthalpies will ordinarily comprise exothermic contributions from specific hydrogen bonding, van der Waals interactions, and from a net coulombic hydration change.

Complete Thermodynamic Profiles

Values for all thermodynamic parameters ($\Delta G°$, ΔH, $T\Delta S$, and ΔV) for the binding of netropsin and distamycin to each helical polynucleotide at 20° are listed in Table II. The binding affinities, K, were obtained from UV melting curves at several NaCl concentrations, using the increase in thermal stability of the complexes relative to the ligand-free polymers.[23] Binding of these minor groove ligands results in K values in excess of 10^8 and corresponds to the high specificity of the ligands to dA · dT base pairs. These binding affinities will include contributions from hydrogen bonding, van der Waals contacts, and electrostatic interactions. The corresponding free energy changes, $\Delta G°$, were derived from $\Delta G° = -RT \ln K$, after which the entropy changes, in terms of $T\Delta S$, were calculated from the Gibbs equation. The last column of Table II corresponds to the release of counterions on binding of the ligand, which reflects the charge on the ligands. Thus, for every mole of bound netropsin there is a release of 2 mol of counterions, and for 1 mol of bound distamycin the release is only 1 mol of counterions, consistent with polyelectrolyte effects.[24]

Although similar free energy changes were obtained for the binding of netropsin and distamycin to this pair of isomeric polymers, the origin of the energy contributions was quite different. For the alternating purine–pyrimidine polymer, poly[(dAT)] · poly[(dAT)], the binding energies were primarily of enthalpic origin, whereas the binding energies for the corresponding homopolymer isomer, poly(dA) · poly(dT), were dominated by the entropic contributions.

Differential Profiles of Minor Groove Binding

To comprehend the observed differences in the thermodynamic parameters obtained from the ligand-binding experiments, we have recast the data in differential terms, by subtracting those obtained in a given polynucleotide reaction from those of the compositional isomer. We obtain similar ligand–DNA binding affinities among the compositional isomers, in spite of the fact that the association of netropsin with all A · T polymers[14] yields completely different thermodynamic values (ΔH and $T\Delta S$) relative to the association of distamycin with the same pair of DNA molecules.[15] However, the differential profiles are similar (see Table III). This strongly suggests that both ligands, regardless of their structural differences, are buried in the minor groove and only partially exposed to solvent. The results are consistent with the intrinsic property of poly(dA) · poly(dT) being heavily hydrated. The small differences in the differential binding enthalpy and entropic contributions between these two ligands reflect primarily differences in hydrogen bonding and van der Waals contacts due to the extra methyl-pyrrole group with no further removal of water molecules. The large net volume expansion (dehydration) is reflected in the differential entropy gain that nearly compensates for the large reduction in the differential exothermic enthalpy.

Inclusion of Mismatch in DNA Duplex

Physical or chemical damage affects the integrity of the duplex structure of DNA, resulting in the formation of non-Watson–Crick base-pair mismatches, unpaired bases (bulges), and other base modifications. These lesions may be associated with different types of mutagenesis if incomplete or erroneous repair takes place.[28] For the complete understanding of the

[28] E. C. Friedberg, G. C. Walker, and W. Siebe, "DNA Repair and Mutagenesis." ASM Press, Washington, DC, 1995.

mechanisms of mutations, detailed knowledge of the thermodynamic contributions of these lesions must be determined. For this reason, we include an additional analysis of the thermodynamics, and particularly the solvation changes, resulting from substituting a dA · dT base pair for a dT · dT mismatch in a 16-mer duplex.[17]

We have measured directly at 20° the change in volume and heat associated with the formation of each molecule from mixing appropriate sets of complementary strands. The results are listed in Table I. We obtained negative ΔV values, which indicates an uptake of water molecules; the different ΔV values imply that the mismatched duplex differs from its respective control duplex significantly in its degree of hydration. The heat of duplex formation is exothermic and corresponds to exothermic contributions from formation of base-pair stacks, hydrogen bonding, and uptake of water molecules. These terms override the endothermic contribution from disrupting base–base stacking interactions of the single strands; the differential uptake of cations normally has a negligible contribution to the enthalpy. The overall magnitude of these enthalpies depends on the nature of the structure formed. Complete thermodynamic profiles obtained from isothermal titration calorimetry and volumetric measurements are presented in Table II. At 20°, similar favorable free energy values are obtained for the formation of each molecule that results from the characteristic partial compensation of exothermic enthalpies and unfavorable entropies. However, differences are seen in the magnitude of the enthalpy and entropy terms that give rise to the overall observed free energy change. The unfavorable entropy values are in good agreement with the volume change measurements, which indicates an increase in order due to the immobilization of water molecules and counterion binding in the formation of each molecule.

Differential Thermodynamic Profiles

If the thermodynamic parameters for the formation of the T · T duplex are subtracted from the corresponding parameters of forming the 16-mer duplex, the contributions from the single strands cancel out almost exactly, because there is a single base difference out of 32, and the resulting differential thermodynamic profiles correspond to the reaction

$$16\text{-Mer duplex} \rightarrow T \cdot T \text{ mismatch duplex} \qquad (6)$$

This reaction actually corresponds to the substitution of a dA · dT base pair for a dT · dT mismatch within a duplex of 16 base pairs. For this reaction, we calculate a $\Delta\Delta G°$ of 2.4 kcal/mol of duplex, which reflects the higher stability of the fully paired duplex, the result of an unfavorable $\Delta\Delta H$ of 9.0 kcal/mol and a favorable $\Delta(T\Delta S)$ of 7.0 kcal/mol. The similar signs of the $\Delta\Delta H$ and $\Delta(T\Delta S)$ terms contribute toward a partial differential compensation. Because the contributions of base-pair stacking and base pairing cancel in the reaction, this result suggests the participation of water molecules, as has been indicated previously.[14,15,29–31] This differential compensation effect is accompanied by a $\Delta\Delta V$ of -19 ml/mol. Furthermore, the opposite signs between $\Delta\Delta V$ and the differential enthalpy–entropy compensation suggest that the substitution of a dA for a dT, to yield a mismatched duplex, actually immobilizes more structural water molecules.[16,30] If the sign of $\Delta\Delta V$ were positive it would have indicated a release of electrostricted water molecules in order to be consistent with the energy needed to remove electrostricted water molecules, which also contributes to the disorder of the system (favorable entropy); most likely these differential terms correspond to a local substitution of electrostricted water for structural (or hydrophobic) water.

Hydration of Parallel versus Antiparallel DNA

Another addition to the family of DNA conformations is that of parallel-stranded DNA (ps-DNA),[32–34] in which both complementary strands have the same 5' to 3' orientation, held together by reverse Watson–Crick base pairs, and forming a double helix with two equivalent grooves in depth and width.[35]

We have studied the formation of two DNA 25-mer duplexes containing exclusively dA · dT base pairs in either parallel (ps-D1 · D2) or antiparallel (aps-D1 · D3) orientation.[18] We have measured at 15° the change in volume and heat associated with the formation of each duplex. The results, listed in Table I, indicate that the formation of each duplex is accompanied by an uptake of water molecules and exothermic heats. The difference in their

[29] D. Rentzeperis, D. W. Kupke, and L. A. Marky, *Biopolymers* **33,** 117 (1993).
[30] L. A. Marky, D. Rentzeperis, N. P. Luneva, M. Cosman, N. E. Geacintov, and D. W. Kupke, *J. Am. Chem. Soc.* **118,** 3804 (1996).
[31] R. Lumry and S. Rajender, *Biopolymers* **9,** 1125 (1970).
[32] J. H. van de Sande, N. B. Ramsing, M. W. Germann, W. Elhorst, B. W. Kalisch, E. V. Kitzing, R. T. Pon, R. C. Clegg, and T. M. Jovin, *Science* **241,** 551 (1988).
[33] T. M. Jovin, *in* "Nucleic Acids and Molecular Biology" (F. Eckstein and D. M. Lilley, eds.), Vol. 5, p. 25. Springer-Verlag, Berlin, 1991.
[34] N. B. Ramsing and T. M. Jovin, *Nucleic Acids Res.* **16,** 6659 (1988).
[35] K. Rippe and T. M. Jovin, *Methods Enzymol.* **211,** 199 (1992).

ΔV and ΔH values suggests that the conformation of the duplex formed reflects primarily on the degree of hydration.

Complete Thermodynamic Profiles of Duplex Formation

Complete thermodynamic profiles at 15° are presented in Table I.[18] At this temperature, favorable free energy values are obtained for the formation of each duplex that results from the characteristic partial compensation of exothermic heats and unfavorable entropies. However, differences are seen in the magnitude of the forces (enthalpy versus entropy) that contribute to the overall observed free energy change. The unfavorable entropy values are in good agreement with the volume change measurements that indicate an increase in order due primarily to the compression of water molecules.

Differential Thermodynamic Profiles

If each of the thermodynamic parameters for the formation of the ps-D1 · D2 duplex is substracted from the corresponding parameters of forming the aps-D1 · D3 duplex, the contributions from the D1 single strands cancel out exactly, and the resulting differential thermodynamic profiles correspond to the substitution reaction

$$\text{aps-D1} \cdot \text{D3} + \text{D2} \rightarrow \text{ps-D1} \cdot \text{D2} + \text{D3} \tag{7}$$

In addition, we can also cancel out the D2 and D3 single strands of Eq. (7) because their sequences are identical but in reverse orientation, and their thermodynamic contributions from base-stacking interactions and hydration may be assumed to be identical. For the resulting reaction, we obtain a $\Delta\Delta G°$ of 9.0 kcal/mol, which is the result of a compensation of an unfavorable $\Delta\Delta H$ of 42.1 kcal/mol and a favorable $\Delta(T\Delta S)$ of 33.1 kcal/duplex, and a $\Delta\Delta V$ of 257 ml/mol of duplex. This reaction is accompanied by a marginal differential counterion uptake, $\Delta\Delta n_{Na^+}$, of 0.19 mol of Na^+/mol of duplex. The positive $\Delta\Delta G$ value of Eq. (7) provides evidence for the greater stability of the more native antiparallel (Watson–Crick) duplexes. This positive enthalpy–entropy compensation is characteristic of processes driven by a differential hydration. Therefore, a Δn_{water} term should be included on the right-hand side of Eq. (7) corresponding to a change in the release or compression of water molecules. This strongly suggests that the ps-D1 · D2 parallel duplex is much less hydrated than its antiparallel counterpart.

Differential Hydration of Bent versus Straight Sequences

Curve DNA plays an important role in the packing of DNA around nucleosomes.[36] Certain DNA sequences are curved, such as A tracts repeated in phase with the DNA helical screw, and these sequences run slowly in gel electrophoretic experiments.[37] For instance, polymers resulting from the ligation of GA_4T_4C oligomers have shown much lower electrophoretic mobility relative to polymers resulting from ligating GT_4A_4C oligomers; the former are bent while the latter are straight.[38] To investigate further the thermodynamic behavior of these sequences and especially their hydration properties, we have studied the interaction of the minor groove ligand netropsin with each of the following oligomer duplexes: $[d(GA_4T_4C)]_2$ and $[d(GT_4A_4C)]_2$.[19]

Complete Thermodynamic Profiles

We have obtained complete thermodynamic profiles for the association of netropsin to each oligomer duplex according to the reaction

$$\text{Duplex} + 2 \text{ netropsin} \rightarrow \text{complex} \tag{8}$$

Values for all thermodynamic parameters at 20° ($\Delta G°$, ΔH, $T\Delta S$, and ΔV) on the binding of netropsin to each of its two sites of each oligomer are listed in Table II. Overall binding affinities were obtained from UV melting curves, using the increase in thermal stability of the saturated 2:1 complexes relative to the ligand free oligomers. Binding of netropsin results in K values in excess of 10^7 for each site of the oligomers and correspond to the strong specificity of the ligands for $dA \cdot dT$ base pairs. These K values were slightly higher for the GT_4A_4C oligomers. The $\Delta G°$ values, derived from $\Delta G° = -RT \ln K$, corresponded to 0.4 kcal/mol more favorable for the GT_4A_4C oligomers.

ITC experiments at 20° yielded the interacting heats and confirmed the 2:1 stoichiometry of the netropsin–oligomer complexes. Molar binding enthalpies were calculated from the fits of the calorimetric binding isotherms; these values are listed for each binding site in Table II. Inspection of the ΔH column shows that these values varied with both the oligomer used and type of site. Binding to the first site of GA_4T_4C was marginally endothermic while all others were exothermic; the ΔH values for each site of GT_4A_4C were highly exothermic. The entropies

[36] C. R. Cantor and P. R. Schimmel, *in* "Biophysical Chemistry," p. 226. W. H. Freeman and Company, New York, 1980.

[37] H.-H. Wu and D. M. Crothers, *Nature (London)* **308,** 509 (1984).

[38] P. J. Hagerman, *Proc. Natl. Acad. Sci. U.S.A.* **81,** 4632 (1984).

of netropsin interacting with each site of the oligomers were calculated using the Gibbs equation and are shown in Table II. Although similar free energy changes were obtained for the binding of netropsin to this pair of isomeric oligomers, the origin of the energy contributions was quite different. For the first site of GA_4T_4C, the binding energy was enthalpic in origin while for the second site the result was more or less equally enthalpy and entropy driven, whereas for each site of GT_4A_4C the binding energy was enthalpic driven.

In the volume change measurements, the two reactants were mixed using batch experiments in two different ways: to form 1 : 1 ligand–oligomer complexes and this was immediately followed by the addition of a second equivalent of ligand, and to form directly the 2 : 1 complexes; both methods yielded similar results shown in the ΔV column of Table II. Binding of netropsin to the first site of each oligomer generated a volume contraction while binding to the second site a volume expansion is observed. The magnitude of both of these volume effects were higher with the GA_4T_4C oligomer than for its GT_4A_4C isomer. Thus, at this salt concentration of 0.1 M, we infer that a net hydration accompanies the total binding to these oligomer duplexes as reflected by the apparent compression of water molecules. The ΔV values indicate that this net hydration is strongly dependent on the DNA sequence and the type of site of the oligomer duplex. Furthermore, the calculated entropy values are in good agreement with the volume change measurements for three of the four sets of thermodynamic profiles; the exception is the binding of netropsin to the first site of GA_4T_4C, which has these parameters with opposite signs.

Differential Profiles of Netropsin Binding to Oligomers

We have recast the thermodynamic data in differential terms by subtracting those obtained in a given oligomer reaction with the ligand from those of the compositional isomer. Each of the thermodynamic parameters for the formation of the netropsin–GA_4T_4C complex is subtracted from the corresponding parameters of forming the netropsin–GT_4A_4C complex. This assumes similar structures of the ligand–oligomer complexes; hence the binding affinities are almost identical. The contributions from the two netropsin molecules and complexes cancel out exactly. The resulting differential thermodynamic profiles correspond to the reaction

$$[d(GA_4T_4C)]_2 \rightarrow [d(GT_4A_4C)]_2 \qquad (9)$$

For the overall reaction, we obtain a small $\Delta\Delta G°$ term of 0.8 kcal/mol, which is the result of a nearly total compensation of an unfavorable $\Delta\Delta H$

of 18 kcal/mol with a favorable $\Delta(T\Delta S)$ of 17.2 kcal/duplex, and a $\Delta\Delta V$ of 5 ml/mol of duplex. The overall results are consistent with $[d(GA_4T_4C)]_2$ being slightly more hydrated. However, the small net volume expansion (dehydration) is not reflected in the magnitude of the differential entropy gain.

To obtain a better understanding of the observed differences in this set of oligomers, we have dissected the overall differential binding profiles into the individual contributions for netropsin binding to the first and second site, respectively, by using thermodynamic cycles (Eq. (9)) that result in parameters that are shown at the bottom of Table III. We obtained marginal $\Delta\Delta G°$ terms for each binding site that result from positive $\Delta\Delta H - \Delta(T\Delta S)$ compensations; binding of netropsin to the first site yielded a negative $\Delta\Delta V$, and for the second site a positive $\Delta\Delta V$ was obtained. These results indicate that in the sequential binding of netropsin there is an initial increase in hydration, due to a hydrophobic or structural effect, that is followed by a dehydration event due to electrostriction effects. These results correlate with those obtained in the analysis of the electrophoretic mobility of the free and netropsin-bound oligomer duplexes (data not shown),[19] in which the mobility of the free $[d(GA_4T_4C)]_2$ is lower than that of the free $[d(GT_4A_4C)]_2$; however, the 1:1 netropsin–oligomer complexes ran with similar mobilities while for the 2:1 complexes the trend is reversed. These results strongly indicate that binding of the first netropsin is straightening the curved $[d(GA_4T_4C)]_2$ duplex, yielding an important correlation with our volumetric studies in that the straightening of this oligomer invokes the participation of structural water.

Molar Volume of Water in Vicinity of Five dA · dT Base Pairs

Two types of processes reported here, (1) binding of minor groove ligands to stretches of 5 dA · dT and (2) the inclusion of an extra dA · dT base pair during duplex formation to yield a stretch of 5 dA · dT base pairs, have one thing in common. They refer to thermodynamic events for the hydration of 5 dA · dT base pairs. As such, we can use our differential thermodynamic profiles to estimate the physical properties of electrostricted water in the vicinity of dA · dT base pairs. To estimate the molar volume for this type of water, we have averaged the differential thermodynamic parameters of both ligand interactions and obtained the following: $\Delta\Delta G° = 1.1$ kcal/mol, $\Delta\Delta H = 13.4$ kcal/mol, $\Delta(T\Delta S°) = 12.3$ kcal/mol, and $\Delta\Delta n_{Na^+} = 0.08$. We use the value of 0.3 (± 0.1) kcal/mol for the enthalpy of releasing 1 mol of water from native DNA[25,26] and noting that $\Delta\Delta H = \Delta(T\Delta S°) = 12.3$ kcal/mol, we arrived at a value of 41 mol of water molecules

TABLE III
DIFFERENTIAL PROFILES FOR APPROPRIATE THERMODYNAMIC CYCLES[a]

Ligand	$\Delta\Delta G°$ (kcal/mol)	$\Delta\Delta H$ (kcal/mol)	$\Delta(T\Delta S)$ (kcal/mol)	$\Delta\Delta V$ (ml/mol)
Inclusion of Extra dA · dT Base Pair				
	−0.5	−5.6	−5.1	−41
Inclusion of Bulges				
	6.4	27.4	21	−222
Poly(dA) · poly(dT) Minus Poly[d(AT)] · poly[d(AT)]				
Netropsin	0.8	11.8	11.0	113
Distamycin	1.5	15.1	13.6	99
Parallel DNA Minus Antiparallel DNA				
	9.0	42.1	33.1	257
Inclusion of Mismatch				
	2.4	9.0	6.6	−19
Straightening Bent Sequences				
Portion of oligomer bound				
Overall	0.8	18.0	17.2	5
First site	0.4	12.3	11.9	−17
Second site	0.4	5.7	5.3	22

[a] All parameters calculated under the experimental conditions indicated in Tables I and II.

per mole of a 5 base-pair binding site. This value suggests that the homopolymer is more hydrated than the alternating one by ~8 mol of water per dA · dT base pair. From the average of the total volume change measurements, $\Delta\Delta V$ = 106 ml/mol, we estimate an average contraction of electrostricted water molecules as 2.6 ml/mol at 20° (106 ml/mol divided by 41 mol of water per mole of 5 base pairs). If instead, we use the differential parameters of duplex formation (Table III), $\Delta\Delta H = \Delta(T\Delta S°) = -5.1$ kcal/mol, we obtained from this heat exchange that the formation of an extra

$dA \cdot dT$ is accompanied by an overall increase in hydration equal to 17 mol of water per mole of base pair and an average volume contraction (41 ml/mol divided by 17 mol/mol of base pair) of 2.4 ml/mol at 20°. This average value of 2.5 ml/mol for the overall contraction in the molar volume of bound water to DNA (15.5 ml/mol for the molar volume) is in excellent agreement with the average molar volume of purely electrostricted water determined by Millero et al.[39] These results are encouraging because they tell us that the approach that we have used, by determining complete thermodynamic profiles for these biomolecular reactions, is a proper one in order to study the hydration of DNA as a function of sequence. Obviously, more sequences must be studied before generalizing these results.

Narrower Minor Grooves Yielding Increased Elecrostricted Hydration

We have shown previously that homoduplexes in the B-conformation are more hydrated, by an average of 3 mol of water per mole of base pair, than homoduplexes in the A-conformation.[10] Thus, the duplex with a narrower minor groove is more hydrated. We have also shown in the previous section that the narrower minor groove of AA/TT base-pair stacks[40] in poly(dA) $\cdot$ poly(dT) is more hydrated, by ~8 mol of water per $dA \cdot dT$ base pair, than the AT/TA (and TA/AT) base-pair stacks in poly[d(AT)] $\cdot$ poly[d(AT)]. A third example is the estimation of the differential hydration of parallel DNA relative to antiparallel DNA. For this estimation, we just divide the resulting differential compensation term, $\Delta\Delta H = \Delta(T\Delta S) = 33.1$ kcal/mol, by the heat of 0.3 kcal/mol of releasing 1 mol of electrostricted water from native DNA, to obtain 110 ± 24 mol of H_2O per mole of duplex or ~4.4 ± 1 mol of H_2O per mole of base pair from this heat exchange experiment. If, however, we use 2.5 ml/mol for the average compression of electrostricted water around the negatively charged DNA of $dA \cdot dT$ base pairs, our $\Delta\Delta V$ of 257 ml/mol of duplex reduces to 103 ± 7 mol of electrostricted H_2O per mole of duplex or ~4 mol of H_2O per mole of base pair from the volume-change measurements. The significant observation is that the antiparallel DNA duplex with a narrower minor groove is more hydrated by an average of four water molecules per mole of base pair.

Thus far, the preceding three comparisons are in agreement that those entities showing the largest electrostricted hydration are the ones

---

[39] F. J. Millero, G. K. Ward, F. K. Lepple, and E. V. Hoff, *J. Phys. Chem.* **78**, 1636 (1974).
[40] H. C. Nelson, J. T. Finch, B. F. Luisi, and A. Klug, *Nature (London)* **330**, 221 (1987).

with the narrowest minor groove. This effect may be explained in terms of the presence of optimum base-pair stacking interactions in the duplexes with narrower minor grooves yielding higher linear charge density parameters. This in turn immobilizes an additional amount of electrostricted water molecules.

Hydrophobic versus Electrostrictive Immobilized Water

We used a combination of densimetric, calorimetric, and temperature-dependent UV spectroscopy techniques to measure complete thermodynamic profiles for two types of processes: (1) the formation of deoxyoligonucleotide duplexes from mixing their respective complementary single strands; and (2) the interaction of ligands with the minor groove of both synthetic polymers and oligomers containing $dA \cdot dT$ stretches. By constructing appropriate thermodynamic cycles, for pairs of reactions, we obtained differential thermodynamic profiles that correspond to the inclusion of base pairs, bulges, or mismatches in DNA oligomer duplexes, and to the comparison of the relative hydration of pairs of DNA molecules in their free states. We find quite sharp correlations with the amounts and kinds of hydrating water. In summary (see Table III), we note that if the sign (plus or minus) of the differential enthalpy–entropy compensation, $\Delta\Delta H - \Delta(T\Delta S)$, is the same as the sign of the differential volume change, $\Delta\Delta V$, electrostricted hydration predominates. If instead, the signs are opposite for $\Delta\Delta H - \Delta(T\Delta S)$ and $\Delta\Delta V$, hydrophobic or structural type of hydration becomes overriding. This conclusion also shows that the sign of the differential free energy, $\Delta\Delta G$, is the same as that of the $\Delta\Delta H - (T\Delta S)$ compensation. Hence, we suggest that direct comparison of the sign of the $\Delta\Delta G$ term with the sign of the $\Delta\Delta V$ term presents immediately the distinction between electrostrictive and hydrophobically immobilized waters.

Another important correlation resulting from these studies is the estimation of the molar volume of electrostricted water, equal to ~15.5 ml/mol, in the vicinity of 5 $dA \cdot dT$ base pairs of B-DNA; in other words, this type of water experiences an electrostatic field that compresses its volume by ~14%. This result can be used to estimate the overall electrostricted hydration of nucleic acid duplexes.

Acknowledgments

The technical assistance of Drs. Dionisios Rentzeperis and Karen Alessi, and of Krzyztof Zieba, is greatly appreciated. This work was supported by Grant GM42223 (L.A.M.) from the National Institutes of Health.

[19] Time-Resolved Fluorescence Methods for Analysis of DNA–Protein Interactions

By DAVID P. MILLAR

Introduction

Fluorescence spectroscopy is an established technique for the analysis of macromolecular structure and dynamics and, in more recent studies, is emerging as a promising method for the study of DNA–protein interactions. The chief advantage of fluorescence-based approaches over the gel electrophoretic and filter-binding methods commonly used to characterize DNA–protein interactions is the ability to monitor binding events in solution under conditions of true thermodynamic equilibrium. In addition, fluorescence measurements provide high detection sensitivity, enabling measurements to be performed at concentrations approaching the K_d values characteristic of many DNA–protein complexes, and can accommodate a wide range of solution conditions, including those that are incompatible with gel-based methods. The most important feature of fluorescence spectroscopy, however, is the characteristic time scale of the emission process itself. The emission of a photon occurs over a time period ranging from picoseconds to tens of nanoseconds, during which a variety of dynamic processes can deactivate the excited-state and compete with the fluorescence. These processes include solvent relaxation, intersystem crossing, charge transfer, resonance transfer of excitation energy, and quenching by exogenous agents. As a result, emission properties are extremely sensitive to the local environment. Moreover, the rapid time scale of the emission process allows for the direct observation of various types of molecular motion, including overall rotational diffusion and internal motion. Measurements of these motional properties provide a sensitive means of monitoring macromolecular complexation.

Because DNA has no measurable intrinsic fluorescence in solution at room temperature, most fluorescence studies of DNA–protein interactions require the attachment of fluorescent probes to DNA. With advances in solid-phase synthesis and labeling methodologies, a variety of fluorescent probes can be readily attached to the 5' or 3' terminus or placed at specific internal positions within a DNA oligonucleotide. The formation of a complex between labeled DNA and a specific DNA-binding protein may be monitored by a change of emission intensity and/or polarization anisotropy of the extrinsic probe, in both steady-state and time-resolved experiments.

Steady-state fluorescence methods provide a relatively simple tool for monitoring DNA–protein complexation and have been used successfully to determine binding affinities for a variety of DNA-binding proteins. These applications have been reviewed elsewhere.[1–4] This chapter focuses instead on the use of time-resolved fluorescence methods for characterizing the interaction between labeled DNA and specific binding proteins.

Time-resolved fluorescence spectroscopy is a powerful technique for probing structural and dynamic features of macromolecular systems.[5,6] There are two types of experiments that can be performed using time-resolved fluorescence techniques. The first involves measurement of the time decay of emission following a brief excitation pulse. This is used to determine the fluorescence lifetime of a fluorophore, which reflects the average time that the molecule remains in the excited singlet state. The fluorescence lifetime is strongly dependent on the molecular properties of the fluorophore environment and can vary from a few picoseconds to tens of nanoseconds. In the case of fluorophore-labeled DNA, lifetime measurements can be used to detect changes in the local environment of specific bases on complexation with a DNA-binding protein. The other type of time-resolved fluorescence experiment is based on measurement of the fluorescence anisotropy decay. The anisotropy decay monitors reorientation of the emission dipole during the excited-state lifetime and can report on local fluorophore motion, segmental motions, or overall rotational diffusion of DNA. The chief advantage of time-resolved anisotropy measurements over steady-state measurements is that the former can be used to separate the effects of protein binding on the overall motion of the DNA and the local rotations of the probe. Changes in the rate or amplitude of local motion on protein binding can provide information on DNA–protein contacts at specific sites. Thus, the information available from fluorescence lifetime and anisotropy decay experiments can be used to infer structural details about the DNA–protein complexes, which is of particular value in the case of large or disordered complexes that are not amenable to direct structural analysis by X-ray crystallographic or nuclear magnetic resonance (NMR) methods. Time-resolved fluorescence and site-specific labeling methods are also useful for analyzing complex DNA–protein interactions involving multiple modes of binding. In favorable cases, time-resolved fluorescence signals can be deconvoluted into contributions from the different binding

[1] D. M. Jameson and W. H. Sawyer, *Methods Enzymol.* **246,** 283 (1995).
[2] T. Heyduk, Y. Ma, H. Tang, and R. H. Ebright, *Methods Enzymol.* **274,** 492 (1996).
[3] J. J. Hill and C. A. Royer, *Methods Enzymol.* **278,** 390 (1997).
[4] J. R. Lundblad, M. Laurance, and R. H. Goodman, *Mol. Endocrinol.* **10,** 607 (1996).
[5] A. R. Holzwarth, *Methods Enymol.* **246,** 334 (1995).
[6] D. P. Millar, *Curr. Opin. Struct. Biol.* **6,** 637 (1996).

modes of the DNA–protein complex, allowing for direct quantitation of the various species present in solution. When used in conjunction with techniques for manipulating the structure of the DNA and/or protein, such measurements can be used to dissect the interactions that stabilize the different modes of binding and to obtain thermodynamic information on the energetics of the interactions.

The purpose of this chapter is to review the application of time-resolved fluorescence spectroscopy to the study of DNA–protein interactions. The chapter begins with a brief overview of general principles, focusing on those time-resolved fluorescence techniques that are most pertinent to the study of DNA–protein interactions. In addition, the overview includes a brief description of the instrumentation required for time-resolved fluorescence measurements and the methods employed in site-specific fluorescent labeling of DNA. A number of representative studies are then presented to demonstrate the different experimental strategies that can be used to characterize DNA–protein complexes. The specific examples presented have also been chosen to illustrate the ability of time-resolved fluorescence techniques to yield both structural and energetic information on DNA–protein interactions. The interested reader is referred elsewhere for more detailed accounts of time-resolved fluorescence spectroscopy and other biochemical applications.[5,6]

General Considerations

Time-Resolved Fluorescence Parameters

The experiments described in this chapter are based on measurements of the time-resolved emission of fluorescent probes attached to DNA and the influence of bound proteins on the fluorescence properties. Before considering specific examples, it is useful to briefly review the time-resolved emission parameters that are most pertinent to the study of DNA–protein interactions.

The fluorescence lifetime of a probe attached to DNA is a useful parameter for the analysis of DNA–protein interactions because it is extremely sensitive to changes in the local environment. This environmental dependence arises because the fluorescence lifetime is governed by a variety of dynamic processes that can deactivate the excited state, as expressed mathematically in Eq. (1):

$$\tau = (\Sigma\, k_r + k_{nr} + k_p + k_{et} + k_Q[Q])^{-1} \tag{1}$$

where τ is the fluorescence lifetime, k_r is the radiative rate constant, k_{nr} is the rate constant of all nonradiative processes including intersystem crossing and internal conversion, k_p is the rate constant for photochemical processes such as excited-state charge transfer and isomerization, k_{et} is the rate constant for energy transfer when an acceptor is present, and k_Q is the rate constant for quenching by an exogenous quenching agent Q with concentration [Q]. Each of the rate constants in Eq. (1) depends on the properties of the molecular environment of the fluorescent probe. Accordingly, changes in the local environment of a fluorescent probe that occur on the binding of a protein to a labeled DNA molecule can be detected through their effect on the fluorescence lifetime, although it is often difficult to interpret the lifetime changes in terms of individual kinetic processes. Nonetheless, the fluorescence lifetime of a probe attached to DNA is an extremely sensitive indicator of protein binding.

The fluorescence lifetime is measured experimentally by analyzing the response of the emission to pulsed or modulated excitation. The instrumentation required for such measurements is discussed briefly in the following section. For a single fluorophore in a homogeneous environment, the intensity decays exponentially with a single lifetime. In biological systems, including labeled DNA oligomers and DNA–protein complexes, the decay is more commonly described by multiple exponentials [Eq. (2)]:

$$I(t) = \sum_i \alpha_i \, \exp(-t/\tau_i) \tag{2}$$

where α_i are the amplitudes associated with each lifetime τ_i. These parameters are determined by nonlinear least-squares fitting of the intensity decay. The heterogeneity underlying Eq. (2) can arise from distinct ground-state conformations of the labeled DNA or DNA–protein complex, in which case the α_i terms reflect the relative populations of each species. In such instances, the average fluorescence lifetime of a probe is defined according to Eq. (3):

$$\tau_{ave} = \sum_i \alpha_i \tau_i \tag{3}$$

Formation of a DNA–protein complex can also alter the motional properties of a fluorescent probe attached to the DNA. The rotational motions of a probe occurring in the picosecond to nanosecond time range can be directly observed in time-resolved fluorescence depolarization experiments. For a probe that is excited by plane-polarized light, different intensity decays will be observed for the emission that is polarized parallel or perpendicular to the excitation polarization. This difference is expressed by the time-dependent fluorescence anisotropy, defined in Eq. (4):

$$r(t) = \frac{I_\parallel(t) - I_\perp(t)}{I_\parallel(t) + 2I_\perp(t)} \tag{4}$$

where $I_\parallel(t)$ and $I_\perp(t)$ are polarized intensity decays observed parallel or perpendicular to the vertical excitation polarization, respectively. The polarized excitation creates an initial anisotropy in the sample that subsequently decays over time as molecules randomize their orientations. For a probe attached to a DNA oligomer, two types of depolarizing motions can be observed, corresponding to local rotation of the probe and overall tumbling of the DNA. Each motional process is associated with a separate exponential decay term in the time-dependent anisotropy [Eq. (5)]:

$$r(t) = \beta_1 \exp(-t/\phi_1) + \beta_2 \exp(-t/\phi_2) \qquad (5)$$

where ϕ_1 is the correlation time for local probe motion, ϕ_2 is the rotational correlation time of the DNA, and β_1 and β_2 are corresponding amplitude factors (the sum of both amplitudes is equal to the limiting anisotropy, an intrinsic property of the particular fluorophore). Binding of a protein to the DNA will generally increase the rotational correlation time, ϕ_2, because the DNA–protein complex has a larger mass than the DNA alone and will therefore tumble more slowly in solution. Complexation with a protein may also restrict the local rotation of the probe, the latter being manifested by a decrease in β_1 and a corresponding increase in β_2. The decrease in β_1 can be interpreted as a reduction in the angular range of motion of the probe at its point of attachment to the DNA. This motional restriction of the probe can be expressed more directly by relating the anisotropy amplitude factors to the cone half-angle describing the angular excursions of the probe [Eq. (6)]:

$$\theta = \cos^{-1}\{\tfrac{1}{2}[(1 + 8S)^{1/2} - 1]\} \qquad (6)$$

where $S = [\beta_2/(\beta_1 + \beta_2)]^{1/2}$. Motional restriction of the probe is directly reflected in the cone half-angle, which may decrease on complexation of the labeled DNA with a protein. The change in tumbling rate that results from protein binding reflects a global property of the DNA–protein complex, whereas the restriction in local motion of the probe is a result of specific steric interactions and is highly dependent on the position of the probe within the three-dimensional structure of the DNA–protein complex.

In view of the above considerations, it is expected that the fluorescence lifetime and/or local rotation of a probe attached to DNA can respond to changes in the microenvironment induced by protein binding. As noted above, these parameters report on different properties of the probe environment. Accordingly, by labeling DNA at different positions within the sequence, fluorescent probes can be used to map the points of contact between a protein and its DNA target and to probe variations in the local microenvironment within the resulting DNA–protein complex. Spectroscopic footprinting experiments based on this concept are discussed later.

Time-resolved fluorescence methods are particularly important for the analysis of complex DNA–protein interactions involving two or more distinct modes of binding. For example, two local probe environments can exist for a labeled DNA molecule that binds to two distinct sites on a protein, giving rise to heterogeneity in the fluorescence lifetime and/or rotational motions of the probe. The applications discussed in this chapter deal with the case in which there is a fraction of long-lived, slowly rotating probes (hereafter termed "buried") mixed in with a bulk population that is more quenched and faster rotating (hereafter termed "exposed"). This situation applies, for example, to dansyl-labeled DNA substrates bound to the separate polymerase and $3' \rightarrow 5'$-exonuclease sites of DNA polymerase (considered later under Applications). The time-dependent anisotropy observed for such a system contains contributions from each probe population [Eq. (7)]:

$$r(t) = f_e(t)r_e(t) + f_b(t)r_b(t) \qquad (7)$$

where $r_e(t)$ and $r_b(t)$ are the anisotropies for the exposed and buried probes, respectively. Both anisotropy decay functions can be represented in terms of local rotation and overall tumbling, as in Eq. (5), but the β terms are quite different for the two populations (β_1 is smaller for the buried probes). Moreover, the contribution of each population to the observed anisotropy changes over time because of the difference in fluorescence lifetimes. These contributions are expressed by the weighting factors $f_e(t)$ and $f_b(t)$, which in the case of the exposed probes can be represented as follows [Eq. (8)]:

$$f_e(t) = \frac{x_e \exp(-t/\tau_e)}{x_e \exp(-t/\tau_e) + x_b \exp(-t/\tau_b)} \qquad (8)$$

where x_e and x_b are the equilibrium mole fractions, and τ_e and τ_b are the fluorescence lifetimes of the exposed and buried probes, respectively (note that $x_e + x_b = 1$). An analogous expression applies for the contribution of the buried probes, $f_b(t)$. The expressions for $f_e(t)$ and $f_b(t)$ can be generalized to include more than one fluorescence lifetime for each probe population.

As a consequence of the different fluorescence lifetimes and mobilities of the exposed and buried probes, the time-dependent fluorescence anisotropy described by Eq. (7) can exhibit a distinctive "dip and rise" pattern consisting of an initial rapid decline, a rising portion at intermediate times, followed by a slow decay at longer times. The precise shape of the anisotropy decay is strongly dependent upon the actual fractions of exposed and buried probes, x_e and x_b, respectively. Provided that the two probe populations can be assigned to specific modes of binding, the analysis of such anisotropy decay patterns can be used to measure the relative amounts of the various bound forms. For example, this method has been used to quantify the partitioning of DNA substrates between the polymerase and $3' \rightarrow 5'$-exonuclease sites of DNA polymerase (discussed under Applications).

Instrumentation

The two most popular methods used to record time-resolved fluorescence data are the impulse-response and harmonic-response methods. The impulse-response method involves direct observation of the time decay of emission following a short excitation pulse. The excitation pulses are typically derived from mode-locked dye lasers or titanium:sapphire lasers, which produce picosecond or subpicosecond pulses and are widely tunable in the visible and ultraviolet regions of the spectrum. Emission decay profiles are recorded by the time-correlated single photon-counting technique using fast photomultiplier tubes, such as those employing microchannel plate design. Unfortunately, such instrumentation is not yet available in an integrated commercial system. However, sophisticated laser-based instrumentation for time-resolved fluorescence measurements is available in a number of national fluorescence user facilities funded by the National Science Foundation or the National Institutes of Health, including the Laboratory for Fluorescence Dynamics at the University of Illinois and the Center for Fluorescence Spectroscopy at the University of Maryland.

In the harmonic-response method, time-resolved fluorescence parameters are deduced by analyzing the response of the emission to sinusoidally modulated excitation. The light modulators and signal processing circuitry required for these measurements can be incorporated into the design of standard spectrofluorimeters and integrated systems are offered by a number of manufacturers (Jobin-Yvon Spex, Rutherford, NJ; Spectronics Instruments, Rochester, NY). In principle, the impulse-response and harmonic-response methods yield equivalent information, although there may be differences in detection sensitivity, time resolution, and the statistical quality of the data depending on the specific details of the instrumentation used. In the author's experience, the use of time-correlated single photon counting with laser excitation and a microchannel plate detector allows for the recording of fluorescence decay profiles with an extremely high degree of precision and a time resolution on the order of 10 psec or less. Detailed descriptions of time-resolved fluorescence instrumentation are presented elsewhere.[7,8]

DNA-Labeling Methods

As noted previously, fluorescence-based studies of DNA–protein interactions require the attachment of fluorescent probes to the DNA. A variety

---

[7] E. W. Small, *Top. Fluoresc. Spectrosc.* **1,** 97 (1991).
[8] J. R. Lakowicz and I. Gryczynski, *Top. Fluoresc. Spectrosc.* **1,** 293 (1991).

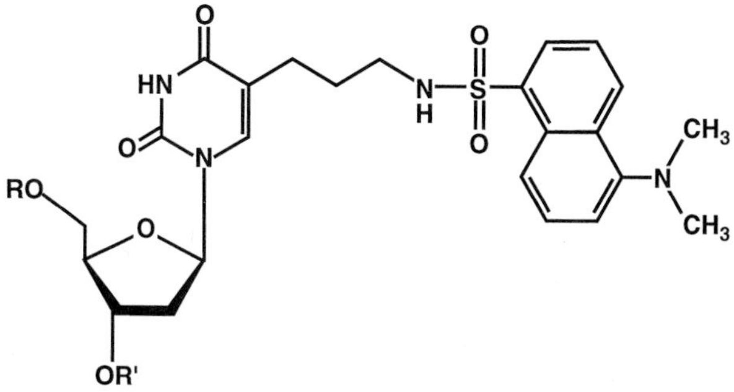

FIG. 1. Fluorescent labeling of deoxyuridine with a dansyl probe.

of methods exist for site-specific fluorescent labeling of DNA oligonucleo-tides, including labeling at the 3′ or 5′ end, internal labeling of bases, and incorporation of fluorescent base analogs. These methods are described in detail elsewhere.[1,9,10] The applications outlined in this review all involve internal labeling of specific bases with extrinsic fluorescent probes. Incorporation of a derivatizable base analog into DNA provides a convenient method of site-specific labeling. For this purpose, 5-aminopropyluridine can be substituted for thymidine during solid-phase oligonucleotide synthesis using a suitably protected phosphoramidite monomer.[11] After incorporation into oligonucleotides and subsequent deprotection of the primary amine group, the oligonucleotide is reacted with a succinimidyl ester, isothiocyanate, or sulfonyl chloride derivative of the dye of interest.[9] The unreacted dye is removed by size-exclusion chromatography and the dye–DNA conjugate is subsequently purified by reversed-phase HPLC. Because conjugation occurs at the 5-position of the uracil ring, the attached probe and linker arm (Fig. 1) project into the major groove of the DNA double helix and have minimal effect on the base pairing or helical structure of the DNA.

Applications

Fluorescence Footprinting

A variety of methods are used to map the binding sites of proteins on DNA, including DNase I footprinting, hydroxyl radical footprinting, and chemical modification interference. These methods are based on the

[9] A. Waggoner, *Methods Enzymol.* **246,** 362 (1995).
[10] D. P. Millar, *Curr. Opin. Struct. Biol.* **6,** 322 (1996).
[11] K. J. Gibson and S. J. Benkovic, *Nucleic Acids Res.* **15,** 6455 (1987).

accessibility of the DNA phosphate backbone to enzymes and small molecule reagents. The resolution of DNase I footprinting is limited by the size of the enzyme and no information is obtained on the local environment of bases within the DNA–protein complex. The resolution of hydroxyl radical and chemical footprinting is greater, but little information is available on the structural and dynamic properties of the microenvironment of particular bases in the DNA–protein complex. An alternative method for DNA footprinting involves the use of fluorescent probes that are attached to specific bases within the DNA sequence. As noted above, fluorescent probes are able to reflect changes in the structure and dynamics of their surrounding environment through measurable alterations in emission properties, particularly in time-resolved experiments. This capability can be exploited for footprinting studies by placing the probe at successive positions along the DNA helix, which is readily achieved using the labeling method described above. Labeled positions that come into contact with the protein on its binding to the DNA are readily identified from the resulting changes in the probe emission. Moreover, information is also obtained on the polarity and rigidity of the microenvironment surrounding specific bases within the DNA–protein complex. Another advantage is that resonance energy transfer strategies can be used to obtain structural information about the DNA–protein complex,[12] a particular advantage in the case of large or disordered complexes that are not amenable to high-resolution structural analysis by X-ray diffraction or NMR spectroscopic methods.

Guest et al.[13] have used time-resolved fluorescence spectroscopy to study the interaction between DNA and the Klenow fragment of DNA polymerase I. DNA Pol I is a multifunctional enzyme involved in DNA replication and repair. In addition to the $5' \rightarrow 3'$-polymerase activity responsible for template-directed synthesis of DNA, the enzyme has $3' \rightarrow 5'$- and $5' \rightarrow 3'$-exonuclease activities located in separate structural domains. The large proteolytic fragment of Pol I, termed the Klenow fragment, retains the polymerase and $3' \rightarrow 5'$ exonuclease activities. This fragment has served as a model system for more complex DNA replication enzymes.[14] The crystal structure of the free protein has been solved, but there is no direct structural data available on the binding of a duplex DNA substrate to the polymerase active site. To study DNA–polymerase interactions by

[12] W. S. Furey, C. M. Joyce, M. A. Osborne, D. Klenerman, J. A. Peliska, and S. Balasubramanian, *Biochemistry* **37,** 2979 (1998).
[13] C. R. Guest, R. A. Hochstrasser, C. G. Dupuy, D. J. Allen, S. J. Benkovic, and D. P. Millar, *Biochemistry* **30,** 8759 (1991).
[14] C. M. Joyce and T. A. Steitz, *Annu. Rev. Biochem.* **63,** 777 (1994).

fluorescence spectroscopy, Guest *et al.* prepared a synthetic DNA primer–template with a dansyl probe attached to a modified uridine residue (Fig. 1) located next to the 3' end of the primer strand. Binding of the labeled DNA to Klenow fragment resulted in a substantial lengthening of the average fluorescence lifetime of the probe and a pronounced reduction in the amplitude of the short correlation time (β_1), which was interpreted as a decrease in the cone half-angle for the motion of the probe at its point of attachment to the DNA (from 40 to 19°). The increase in fluorescence lifetime was attributed to a less polar environment within the polymerase active site, based on the known photophysical properties of the dansyl group in different solvents. The large decrease in the cone half-angle indicated that the dansyl group was rotationally restricted within the polymerase active site. On the basis of these changes, it was concluded that the polymerase was in close contact with the labeled base adjacent to the primer 3' terminus and that water molecules were excluded from the active site.

Nucleoside triphosphates were then incorporated to move the probe to positions 2, 4, and 7 bases upstream from the primer 3' terminus. The fluorescence lifetime and local rotation of the dansyl probe at positions 2 and 4 were also indicative of direct DNA–protein contacts, showing that the footprint of the polymerase extended over this region of the DNA, although the local motion was less restricted (cone half-angles of 27 and 31°, respectively) and the surrounding environment was less hydrophobic than observed in the original complex with the probe adjacent to the primer terminus. In contrast, the properties of the probe at position 7 were indicative of an exposed, aqueous environment, although a small fraction of buried probes was also observed. As noted in the following section, these buried probes arise from a minor population of primer templates bound at the 3'→5'-exonuclease site of the polymerase. Thus, these results indicated that 5 or 6 bp of duplex DNA upstream from the primer 3' terminus are in contact with the enzyme when the primer terminus is at the polymerase active site and that there are significant variations in the local microenvironment within this contact region.

Fluorescence footprinting methods are not limited to the study of DNA polymerases, however. Bailey *et al.*[15] have used similar methods to examine the interaction between the regulatory protein TyrR and a 42-bp oligonucleotide containing a centrally located binding site (TyrR box). A series of oligonucleotides was prepared in which a fluorescein probe was conjugated separately to positions within and adjacent to the TyrR box (Fig. 2). The labeling scheme was similar to that shown in Fig. 1, except that the

[15] M. Bailey, P. Hagmar, D. P. Millar, B. E. Davidson, G. Tong, J. Haralambidis, and W. H. Sawyer, *Biochemistry* **34**, 15802 (1995).

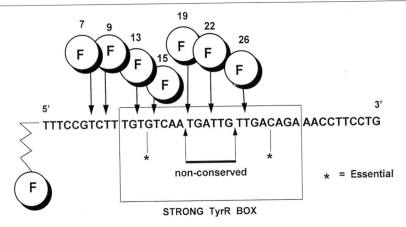

Fig. 2. Design of oligonucleotides labeled at specific bases within and adjacent to the TyrR box. F denotes a fluorescein probe, and the numbers refer to the position of the labeled residue from the 5' end. Nonconserved bases in the TyrR consensus sequence are indicated, together with the positions of essential C and G residues (*). [Reproduced from M. Bailey et al., Biochemistry **34,** 15802 (1995), with permission.]

probe was attached to the uracil ring via an aminopropynyl rather than an aminopropyl linkage. As in the case of the DNA polymerase study, binding of TyrR to the labeled DNAs resulted in significant changes in both the fluorescence lifetime and local rotational motions of the probe, and these changes were dependent on the position of the probe within the DNA sequence (Fig. 3). An interesting aspect of this study was that the DNA–protein interactions were examined for two different oligomeric states of the protein. In the absence of coeffectors, the TyrR protein exists in solution as a dimer composed of 57.6-kDa subunits. On binding of the corepressor tyrosine, however, the molecule undergoes self-association from a dimer to a hexamer. Thus, Bailey et al. were able to characterize separately the binding of the TyrR dimer or the TyrR hexamer to DNA by carrying out the fluorescence measurements in the absence or presence of tyrosine, respectively. The effects of dimer binding were most obvious for oligonucleotides labeled at positions 13, 15, and 26 within the TyrR box, as reflected by an increase in the average fluorescence lifetime of the probe and a decrease in the anisotropy amplitude associated with local probe motion (Fig. 3). These spectroscopic changes indicated that the TyrR dimer makes close contacts with these bases within the TyrR box. Interestingly, positions 19 and 22 within the central nonconserved 6-bp segment were relatively insensitive to dimer binding but showed large changes on binding of the TyrR hexamer. Similarly, positions 7 and 9 adjacent to the TyrR box were sensitive to hexamer binding but not to dimer binding. These

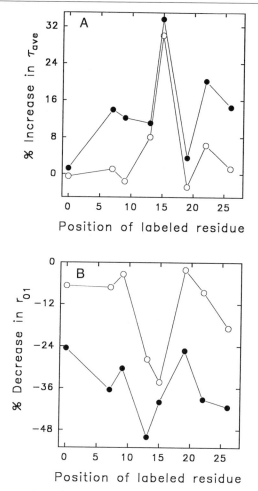

Fig. 3. Fluorescence footprinting of TyrR bound to DNA. (A) Effect of the labeling position on the change in the average fluorescence lifetime of fluorescein-labeled DNA in the presence of 2.4 μM TyrR dimer (open symbols) or 2.4 μM TyrR plus 470 μM tyrosine (filled symbols). The TyrR protein self-associates from dimer to hexamer in the presence of tyrosine. (B) Change in the amplitude of the fast anisotropy decay component (r_{01}), plotted as a function of the probe position within the DNA sequence. Note that r_{01} is equivalent to β_1 in Eq. (5). [Reproduced from M. Bailey, *et al.*, *Biochemistry* **34,** 15802 (1995), with permission.]

results indicated that the hexameric form of TyrR makes more extensive contacts with the DNA than does the TyrR dimer, either by providing extra protein surface with which the flanking and internal regions of the DNA can make contact, or distorting the DNA so that spatially distant regions are brought into contact with the protein surface.

Spectroscopic Studies of DNA Polymerase Proofreading

As noted above, DNA polymerase I has $3' \rightarrow 5'$- and $5' \rightarrow 3'$-exonuclease activities in addition to the basic $5' \rightarrow 3'$-polymerase activity. The $3' \rightarrow 5'$-exonuclease activity performs a proofreading function by selectively removing misincorporated nucleotides from the $3'$ end of the primer strand. The polymerase and $3' \rightarrow 5'$-exonuclease activities must be tightly coordinated to ensure that mismatches are efficiently removed from the DNA without wasteful excision of correct nucleotides. Crystallographic and mutational studies of Klenow fragment have revealed that the $3' \rightarrow 5'$-exonuclease active site is separated from the polymerase active center by 25–30 Å,[14] raising the question of how the two sites work together during DNA replication. Although crystal structures exist for DNA–protein complexes with DNA bound at the $3' \rightarrow 5'$-exonuclease active site,[14] a solution method is needed to investigate the mechanisms that govern the competition between the two sites for binding the DNA substrate and that control movement of DNA between sites.

Carver et al.[16] have used time-resolved fluorescence anisotropy decay to monitor the partitioning of dansyl-labeled DNA substrates between the polymerase and $3' \rightarrow 5'$ exonuclease sites (hereafter called pol and exo sites) of Klenow fragment. The dansyl probe was attached to a uridine residue 7 bases upstream from the primer $3'$ terminus, the position of greatest sensitivity to the binding site on the polymerase, as evidenced by the fluorescence footprinting study of Guest et al. outlined in the previous section. The time-resolved fluorescence anisotropy decay profile for the DNA–protein complex exhibited an unusual "dip and rise" shape indicative of two different environments for the dansyl probe (Fig. 4). The two environments could be assigned to DNA primer templates bound at the pol site or the exo site (Fig. 5), on the basis of experiments with mismatched DNA substrates and with chemically modified DNA substrates that bound tightly to the pol site.[13] The anisotropy decays recorded for a wide variety of matched and mismatched DNA substrates bound to Klenow fragment could be uniquely analyzed in terms of a two-state model of exposed and buried dansyl probes [Eq. (7)], using a common set of lifetime and rotational parameters to describe each probe population. Examples of these global fits for a few representative DNA sequences are shown in Fig. 4. Thus, the relative fractions of primer termini bound at either the pol site or the exo site could be measured by this method. This information immediately yields the equilibrium constant, K_{pe}, describing the partitioning of a DNA primer template between the two active sites of the polymerase (Fig. 5).

---

[16] T. E. Carver, R. A. Hochstrasser, and D. P. Millar, Proc. Natl. Acad. Sci. U.S.A. **91**, 10670 (1994).

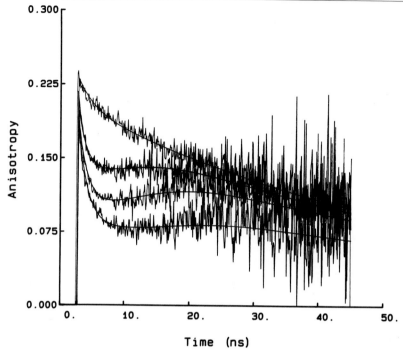

FIG. 4. Fluorescence anisotropy decays of dansyl-labeled primer-template DNA substrates bound to the Klenow fragment of DNA polymerase I. The DNA substrates contain zero to four mismatches, as follows (from bottom to top): matched DNA sequence; single G · G mismatch at the primer 3' terminus; an A · A mismatch 4 bases from the 3' terminus; four consecutive purine–purine mismatches at the primer terminus. The solid lines are from a global fit to a two-state model of exposed and buried probes [Eq. (7)]. [Reproduced from T. E. Carver, R. A. Hochstrasser, and D. P. Millar, *Proc. Natl. Acad. Sci. U.S.A.* **91,** 10670 (1994), with permission. Copyright (1994) National Academy of Sciences, U.S.A.]

This method of examining partitioning of a DNA substrate between the pol and exo sites has the advantage that the reporter group is distant from the primer 3' terminus. Thus, sequence changes can be introduced at the primer terminus without directly affecting the dansyl probe. As a result, the method is ideal for characterizing the effects of mismatches, frameshifts, and other mutagenic phenomena that exert their effect at the 3' end of the DNA primer. This approach was used to characterize the interaction of Klenow fragment with a variety of different mispaired DNA substrates in order to assess the contribution of mismatched base pairs to the energetics of DNA proofreading. Whereas matched sequences bound predominantly at the pol site of the enzyme, addition of an increasing number of mismatches caused the DNA to partition in favor of the exo site (Table I). These observations support the idea that the occupancy of the exo site is

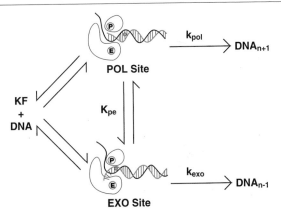

FIG. 5. Simplified scheme for a single round of DNA processing by the Klenow fragment. k_{pol} is the rate constant for addition of the next nucleotide to DNA bound at the polymerase active site and k_{exo} is the rate constant for exonucleolysis of DNA bound at the $3' \rightarrow 5'$-exonuclease site. The circle within the DNA helix represents a dansyl probe. The probe is located in a solvent-exposed environment when the primer $3'$ terminus is at the polymerase site and is buried in the protein interior when the terminus is at the $3' \rightarrow 5'$-exonuclease site. K_{pe} is the equilibrium constant describing the distribution of DNA substrates between the polymerase site and the $3' \rightarrow 5'$-exonuclease site. [Reproduced from T. E. Carver, R. A. Hochstrasser, and D. P. Millar, *Proc. Natl. Acad. Sci. U.S.A.* **91**, 10670 (1994), with permission. Copyright (1994) National Academy of Sciences, U.S.A.]

correlated with the melting capacity of the DNA terminus, which increases with mismatching in the duplex sequence. This, in turn, is consistent with biochemical data indicating that localized melting and unwinding of the primer $3'$ terminus is required for exonuclease activity on a duplex DNA substrate.

Close inspection of the results in Table I reveals that other effects can also influence the partitioning of a DNA substrate between the polymerase and $3' \rightarrow 5'$-exonuclease sites. For example, the $\Delta\Delta G$ values describing the change in the partitioning free energy due to mismatches (Table I) are larger than the differential change in the melting energy due to a mismatch, indicating that the melting capacity of the primer terminus is not the sole determinant governing the partitioning equilibrium.[16] In addition, DNA substrates containing embedded mismatches up to 4 bp from the primer terminus induce greater partitioning of DNA into the exo site than equivalent mismatches at the primer terminus itself (Table I). These observations suggest that unfavorable interactions between duplex DNA and the polymerase domain of the enzyme can also contribute to preferential partitioning of the primer $3'$ terminus into the exo site. Studies of polymerase

TABLE I

PARTITIONING OF DNA BETWEEN POLYMERASE AND $3'\rightarrow5'$-EXONUCLEASE SITES

DNA sequence[a]	K_{pe}[b]	$\Delta\Delta G$[c] (kJ mol^{-1})
Matched		
5'-TCGCAGCCG**U**CCAAGGG 3'-AGCGTCGGCAGGTTCCCATATAGCCGA	0.07	0
Single mismatch		
5'-TCGCAGCCG**U**CCAAGG<u>G</u> 3'-AGCGTCGGCAGGTTCC<u>T</u>ATATAGCCGA	0.18	−2.3
5'-TCGCAGCCG**U**CCAAGG<u>G</u> 3'-AGCGTCGGCAGGTTCC<u>A</u>ATATAGCCGA	0.22	−2.8
5'-TCGCAGCCG**U**CCAAGG<u>G</u> 3'-AGCGTCGGCAGGTTCC<u>G</u>ATATAGCCGA	0.22	−2.8
Internal mismatches		
5'-TCGCAGCCG**U**CCAAG<u>G</u>G 3'-AGCGTCGGCAGGTTC<u>G</u>CATATAGCCGA	0.63	−5.3
5'-TCGCAGCCG**U**CCAA<u>G</u>GG 3'-AGCGTCGGCAGGTT<u>G</u>CCATATAGCCGA	0.46	−4.6
5'-TCGCAGCCG**U**CCA<u>A</u>GGG 3'-AGCGTCGGCAGGT<u>A</u>CCCATATAGCCGA	0.70	−5.6
Multiple mismatches		
5'-TCGCAGCCG**U**CCAAG<u>GG</u> 3'-AGCGTCGGCAGGTTC<u>GG</u>ATATAGCCGA	≥19	≤−13.6
5'-TCGCAGCCG**U**CCAA<u>GGG</u> 3'-AGCGTCGGCAGGTT<u>GGG</u>ATATAGCCGA	≥19	≤−13.6
5'-TCGCAGCCG**U**CCA<u>AGGG</u> 3'-AGCGTCGGCAGGT<u>AGGG</u>ATATAGCCGA	≥19	≤−13.6
A-tract at primer terminus		
5'-TCGCAGCCG**U**CCAATTT 3'-AGCGTCGGCAGGTTAAAATATAGCCGA	1.33	0
5'-TCGCAGCCG**U**CCAATT<u>T</u> 3'-AGCGTCGGCAGGTTAA<u>T</u>ATATAGCCGA	0.35	3.3

[a] **U** denotes a modified uridine residue labeled with a dansyl probe (Fig. 1). Mismatched base pairs are underlined.

[b] Equilibrium constant for partitioning of DNA from the polymerase site to the $3'\rightarrow 5'$-exonuclease site.

[c] Change in free energy of partitioning relative to the corresponding matched DNA substrate.

TABLE II
EFFECTS OF MUTATIONS ON PARTITIONING OF
DNA BETWEEN POL AND EXO SITES

Mutation[a]	$\Delta\Delta G^b$ (kJ mol^{-1})
Leu361Ala	3.9
Phe473Ala	3.2
Glu357Ala	2.6
Tyr497Ala	−0.8
His660Ala	1.8

[a] Mutant enzymes also contained a Asp424Ala mutation to prevent substrate hydrolysis.
[b] Change in the free energy of partitioning of DNA from the polymerase site to the $3' \to 5'$-exonuclease site due to the specified mutation. The DNA substrate contained a $G \cdot T$ mismatch at the primer $3'$ terminus.

recognition of sequence-directed DNA structure reinforce this conclusion.[17] DNA substrates containing an A-tract motif (5'-AATTT-3') flanking the primer 3' terminus were observed to partition in favor of the exo site, despite the absence of mismatched base pairs within the DNA. In fact, A-tract DNA exhibited a greater preference for binding at the $3' \to 5'$-exonuclease site than a corresponding mismatched sequence (Table I). These results were interpreted in terms of unfavorable binding of the duplex DNA to the polymerase domain, rather than enhanced local melting, an effect that was attributed to the unusual helix geometry and intrinsic curvature of A-tract motifs. On the basis of these results, it appears that the polymerase domain of Klenow fragment prefers to bind straight segments of duplex DNA that conform to normal B-form geometry.

The time-resolved fluorescence anisotropy technique has also been used to analyze the effects of protein mutations on the partitioning of DNA substrates between the pol and exo sites of Klenow fragment.[18] Mutations (alanine replacements) were introduced into amino acid side chains that are seen by X-ray crystallography to be in close proximity to the 3' terminus of a DNA substrate bound to the exo site. The residues examined are those involved in direct contacts with the terminal bases of the primer strand, that serve as ligands to two divalent metal ions at the active site, or that

[17] T. E. Carver, Jr. and D. P. Millar, *Biochemistry* **37,** 1898 (1998).
[18] W.-C. Lam, E. J. C. Van der Schans, C. M. Joyce, and D. P. Millar, *Biochemistry* **37,** 1513 (1998).

interact with the sugar–phosphate backbone upstream from the point of hydrolysis. Each mutation was observed to have a different effect on the partitioning of a DNA substrate between two active sites of the polymerase, reflecting the loss of the binding energy contributed by the wild-type side chain at the 3′→5′-exonuclease site (Table II). The largest effects were observed for alanine replacements of Leu-361 and Phe-473, which destabilized binding of a mismatched DNA substrate to the exo site by 3.9 and 3.2 kJ mol^{-1}, respectively. The side chains of these residues are evidently important for binding DNA within the exonuclease domain, consistent with structural data showing that these residues make intimate contacts with the penultimate and terminal bases.[19] Thus, the DNA partitioning measurements can be used to evaluate the energetic contributions associated with structurally defined interactions. Moreover, information is obtained on the energetics of DNA–protein interactions at just one of the active sites of the polymerase, in contrast to standard binding measurements, which would simply yield an average of both sites. Another interesting observation is that the Tyr-497Ala mutation actually increased binding of DNA to the exo site (Table II), suggesting that the side chain of tyrosine interfered with binding. This observation indicates that this residue plays a more subtle role in the exonuclease reaction, possibly acting to strain the DNA substrate toward the transition state.

Acknowledgments

It is a pleasure to acknowledge the colleagues and collaborators who have contributed to the research described in this chapter: Ted Carver, Remo Hochstrasser, Wai-Chung Lam, Christopher Guest, Elizabeth Thompson, Edwin Van der Schans, Catherine Joyce, Stephen Benkovic, Lawrence Sowers, William Sawyer, and Michael Bailey. Work in the author's laboratory was supported by NIH Grant GM44060.

[19] L. Beese and T. A. Steitz, *EMBO J.* **10**, 25 (1991).

[20] Use of Fluorescence Spectroscopy as Thermodynamics Tool

By MAURICE R. EFTINK

Introduction

This chapter discusses the various ways that fluorescence techniques can be used to obtain thermodynamics information about biological macromolecules. Whereas one might think of fluorescence as a spectroscopic method, which of course it is, some of the most valuable applications of fluorescence are to determine thermodynamic and kinetic information about proteins, nucleic acids, and membrane systems.

Some of the advantages of fluorescence, over other experimental methods such as scanning or titration microcalorimeter or plasmon resonance, are that (1) fluorescence permits study with an extremely wide range of macromolecule concentrations, (2) fluorescence measurements are amenable to a wide range of solution conditions (i.e., a range of salt concentration, pH, temperature, pressure, etc.), (3) it is a rapid monitoring method, which makes it possible to have a rapid throughput in data collection, and (4) fluorescence can provide selectivity by focusing on a relatively few (or single) fluorescing centers within a macromolecule. This latter feature can be important, because it helps in our understanding of what molecular events are happening and it permits some thermodynamics studies to be carried out even when the samples are not ~100% pure. Many biological macromolecules contain intrinsic fluorescing groups, or these groups can be attached as extrinsic probes. Also, in cases involving a small molecule as a ligand, the ligand may be the fluorescing species.

Choice of Fluorescence Signal

Thermodynamic studies involve tracking the population of states under equilibrium conditions. The fundamental relationship that makes fluorescence useful in this regard is that fluorescence intensity, F, is directly proportional to molar concentration over a wide concentration range, i.e., $F_i \propto [X_i]$, a Beer's law type of relationship. If the total concentration of the fluorescing species remains constant in an equilibrium study, then one can write

$$F = \sum X_i F_i \tag{1}$$

where X_i is the mole fraction of species i and F_i is the relative fluorescence intensity of species i at some experimental condition (excitation and emission wavelengths, temperature, etc).* So long as the F_i of two or more

---

* Actually, Eq. (1) should include the possibility of different absorption coefficients at the excitation wavelength for the two or more species in equilibrium.[1] However, changes in the absorbance of a biological macromolecule, on some conformational transition, are usually much smaller than changes in fluorescence intensity. The discussion assumes that any fluorescence changes are due to changes in the fluorescence quantum yield, rather than to changes in absorbance.

molecular species is discernibly different, fluorescence intensity measurements can be used to track the relative population of the species. The same applies for measurements of fluorescence quantum yield, Φ, because the yield is directly proportional to fluorescence intensity.[1]

When the quantum yield (or intensity) is not significantly different for different thermodynamic states under a particular condition, this difference can sometimes be enhanced by adding a solute quencher or substituting deuterium oxide as solvent, because these agents can change the fluorescence intensity of solvent-exposed fluorophores more than internal fluorophores. Also, fluorescence intensity can be influenced by resonance energy transfer between donor and acceptor groups and any difference in energy transfer efficiency between different thermodynamic states can be used to advantage in fluorescence methods.

The practical range of fluorescence intensity measurements is hard to state, because the upper and lower range will depend on many factors, including the light source intensity, the effective cell path length, the quantum yield of the fluorophore, the magnitude of the fluorescence change, the photon detection efficiency of the instrument, and the extinction coefficient of the fluorophore. Even for tryptophan, the intrinsic probe often used in studies with proteins, with its modest extinction coefficient and quantum yield and with the lower light intensity at 280–300 nm for most excitation sources, a concentration range of 10^{-8} to 10^{-4} M can usually be observed with adequate precision. With extrinsic fluorescence probes such as fluorescein, which have higher extinction and yield and a better match with the output of laser excitation sources, the lower limit can usually be reduced by two or more orders of magnitude. The upper concentration limit is usually determined by the optical path length (e.g., optical thickness at the excitation and/or emission wavelengths) or practical considerations (e.g., solubility limits).

Whereas fluorescence intensity measurements are usually dependable in applications to thermodynamic studies, there are other fluorescence methods that should be used with caution. Time-resolved fluorescence lifetime measurements can in principle be used to track the population of states in equilibrium, but it is important to realize that it is the preexponentials associated with decay times, not the decay times themselves, that are related to a population of species (and the analysis of preexponentials can be complicated by excited state reactions). As an example of this application, and the complexity of trying to analyze fluorescence decay times, we have presented a study of the use of time-resolved fluorescence to track

[1] M. R. Eftink, *Biophys. J.* **66**, 482 (1994).

the thermal and chemical denaturant-induced unfolding of a protein.[2] Because it is difficult to correlate preexponential factors with the population of species, and because it is moderately inconvenient to make time-resolved fluorescence measurements, as compared with simple intensity measurements, time-resolved methods are recommended only when a direct relationship between fluorescence intensity and species concentration is not useful. The latter problem might exist when there is not a constant thickness of the illuminated sample, as might be the case in some studies with *in vitro* biological samples. In such experiments, changes in fluorescence decay profiles might be used to track the population of species in equilibrium, with the application known as fluorescence lifetime sensing or imaging.[3] A clever application of a time-resolved fluorescence method is the use of anisotropy decay data to estimate the partitioning of a DNA–protein complex between two possible binding modes, one of which limits the motional freedom of the fluorescence probe more than does the other binding mode.[4]

Measurements of emission maxima or the center of mass of emission spectra can be problematic. If the fluorescence maximum of two (or more) thermodynamic species is significantly different (e.g., one is red and one is blue), then it might appear that the emission maximum or center of mass would track the population of species. But this works well only in a quantitative sense when the emission intensity (or quantum yield) of the species is about the same. Otherwise, the emission maximum will be weighted toward the dominantly emitting species. As has been shown by simulations of protein-unfolding studies, this skewing can easily lead to significant misinterpretation of the thermodynamic parameters for a transition.[1] This problem exists for the spectral center of mass as well. A more thermodynamically rigorous approach is to use the basis spectra for the thermodynamic species to aid in fitting spectra to determine the relative concentration of species under equilibrium conditions. Of course, if the fluorescence intensity of the thermodynamic species is about the same, yet they show different emission maxima, then measurements of apparent emission maxima can be analyzed meaningfully to track a transition.

Steady state fluorescence anisotropy is another method that can be problematic to analyze, unless the intensity of the thermodynamic species is about the same. When the intensities are different, then the apparent anisotropy, *r*, will be weighted toward the more dominantly emitting spe-

[2] M. R. Eftink, I. Gryczynski, W. Wiczk, G. Laczko, and J. R. Lakowicz, *Biochemistry* **30**, 8945 (1991); M. R. Eftink and Z. Wasylewski, *SPIE Proc.* **1640**, 579 (1992).
[3] J. R. Lakowicz, P. A. Koen, H. Szmacinski, I. Gryczynski, and J. Kusba, *J. Fluoresc.* **4**, 117 (1994).
[4] C. R. Guest, R. A. Hochstrasser, C. G. Dupuy, D. J. Allen, S. J. Benkovic, and D. P. Millar, *Biochemistry* **30**, 8759 (1991).

cies.[1] That is, whereas the total fluorescence intensity is a mole fraction weighted signal ($F = \Sigma X_i F_i$), the relationship for the apparent fluorescence anisotropy is

$$r_i = \frac{\Sigma X_i \phi_i r_i}{\Sigma X_i \phi_i} \qquad (2)$$

If the fluorescence quantum yield (or relative intensity) of each species i is known or is the same for the species in equilibrium, then anisotropy measurements can be used to track the population of species just as can fluorescence intensity. Anisotropy measurements will have less precision than intensity measurements, but anisotropy is independent of the intensity of the illuminating source and there is always the possibility that two (or more) thermodynamic species will have the same fluorescence intensity yet will have different anisotropy values (due to different rotational motion for the species).

Thermodynamic Models

The types of thermodynamic systems to be investigated can be crudely divided into two categories: unimolecular isomerizations processes and bi- (or multi-) molecular association processes. A third type of thermodynamic study involves transmembrane partitioning; the focus here is on the first two types of thermodynamic studies.

Isomerization (A ⇌ B) Processes

Examples of isomerization processes are protein conformational changes (e.g., a transition between two different folded states of a protein), protein-unfolding reactions (e.g., a transition between a native and an unfolded state), and phase transitions of phospholipid bilayers (e.g., a transition between liquid crystalline and gel-phase states). In general, for a transition between two states, A ⇌ B, Eqs. (a)–(d) in Table I apply for transitions induced by temperature, pressure, or chemical denaturant. The thermodynamic parameters of interest are the enthalpy change, ΔH, the entropy change, ΔS, the volume change, ΔV (for pressure studies), and the sensitivity to chemical denaturant, m ($= -\delta \Delta G / \delta [d]$, where $[d]$ is the molar concentration of chemical denaturant). For thermally induced transitions, there may also be a nonzero change in heat capacity, ΔC_p ($= \delta \Delta H / \delta T$), for the transition. If this is the case, then the more complete equation for thermal induced transitions is given by Eq. (b) in Table I, which includes an enthalpy change and entropy change at some reference temperature, T_0. The most widely accepted relationship for chemical denaturant-induced unfolding of a protein is the so-called linear extrapolation model [Eq. (c) in Table I].

TABLE I
RELATIONSHIPS DESCRIBING TWO-STATE TRANSITIONS IN PROTEINS[a]

Temperature

(a) $\Delta G_{un}(T) = \Delta H_{un} - T\Delta S_{un}$

(b) $\Delta G_{un}(T) = \Delta H_{0,un}^{\circ} + \Delta C_p(T - T_0) - T[\Delta S_{0,un}^{\circ} + \Delta C_p \ln(T/T_0)]$

$\Delta H_{0,un}^{\circ}$ is the enthalpy change at $T = T_0$

$\Delta S_{0,un}^{\circ}$ is the entropy change at $T = T_0$;

ΔC_p is the change in heat capacity on unfolding.

Chemical denaturants

(c) $\Delta G_{un}([d]) = \Delta G_{0,un}^{\circ} - m[d]$ (Linear extrapolation model)

$\Delta G_{0,un}^{\circ}$ is the free energy change in the absence of d

$m = \delta\Delta G_{un}/\delta[d]$.

Pressure

(d) $\Delta G_{un}(P) = \Delta G_{0,un}^{\circ} - \Delta V_{un}(P_0 - P)$

ΔV_{un} is the volume change for N $\rightleftharpoons$ U transition

P_0 is the reference pressure.

---

[a] For a two-state transition, A $\rightleftharpoons$ B (or N $\rightleftharpoons$ U for the unfolding of a native, N, to an unfolded, U, state of a protein), the mole fraction of the N and U states are given as $X_N = 1/Q$, $X_U = \exp(-\Delta G_{un}/RT)/Q$, where $Q = 1 + \exp(-\Delta G_{un}/RT)$ and the function for ΔG_{un} is taken from above; the average fluorescence signal $F_{calc} = \sum X_i(F_i + x \cdot \delta F_i/\delta x)$, where x is a generalized perturbant.

For an unfolding transition of a biological macromolecule, whether induced by temperature, pressure, or chemical denaturant, the thermodynamic parameter of primary interest is the free energy change at some reference condition (such as 20°, neutral pH, and 1 atm). The latter free energy change, $\Delta G_{0,un}^{\circ}$, can serve as a measure of the thermodynamic stability of the "B" state with respect to the "A" state; if these are native and unfolded states, respectively (i.e., N $\rightleftharpoons$ U), then $\Delta G_{0,un}^{\circ}$ is a measure of the stability of the native state, for which comparisons can be made for a set of mutant proteins, etc. This strategy and the length and nature of the extrapolations to the reference conditions [e.g., the extrapolations inherent to Eqs. (a)–(d) in Table I to obtain $\Delta G^{\circ}_{0,un}$] are discussed in various writings on the thermodynamics of protein unfolding.[5–11]

A typical example of such an isomerization process is the urea-induced

[5] J. A. Schellman, *Biopolymers* **17**, 1305 (1978).

[6] W. J. Becktel and J. A. Schellman, *Biopolymers* **26**, 1859 (1987).

[7] C. N. Pace, *Methods Enzymol.* **131**, 266 (1986).

[8] C. N. Pace, B. A. Shirley, and J. A. Thomson, *in* "Protein Structure and Function: A Practical Approach" (T. E. Creighton, ed.), pp. 311–330. IRL Press, Oxford, 1989.

[9] M. M. Santoro and D. W. Bolen, *Biochemistry* **27**, 8063 (1988).

[10] P. L. Privalov, *Annu. Rev. Biophys. Biophys. Chem.* **18**, 47 (1989).

[11] M. R. Eftink and R. Ionescu, *Biophys. Chem.* **64**, 175 (1997).

unfolding of the protein HPr F22W (a single tryptophan-containing mutant of *Escherichia coli* histidine-containing phosphocarrier protein). Figures 1 and 2 present data for the unfolding of the native state to the unfolded state of this protein. This is a small protein that shows reversible unfolding. The data illustrate some of the points raised in the previous section. The fluorescence intensity of the native state is larger and blue-shifted with respect to that of the unfolded state and this provides a convenient way to directly track the transition. Fitting the intensity data with a combination of Eq. (c), along with the fundamental two-state equations given in Table I, yields the shown fit and a value of $\Delta G^{\circ}_{0,\mathrm{un}}$ = 3.25 kcal/mol in the absence of denaturant and a value of m = 1.15 kcal/mol $\cdot$ M. Because the fluorescence intensity of the native state is greater than that of the unfolded state, measurement of either the emission maximum or the steady state fluorescence anisotropy will be skewed, with the native state appearing to be more populated than should be the case, for reasons discussed above. An attempt to fit the apparent emission maximum or anisotropy data to a two-state model via Eqs. (1) and (c) in Table I yields apparent $\Delta G^{\circ}_{0,\mathrm{un}}$ and m values that are significantly larger than the preceding values from analysis of intensity data (see Fig. 2). However, as shown in Fig. 2 the "correct" thermodynamic parameters can be recovered from the anisotropy data if

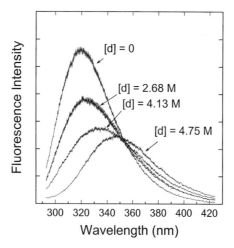

FIG. 1. The fluorescence spectrum of HPr F22W in the presence of 0, 2.68, 4.13, and 4.75 M urea. Conditions: 20°, excitation at 290 nm, 4-nm slits, pH 7.1, 0.02 M phosphate buffer, spectra acquired with a diode array detector. The spectra at 0 and 4.75 M urea were assumed to represent the spectra of the N and U states, respectively. The spectra for the other denaturant concentrations were fitted as a sum of these two basis spectra. For example, the spectrum for 2.68 M denaturant is overlaid with a fitted spectrum with X_{U} = 0.422.

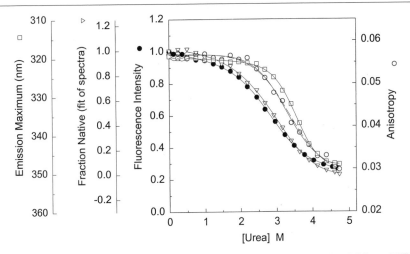

FIG. 2. Use of various fluorescence signals to track the urea-induced unfolding of HPr F22W. (●) Fluorescence intensity at 320 nm; (○) anisotropy; (□) emission maximum; (▽) mole fraction of N state (from a fitting of the spectra in Fig. 1 to a sum of two component spectra). The signals were individually fitted (solid curves) with Eq. (1) and (c) from Table I (and the two-state model given in Table I) to obtain the following estimates of apparent $\Delta G_{0,un}^{\circ}$ (in units of kcal/mol), m (in units of kcal/mol/M), $[d]_{1/2}$ (= $\Delta G_{0,un}^{\circ}/m$, in molarity).

Parameter	$\Delta G_{0,un}^{\circ}$	m	$[d]_{1/2}$
Fluorescence intensity	3.25	1.15	2.83
Anisotropy	4.18	1.29	3.24
Emission maximum	6.58	1.87	3.54
X_i from spectra fitting	3.15	1.08	2.92

A global analysis of intensity and anisotropy data is shown as the dashed curves through these points (difficult to see, because it overlays almost perfectly with the individual fits); the global fit is with $\Delta G_{0,un}^{\circ}$ = 3.32 kcal/mol, m = 1.17 kcal/mol/M, and $[d]_{1/2}$ = 2.84.

one uses Eq. 2 along with known relative fluorescence intensity values for the native and unfolded states. Likewise, by fitting the spectra for intermediate conditions (i.e., in which both native and unfolded species coexist) as a linear combination of the basis spectra for the native and unfolded states, the population of the native and unfolded states can be determined as a function of urea concentration. This allows thermodynamic parameters to be obtained, which are consistent with the parameters obtained from the fit of intensity versus $[d]$ data.

The above example, and Eqs. (a)–(d) in Table I, are for a two-state model for a conformational transition. Whether or not the two-state model is valid for any particular conformational transition is a major question.[11,12] It is possible that a transition involves one or more equilibrium intermediates, in which case the model should be three-state or multistate. As discussed elsewhere, it will often be difficult to experimentally determine whether a transition is two-state or multistate, because even the latter can easily appear to be a smooth sigmoidal transition with respect to the perturbant axis.[11] A classic test of the validity of the two-state model is to monitor the transition using two or more structural probes and then check for coincidence in the trackings.[12] If there is a divergence between two or more tracking methods, this is an indication of a multistate transition. In applying this test it is important that the two or more methods respond to different structural features. Fluorescence is a good choice for one of these tracking methods, because it focuses on one (or a small number of) fluorescing centers in the macromolecule. A combination of fluorescence and circular dichroism methods is particularly useful when studying the unfolding of globular proteins, because the latter method senses changes in secondary structure throughout the native state of the protein.[13]

Another valuable application of fluorescence to monitor a conformational transition is the measurement of the steady state fluorescence anisotropy of diphenylhexatriene (DPH) in studies of the gel to liquid–crystalline phase transition of model bilayer membranes.[14] This is a frequently used application, which takes advantage of the fact that the probe DPH, being a long and rigid molecule, has greater rotational freedom in the liquid–crystalline phase state of the phospholipid bilayer. Consequently, the anisotropy of DPH is lower above the phase transition temperature.

$M + L \rightleftharpoons ML$ and $M + M \rightleftharpoons M_2$ Type Association Processes

For association processes, there are a number of possible thermodynamic models: a simple one-to-one (stoichiometry, $n = 1$) macromolecule–ligand system, or the independent and identical binding of multiple ligands ($n > 1$ with a single microscopic K for binding), or independent and nonidentical binding of multiple ligands ($n > 1$ with different K_i for each class of binding site), or the cooperative binding of multiple ligands ($n > 1$ with a single K and a cooperativity index $\neq 1$). For sake of generality, the ligand can either be a small molecule or a macromolecule. Also, we could consider the self-association of the macromolecule: whether or not this self-association is linked to the binding of a ligand. There are a variety of association

---

[12] R. Lumry, R. Biltonen, and J. F. Brandts, *Biopolymers* **4,** 917 (1966).

[13] G. D. Ramsay, R. Ionescu, and M. R. Eftink, *Biophys. J.* **69,** 701 (1995).

[14] J. Suurkuusk, B. R. Lentz, Y. Barenholz, R. L. Biltonen, and T. E. Thompson, *Biochemistry* **15,** 1393 (1976).

models and it is useful to refer to a previous chapter in this series,[15] where a large number of such models and equations are given to relate to the analysis of fluorescence intensity data.

The analysis of fluorescence data for a simple one-to-one binding system is fairly straightforward. For the reaction $M + L \rightleftharpoons ML$, where M is macromolecule, L is ligand, and ML is a binary complex formed with association constant $K = [ML]/([M][L])$, the following relationship applies:

$$F_{rel} = (1 + K[L]F_{ML})/Q \tag{3}$$

where Q is $1 + K[L]$. In Eq. (3) F_{ML} is the relative fluorescence intensity of the complex (i.e., relative to a value of 1.0 for the free M, assuming that the fluorescence of the macromolecule is being observed). A plot of F_{rel} versus [L] should have a hyperbolic shape and should approach a value of F_{ML} at saturating ligand concentration. A difficulty with such systems is that the ligand concentration in Eq. (3) is the free concentration, not the total concentration. The free ligand concentration will depend on both the total ligand concentration, $[L]_0$, and the total macromolecule concentration, $[M]_0$. An analytical solution for [L] is possible, however (as a solution to a quadratic equation containing $[L]_0$, $[M]_0$, and K), making it possible to perform fits of Eq. (3) with only two fitting parameters, K and F_{ML} (see Ref. 15).

However, it is much more difficult to handle the more complicated models. The basic problem is that the number of thermodynamic fitting parameters and spectral fitting parameters can easily be too large to allow convergence in a fit with a particular model. Even for a model having two independent and nonidentical binding sites, there can be five fitting parameters (i.e., two association constants and the relative fluorescence intensities for the two different singly ligated species and the relative intensity for the doubly ligated species). This is to say nothing of the difficulty of determining the free ligand concentrations from the total ligand concentrations. As we have shown via simulations,[15] it is possible to observe binding isotherms that have a variety of appearances, including some that appear to be sigmoidal, even when there is no cooperativity between binding sites. This is because the relative fluorescence intensity of one of the ML species can either be higher or lower than the free M or other ML species.

Fluorescence ligand-binding studies are not particularly useful for determining binding stoichiometries, because of the correlation that may exist between the binding site number, n, and the relative signal, F_{MLi}, for a multiligand system. If the product of the association constant, K, times the total macromolecule concentration, $[M]_0$, is large, then end-point titrations

[15] M. R. Eftink, *Methods Enzymol.* **278**, 221 (1997).

can reveal the binding site number. Such experiments work best when the fluorescence being monitored comes from the ligand instead of the macromolecule.[15]

Bujalowski and Lohman[16] have developed a strategy for overcoming these problems. Their approach basically involves obtaining binding isotherms as a function of both total ligand and total macromolecule concentration and interpolating the data sets to obtain $[M]_0$ and $[L]_0$ data pairs that have the same $\Delta F/[M]_0$. From sets of these data pairs, the degree of saturation of the binding sites can be calculated as a function of free ligand concentration, thus enabling the characterization of the binding process.

When the binding stoichiometry and binding model are known from structural considerations, then fluorescence titrations can be used effectively to determine binding affinities. For example, there are a variety of applications of antibody–antigen fluorescence binding studies, with either the fluorescence of the antibody (e.g., intrinsic tryptophan or extrinsic fluorescein labels) being quenched by hapten binding or with a fluorescence labeled hapten/antigen being quenched on association with antibody.[17-19] Effector-receptor binding reactions are also conveniently studied by fluorescence measurements.[20] Another general example of binding reactions with a defined stoichiometry, for which there are convenient fluorescence titration methods, is fluorescent indicator (e.g., Fura-2) complexation with calcium ions, hydrogen ions, or other inorganic ions.[3,21,22] Because the association constant is often known for the latter complexes, the application of the method is usually to "sense" the unknown concentration of the analyte, for example within a cell.

Transmembrane Potentials

A third general type of thermodynamic measurement, which does not fall into the categories of conformational isomerization or an association process, is that of the transmembrane potential, for example, across biologi-

[16] W. Bujalowski and T. M. Lohman, *Biochemistry* **26**, 3009 (1987).

[17] E. V. Voss, Jr., R. M. Watt, and G. Weber, *Mol. Immunol.* **17**, 505 (1980).

[18] J. Erickson, P. Kane, D. Goldstein, and B. Baird, *Mol. Immunol.* **23**, 769 (1986).

[19] J. Erickson, R. Posner, B. Goldstein, D. Holowaka, and B. Baird, *in* "Biophysical and Biochemical Aspects of Fluorescence Spectroscopy" (T. G. Dewey, ed.), pp. 169–195. Plenum Press, New York, 1991.

[20] W. J. Phillips and R. A. Cerione, *in* "Biophysical and Biochemical Aspects of Fluorescence Spectroscopy" (T. G. Dewey, ed.), pp. 135–167. Plenum Press, New York, 1991.

[21] R. Y. Tsien, *Biochemistry* **19**, 2396 (1980).

[22] R. P. Haugland, "Handbook of Fluorescent Probes and Research Chemicals," 6th Ed., Chaps. 22, 23, and 24. Molecular Probes, Eugene, Oregon, 1996.

cal membranes. There are a variety of fluorescent probes for transmembrane potentials, as has been reviewed.[23,24]

Problems in Using Fluorescence to Obtain Thermodynamic Parameters

Probably the biggest concern about the use of fluorescence techniques in studying equilibrium processes in biological macromolecules is that there may be interferences with the signal being measured and/or that baseline trends (e.g., as temperature, pH, ligand concentration, etc., is varied) may dominate a fluorescence signal change due to a transition or binding process.

It is well known that the fluorescence intensity of various fluorophores decreases in a slightly nonlinear manner with increasing temperature, because of what can be called thermal quenching.[25] Because it is common to assume a linear temperature dependence for baseline trends,* this intrinsic nonlinearity can lead to a small deviation of the fitted thermodynamic parameters from their true values. We have demonstrated this by a simulation for the unfolding of a protein.[26] However, this nonlinearity of a fluorescence signal is of minor importance in most studies.

Fluorescence intensity can also depend on the concentration of chemical denaturants, including urea and guanidine hydrochloride.[1,27] This dependence is reasonably linear and shallow, making it possible to easily handle the baselines for protein-unfolding reactions. It is our experience that the

---

[23] L. W. Loew, *in* "Fluorescent and Luminescent Probes for Biological Activity" (W. T. Mason, ed.), pp. 150–160. Academic Press, San Diego, California, 1994.

[24] J.-Y. Wu and L. B. Cohen, *in* "Fluorescent and Luminescent Probes for Biological Activity" (W. T. Mason, ed.), pp. 389–404. Academic Press, San Diego, California, 1994.

[25] J. Eisinger and G. Navon, *J. Chem. Phys.* **50,** 2069 (1969).

* By baseline slope we mean that the F_i value in Eq. (1) is intrinsically dependent on the perturbing condition. For example, a linear dependence of the fluorescence intensity of state A on temperature would be expressed as $F_A = F_{A,0} + T \, \delta F_A/\delta T$, where the $\delta F_A/\delta T$ is the baseline slope and $F_{A,0}$ is the fluorescence intensity of species A at a reference temperature, such as $0°$. Including such baseline slopes expands Eq. (1) to $F = \Sigma \, X_i(F_{i,0} + T \, \delta F_i/\delta T)$ in the case of a thermal transition. For a two-state process, the existence of baselines slopes for the two states increases the number of fitting parameters by two. A similar expression for the fluorescence baseline can be written for denaturant-induced transitions or for a ligand-binding process, where the "baseline slope" in the latter case would reflect the effect of the ligand on the intrinsic fluorescence of the macromolecule in the absence of a binding process.

[26] M. R. Eftink, *Biochemistry (Moscow)* **63,** 327 (1998).

[27] F. X. Schmid, *in* "Protein Structure: A Practical Approach" (T. E. Creighton, ed.), pp. 251–285. IRL Press, Oxford, 1989.

baseline slope for an unfolded protein is greater than that for a folded protein.

pH titrations of macromolecules can result in either unfolding reactions, more subtle conformational changes, or proton binding to specific side chains, such as histidine residues. It is important to realize that pH-induced changes in the fluorescence of a macromolecule do not always indicate that there is a conformational change coupled to the proton-binding process. For example, protonated histidine residues are known to be better quenchers of tryptophan fluorescence in proteins, as compared with the neutral histidine ring.[28] So a drop in the fluorescence of a protein between pH 7 and pH 5 may indicate only the protonation of a residue near a tryptophan. Also, deprotonation of tyrosine residues will produce tyrosinate, which can act as a quencher of tryptophan fluorescence by resonance energy transfer. In some cases fluorescence from tyrosinate can interfere with the fluorescence of tryptophan in a protein.

The linearity of a fluorescence signal with analyte concentration can be compromised by inner filter effects as the absorbance increases with increasing concentration. Besides the use of reduced path length cells, there are corrections that work reasonably well up to an absorbance of ~ 0.5 at the excitation wavelength.[29]

The linear relationship between fluorescence intensity and analyte concentration can potentially be complicated if the rate constant for the conformational transition (or association/dissociation processes) occurs on the same time scale as does the radiative emission of a photon from the fluorescing center.[30] However, the fluorescence decay time of most intrinsic and extrinsic probes is on the tens of nanoseconds time scale, whereas the effective relaxation time for most protein conformational changes and ligand-binding reactions is on the milliseconds to seconds (or longer) time scale, thus eliminating coupling between the fluorescence decay and the chemical kinetics processes in most cases of biochemical interest. A case in which such coupling is a concern is for proton binding to chromophores.[30] Yet another phenomenon that can lead to a slight deviation from the expected linear dependence of a fluorescence intensity measurement on the concentration of analyte is related to the polarization bias caused by the photoselection that occurs when either naturally polarized or linearly polarized excitation light is used.[31] Because the standard detection geometry is 90° with respect to the excitation beam, the photoselection bias will be

[28] Y. Chen and M. D. Barkley, *Biochemistry* **37**, 9976 (1998).
[29] B. Birdsall, R. W. King, M. R. Wheeler, C. A. Lewis, S. R. Goode, R. B. Dunlap, and G. C. K. Roberts, *Anal. Biochem.* **132**, 353 (1983).
[30] A. Kowalczyk, N. Boens, and M. Ameloot, *Methods Enzymol.* **278**, 94 (1997).
[31] M. Shinitzky, *J. Chem. Phys.* **56**, 5979 (1972).

negligible for small molecules, because their emission is depolarized due to rotation. However, if a fluorescing small molecule binds to a large macromolecule, the fluorescence of the ligand can be attenuated due to this photoselection, even when using natural light for excitation. This photoselection detection bias is almost always ignored in steady state measurements and probably causes only a few percent change in the fluorescence signal of such a ligand on binding to a macromolecule.

A major concern in the use of fluorescence, or any other method, to track a transition is that a thermodynamic analysis is valid only if the proper model is being used to fit the data. We have discussed this point with simulations for protein-unfolding transitions.[11,26] For example, if the unfolding of a protein is a multistate process, then fitting a two-state model to the data will lead to apparent thermodynamic parameters that have essentially no meaning. The same general warning also applies to ligand-binding interactions; if an invalid model is used to fit the data, any recovered thermodynamic parameters, likewise, will be dubious. One area where this concern has surfaced is the chemical denaturant-induced unfolding of a series of globular proteins. If one assumes the two-state model to be valid for a set of mutant proteins, then one would likely find a range of "m" values for these proteins. Alternatively, the range of m values may be an indication that the unfolding is not two-state for these proteins.

A final concern regarding the use of fluorescence in thermodynamics studies is that one must be careful to make sure that a sample is truly at chemical equilibrium when its signal is being measured. This requires attention to the experimental design. Because fluorescence measurements can be made rapidly and can be easily automated and computer interfaced, it is important to ensure that chemical equilibrium has been reached in conformational transitions and titration experiments. If, for example, a thermal scanning measurement is performed rapidly (compared with the time required for chemical equilibration), then the transition curve can have a distorted shape, leading to the recovery of incorrect thermodynamic parameters. The simulations by Lepock and co-workers,[32] although involving differential scanning calorimetry, are applicable to other types of automated transitions (e.g., thermal scanning or automated titrations) where an experimental scan rate is involved. This concern is mentioned not to discourage automated scanning or titration experiments, but to emphasize that it is important to ensure that the chemical reaction has reached equilibrium at each point in order to "do thermodynamics."

[32] J. R. Lepock, K. P. Ritchie, M. C. Kolios, A. M. Rodahl, K. A. Heinz, and J. Kruuv, *Biochemistry* **31,** 12706 (1992).

Concluding Remarks

As with any method, there are limitations and experimental pitfalls in the use of fluorescence as a thermodynamic tool. However, the numerous advantages listed in the first section make fluorescence spectroscopy a tool that is almost unparalleled in its breadth of applications.

[21] Microsecond Dynamics of Biological Macromolecules

By Joseph R. Lakowicz, Ignacy Gryczynski, Grzegorz Piszczek, Leah Tolosa, Rajesh Nair, Michael L. Johnson, and Kazimierz Nowaczyk

Fluorescence spectroscopy is widely used to study the nanosecond time-scale dynamics of biological macromolecules. The spectral observables are sensitive to nanosecond dynamics because emission also occurs on the nanosecond time scale. While nanosecond biopolymers dynamics are important, these rapid processes reflect mostly local fluorophore motions and its interactions with the immediate environment. However, biological macromolecules also display structural changes on the microsecond time scale. Processes that occur on the microsecond time scale include domain flexing in proteins and lateral diffusion in membranes. It is also likely that nucleic acid junctions and structured RNAs display microsecond motions.

Fluorescence is now capable of detecting microsecond dynamics. This change in time scale is made possible by the development of metal–ligand complexes (MLCs), which display decay times ranging from 10 nsec to more than 10 μsec. The MLCs display several spectral characteristics that make them useful probes, including high photostability, a large Stokes shift, and polarized emission. The use of MLCs to measure microsecond dynamics is just beginning. In this overview chapter we present data and simulations showing the possibility of measuring protein domain flexing, lateral diffusion in membranes, and microsecond rotational correlation times. Because the lanthanides display millisecond decay times, fluorescence is no longer trapped on the nanosecond time scale, and can be used to quantify dynamic processes from nanoseconds to microseconds to milliseconds.

Introduction

A favorable aspect of fluorescence is that it occurs on the nanosecond time scale. After excitation the fluorophore emission is sensitive only to

Copyright © 2000 by Academic Press
0076-6879/00 $30.00

processes that occur prior to emission. As a consequence fluorescence spectroscopy, and especially time-resolved fluorescence, has been widely used to study the local dynamics of amino acid side chains in proteins, motions of probes in membranes, hydrodynamics of moderate-size proteins, and torsional motions of DNA.[1-5] The results of such measurements are sometimes compared with molecular dynamics calculations on similar systems.[6-9]

It is possible that fluorescence has been trapped on the nanosecond time scale. While thousands of fluorophores are known, most display decay times from 1 to 10 nsec. Hence, the emission has largely decayed after 30 nsec, so that slower processes cannot be observed. This linkage between the fluorescence lifetime and observable motions can be seen from the Perrin equation,

$$r = \frac{r_0}{1 + \tau/\theta} \tag{1}$$

In this expression r is the steady state anisotropy and r_0 is the anisotropy of the fluorophore in the absence of rotational motion. If the lifetime (τ) is much longer than the rotational correlation time (θ), the anisotropy (r) is zero. If the lifetime is much shorter than the correlation time, the anisotropy is r_0. The anisotropy (r) is sensitive to the rate of rotational diffusion only if θ is comparable to the lifetime τ. This relationship shows why the anisotropy decays of membrane-bound fluorophores reveal the local

[1] A. P. Demchenko, Fluorescence and dynamics in proteins. *In* "Topics in Fluorescence Spectroscopy," Vol. 3: "Biochemical Applications" (J. R. Lakowicz, ed.), pp. 65–111. Plenum Press, New York, 1992.

[2] J. B. A. Ross, W. R. Laws, K. W. Rousslang, and H. R. Wyssbrod, Tyrosine fluorescence and phosphorescence from proteins and polypeptides. *In* "Topics in Fluorescence Spectroscopy," Vol. 3: "Biochemical Applications," pp. 1–63. Plenum Press, New York, 1992.

[3] C. D. Stubbs and B. W. Williams, Fluorescence in membranes. *In* "Topics in Fluorescence Spectroscopy," Vol. 3: "Biochemical Applications," pp. 231–271. Plenum Press, New York, 1992.

[4] P. Auffinger and E. Westhof, *Curr. Opin. Struct. Biol.* **8,** 227 (1998).

[5] R. M. Hochstrasser, *J. Chem. Ed.* **75,** 559 (1998).

[6] J. M. Schurr, B. S. Fujimoto, P. Wu, and L. Song, Fluorescence studies of nucleic acids: Dynamics, rigidities, and structures. *In* "Topics in Fluorescence Spectroscopy," Vol. 3: "Biochemical Applications," pp. 137–229. Plenum Press, New York, 1992.

[7] J. A. McCammon and S. C. Harvey, "Dynamics of Proteins and Nucleic Acids," p. 234. Cambridge University Press, New York, 1987.

[8] H. Merlitz, K. Rippe, K. V. Klenin, and J. Langowski, *Biophys. J.* **74,** 773 (1998).

[9] D. Sprous, M. A. Young, and D. L. Beveridge, *J. Mol. Biol.* **265,** 1623 (1999).

motions of probes within the bilayers,[10,11] but does not reveal the microsecond time-scale rotational diffusion of the lipid vesicles themselves. The vesicles themselves do not undergo significant rotational motion during a nanosecond intensity decay. Similarly, the anisotropy decays of probes bound to DNA reveal mostly the local torsional motions of the DNA and contain less information on the slower bending motions.[12,13] For proteins the upper molecular mass limit for measuring rotational diffusion is near 50 kDa. A protein the size of human serum albumin (HSA), 66 kDa, has a rotational correlation time near 30 nsec.[14] It is difficult to measure slower motions because the nanosecond emission has decayed before the slower rotation has displaced the transition moment of the fluorophore and decreased the anisotropy.

Another example of the linkage between lifetime and dynamics is the effect of donor-to-acceptor diffusion on resonance energy transfer (RET). The mean distance a molecule diffuses during a donor decay time τ is given by

$$(\Delta x^2)^{1/2} = 2D\tau \tag{2}$$

where D is the donor (D)-to-acceptor (A) diffusion coefficient. To avoid complex nomenclature, we use D to indicate either the donor or the diffusion coefficient, as the meaning will be clear from the content. For moderately slow diffusion of $D = 10^{-8}$ cm^2/sec the root mean square distances a molecule diffuses in 1 nsec and 1 μsec are 0.45 and 14 Å, respectively. Hence diffusion will not affect the RET efficiency of a 1-nsec probe, but will result in increased RET efficiency from a 1-μsec donor. Because of its distance dependence RET is frequently used to measure distances between sites on proteins. Because of the nanosecond decay times of most fluorophores RET has not been used to detect domain motions in proteins on the time scale of these motions.

There are several methods to circumvent the nanosecond time scale of fluorescence. These include the use of phosphorescence,[15,16] fluorescence

[10] D. C. Mitchell and B. J. Litman, *Biophys. J.* **74,** 879 (1998).

[11] D. A. van der Sijs, E. E. vann Faassen, and Y. K. Levine, *Chem. Phys. Lett.* **216,** 559 (1993).

[12] M. Collini, G. Chirico, and G. Baldini, *Biophys. Chem.* **53,** 227 (1995).

[13] M. L. Barcellona and E. Gratton, *Biophys. J.* **70,** 2341 (1996).

[14] J. R. Lakowicz, "Principles of Fluorescence Spectroscopy," 2nd Ed. Plenum Press, New York, 1999.

[15] J. M. Vanderkooi, Tryptophan phosphorescence from proteins at room temperature. *In* "Topics in Fluorescence Spectroscopy," Vol. 3: "Biochemical Applications," pp. 113–136. Plenum Press, New York, 1992.

[16] J. A. Schauerte, D. G. Steel, and A. Gafni *Methods Enzymol.* **278,** 49 (1997).

recovery after photobleaching (FRAP),[17,18] polarized photobleaching,[19–21] and the use of lanthanides with millisecond decay times.[22] We do not wish to criticize the creative implementations of these techniques, but all techniques have limitations. Relatively few probes such as eosin display useful phosphorescence in aqueous solution, and detection of phosphorescence typically requires the complete exclusion of oxygen. FRAP and polarized photobleaching require somewhat intense illumination to photobleach the probes, and there has been controversy about interpretation of the FRAP data. The lanthanides display millisecond lifetimes, which are due to forbidden transitions between shielded atomic orbitals. These lifetimes are not sensitive to the local environment and are not easily changed to the microsecond time scale. In many cases the millisecond decay times result in complete spatial averaging to the "rapid diffusion limit,"[23,24] with loss of information of motions on the microsecond time scale. Also, lanthanide emission is not known to be polarized and thus not useful for anisotropy measurements. As an overall conclusion, fluorescence has not been useful on the microsecond time scale, and is less than optimal on the millisecond time scale.

The importance of microsecond time-scale measurements is seen from the large number of studies reporting biopolymer dynamics on the microsecond time scale. For instance, the lateral diffusion coefficients of lipids in model membranes are reported to be near 1×10^{-8} to 10×10^{-8} cm^2/sec,[25–27] and membrane-bound proteins diffuse about 10-fold slower.[28,29] Using an estimated diffusion coefficient of 5×10^{-8} cm^2/sec for lipids in membranes, and using $\Delta x^2 = 4D\tau$ for diffusion in two dimensions, one can calculate a displacement of 30 Å in 1 μsec but only 1.4 Å in 1 nsec. As is shown below, such diffusion coefficients in membranes are expected to be

[17] N. Periasamy and A. S. Verkman, *Biophys. J.* **75**, 557 (1998).
[18] J. Lippincott-Schwartz, J. F. Presley, K. J. M. Zaal, K. Hirschberg, C. D. Miller, and J. Ellenberg, *Methods Cell Biol.* **58**, 261 (1999).
[19] M. Velez and D. Axelrod, *Biophys. J.* **53**, 575 (1988).
[20] T. M. Yoshida and B. G. Barisas, *Biophys. J.* **50**, 41 (1986).
[21] B. A. Scalettar, P. R. Selvin, D. Axelrod, J. E. Hearst, and M. P. Klein, *Biophys. J.* **53**, 215 (1988).
[22] N. Sabbatini, M. Guardigli, and I. Manet, *Adv. Photochem.* **23**, 213 (1997).
[23] D. D. Thomas, W. F. Carlsen, and L. Stryer, *Proc. Natl. Acad. Sci. U.S.A.* **75**, 5746 (1978).
[24] L. Stryer, D. D. Thomas, and C. F. Meares, *Annu. Rev. Biophys. Bioeng.* **11**, 203 (1982).
[25] G. J. Schötz, H. Schindler, and T. Schmidt, *Biophys. J.* **73**, 1073 (1997).
[26] S. Ladha, A. R. Mackie, L. J. Harvey, D. C. Clark, E. J. A. Lea, M. Brullemans, and H. Duclohier, *Biophys. J.* **71**, 1364 (1996).
[27] R. Gilmanshin, C. E. Creutz, and L. K. Tamm, *Biochemistry* **33**, 8225 (1994).
[28] S. Nelson, R. D. Horvat, J. Malvey, D. A. Roess, B. G. Barisas, and C. M. Clay, *Endocrinology* **140**, 950 (1999).
[29] K. Kitani, S. Tanaka, and I. Z. Nagy, *Arch. Gerontol. Geriat.* **26**, 257 (1998).

readily detectable using RET donors with microsecond decay times. Lateral diffusion in membranes does not have a significant effect on RET of nanosecond membrane-bound donors.

We now consider domain motions in proteins, which appear to be important for signaling and enzyme regulation.[30–33] Because domain motions often occur on substrate binding, domain motions are also thought to be involved in catalysis. However, attempts to use RET to quantify the rates of domain motions have been largely unsuccessful.[34,35] For instance, RET between labeled subunits was used to measure the distribution of distances between two sites on phosphoglycerate kinase.[34] The time–domain data revealed a range of distances, that is, a distance distribution, but the data provided no information on the rate of interchange between the distances and no information on the time scale of the domain motions. In contrast, domain-to-domain motions are expected to be detectable using microsecond decay time donors. Furthermore, the common structural motifs of proteins appear to form on the microsecond time scale. For instance, α helices appear to fold and unfold in about 100 μsec,[36,37] and β hairpins fold in about 6 μsec.[38,39] These structural changes result in changes in the distance between sites on the peptides.

Finally, we note there have been numerous reports of new structural motifs in nucleic acids. RET has already provided considerable information about the solution conformation of double and triple helices,[40,41] three- and four-way junctions,[42] and curved DNA.[40–48] However, the nanosecond

[30] M. Gerstein, A. M. Lesk, and C. Chothia, *Biochemistry* **33**, 6738 (1994).

[31] G. E. Schulz, *Curr. Opin. Struct. Biol.* **1**, 883 (1991).

[32] W. Wriggers, E. Mehler, F. Pitici, H. Weinstein, and K. Schulten, *Biophys. J.* **74**, 1622 (1998).

[33] K. Hinsen, A. Thomas, and M. J. Field, *Proteins Struct. Funct. Genet.* **34**, 369 (1999).

[34] G. Haran, E. Hass, B. K. Szpikowska, and M. T. Mas, *Proc. Natl. Acad. Sci. U.S.A.* **89**, 11764 (1992).

[35] M. Miki and T. Kouyama, *Biochemistry* **33**, 10171 (1994).

[36] S. Williams, T. P. Causgrove, R. Gilmanshin, K. S. Fang, R. H. Callender, W. H. Woodruff, and R. B. Dyer, *Biochemistry* **35**, 691 (1996).

[37] C. L. Brooks, *J. Phys. Chem.* **100**, 2546 (1996).

[38] V. Muñoz, P. A. Thompson, J. Hofrichter, and E. A. Eaton, *Nature (London)* **390**, 196 (1997).

[39] F. Blanco, M. Ramirez-Alvarado, and L. Serrano, *Curr. Opin. Struct. Biol.* **8**, 107 (1998).

[40] E. A. Jares-Erijman and T. M. Jovin, *J. Mol. Biol.* **257**, 597 (1996).

[41] C. Gohlke, A. I. H. Murchie, D. M. Lilley, and R. M. Clegg, *Proc. Natl. Acad. Sci. U.S.A.* **91**, 11660 (1994).

[42] F. Walter, A. I. H. Murchie, D. R. Duckett, and D. M. J. Lilley, *RNA* **4**, 719 (1998).

[43] J.-L. Mergny, T. Garestier, M. Rougee, A. V. Lebedev, M. Chassignol, N. T. Thuong, and C. Hélène, *Biochemistry* **33**, 15321 (1994).

[44] R. M. Clegg, A. I. H. Murchie, Z. Zechel, and D. M. J. Liley, *Proc. Natl. Acad. Sci. U.S.A.* **90**, 2994 (1993).

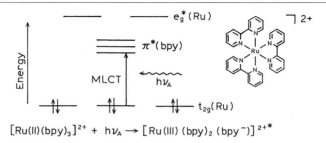

$$[Ru(II)(bpy)_3]^{2+} + h\nu_A \rightarrow [Ru(III)(bpy)_2(bpy^-)]^{2+*}$$

Fɪɢ. 1. Jablonski diagram for metal–ligand complexes (MLCs). MLCT, Metal-to-ligand charge transfer. [Reprinted with permission from Kluwer Academic/Plenum Publishing.]

decay times of the donors have precluded measurements of the solution dynamics of these nucleic acid structures.

The use of RET with microsecond decay time donors should allow measurement of the rates of motions from their effects on the intensity decays of the donor. In the following section we show how microsecond dynamic processes can be observed using microsecond decay time luminophores.

Metal–Ligand Complexes

Prior to describing the measurements possible with microsecond decay times it is informative to describe the photophysics and spectral properties of the metal–ligand complexes (MLCs). By metal–ligand complexes we are referring to luminophores that contain rhenium (Re), ruthenium (Ru), or osmium (Os) and typically at least one diimine ligand. The best known example is $[Ru(bpy)_3]^{2+}$, where bpy is 2,2′-bipyridine (Fig. 1). At first glance one might expect $[Ru(bpy)_3]^{2+}$ to behave like a metal chelate. However, the metal-to-ligand bonds behave almost like covalent bonds, and there is no dissociation of the ligands from the metal unless the MLC is placed under extreme nonphysiological conditions. The optical properties of the MLCs are also characteristic of a single molecular species, rather than a chelate.

The photophysical properties of the metal–ligand complexes have been

[45] P. S. Eis and D. P. Millar, *Biochemistry* **32**, 12852 (1993).
[46] R. M. Clegg, A. I. H. Murchie, and D. M. J. Lilley, *Biophys. J.* **66**, 99 (1994).
[47] F. Walter, A. I. H. Murchie, D. R. Duckett, and D. M. J. Lilley, *RNA* **4**, 719 (1998).
[48] K. Toth, V. Sauermann, and J. Langowski, *Biochemistry* **37**, 8173 (1998).

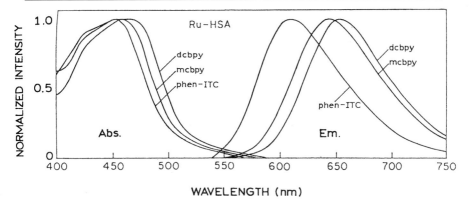

F𝗂G. 2. Absorption and emission spectra of [Ru(bpy)₂(dcbpy)], [Ru(bpy)₂(mcbpy)]⁺, and [Ru(bpy)₂phen-ITC]⁺ conjugated to HSA. Excitation wavelength 460 nm, 20°. Structures are shown in Fig. 3. dcbpy, 4,4′-Dicarboxylic acid-2,2′-bipyridine; mcbpy, 4-methyl-4′-carboxylic acid-2,2′-bipyridine; phen-ITC, isothiocyanate derivative of 5-amino-1,10-phenanthroline.

extensively studied.[49–52] Much of this research was directed toward using these molecules to capture solar energy. The basic idea is illustrated by the excited state, in which an electron is transferred from the metal to one of the ligands (Fig. 1). The excited state is simultaneously a strong oxidant and reductant, and it was hoped that the energy of the excited state would split water to oxygen and hydrogen. Fortunately for biophysics, but unfortunately for solar energy, the excited states of the metal–ligand complexes do not catalyze chemical reactions under normal experimental conditions. There are some examples in which MLCs were used to cleave DNA, but the cleavage requires a sacrificial amine, copper, and H_2O_2.[53] MLC probes have now been used bound to proteins, lipids, and nucleic acids without observable photodamage.

An important property of the MLCs is their long-wavelength absorption, which is distinct from the absorption of the metal or ligand alone. Typical spectra for Ru MLCs are shown in Fig. 2. In this case the MLCs have been covalently linked to human serum albumin. The metal-to-ligand charge

[49] K. Kalyanasundaram, "Photochemistry of Polypyridine and Porphyrin Complexes," p. 626. Academic Press, New York, 1992.

[50] D. J. Stufkens and A. Vlcek, *Coord. Chem. Rev.* **177,** 127 (1998).

[51] A. Juris, V. Balzani, F. Barigelletti, S. Campagna, P. Belser, and A. von Zelewsky, *Coord. Chem. Rev.* **84,** 85 (1988).

[52] J. N. Demas and B. A. DeGraff, Design and application of highly luminescent transition metal complexes. *In* "Topics in Fluorescence Spectroscopy," Vol. 4: "Probe Design and Chemical Sensing" (J. R. Lakowicz, ed.), pp. 71–107. Plenum Press, New York, 1994.

[53] L. A. Basile and J. K. Barton, *J. Am. Chem. Soc.* **109,** 7548 (1987).

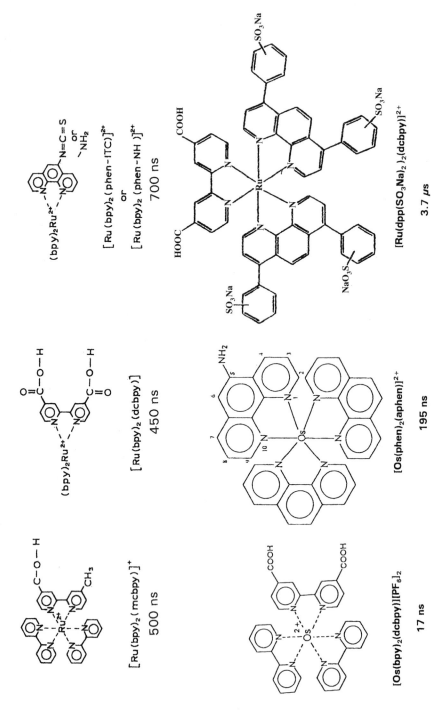

FIG. 3. Representative ruthenium MLCs synthesized in the laboratory. Aphen, aminophenanthroline; dpp, 4,7-diphenyl-1,10-phenanthroline.

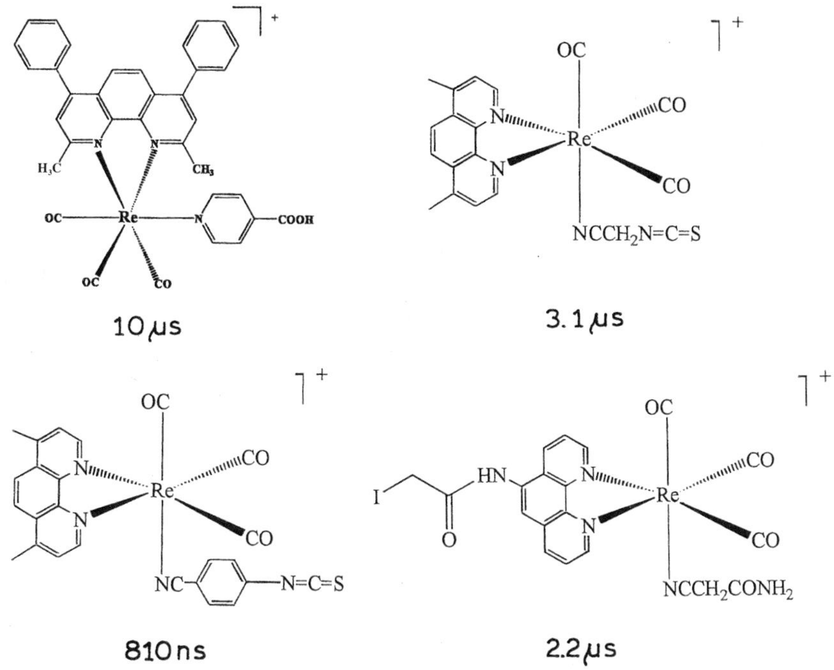

FIG. 4. Rhenium MLCs synthesized in the laboratory. The decay times are in deoxygenated organic solvents.

transfer (MLCT) absorption is the broad emission centered at 450 nm. Absorption of these wavelengths results in creation of the charge transfer state in less than 300 fsec.[54]

A large number of MLCs have been characterized. Figures 3 and 4 show some typical MLCs that have been conjugated with macromolecules and used as biophysical probes.[55-68] Emission maximum for the Ru MLCs

[54] N. H. Damrauer, G. Cerullo, A. Yeh, T. R. Boussie, C. V. Sharnk, and J. K. McCusker, *Science* **275,** 54 (1997).

[55] E. Terpetschnig, H. Szmacinski, H. Malak, and J. R. Lakowicz, *Biophys. J.* **68,** 342 (1995).

[56] E. Terpetschnig, H. Szmacinski, and J. R. Lakowicz, *Anal. Biochem.* **227,** 140 (1995).

[57] H. Szmacinski, E. Terpetschnig, and J. R. Lakowicz, *Biophys. Chem.* **62,** 109 (1996).

[58] L. Li, H. Szmacinski, and J. R. Lakowicz, *Anal. Biochem.* **244,** 80 (1997).

[59] L. Li, H. Szmacinski, and J. R. Lakowicz, *Biospectroscopy* **3,** 155 (1997).

[60] Z. Murtaza, Q. Chang, G. Rao, H. Lin, and J. R. Lakowicz, *Anal. Biochem.* **247,** 216 (1997).

[61] E. Terpetschnig, J. D. Dattelbaum, H. Szmacinski, and J. R. Lakowicz, *Anal. Biochem.* **251,** 241 (1997).

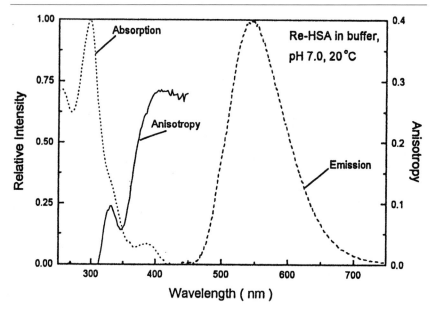

FIG. 5. Absorption, emission, and anisotropy spectra of [Re(bcp)(CO)$_3$(4-COOHPy)]$^+$ (the structure shown in Fig. 4, upper left). [Reprinted with permission from the American Chemical Society.]

is typically near 650 nm, and emission from the Re MLCs is typically near 550 nm. The shorter absorption and emission spectra of the Re MLCs (Fig. 5) are typically associated with longer decay times and higher quantum yields. This dependence of quantum yield on the energy of the excited state is due to the energy gap law,[49] which states that the nonradiative decay rates become faster as the energy of the excited state becomes closer to that of the ground state. Osmium MLCs typically display emission maxima near 750 nm. Because of the energy gap law the nonradiative decay rates are large, the lifetimes are short (5–30 nsec), and the quantum yields

[62] F. N. Castellano, L. Li, and J. R. Lakowicz, *Biophys. Chem.* **71**, 51 (1998).

[63] X.-Q. Guo, F. N. Castellano, L. Li, H. Szmacinski, J. R. Lakowicz, and J. Sipior, *Anal. Biochem.* **254**, 179 (1997).

[64] X.-Q. Guo, F. N. Castellano, and J. R. Lakowicz, *Anal. Chem.* **70**, 632 (1998).

[65] Z. Murtaza, P. Herman, and J. R. Lakowicz, *Biophys. Chem.* **80**, 143 (1999).

[66] H. Szmacinski, F. N. Castellano, E. Terpetschnig, J. D. Dattelbaum, J. R. Lakowicz, and G. J. Meyer, *Biochim. Biophys. Acta* **1383**, 151 (1998).

[67] F. N. Castellano and J. R. Lakowicz, *Photochem. Photobiol.* **67**, 179 (1998).

[68] L. Li, F. N. Castellano, I. Gryczynski, and J. R. Lakowicz, *Chem. Phys. Lipids* **99**, 1 (1999).

are low. The lifetimes can be increased by using phenanthroline or ter-pyridyl ligands in place of the bipyridyl ligands, as shown for [Os(phen)$_2$aphen]$^{2+}$ in Fig. 3. However, such long lifetimes near 200 nsec are exceptional for osmium MLCs.

The metal–ligand complexes display a number of characteristics that make them versatile biophysical probes. The Stokes shifts are large, as seen for Ru and Re MLCs in Figs. 2 and 5, respectively. Because of the large Stokes shift the MLCs do not display significant radiative or nonradiative homo transfer. Consequently, they do not self-quench even for a protein randomly labeled with several MLCs. In addition, the MLCs display good water solubility and are not prone to self-association. We have occasionally observed aggregation with MLC-labeled proteins, but this is not common and may not be due to the MLC probes themselves. The lack of self-association is probably due to the somewhat spherical structures of the MLCs, which provide less opportunity for stacking interactions than for more planar fluorophores.

Another favorable property of the MLCs is their high photostability. While some MLCs are known to undergo light-dependent dissociation, these are the exceptions. Most MLCs appear to be more stable then fluorescein by a factor of 100 or more. We have had an aqueous sample of [Ru(bpy)$_2$(dcpy)]$^{2+}$ in the laboratory room light for several years with no apparent change in absorption, emission, or lifetime.

Perhaps the most important property of the MLCs are their long lifetimes. The lifetimes of Ru MLCs are typically near 200–1000 nsec, and the Re MLCs typically have lifetimes from 1 to 10 μsec (Fig. 6). Osmium MLCs typically have shorter decay times. The decay time can be varied by choice of the diimine ligand, and there is extensive literature on this topic.[49–51] The dependence of the lifetimes on the emission maximum and the ligand can be mostly explained by the energy gap law. This law works well within a homologous class of MLCs, but less well when comparing less similar structures. The long lifetime of the MLCs is the result of a mixed singlet–triplet excited state. For this reason it is not clear if the emission is fluorescence or phosphorescence. While the lifetimes of the MLCs are long, they are considerably shorter than the millisecond decay time characteristic of phosphorescence. For simplicity we refer to the emission as fluorescence. We typically perform the measurements on samples equilibrated with atmospheric oxygen. Ruthenium and Re MLCs, when bound to macromolecules, are typically quenched less than 30% because of dissolved oxygen. In contrast, measurement of phosphorescence typically requires the complete exclusion of oxygen, which is sometimes not practical during studies of macromolecular or cellular systems.

And, finally, the MLCs display polarized emission, as can be seen from

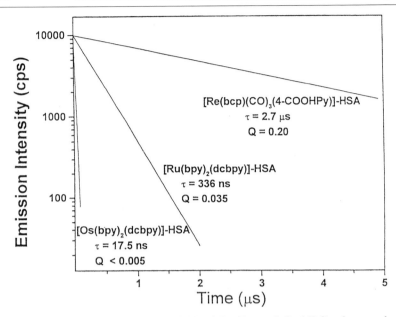

FIG. 6. Decay times and quantum yields of Re, Ru, and Os MLCs when covalently linked to human serum albumin. The actual intensity decays are somewhat multiexponential. The single exponential intensity decays are to illustrate the time scale of the decays. bcp, 2,9-Dimethyl-4,7-diphenyl-1,10-phenanthroline. [Reprinted with permission from Kluwer Academic/Plenum Publishing.]

the excitation anisotropy spectra in Figs. 5 and 7. This was a somewhat unexpected observation because molecules like $[Ru(bpy)_3]^{2+}$ appear to be somewhat spherical, and the emission from the lanthanides is not polarized. Also, it was assumed that the excited state was randomized among the diimine ligands. We now know that an electron-withdrawing ligand such as dcbpy (4,4'-dicarboxylic acid-2,2'-bipyridine) results in a high fundamental anisotropy (Fig. 7), which we interpret as due to directed electron transfer from Ru to the dcbpy ligand. The fundamental anisotropies decrease with less electron-withdrawing ligands, as seen for the Ru complexes with the monocarboxylbipyridine and the aminophenanthroline ligands (Fig. 7). The Re MLCs typically have a single chromophoric diimine ligand, and all the Re MLCs we have studied display high r_0 values. Presumably the electron can be transferred only to this single diimine ligand.

In summary, the Ru and Re MLCs are generally useful fluorescence probes that display long decay times and polarized emission. Consequently, they can be used to detect translational or rotational diffusion on the microsecond time scale.

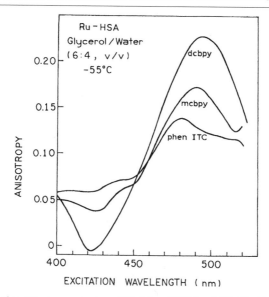

FIG. 7. Excitation anisotropy spectra of $[Ru(phen\text{-}ITC)(bpy)_2]^{2+}$, $[Ru(dcbpy)(bpy)_2]$, and $[Ru(mcbpy)(bpy)_2]^+$ conjugated to HSA. The emission wavelength was 650 nm for dcbpy and mcbpy, and 605 nm for phen-ITC. The structures are shown in Fig. 3. [Reprinted with permission from Elsevier Science, Inc.]

Microsecond Dynamics

Microsecond Rotational Diffusion

The long decay times and polarized emission from the MLCs make it possible to measure microsecond correlation times. The difficulty of measuring such correlation times with nanosecond decay time fluorophores is shown in Fig. 8, which shows the dependence of anisotropy on the molecular weight according to Eq. (1). For these calculations we used the Stokes–Einstein equation to calculate the correlation time θ,

$$\theta = \frac{\eta V}{RT} = \frac{\eta M}{RT}(\bar{v} + h) \tag{3}$$

where η is the viscosity in poise, $\bar{v}$ is the specific volume, T is the temperature in degrees kelvin, and R is the gas constant. The volume (V) of the protein is calculated from its molecular weight (M) with the typical assumption of 20% hydration $(h = 0.20)$.[14] For a protein the size of HSA (66 kDa) labeled with a 4-nsec decay time fluorophore the emission is expected to be completely polarized due to the correlation time near 30

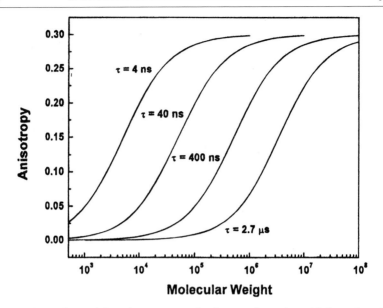

FIG. 8. Dependence of the anisotropy on molecular mass for probes with decay times from 4 to 2700 nsec. [Reprinted with permission from the American Chemical Society.]

nsec. Changes in the correlation time due to association or other phenomena will have a minimal effect on the anisotropy, which is already near the maximum value.

At present our longest lived probe is $Re(bcp)(CO_3)(COOH-py)$ (shown in the upper half of Fig. 4). This MLC shows a decay time near 3 μsec in aqueous solution in the presence of dissolved oxygen. Hence we now have MLC probes with decay times that are nearly 1000-fold larger than those of typical fluorescence probes. The use of long-lifetime MLC probes allows measurement of microsecond correlation times. For decay times of 400 nsec or 4 μsec the emission from a 66-kDa protein would be nearly completely depolarized, allowing detection of association reactions which increase its correlation time. Stated conversely, the rotational correlation times of much larger proteins can be measured with the microsecond probes. For a decay time of 2.7 μsec the anisotropy will be at one-half the maximum value for a molecular mass of 4×10^6 daltons.

Long-lifetime metal–ligand complexes have already been used to measure correlation times ranging from 30 nsec to 7 μsec. Figure 9 shows the frequency-domain intensity decays of the Ru complexes containing the dcbpy or phen-ITC ligands (Fig. 3, upper row). The frequency responses are reasonably approximated by a single decay time, but are weakly multiexponential. Such intensity decays are typical of macromolecules labeled with

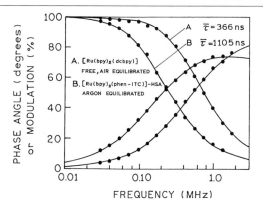

FIG. 9. Frequency-domain intensity decay of HSA labeled with Ru MLCs. [Reprinted with permission from Elsevier Science.]

MLCs. One notices that the frequency responses are centered near 0.3 MHz. As a result the time-resolved measurements are technically simple and can even be measured with directly modulated light-emitting diodes (LEDs) as the light source.[69,70] The ability to directly modulate the LED output eliminates the expensive and cumbersome electro-optic modulators found in most frequency-domain (FD) instruments. The use of LED and laser diode light sources in the FD instruments is expected to result in a new generation of simpler and lower cost instruments, and these instruments will be ideal for studies of microsecond time-scale processes.

The usefulness of the MLC probes for measuring microsecond rotational correlation times is shown in Fig. 10. These data are the frequency-domain anisotropy decays of HSA covalently labeled with [Ru(bpy)$_2$(dcbpy)]. Also shown are the FD anisotropy decay with increasing amounts of a polyclonal antibody directed against HSA. Least-squares analysis of these data yielded two rotational correlation times, near 36 nsec and 5 μsec. Similar long and short correlation times were found for HSA labeled with another Ru MLC (Table I). Examination of Table I shows that the amplitude (r_2) of the long correlation times (θ_2) increases on binding of HSA to anti-HSA. These results demonstrated that microsecond protein motions can be measured using the MLC probes. One can now imagine the routine measurement of the anisotropy decays of large macromolecular complexes.

A second example of slow anisotropy decays is shown for DNA labeled with [Ru(bpy)$_2$(dppz)]$^{2+}$ or [Ru(phe)$_2$(dppz)]$^{2+}$, where dppz is dipyri-

[69] J. Sipior, G. M. Carter, J. R. Lakowicz, and G. Rao, *Rev. Sci. Instrum.* **68,** 2666 (1997).
[70] J. Sipior, G. M. Carter, J. R. Lakowicz, and G. Rao, *Rev. Sci. Instrum.* **67,** 3795 (1996).

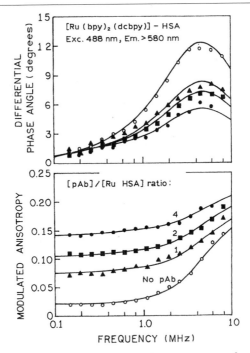

FIG. 10. Anisotropy decay of HSA labeled with $[Ru(bpy)_2(dcbpy)]^{2+}$ in aqueous buffer at 24°. The ratios are the molar ratios of polyclonal antibody (pAC) to Ru-labeled HSA.[57] [Reprinted with permission from Elsevier Science.]

TABLE I
GLOBAL CORRELATION TIMES AND AMPLITUDES RECOVERED FOR
RU-LABELED HSA IN PRESENCE OF ANTI-HSA POLYCLONAL ANTIBODY

Molar ratio (anti-HSA/HSA)	r_{01}	r_{02}	θ_1 (nsec)	θ_2 (nsec)[a]
Ru(bpy)$_2$(dcbpy)-HSA				
0	0.173[b]	0.008	36	5115
4.0	0.078	0.145	36	5115
Ru(bpy)$_2$(phen-ITC)-HSA				
0	0.102	0.0006	68	2023
4.0	0.044	0.086	68	2023

[a] The correlation times were global while fitting the data at different anti-HSA/HSA mole ratios.
[b] The anisotropy decay data were fit to $r(t) = \sum_j r_{0j} \exp(-t/\theta_j)$.

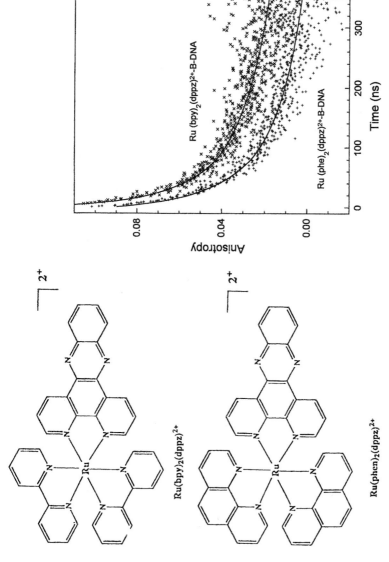

Fig. 11. Anisotropy decay of DNA labeled with $[Ru(bpy)_2(dppz)]^{2+}$ or $[Ru(phe)_2(dppz)]^{2+}$, whose structures are shown. [Reprinted with permission from Kluwer Academic/Plenum Publishing.]

do[3,2-a:2′,3′-c]phenazine (Fig. 11). Ru metal–ligand complexes containing the dppz ligand are known to be nonfluorescent in water but emit when the dppz ligand is intercalated into a DNA helix.[71] The emission of dppz-containing MLCs is typically quenched when the nitrogens on the dppz are exposed to water or polar solvents. These nitrogens are shielded from water when the dppz intercalates into double-helical DNA and the complex becomes fluorescent. The intensity decays for the DNA-bound MLCs are multiexponential (not shown) with decay components ranging from 22 to 138 and mean decay times from 97 to 159 nsec. These decay times allowed measurement of the DNA anisotropy decays up to 340 nsec. Longer data acquisition with these probes should allow the anisotropy data to be useful to about 1 μsec. Such data should allow measurement of the bending motions of DNA, in contrast to nanosecond probes, which report mostly on local torsional motions of the probes or of nearby base pairs. One can imagine the use of such long-lifetime DNA probes bound to DNA junctions, tRNA, or to ribozymes to measure the microsecond dynamics of these structures.

As a final example of measuring microsecond correlation times we show data that reveal the overall correlation time of dipalmitoyl-L-α-phosphatidyl glycerol (DPPG) vesicles. Several lipid MLC probes have been synthesized with some displaying decay times as long as 4.7 μsec (Fig. 12). Frequency-domain anisotropy decays for DPPG vesicles labeled with Ru(bpy)$_2$(dcbpy)-PE$_2$ are shown in Fig. 13. This MLC complex is covalently linked to two molecules of dipalmitoyl-L-α-phosphatidylethanolamine (PE). This probe displays decay times from 620 to 405 nsec in DPPG vesicles at temperatures ranging from 2 to 53°.[58] The anisotropy decay data (Fig. 13) were fit to a multicorrelation time model. At each temperature the Ru–lipid probe resulted in a correlation time near 150 nsec and a second longer correlation time from 5 to 10 μsec. These long correlation times are consistent with those expected for overall rotational diffusion of lipid vesicles with radii near 250–400 Å, which are expected to display microsecond correlation times.

It is interesting to examine the time-domain (TD) anisotropy decays of the labeled vesicles (Fig. 14). The anisotropies show a rapid decrease at times below 50 nsec, followed by a much slower loss of the remaining anisotropy at longer times. We attribute this slower anisotropy decay component to overall rotational motion of the lipid vesicles. This capability is in stark contrast to the usual result with nanosecond probes, which reveal only the local probe motions within the membrane. One can now imagine a new class of experiments that monitor the interaction of lipid vesicles

[71] C. Turro, S. H. Bossmann, Y. Jenkins, J. K. Barton, and N. J. Turro, *J. Am. Chem. Soc.* **117**, 9026 (1995).

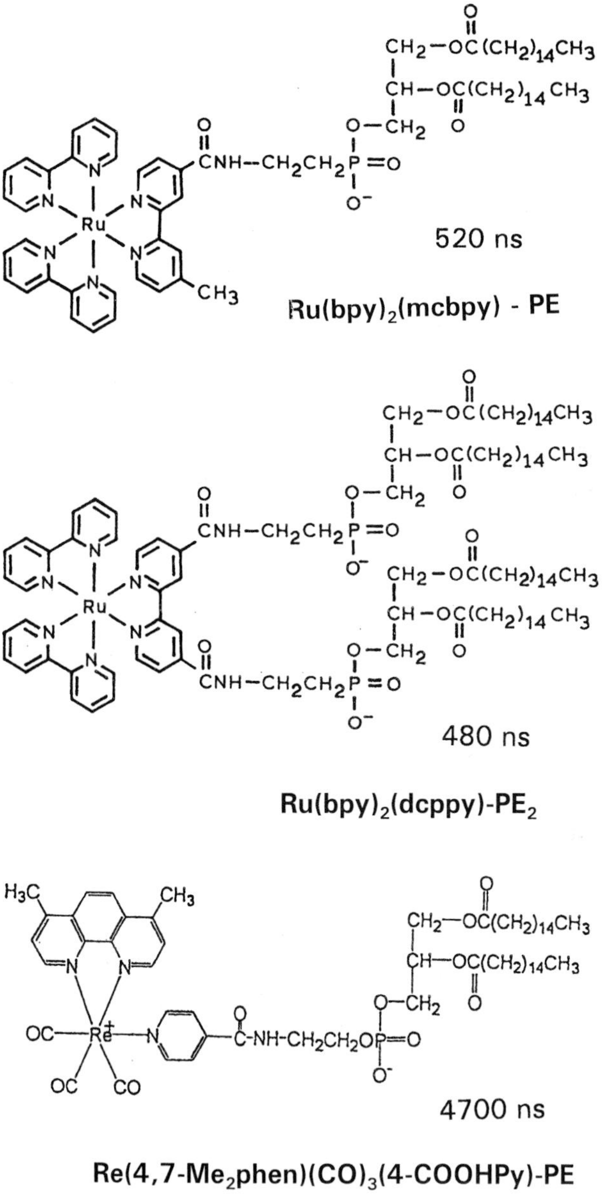

520 ns

Ru(bpy)$_2$(mcbpy) - PE

480 ns

Ru(bpy)$_2$(dcppy)-PE$_2$

4700 ns

Re(4,7-Me$_2$phen)(CO)$_3$(4-COOHPy)-PE

FIG. 12. Structure of MLC-labeled lipids.

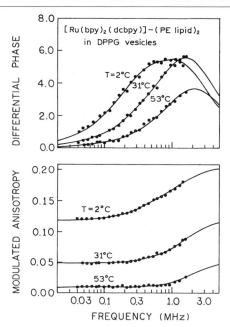

FIG. 13. Rotational diffusion of DPPG vesicles observed using Ru(bpy)$_2$(dcbpy)-PE$_2$.[58] [Reprinted with permission from Academic Press.]

with proteins or other membranes, as revealed by the vesicle correlation times.

Microsecond Translational Motions

In the preceding sections we described the use of the MLC probes to study slow rotational motions. There are also numerous opportunities for the use of resonance energy transfer (RET) with the microsecond probes to study translational diffusion or site-to-site motions in macromolecules. The MLCs are known to display RET to suitable acceptors. We presently believe that the metal-to-ligand charge transfer (MLCT) state undergoes Förster transfer just like a nanosecond fluorophore. However, we are not aware of a detailed study that demonstrates the distance dependence of RET from the MLC probes. To date, the measured Förster distance agrees with the values calculated from the spectral overlap.

The usefulness of the MLC probes for measuring translational motions can be seen from the dependence of the RET efficiency on the mutual diffusion coefficient (D) of the donor (D) and acceptor (A). It is well known that diffusion increases the transfer efficiency of linked and unlinked

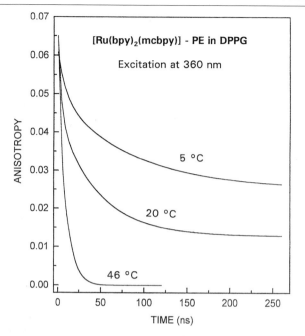

Fig. 14. Time-domain anisotropy, decays of DPPG vesicles labeled with [Ru(bpy)$_2$ (mcbpy)]-PE.[59] [Reprinted with permission from John Wiley & Sons.]

donors and acceptors.[72,73] The relationship between detectable translational diffusion coefficients and the donor decay times was elegantly presented by Stryer and co-workers.[74,75] Figure 15 shows the efficiency of resonance energy transfer between freely diffusing donor and acceptors in a three-dimensional solution. For nanosecond decay times there is no increase in RET efficiency for diffusion coefficients smaller than 10^{-5} cm^2/sec. For a microsecond donor decay time the transfer efficiency is sensitive to the diffusion coefficient because the donor (D) and acceptor (A) can diffuse within the Förster distances (R_0) during the excited state lifetime. For millisecond decay times the extent of transfer becomes limited by the distance of closest D–A approach. The RET efficiency is no longer sensitive

[72] I. Z. Steinberg and E. Katchalski, *J. Chem. Phys.* **48,** 2404 (1968).

[73] E. Katchalski-Katzir, E. Haas, and I. Z. Steinberg, Study of conformation and intramolecular motility of polypeptides in solution by a novel fluorescence method. *In* "Luminescence from Biological and Synthetic Macromolecules" (H. Morawetz and I. Z. Steinberg, eds.), pp. 44–61. Annals of the New York Academy of Sciences, New York, 1981.

[74] D. D. Thomas, W. F. Carlsen, and L. Stryer, *Proc. Natl. Acad. Sci. U.S.A.* **75,** 5746 (1978).

[75] L. Stryer, D. D. Thomas, and C. F. Meares, *Annu. Rev. Biophys. Bioeng.* **11,** 203 (1982).

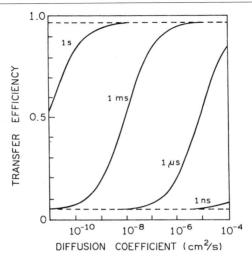

FIG. 15. Transfer efficiency for donor and acceptors in solution. $R_0 = 50$ Å and [acceptor] $= 0.1$ mM. [Revised from. Ref. 75.]

to the diffusion coefficients near 10^{-8} cm^2/sec because the millisecond decay times result in complete spatial averaging to the distance of closest approach (upper dashed line, Fig. 15). Because of this averaging, diffusion-enhanced energy transfer has been used in experiments that require a closest approach measurement, such as the depths of probes in membranes, depths of chromophores in visual proteins,[76,77] and the distance of closest approach of charged donors to dyes intercalated in DNA.[78] Importantly, Fig. 15 shows that the transfer efficiency is sensitive to D–A diffusion coefficients near 10^{-8} cm^2/sec when the lifetimes are near 1 μsec. Such D–A diffusion coefficients are expected for domain motions in proteins and for lateral diffusion in membranes. As a result, the time-resolved donor decays are expected to be sensitive to translational diffusion of proteins, membranes, and nucleic acids.

Domain Motions in Proteins

The long-lifetime MLC probes can be used to study domain motions in proteins or in DNA structural motifs. The difficulty of measuring domain motions was described above for RET studies of donor and acceptor-labeled phosphoglycerate kinase (PGK).[34] The structure of PGK in its open

[76] S. M. Yeh and C. F. Meares, *Biochemistry* **19,** 5057 (1980).
[77] D. D. Thomas and L. Stryer, *J. Mol. Biol.* **154,** 145 (1982).
[78] T. G. Wensel, C.-H. Chang, and C. F. Meares, *Biochemistry* **24,** 3060 (1985).

and closed conformations is shown in Fig. 16. A donor and acceptor were placed at positions 135 and 290. Opening and closing of the domains is expected to change the D-to-A distance from 42 to 38 Å. The time-resolved donor decays for the PGK mutant revealed a distribution of D-to-A distances, as expected for two linked domains that were flexing in solution. However, the data using the 5-((((2-iodoacetyl)amino)ethyl)amino)naphthalene-1-sulfonic acid (IAEDANS) donor with a 15-nsec decay time did not provide any information on the rate of domain flexing.[34]

We questioned whether domain-to-domain motions could be detected using a donor decay time of 2 μsec. We simulated data for a D–A pair with a distance distribution comparable to that found for PGK, with a mean distance of (R_{AV}) of 35 Å and half-width (hw) of 20 Å.[34] The FD data were simulated in the presence (solid line, Fig. 17) and absence (dashed line, Fig. 17) of diffusion. For a 20-nsec donor lifetime there is essentially no effect of diffusion (Fig. 17, top). By diffusion we mean the donor-to-acceptor motions that are the result of domain flexing of the protein. We are not referring to rotational diffusion of the protein, nor are we referring to the effects of diffusion on RET between donors and acceptors on different protein molecules. With a 2-μsec lifetime, and a site-to-site diffusion

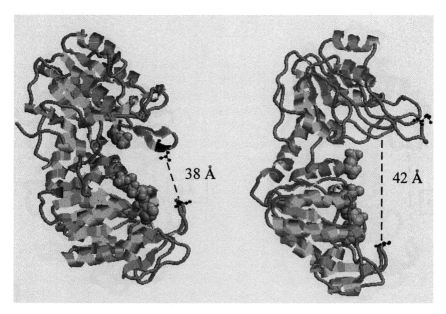

FIG. 16. Ribbon structures of phosphoglycerate kinase in the open (*left*) and closed (*right*) conformations. The positions labeled with the nanosecond donor and acceptor, as residues 135 and 290,[34] are shown as ball-and-stick models. The substrates 1,3-bisphosphoglycerate and ADP are shown as space-fill models.

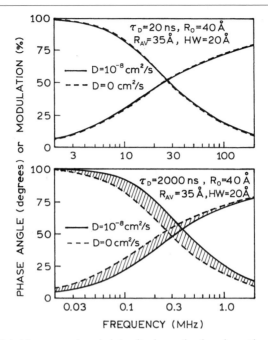

FIG. 17. Simulated frequency-domain intensity decays for domain motions in proteins. The shaded area shows the effect of D-to-A diffusion on RET.

coefficient of 10^{-8} cm^2/sec over the range of distances found for PGK, the frequency response shifted to higher frequencies (Fig. 17, bottom, solid line), due to the domain motions and a higher RET efficiency. The dashed line shows the donor decay expected in the absence of D–A diffusion. The shaded area shows the effect of D–A diffusion, which demonstrates that protein domain flexing on this time scale can be detected using RET with microsecond decay time probes.

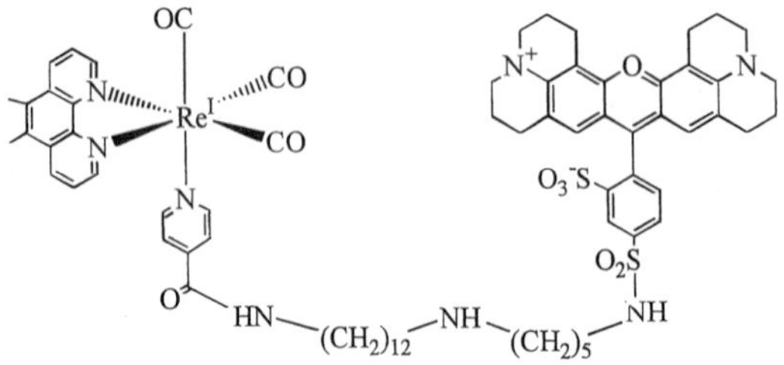

Re – C$_{17}$– Tr , Donor – Acceptor Pair

FIG. 18. Covalently linked Re MLC–Texas Red D–A pair.[79]

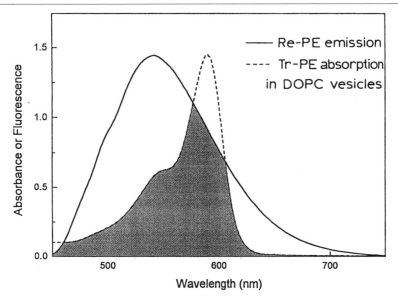

FIG. 19. Re–PE emission and Texas Red–PE absorption spectra.[80] [Reprinted with permission from Chemistry and Physics of Lipids, Elsevier Science.]

We have not yet obtained a suitably labeled domain protein. However, we were able to synthesize a covalently linked donor–acceptor pair (Fig. 18). The Re MLC donor displayed a decay time of 2.15 μsec in the absence of acceptor.[79] The extent of spectral overlap (Fig. 19) resulted in a Förster distance of $R_0 = 35.5$ Å.[80] The frequency-domain donor intensity decay shifts to higher frequency and becomes nonexponential in the presence of the covalently linked acceptor (Fig. 20). The shift toward higher frequency between the dashed line for no diffusion and the solid line with diffusion shows that D–A diffusion makes a substantial contribution to shortening the intensity decay of the donor.

The donor and donor–acceptor decays were analyzed according to a previously developed model that fits the data in terms of mean distance ($\bar{r}$) and half-width (hw) of the distance distribution, and the D–A diffusion coefficient (D),[81,82] resulting in values of $\bar{r} = 10.5$ Å, hw = 26 Å, and $D = 1.6 \times 10^{-8}$ cm²/sec at 20° in propylene glycol (Fig. 20). We questioned the

[79] R. Nair, G. Pisczek, I. Gryczynski, and J. R. Lakowicz, *Photochemistry and Photobiology* **71**(2), 157 (2000).
[80] L. Li, I. Gryczynski, and J. R. Lakowicz, *Chem. Phys. Lipids* **101**, 243 (1999).
[81] J. R. Lakowicz, J. Kusba, I. Gryczynski, W. Wiczk, H. Szmacinski, and M. L. Johnson, *J. Phys. Chem.* **95**, 9654 (1991).
[82] J. R. Lakowicz, J. Kusba, W. Wiczk, and I. Gryczynski, *Chem. Phys. Lett.* **173**, 319 (1990).

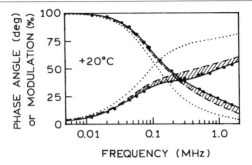

Fig. 20. Frequency-domain donor decay of the Re MLC–Texas Red D–A pair in propylene glycol at 20° (Fig. 18). The dotted line represents donor without acceptor. The solid line shows the intensity decay of the D–A pair (●) fit to $\bar{r} = 10.5$ Å, hw = 26 Å, and $D = 1.6 \times 10^{-8}$ cm²/sec. The dashed line is for the same values of $\bar{r}$ and hw but with $D = 0$.

uncertainty in the recovered diffusion coefficient. This uncertainty is not accurately recovered from the least-squares statistics, which typically assumes there is no correlation between the parameters ($\bar{r}$, hw, and D). There was correlation, which means that if one parameter value changes, the other parameter value can change to compensate and still yield an acceptable value of the goodness-of-fit parameter χ_R^2. This results in a wider range of parameter values that are consistent with the data than in the absence of correlation, and possible overestimation of the resolution of the experiments. The range of values consistent with data can be found from the χ_R^2 surface. This surface is obtained by holding one parameter value fixed at the desired value and minimizing χ_R^2 by adjustment of the remaining parameters. The χ_R^2 surface for the diffusion coefficient is shown in Fig. 21,

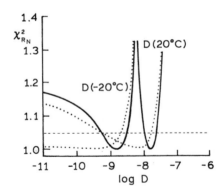

Fig. 21. χ_R^2 surface for the Re MLC to Texas Red diffusion coefficient from Fig. 20 from a global fit of −20 and 20°. The solid lines are for a global analysis at −20 and 20°. The dotted lines are for analysis at a single temperature.

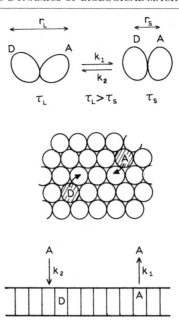

FIG. 22. Jump models for domain motions in proteins, lateral diffusion in membranes, and intercalation rates for dyes in DNA.

along with the χ_R^2 elevation, which is not acceptable (dashed line, Fig. 21).[14,83] This surface shows good resolution of the diffusion coefficient even when correlation is rigorously considered. It appears that the data will be adequate to recover diffusion coefficients larger than 10^{-8} cm²/sec, and diffusion coefficients as low as 10^{-9} cm²/sec could be resolved if some constraints were placed on the time-independent distance distribution.[79] In practice such constraints are readily available using shorter lifetime donors or by quenching a long-lifetime donor. These data for a simple D–A pair confirm the usefulness of the microsecond MLC probes for measurement of domain motions in proteins.

Jump Models for Microsecond Dynamics

It is likely that many domain proteins exist in a limited number of conformations, such as the open and closed conformations shown schematically in Fig. 22. In these cases it may be informative to consider the effect of conformational jumps on the RET data. One can also imagine jump

[83] M. L. Johnson and L. M. Faunt, Methods Enzymol. **210**, 1 (1992).

models for lipid diffusion in membranes (Fig. 22, middle) and for the on and off rates of dyes intercalated into DNA (Fig. 22, bottom).

We used simulations to determine the effect of conformational jumps on the donor decay of a D and A-labeled protein. We assumed that the protein could exist in two states such that the donor decay times were $\tau_S = 200$ nsec and $\tau_L = 2000$ nsec.

$$
\begin{array}{ccc}
\underset{\tau_S = 200\ nsec}{D\ A} & \underset{k_2}{\overset{k_1}{\rightleftharpoons}} & \underset{\tau_L = 2000\ nsec}{D\diagdown A}
\end{array}
\tag{4}
$$

One can readily obtain an analytical solution for the time-dependent donor decays assuming that both species are excited in proportion to their fractional ground state populations.[84] These equations are different from those of an excited state reaction because both species are excited, whereas in the excited state reaction only one species is typically excited. For this model the emission from both states is at the same wavelength, and it is not possible to independently observe the emission from each state, and we can observe only the total emission.

We used these expressions[84] to model the TD and FD decay expected for a protein jumping between two conformations. In the time domain the intensity decay with a slow jump rate (10^4 sec^{-1}) is a double exponential decay with decay times of 200 and 2000 nsec. Of course, the system is then in slow exchange so that one observes the sum of two single exponential decays, with one decay time for each state.

The shape of the intensity decay changes dramatically as the interchange rate increases. When the exchange rate is much faster than the reciprocal lifetime the intensity decay becomes a single exponential with a decay time of 364 nsec, which is given by

$$
\frac{1}{\tau} = \frac{1}{2}\left(\frac{1}{\tau_1} + \frac{1}{\tau_2}\right)
\tag{5}
$$

In this case the system behaves as if all the D–A pairs are at a shorter distance because the jump occurs many times during the donor lifetime. When the interchange rates are comparable to the reciprocal lifetime then the shape and mean decay time depends on k_1 and k_2 (Fig. 23). This dependence should allow determination of k_1 and/or k_2 from the time-resolved data. Of course it may be necessary to use an independent determined value of the equilibrium constant to calculate both k_1 and k_2 from a single set of data.

[84] K. Nowaczyk, M. L. Johnson, and J. R. Lakowicz, unpublished observations (1999).

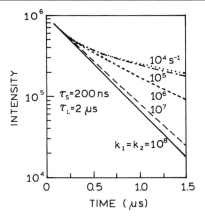

FIG. 23. Simulated time-domain intensity decays for the jump model shown in Eq. (4). From left to right the interchange rates are $k_1 = k_2 = 10^4$ to 10^8/sec.

The frequency-domain data for the two-state jump model also depend on the values of k_1 and k_2 (Fig. 24). In this case the frequency response shows maximum heterogeneity for low values of k_1 and k_2 (dotted line, Fig. 24), where the intensity decay is the sum of two independent decay times. As the values of k_1 and k_2 increase the frequency response shifts toward higher frequency and becomes more similar to single exponential.

To test the possibility of recovering k_1 and k_2 from the time-resolved data we analyzed the frequency response for the MLC D–A pair (Fig. 18) in terms of the jump model.[84] This analysis resulted in apparent jump rates

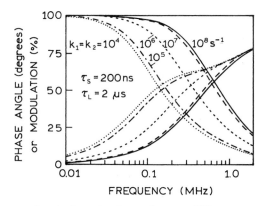

FIG. 24. Frequency-domain intensity decays for $\tau_S = 200$ nsec, $\tau_L = 2$ μsec, and the indicated k_1 and k_2 values.

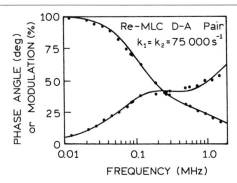

FIG. 25. Analysis of the FD intensity decay data for the Re MLC D–A pairs (FIG. 18) in terms of the jump model.

of 75,000 sec^{-1} (Fig. 25). These results are preliminary but demonstrated that jumps on the order of once every 12 μsec can be recovered using RET and a microsecond MLC donor.

Lateral Diffusion in Membranes

Another opportunity for the microsecond probes is the direct measurement of lateral diffusion in membranes. Numerous publications have appeared on RET in membranes.[85–88] Most reports consider only randomly distributed donors and acceptors, and none of the reports consider the effects of lateral diffusion on RET. Hence we used simulations to predict the effects of lateral diffusion on the RET efficiency for donors and acceptors randomly distributed in two dimensions for various assumed lateral diffusion coefficients. For a typical 10-nsec probe, the RET efficiency is not sensitive to diffusion coefficients below 10^{-5} cm^2/sec (Fig. 26). As the lifetime increases the RET efficiency increases due to the changing D–A distance during the donor decay. If the decay time is 1 msec the RET efficiency is at the diffusion-controlled limit (top dashed line, Fig. 26), and sensitive only to the distance of closest approach ($R_{\min}$). For decay times for 1 to 100 μsec the RET efficiency is sensitive to the lateral diffusion coefficient.

To demonstrate further that lateral diffusion in membranes will be detectable, we simulated frequency responses for the donors in donor and

[85] B. K.-K. Fung and L. Stryer, *Biochemistry* **17**, 5241 (1978).

[86] D. E. Wolf, A. P. Winiski, A. E. Ting, K. M. Bocian, and R. E. Pagano, *Biochemistry* **31**, 2865 (1992).

[87] A. Blumen, J. Klafter, and G. Zumofen, *J. Chem. Phys.* **84**, 1307 (1986).

[88] H. Kellerer and A. Blumen, *Biophys. J.* **46**, 1 (1984).

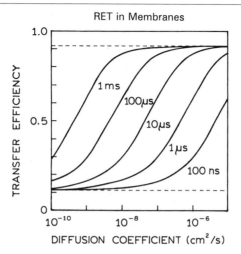

FIG. 26. Simulated RET efficiency for diffusion in two dimensions. The acceptor density was 5×10^{-3} acceptors/lipid and the area per lipid was 75 Å^2/lipid, $R_0 = 25$ Å, and $R_{min} = 7$ Å.

acceptor-labeled membranes. For $D = 5 \times 10^{-8}$ cm^2/sec there is little effect of diffusion on a 300-nsec donor (Fig. 27, top). However, if the donor decay time is increased to 3 μsec, the effect of diffusion is substantial (Fig. 27, shaded area, bottom), demonstrating the possibility of measuring lateral diffusion in membranes.

In preliminary studies we have already experimentally demonstrated that lateral diffusion in membranes can be detected with a microsecond donor.[89] A long-lifetime Re MLC was covalently linked to PE, as was the Texas Red acceptor (Fig. 28). The Förster distance of the D–A pair is 35.3 Å.[80] These probes were placed into dioleoylphosphatidylcholine (DOPC) vesicles by the usual procedure of cosonication. In the absence of acceptor Re–PE displays intensity decay components as long as 2–3 μsec. The Re–PE donor decay shifts to higher frequencies in the presence of acceptor (Fig. 29). We used our software for RET in two dimensions without lateral diffusion to predict the frequency response in the absence of diffusion (Fig. 29, dashed line). The measured decay (Fig. 29, ■) is shifted to much higher frequencies than the diffusionless decay (Fig. 29, dashed line). The intensity decay of the D–A pair shows shorter decay times and is more like a single exponential. These results show a substantial if not dominant contribution of diffusion to the RET at this acceptor density (shaded area). These results

[89] M. Collini, G. Chirico, G. Baldini, and M. E. Bianchi, *Biopolymers* **36,** 211 (1995).

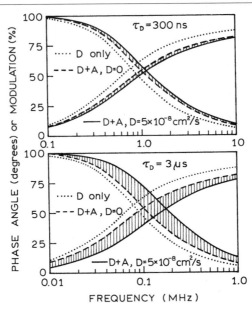

FIG. 27. Simulated donor decays for D and A in two dimensions; $R_0 = 35$ Å, 10^{-2} acceptors/lipid.

demonstrated that lateral diffusion contributes to RET from long-lifetime lipid donors.

We used the theory for RET without diffusion to recover the "apparent" value of R_0 for the membrane-bound D–A pair. It is well known that D-to-A diffusion results in larger apparent R_0 values if the data are analyzed without consideration of diffusion.[72] This analysis resulted in an apparent value of $R_0 = 65$ Å, rather than the known value of 35.3 Å (Fig. 30). One can use this increase in R_0 by 30 Å to estimate the lateral diffusion coefficient. Using the $\Delta x^2 = 4D\tau$ for diffusion in two dimensions yields an approximate value of $D = 2 \times 10^{-8}$ cm²/sec. This value is comparable to that found by FRAP studies of fluid bilayers.[25,26] Smaller values of D have also been reported, particularly in the presence of membrane-bound proteins.[27] These preliminary data clearly demonstrate that the FD time-resolved data, with a 3-μsec donor, will be adequate to recover the expected diffusion coefficients of lipids in membranes.

Microsecond DNA Dynamics

Microsecond probes are also likely to be useful in studies of the structural dynamics of DNA and RNA. The DNA double helix can be bent or

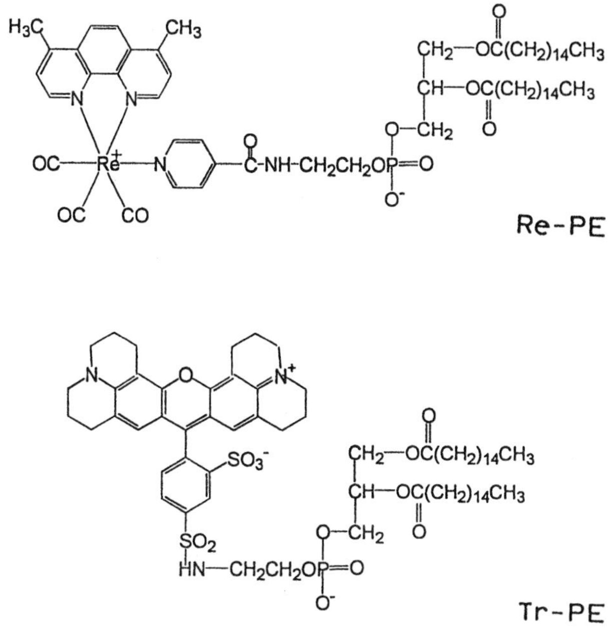

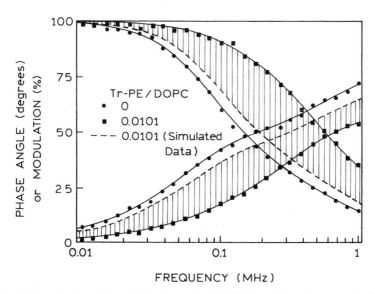

FIG. 28. MLC donor and Texas Red (Tr) acceptor-labeled lipids. [Reprinted with permission from Chemistry and Physics of Lipids, Elsevier Science.]

FIG. 29. Contribution of lateral diffusion to the Re-PE donor decay shown as the shaded area. [Data from Ref. 80.]

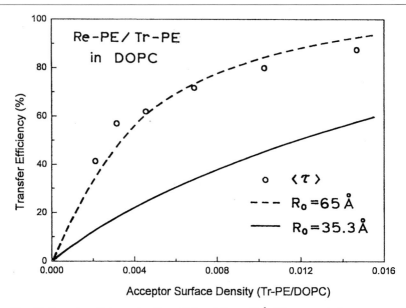

FIG. 30. Transfer efficiency for $R_0 = 35.3$ and $R_0 = 65$ Å. The open circles represent the experimental data. [Data from Ref. 80.]

curved (Fig. 31) by the introduction of specific base sequences.[89–91] In such cases it should be possible to determine the rates and amplitudes of DNA bending. Similarly, structural fluctuations in tRNA[92,93] should now be accessible to direct measurements. It is also possible that long lifetimes may be used for studies of plasmid dynamics. Figure 32 shows solution conformations of a plasmid in the relaxed, intermediate, and supercoiled states.[94] If the flexing rates are fast, these motions may contribute to the extent of RET from long-lifetime donors.

Long-Range Quenching

Fluorescence quenching is usually regarded as a short-range interaction that requires molecular contact between the fluorophore and the quenchers. Although this is true for nanosecond fluorophores, quenching can be a

[90] C. Gohlke, A. I. H. Murchie, D. M. Lilley, and R. M. Clegg, *Proc. Natl. Acad. U.S.A.* **91,** 11660 (1994).

[91] S. Diekmann and D. Porschke, *Biophys. Chem.* **26,** 207 (1987).

[92] P. Auffinger, S. Louise-May, and E. Westhof, *Biophys. J.* **76,** 50 (1999).

[93] M. W. Friederich, E. Vacano, and P. J. Hagerman, *Proc. Natl. Acad. Sci. U.S.A.* **95,** 3572 (1998).

[94] R. K.-Z. Tan, D. Sprous, and S. C. Harvey, *Biopolymers* **39,** 259 (1996).

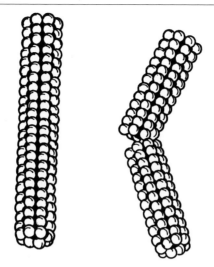

FIG. 31. Structure of a DNA fragment with 50 base pairs and a bent DNA with five unpaired adenines.[91]

longer range through-space interaction for longer lifetime fluorophores. For instance, electron exchange interactions are known to occur through a distance-dependent interaction according to

$$k(r) = k_a \exp\left(\frac{r - a}{r_e}\right) \qquad (6)$$

In this expression $k(r)$ is the quenching rate at a distance r, a is the distance of closest approach, $a = 7$ Å, k_a is the quenching rate at the closest distance,

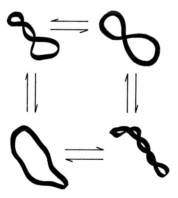

FIG. 32. Possible conformation of a circular DNA plasmid. [Modified from Ref. 94.]

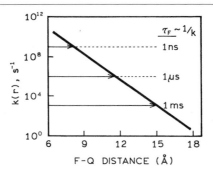

FIG. 33. Dependence of the quenching rate on the fluorophore-to-quencher distance, according to Eq. (6) with $k_a = 10^{10}$ sec^{-1} and $r_e = 0.5$ Å.

and r_e is the characteristic distance. The rates of electron transfer or exchange interaction quenching are fundamentally different from the rates for RET. In the case of RET the transfer rate is proportional to the emission rate of the donor. This means that for a D-to-A distance of $r = R_0$ the donor will be 50% quenched irrespective of the donor lifetime. In the case of electron transfer the distance at which the donor-quenching pair is 50% quenched is when $k(r)$ is equal to the reciprocal of the lifetime.[95] As the lifetime of the donor becomes longer then D–A distances for 50% quenching increases. This suggests that quenching of a 1-μsec donor should occur for a D–Q distance of 12 Å (Fig. 33). Hence, quenching can potentially be used as a structural probe to measure distances in a manner comparable to RET.

Conclusions

In the preceding sections we demonstrated that fluorescence spectroscopy with long-lifetime MLCs provides information on microsecond dynamics. While not the emphasis of this review, translational motions on the millisecond time scale are also directly observable. Such measurements are possible with the luminescent lanthanides, which display decay times near 2 msec.[96–100] This sensitivity over the time scale from picoseconds to millisec-

[95] J. M. Vanderkooi, S. W. Englander, S. Papp, W. W. Wright, and C. S. Owen, *Proc. Natl. Acad. Sci. U.S.A.* **87,** 5099 (1990).
[96] P. R. Selvin and J. E. Hearst, *Proc. Natl. Acad. Sci. U.S.A.* **91,** 10024 (1994).
[97] P. R. Selvin, T. M. Rana, and J. E. Hearst, *J. Am. Chem. Soc.* **116,** 6029 (1994).
[98] P. R. Selvin, *Methods Enzymol*, 301 (1995).
[99] N. Sabbatini, M. Guardigli, and I. Manet, *Adv. Photochem.* **23,** 213 (1997).
[100] J. B. Lamture, B. Iverson, and M. E. Hogan, *Tetrahedron Lett.* **37,** 6483 (1996).

onds is unique among spectroscopic techniques, and is likely to result in increased understanding of the dynamics of macromolecules and their assemblies.

Acknowledgments

This work was supported by a grant from the NIGMS (GM-35154) and from the NCRR (RR-08119).

Author Index

Numbers in parentheses are footnote reference numbers and indicate that an author's work is referred to although the name is not cited in the text.

A

Ackerman, A. L., 67, 74, 75(11)
Ackermann, F., 208, 227(4), 228(4)
Ackers, G. K., 126, 127(7), 128(7), 130(7), 131(7), 132, 132(7), 134(7, 14), 135, 135(7, 14, 15), 136(7), 137(24), 138(24), 140(22–24), 141(23, 24), 142, 142(7), 144, 144(7), 145(7, 33), 146, 146(7, 14, 15, 22, 23, 33), 147, 147(34), 148(26, 38, 41), 149, 149(7, 24, 34), 150(7, 14, 15, 31, 41), 151(33, 41), 152(26, 41), 153(7), 154, 162, 165, 229, 263
Adair, B. D., 66, 76(6)
Adair, G. S., 126, 129(4), 153(4), 156
Adair, J. R., 210, 230
Adams, G. P., 315, 316(39)
Adams, P. D., 67
Air, G. M., 227
Alberts, J., 209
Alegre, M. L., 210
Alessi, K., 421, 438(19)
Allen, D. J., 450, 454(13), 462
Allen, M. J., 224, 225(35), 227, 228, 228(52)
Almeida, P. F. F., 341, 343(16), 346(16), 347(16), 348(16), 350(16)
Altin, M., 306
Amegadzie, B. Y., 177
Ameloot, M., 471
Amos, L. A., 94, 95(38)
Amzel, L. M., 167, 167(4–6), 168, 170(4), 174(4), 175(1, 6), 176(1), 222, 224(27), 225(27)
Anaguchi, H., 201, 202(51), 204(51)
Anand, N. N., 207
Anchin, J. M., 227
Anderson, C. F., 375, 406, 407, 407(1, 2, 4), 410(5), 411, 415(26), 417(26), 425, 432(24)
Anderson, D., 209

Andersson-Teleman, A., 258
Andreu, J. M., 94, 95(43)
Angal, S., 210
Anthony, F. H., 370
Aoki, K. H., 180, 180(21), 181, 186(16), 195(21), 196(21), 197, 198, 201(44), 202(44)
Aplin, R. T., 119
Arakawa, T., 180(21), 181, 195(21), 196(21), 198, 201(44), 202(44)
Argaet, V. P., 237, 240(11), 242(11), 250(10, 11), 253(11)
Argyropoulos, V. P., 231, 242(3)
Arnold, G. E., 201
Arrington, C. B., 104, 105, 110(9), 119(9), 120(9), 121
Arthos, J., 210
Arulanantham, P. G., 326, 327(6)
Arulanantham, P. R., 193
Atha, D. H., 135, 137(28), 138(28), 141(28), 144(28), 165
Auffinger, P., 474, 506
Austin, S., 253
Axel, R., 210
Axelrod, D., 476
Aymamí, J., 399, 401(51), 402(51), 403(51), 405(51)

B

Babu, Y. S., 258
Babul, J., 212
Bai, R., 96, 100(51)
Bai, S. S., 96, 100(51)
Bai, Y., 104, 105, 108(2), 109(2)
Bailey, M., 243, 451, 453
Bailey, M. F., 240, 241(14), 245(14), 246(14), 247(14), 249(14), 250(14)

511

X

Y

Z

Subject Index

A

Accessible surface area
  corresponding enthalpy and entropy
    changes, 225–226
  determination for proteins, 59
  DNA–drug interactions
    free energy relationship, 379–380
    heat capacity change calculations, 399,
      403
    Hoechst 33258 binding calculations,
      399–403
    model structure generation for calcula-
      tion, 396–397
    software for calculation, 397–399
    structure sources for calculation,
      395–396
  erythropoietin–receptor complex,
    197–198
Adair model, hemoglobin cooperativity,
  125, 129–132, 156–162, 165–166, 298
Amide hydrogen exchange, *see* Hydrogen
  exchange
ASA, *see* Accessible surface area

B

Backbone entropy, calculations, 172–173
Barnase
  absolute partial heat capacity measure-
    ment, 41
  calorimetry studies of folding cooperativ-
    ity, 35
BIACORE, *see* Surface plasmon resonance
Binding polynomial, complex ligand-bind-
  ing formulas, 158, 160, 164, 167
Bohr effect, hemoglobin, 124, 126
BoxA, NusB/E complex, 27–28
BoxB
  cooperative interactions with NusA,
    26–27
  promotion of RNA polymerase interac-
    tion with N, 13–18, 22–24

C

Calmodulin
  calcium affinities of homologous do-
    mains, 258–260, 301
  calcium titration simulations
    equal-affinity sites, 266
    four-ligand binding, 266–267
    pairs of sites with different affinities,
      264–266
    two-ligand binding, 264–265
  combinations of vacant and filled calcium
    sites, 260–261
  cooperative binding of calcium, 256–258
  domains, 256
  mutagenesis, 263–264, 301
  proteolytic footprinting titration
    application with other proteins, 268
    binding constant resolution
      fitting for end points, 298–299
      fractional population of single spe-
        cies, 296
      fractional saturation, 296–298
      pairwise cooperativity, 299–300
    calcium titrations
      equilibrium titration, 289
      overview, 287–288
      stoichiometric titration, 288–289
    classes of observed susceptibility pro-
      files, 281, 283, 285, 287
    comparative studies, 300–301
    denaturing gel electrophoresis, 277–279
    interpretation
      absolute susceptibility profiles,
        295–296
      fractional area of peptides from chro-
        matograms, 294
      normalized susceptibility profiles,
        296
      precision of quantification and back-
        ground correction, 294–295
    interval of proteolysis, 275–276
    overview, 262–263
    protease selection

Hill equation, hemoglobin cooperativity,
127, 129, 157
Hoechst 33258, *see* DNA–drug interactions
HPLC, *see* High-performance liquid chromatography
Hüfer model, hemoglobin cooperativity,
125–126
Hydrogen exchange
amide hydrogen exchange in protein conformation analysis
EX1 exchange, 107
EX2 exchange, 106–107
exchange rate calculation for residues
in unstructured peptides, 107–110
observed rate constant for exchange,
106–107
overview, 104–105
pH dependence, 107
two-state model of exchange, 106
nuclear magnetic resonance analysis of
protein amide groups
advantages, 104–105
EX1 versus EX2 exchange processes,
118–120, 124
free energy of opening calculation, 116
kinetic parameter determinations,
118–121
localized fluctuations in native conformation, 116–118
normalization of peaks, 115–116
observed rate constant for exchange determination, 115–116
pH dependence
least-squares analysis, 121–122
protein stability considerations,
122–124
principle, 105, 110
quenched hydrogen exchange measurement, 113–115
real-time hydrogen exchange measurement, 110–113
resonance assignment, 110

I

IL-2, *see* Interleukin-2
Interleukin-2, receptor α subunit interactions studied with BIACORE

assay design, 327
binding conditions, 327–328
data processing
double referencing, 330
overlay of all analyte responses,
330
reference subtraction, 328
replication overlays, 328, 330
zeroing, 328
immobilization of receptor, 327
kinetic data analysis
assessing residuals, 332–333
equilibrium analysis, 333–334
global fitting, 331–332
model, 330–331
rate constants, 333
materials, 326
thermodynamic analysis
calorimetry comparison, 339
Erying analysis, 338
free energy, 334
transition state free energy, 334–335
van't Hoff analysis, 335–339
Isothermal titration calorimetry
CD4–monoclonal antibody interactions,
211, 215–216, 219–223, 229
DNA–drug interactions
binding constant determination, 386
DNA preparation
buffers, 382–383
concentration in experiments,
386–387
purification, 382
quantification, 383–384
sequence, 384–385
structural equilibrium, 384–385
error sources, 389
ethidium bromide binding experiment,
392, 394–395
excess binding site titration, 389–390,
392, 394
Hoechst 33258 binding experiment,
390–392, 394–395
instrumentation
calibration, 388–389
cleaning and degassing, 388
combined microcalorimetry system,
381–382
ligand

M

Magnetic suspension densitometry, volume change measurements of DNA hydration, 423–424

Mass spectrometry, protein proton exchange analysis, 104

Membrane, *see* Lipid bilayer

Monoclonal antibody–protein interactions, *see* CD4–monoclonal antibody interactions; Sedimentation velocity analysis

Monod, Wyman, and Changeux model, hemoglobin cooperativity, 142–144, 146, 149, 151, 162–163

Monte Carlo simulation, *see* Lipid bilayer

MWC model, *see* Monod, Wyman, and Changeux model

N

N, antitermination complex
activation energy of N contribution to antitermination, 9–10
assembly, 3
biological significance of λ-mediated antitermination, 2–3
kinetic competition between elongation and termination
overview, 7
perturbation by antitermination, 7, 9–10
layers of specificity
application to other multilayered regulatory processes, 29–31
concept of accessory factor contributions, 1, 5–6, 28–29
RNA polymerase interaction with N
accessory factor mechanisms of promoting interaction, 12–13
binding energy, 10–11, 21–22
controlling binding equilibria, 18
dissociation constants, 11, 14, 17, 21
effective concentration of N at polymerase, 16
fraction of RNA polymerase bound by N, 12
NusA effects
cooperative interactions with boxB, 26–27

enhancement of efficacy of N-dependent modification of polymerase, 25
inhibition of nonspecific interactions of N with RNA, 25–26
N binding affinity, 25
range of antitermination, 26
termination efficiency, 11–12, 25
NusB/E–boxA effects
overview, 27
range of antitermination, 27–28
RNA looping
boxB effects, 22–24
efficacy of antitermination modification, 19–20
fraction of N bound to transcript, 20, 22
promotion of interaction, 13–18, 22–23
range of antitermination modification along template, 21, 23–24
termination efficiency, 9–11, 21
terminator specificity, 2–3
in vitro characterization of complexes, 3–5, 29

NMR, *see* Nuclear magnetic resonance

Nuclear magnetic resonance, amide hydrogen exchange analysis in proteins
advantages, 104–105
EX1 versus EX2 exchange processes, 118–120, 124
free energy of opening calculation, 116
kinetic parameter determinations, 118–121
localized fluctuations in native conformation, 116–118
normalization of peaks, 115–116
observed rate constant for exchange determination, 115–116
pH dependence
least-squares analysis, 121–122
protein stability considerations, 122–124
principle, 105, 110
quenched hydrogen exchange measurement, 113–115
real-time hydrogen exchange measurement, 110–113
resonance assignment, 110

[22] Errata

Methods in Enzymology, Volume 295: Energetics of Biological Macro-molecules, Part B

Article 17: Theoretical Aspects of Isothermal Titration Calorimetry by Lawrence Indyk and Harvey F. Fisher, p. 353

Equations 10 and 11 should read

$$[L]^2 + ([M_T] - [L_T] + 1/K_B)[L] - [L_T]/K_B = 0 \tag{10}$$

$$[L] = \frac{[L_T] - [M_T] - 1/K_B + \sqrt{([L_T] - [M_T] - 1/K_B)^2 + 4[L_T]/K_B}}{2} \tag{11}$$

The authors are indebted to Brian Feldman of Dartmouth College for bringing these errors to their attention.

ISBN 0-12-182224-9

90038

9 780121 822248